Selected Titles in This Series

208 **Michel L. Lapidus, Lawrence H. Harper, and Adolfo J. [illegible],** Harmonic analysis and nonlinear differential equations: A volume in honor of Victor L. Shapiro, 1997

207 **Yujiro Kawamata and Vyacheslav V. Shokurov, Editors,** Birational algebraic geometry: A conference on algebraic geometry in memory of Wei-Liang Chow (1911–1995), 1997

206 **Adam Korányi, Editor,** Harmonic functions on trees and buildings, 1997

205 **Paulo D. Cordaro and Howard Jacobowitz, Editors,** Multidimensional complex analysis and partial differential equations: A collection of papers in honor of François Treves, 1997

204 **Yair Censor and Simeon Reich, Editors,** Recent developments in optimization theory and nonlinear analysis, 1997

203 **Hanna Nencka and Jean-Pierre Bourguignon, Editors,** Geometry and nature: In memory of W. K. Clifford, 1997

202 **Jean-Louis Loday, James D. Stasheff, and Alexander A. Voronov, Editors,** Operads: Proceedings of Renaissance Conferences, 1997

201 **J. R. Quine and Peter Sarnak, Editors,** Extremal Riemann surfaces, 1997

200 **F. Dias, J.-M. Ghidaglia, and J.-C. Saut, Editors,** Mathematical problems in the theory of water waves, 1996

199 **G. Banaszak, W. Gajda, and P. Krasoń, Editors,** Algebraic K-theory, 1996

198 **Donald G. Saari and Zhihong Xia, Editors,** Hamiltonian dynamics and celestial mechanics, 1996

197 **J. E. Bonin, J. G. Oxley, and B. Servatius, Editors,** Matroid theory, 1996

196 **David Bao, Shiing-shen Chern, and Zhongmin Shen, Editors,** Finsler geometry, 1996

195 **Warren Dicks and Enric Ventura,** The group fixed by a family of injective endomorphisms of a free group, 1996

194 **Seok-Jin Kang, Myung-Hwan Kim, and Insok Lee, Editors,** Lie algebras and their representations, 1996

193 **Chongying Dong and Geoffrey Mason, Editors,** Moonshine, the Monster, and related topics, 1996

192 **Tomek Bartoszyński and Marion Scheepers, Editors,** Set theory, 1995

191 **Tuong Ton-That, Kenneth I. Gross, Donald St. P. Richards, and Paul J. Sally, Jr., Editors,** Representation theory and harmonic analysis, 1995

190 **Mourad E. H. Ismail, M. Zuhair Nashed, Ahmed I. Zayed, and Ahmed F. Ghaleb, Editors,** Mathematical analysis, wavelets, and signal processing, 1995

189 **S. A. M. Marcantognini, G. A. Mendoza, M. D. Morán, A. Octavio, and W. O. Urbina, Editors,** Harmonic analysis and operator theory, 1995

188 **Alejandro Adem, R. James Milgram, and Douglas C. Ravenel, Editors,** Homotopy theory and its applications, 1995

187 **G. W. Brumfiel and H. M. Hilden,** $SL(2)$ representations of finitely presented groups, 1995

186 **Shreeram S. Abhyankar, Walter Feit, Michael D. Fried, Yasutaka Ihara, and Helmut Voelklein, Editors,** Recent developments in the inverse Galois problem, 1995

185 **Raúl E. Curto, Ronald G. Douglas, Joel D. Pincus, and Norberto Salinas, Editors,** Multivariable operator theory, 1995

184 **L. A. Bokut′, A. I. Kostrikin, and S. S. Kutateladze, Editors,** Second International Conference on Algebra, 1995

183 **William C. Connett, Marc-Olivier Gebuhrer, and Alan L. Schwartz, Editors,** Applications of hypergroups and related measure algebras, 1995

182 **Selman Akbulut, Editor,** Real algebraic geometry and topology, 1995

181 **Mila Cenkl and Haynes Miller, Editors,** The Čech Centennial, 1995

(*Continued in the back of this publication*)

Harmonic Analysis and Nonlinear Differential Equations

Victor L. Shapiro

CONTEMPORARY MATHEMATICS

208

Harmonic Analysis and Nonlinear Differential Equations

A Volume in Honor of Victor L. Shapiro

November 3–5, 1995
University of California, Riverside

Michel L. Lapidus, Coordinating Editor

Lawrence H. Harper
Adolfo J. Rumbos
Editors

American Mathematical Society
Providence, Rhode Island

This volume is a collection of papers dealing with harmonic analysis and nonlinear differential equations and stems from a conference in honor of Victor L. Shapiro. This conference was held in conjunction with the fall meeting of the Southern California Analysis and Partial Differential Equations Conference, November 3–5, 1995, in Riverside, CA, and was jointly sponsored by the National Science Foundation and the University of California, Riverside.

1991 *Mathematics Subject Classification.* Primary 35Jxx, 35Pxx, 35Qxx, 42Axx, 42Bxx, 42Cxx, 76D05.

Library of Congress Cataloging-in-Publication Data

Harmonic analysis and nonlinear differential equations : a volume in honor of Victor L. Shapiro : November 3–5, 1995, University of California, Riverside / Michel L. Lapidus, coordinating editor ; Lawrence H. Harper, Adolfo J. Rumbos, editors.

p. cm. — (Contemporary mathematics, ISSN 0271-4132 ; 208)

Includes bibliographical references.

ISBN 0-8218-0565-7 (alk. paper)

1. Harmonic analysis—Congresses. 2. Differential equations, Nonlinear—Congresses. I. Shapiro, Victor L. (Victor Lenard), 1924– . II. Lapidus, Michel L. (Michel Laurent), 1956– . III. Harper, Lawrence H. (Lawrence Hueston), 1938– . IV. Rumbos, Adolfo J., 1962– . V. Series: Contemporary mathematics (American Mathematical Society) ; v. 208.

QA403.H223 1997
515′.2433—dc21

97-8483
CIP

∞ The paper used in this book is acid-free and falls within the guidelines
established to ensure permanence and durability.
Visit the AMS homepage at URL: http://www.ams.org/

10 9 8 7 6 5 4 3 2 1 02 01 00 99 98 97

Table of Contents

Dedication

This volume is dedicated to Victor L. Shapiro in recognition of his long and productive mathematical career, particularly his many valuable contributions to harmonic analysis and quasilinear partial differential equations.

Victor Lenard Shapiro was born in Chicago, Illinois, on October 16, 1924. In September 1943, he entered the United States army at the height of World War II. In September 1944, he was awarded the expert infantry badge for action against the enemy on the South Pacific island of Bougainville. In April 1945, serving as a combat medic with the 132nd infantry regiment, Vic was in the 6th wave of a beachhead landing on Cebu, Phillipine Islands, and saw much action in the ensuing campaign.

After the war, in June 1946, Vic entered the University of Chicago with the help of the G. I. Bill. He received his Ph.D. degree in June 1952. His doctoral dissertation, written under the direction of A. Zygmund, dealt with the localization theory of multiple trigonometric series.

Shapiro was subsequently a professor at Rutgers University (8 years), the University of Oregon (4 years) and the University of California, Riverside (32 years+). In between, he spent a total of 3 years as a member of the Institute for Advanced Study at Princeton, part of the time as a post-doctoral National Science Foundation fellow.

V. L. Shapiro is mainly known for having solved a number of difficult problems in analysis:

(i) He obtained a best possible result on the uniqueness of Abel sums of multiple trigonometric series, establishing as a consequence the analog of Cantor's famous uniqueness theorem for double trigonometric series under circular convergence with a side condition. The side condition was later shown to hold by R. Cooke. Zygmund himself had emphasized the significance of this problem and it had confounded some of the best analysts of the day. Recently, J. Bourgain has extended the theorem to all dimensions $d \geq 3$.

(ii) He proved the analyticity of the Cauchy-Riemann equations under pointwise L^1-total-differentials, which had also been an open problem for many years.

(iii) He proved (in collaboration with his student, R. Crittenden) an analog of Cantor's theorem for Walsh-Fourier series. This had also been a long-standing open problem.

(iv) He did fundamental work on removable singularities and exceptional sets for the Navier-Stokes equations.

(v) Later, shifting over to more general quasilinear differential equations, he wrote a fundamental paper on resonance and the first eigenvalue for such equations. His most recent work involves time-periodic boundary value problems and reaction-diffusion equations.

In 1964, Victor was given an award by the National Science Foundation for his expository paper "Fourier Series in Several Variables" which appeared in the *Bulletin of the American Mathematical Society.* In 1978, Victor was elected the Faculty Research Lecturer at the University of California, Riverside (UCR). In his lecture, entitled "Exceptional Sets and the History of Mathematics," he showed how the apparently arcane technicalities of exceptional sets for trigonometric series had led Cantor to the basic results of set theory, thereby profoundly influencing the course of mathematics.

Victor has also been recognized as an outstanding teacher at the upper division and graduate level, mainly teaching courses on applied mathematics. Bill Wade, one of the 20 students who received their Ph.D. degrees under Victor's tutelage, began his Shapirofest lecture with the following remarks about Vic as a teacher:

"Vic's style was a revelation to me. Before (taking his class) I was convinced that algebra was the only rigorous branch of mathematics, and I was going to be an algebraist. But Vic put in every detail. There was no hand waving about orders of growth, no vague phrases about "as small as you like," no appealing to mysterious principles from physics – just good solid mathematics based on inequalities and estimates. The estimates were carried out in precise, complete detail, in steps which were easy to follow and could not be refuted. Moreover, he shared enough of the motivation behind the results that we could almost anticipate the next result he planned to prove. All of a sudden I realized that analysis was an elegant, natural, almost simple branch of mathematics. And I decided I wanted to work in this area for the rest of my life."

Altogether, Vic, and the doctoral theses of the students working under him, have accounted for 18 papers published in the *Transactions of the American Mathematical Society.*

On his own initiative, Vic created a course at UCR on the History of Mathematics and taught it successfully for a number of years. Recently he did the same for Mathematical Biology. His hobbies are ballroom dancing with his wife, Florence, and hiking with their four children and eight grandchildren.

Because of his distinguished research work, inspirational teaching and exemplary humanity, we, as his students, colleagues and collaborators, dedicate this volume to Victor L. Shapiro.

Lawrence H. Harper, Michel L. Lapidus and Adolfo J. Rumbos

Preface

This volume is a collection of papers dealing with harmonic analysis and nonlinear differential equations and stems from a conference in honor of Victor L. Shapiro specifically on these two mathematical areas and their interface. This conference was held in conjunction with the fall meeting of the Southern California Analysis and Partial Differential Equations Conference, November 3–5, 1995, in Riverside, California, and was jointly sponsored by the National Science Foundation and the University of California, Riverside.

Most of the papers presented in this book are from individuals who spoke at the conference and are closely related to the various areas of analysis in which Vic Shapiro has worked during the last four and one-half decades.

In particular, there are four papers dealing directly with the interface between harmonic analysis and nonlinear partial differential equations, one on the d-torus and the nonlinear Schrödinger equation, two using harmonic analysis to study periodic solutions of the Navier-Stokes equations, one of which has applications to turbulence theory, and a fourth dealing with double Fourier series and time-periodic solutions of the semilinear wave equation. Furthermore, there are two up-to-date overviews of the recent developments in two important branches of harmonic analysis, namely one dealing with the latest spectacular results in the uniqueness theory of multiple trigonometric series and another covering the current work in dyadic harmonic analysis (i.e., Walsh-Fourier series). In addition, there is another overview of important recent work on shock-waves, black holes, and general relativity, as well as a survey (and research) article of recent work on harmonic analysis and PDEs on fractals, and their relationships with noncommutative geometry, of relevance, for example, in condensed matter physics, particularly the study of diffusions and wave propagation in porous media or percolation networks.

There also are several research articles that deal primarily with nonlinear differential equations, one using variational methods to solve higher order quasilinear elliptic equations, another concerned with new results in dynamical systems which have applications to neural networks, a third dealing with the structure of the solution set of semilinear equations involving the Laplacian in $\mathbb{R}^n$, and a fourth concerned with an abstract evolution equation with applications to the blow up of solutions to the nonlinear wave equation. There is another paper in this volume containing recent historical work on the Mary Cartwright-J. E. Littlewood collaboration on the Van der Pol equation, as well as one that deals with the application of dyadic harmonic analysis to coding theory. There is also a paper which gives an overview of recent work in the q-theory of special functions with applications to some classical trigonometric expansions.

A number of the above papers contain open problems in their respective areas.

In closing this foreword, the editors would like to express their appreciation to a number of individuals and sponsors who made the conference and this volume possible.

First, we would like to thank Professor Albert R. Stralka, former Chair of the Department of Mathematics at UCR, for his initial suggestion to organize this symposium in honor of Professor Shapiro. We are also grateful to Professors M. Salah Baouendi and Linda Preiss Rothschild for their advice during the initial preparation of the conference. Further, we wish to thank Cheryl Ann Griffith, Harry Watson, Jr. and Duane Allen for generously donating their time during the conference, as well as Flo Shapiro for her never failing optimism and her help in setting-up the banquet in honor of her husband. Our thanks also go to Jan Patterson for typing-up parts of the manuscripts, to Julie Drouyor for her help in collecting the manuscripts and to Linda Terry for her assistance in planning the conference. The coordinating editor (M. L. L.) is grateful to Professor Donald G. Babbitt, Director of Publications at the American Mathematical Society, for a long phone conversation regarding the possible publication of this volume in *Contemporary Mathematics*, shortly after the conference was held.

Finally, we wish to acknowledge the financial support of the National Science Foundation and the University of California, Riverside (UCR), more specifically, the Department of Mathematics and the Dean of the College of Natural Sciences and Agriculture, as well as the Jones' Chair of Mathematics at UCR.

Michel L. Lapidus, Lawrence H. Harper and Adolfo J. Rumbos

Contemporary Mathematics
Volume **208**, 1997

From Reaction-Diffusion to Spherical Harmonics

Victor L. Shapiro

ABSTRACT. With the concentration on examples and counterexamples, a number of recent results in quasilinear parabolic and elliptic partial differential equations are presented. Also, several illustrations of the close connection between harmonic analysis and the Navier-Stokes system are given. The closing topic gives the background material for a problem open for over seventy-five years in the theory of spherical harmonics.

At the request of the editors of this volume, I give here a survey of some of my work in nonlinear differential equations and harmonic analysis with the emphasis on examples and counterexamples. This review starts with those papers of most recent vintage (Quasilinear Parabolic PDE and Reaction-Diffusion Equations) and works backward chronologically in time to spherical harmonics where a problem open for many years and still unsolved is described. Other papers surveyed deal with various aspects of quasilinear elliptic PDE, in particular the stationary Navier-Stokes system. Throughout the connection between harmonic analysis and differential equations is emphasized. Also, surveyed is a paper dealing with the relationship between number theory, certain fractal sets, and trigonometric series. There is a total of twelve different topics in all.

1. Resonance and Quasilinear Parabolic Partial Differential Equations

Resonance and Quasilinear Parabolic Partial Differential Equations (joint with Lew Lefton), Journal of Differential Equations, Vol. 101 (1993), pp 148–177.

This paper is concerned with using harmonic analysis in combination with Galerkin techniques to obtain results for a time-periodic boundary value problem involving the quasilinear parabolic partial differential equation

$$\partial u/\partial t + Qu = f(x,u) + g(x,t,u) + h(x,t) \tag{1.1}$$

for $(x,t) \in \Omega \times (-\pi,\pi)$ where $\Omega \subset \mathbb{R}^N$ is a bounded open set. Here Q is a quasilinear elliptic operator of order $2m$ and is given by

$$Qu = \sum_{|\alpha|=m} (-1)^{|\alpha|} D^\alpha A_\alpha\big(x,\xi(u)\big) + a_0(x,u)u \tag{1.2}$$

1991 *Mathematics Subject Classification.* Primary: 35K55, 35K35, 76D05, 35Q10, 42B10; Secondary: 33A45, 33A65.

Key words and phrases. Capacity, Fourier series, Navier-Stokes, quasilinear elliptic, quasilinear parabolic, reaction-diffusion, resonance, spherical harmonics, time-periodic.

This paper is in final form and no version of it will be submitted for publication elsewhere.

where $\xi(u(x)) = (D^\alpha u(x))_{|\alpha|=m}$ is a vector in $\mathbb{R}^{s_m}$ and $D^\alpha u = D_1^{\alpha_1} \cdots D_N^{\alpha_N} u$ with $D_j^{\alpha_j} = \partial^{\alpha_j}/\partial x_j^{\alpha_j}$. (The positive integer s_m is the number of N-dimensional multi-indices α of order m. Thus when $N = 2$ and $m = 3$, $s_m = 4$.)

In this paper, we are interested in obtaining results for (1.1) under zero Dirichlet boundary conditions for x and 2π periodic boundary conditions for t. In order to do this we introduce the pre-Hilbert space

$$\mathcal{A} = \{v(x,t) \in C^\infty(\Omega \times \mathbb{R}) \, : \, v(x,t) \text{ is real-valued and satisfies (A-1) and (A-2) } \},$$

where

(A-1) $v(x,t) = v(x, t+2\pi)$, $\forall x \in \Omega$ and $\forall t \in \mathbb{R}$,
(A-2) $\exists E$, a compact subset of Ω, such that $v(x,t) = 0 \ \forall x \in \Omega \setminus E$ and $\forall t \in \mathbb{R}$.

Set $\tilde{\Omega} = \Omega \times \mathbb{T}$, with $\mathbb{T} = (-\pi, \pi)$ and introduce the following inner product in $\mathcal{A}$

$$\langle u, v\rangle_{\tilde{H}} = \langle D_t u, D_t v\rangle_{L^2(\tilde{\Omega})} + \sum_{|\alpha| \le m} \langle D^\alpha u, D^\alpha v\rangle_{L^2(\tilde{\Omega})}. \tag{1.3}$$

The Hilbert space that is obtained by completing $\mathcal{A}$ by the method of Cauchy sequences using the inner product in (1.3) we call $\tilde{H}$. It is clear that functions in $\tilde{H}$ satisfy the boundary conditions stated at the outset in a natural manner.

To state the precise hypothesis on our quasilinear operator Q introduced in (1.1), we proceed as follows. We leave $\xi = (\xi_\alpha)_{|\alpha|=m}$; so ξ is a vector which lies in $\mathbb{R}^{s_m}$. Also, for $u \in \tilde{H}$, we set $\xi_\alpha(u(x,t)) = D^\alpha u(x,t)$ and $\xi(u) = (\xi_\alpha(u))_{|\alpha|=m}$. Now the principal coefficients of Q will be of the form $A_\alpha(x,\xi) \colon \Omega \times \mathbb{R}^{s_m} \to \mathbb{R}$, where $|\alpha| = m$.

For these coefficients we shall assume the following where $|\xi|^2 = \sum_{|\alpha|=m} |\xi_\alpha|^2$.

(Q-1) $A_\alpha(x,\xi)$ is a Caratheodory function on $\Omega \times \mathbb{R}^{s_m}$ (i.e., $A_\alpha(x,\xi)$ is measurable in x, $\forall \xi \in \mathbb{R}^{s_m}$ and continuous in ξ for a.e. $x \in \Omega$) for $|\alpha| = m$.
(Q-2) $\exists c_0 > 0$ s.t. $\sum_{|\alpha|=m} A_\alpha(x,\xi)\xi_\alpha \ge c_0 |\xi|^2$, $\forall \xi \in \mathbb{R}^{s_m}$ and a.e. $x \in \Omega$.
(Q-3) $\exists c_1 > 0$ and $h^\sharp \in L^2(\Omega)$ with $h^\sharp$ nonnegative s.t.

$$|A_\alpha(x,\xi)| \le h^\sharp(x) + c_1|\xi|, \quad \forall \xi \in \mathbb{R}^{s_m}, \ |\alpha| = m, \text{ and a.e. } x \in \Omega.$$

(Q-4) $\sum_{|\alpha|=m} [A_\alpha(x,\xi) - A_\alpha(x,\xi')](\xi_\alpha - \xi'_\alpha) > 0$, $\forall \xi, \xi' \in \mathbb{R}^{s_m}$ with $\xi \ne \xi'$ and for a.e. $x \in \Omega$.
(Q-5) $\int_{\tilde{\Omega}} \sum_{|\alpha|=m} A_\alpha(x, \xi(v)) D^\alpha D_t v = 0$, $\forall v \in \mathcal{A}$.

The first four conditions listed above are the usual ones listed for quasilinear elliptic operators (see [Br] and [LL]). The fifth condition, namely (Q-5), evidently was stated for the first time in the paper under review. It appears to be the condition that makes everything work in the parabolic quasilinear case being considered here and is one of the main conditions to be checked for Q in order to apply the results obtained in the paper. In the sequel, we shall give several examples where condition (Q-5) holds.

For the principal coefficient of Q, $a_0(x,s) \colon \Omega \times \mathbb{R} \to \mathbb{R}$, we shall suppose

(Q-6) $a_0(x,s)$ is a Caratheodory function on $\Omega \times \mathbb{R}$ and $a_0(x,s) \in L^\infty(\Omega \times \mathbb{R})$.

For $u, v \in \tilde{H}$, we introduce the following two-form

$$\mathcal{Q}(u,v) = \sum_{|\alpha|=m} \int_{\tilde{\Omega}} A_\alpha(x, \xi(u)) D^\alpha v + \int_{\tilde{\Omega}} a_0(x,u) uv \tag{1.4}$$

and then define

$$\lambda_1^* = \liminf_{\|u\|_{L^2}\to\infty} \mathcal{Q}(u,u)/\|u\|_{L^2}^2 \qquad u \in \tilde{H}. \tag{1.5}$$

It will be understood from now on, and in particular in (1.5), that L^p and L^2 unqualified will imply the function space over $\tilde{\Omega}$.

λ_1^*, as a "principal" eigenvalue for $\mathcal{Q}$ in conjunction with parabolic differential equations, was evidently defined here for the first time. It was based on the definition given previously for quasilinear elliptic partial differential equations by Shapiro [Sh1].

For the functions on the right-hand side of (1.1), we made the following assumptions where $f\colon \Omega\times\mathbb{R}\to\mathbb{R}$, $g\colon \Omega\times\mathbb{T}\times\mathbb{R}\to\mathbb{R}$, and $h\colon \Omega\times\mathbb{R}\to\mathbb{R}$:

- (f-1) $f(x,s)$ is a Caratheodory function;
- (f-2) $\exists\, c_2 > 0$ and $h_1 \in L^{p'}(\Omega)$ with $h_1 \geq 0$ a.e. in Ω s.t. $|f(x,s)| \leq h_1(x) + c_2|s|^{p-1}$ for a.e. $x\in\Omega$, $\forall\, t\in\mathbb{R}$, where $p = 2(N+1)/(N+1-2)$ for $2 < N+1$, $p > 2$, for $2 \geq N+1$, and $p' = p/(p-1)$;
- (f-3) $\exists\, \Gamma \in \mathbb{R}$ and $h^* \in L^2(\Omega)$ with $h^* \geq 0$ a.e. in Ω s.t. $sf(x,s) \leq \Gamma s^2 + h^*(x)|s|$, $\forall\, s\in\mathbb{R}$ and a.e. $x\in\Omega$.

- (g-1) $g(x,t,s)$ is a Caratheodory function on $\tilde{\Omega}\times\mathbb{R}$;
- (g-2) $\exists\, c_3 > 0$ and $h_2 \in L^2(\tilde{\Omega})$ s.t.

$$|g(x,t,s)| \leq c_3|s| + |h_2(x,t)|, \quad \forall\, s\in\mathbb{R} \text{ and a.e. } (x,t)\in\tilde{\Omega}$$

- (g-3) $\exists\, \Gamma_1 > 0$ and $\tilde{h} \in L^2(\tilde{\Omega})$ with $\tilde{h} \geq 0$ a.e. in $\tilde{\Omega}$ s.t. $sg(x,t,s) \leq \Gamma_1 s^2 + \tilde{h}(x,t)|s|$ a.e. in $\tilde{\Omega}$.

- (h-1) $h \in L^2(\tilde{\Omega})$

Using harmonic analysis in conjunction with Galerkin techniques, we then obtain the following nonresonant result:

THEOREM 1.1. *Let $\Omega \subset \mathbb{R}^N$ be a bounded open set with $\tilde{\Omega} = \Omega\times\mathbb{T}$. Suppose that the coefficients of Q satisfy* (Q-1)–(Q-6), *that f satisfies* (f-1)–(f-3), *g satisfies* (g-1)–(g-3), *and h satisfies* (h-1). *Suppose furthermore that $\Gamma + \Gamma_1 < \lambda_1^*$. Then $\exists\, u \in \tilde{H}_1$ which is a weak solution of* (1.1) *on $\tilde{\Omega}$.*

What we mean by a weak solution of (1.1) is

$$\int_{\tilde{\Omega}} vD_t u + \mathcal{Q}(u,v) = \int_{\tilde{\Omega}} f(x,u)v + \int_{\tilde{\Omega}} g(x,t,u)v + \int_{\tilde{\Omega}} hv \quad \forall\, v\in\tilde{H}.$$

(For a discussion of resonance and nonresonance, see [Laz].) The above result is called nonresonant because $\Gamma+\Gamma_1 < \lambda_1^*$. It is partially superlinear because $p > 2$ in (f-2) and the condition (f-3) is only a one-sided restriction.

A resonance result, i.e. when $\Gamma + \Gamma_1 = \lambda_1^*$, is also obtained in Theorem 2 of this paper, but we won't comment here about this result. Instead we will give two examples of a quasilinear Q that meet the conditions stated in (Q-1)–(Q-5).

For clarity of understanding, we give as our first example a fairly simple second order partial differential equation, (i.e., $m = 1$). For a more sophisticated second order example we refer the reader to page 155 of the paper.

With $\Omega \subset \mathbb{R}^N$, a bounded domain, we define

(1.6)
$$Qu = \sum_{i=1}^{N} D_i \{1 + [1 + b_1(x)|D_1 u|^2 + \cdots + b_N(x)|D_N u|^2]^{-1/2}\} b_i(x) D_i u$$

where

(1.7) $$b_i(x) \in C^0(\overline{\Omega}) \text{ and } b_i(x) > 0 \quad \forall x \in \overline{\Omega},\ i = 1, \ldots, N.$$

Identifying every α with $|\alpha| = 1$, as an i, we see that

(1.8) $$A_i(x, \xi) = \{1 + [1 + \sum_{j=1}^{N} b_j(x)\xi_j^2]^{-1/2}\} b_i(x)\xi_i$$

With $\xi = (\xi_1, \ldots, \xi_N)$, it is clear from (1.7) and (1.8) that (Q-1)–(Q-3) are satisfied. It remains to show that (Q-4) and (Q-5) are satisfied. It is easy to see that for each fixed $x \in \Omega$, $\left[1 + \sum_{i=1}^{N} b_i(x)\xi_i^2\right]^{1/2}$ is a convex function of ξ on $\mathbb{R}^N$. Since the gradient of a convex function is monotone (see [KS, p. 16]) and

$$\partial \left[1 + \sum_{j=1}^{N} b_j(x)\xi_j^2\right]^{1/2} \Big/ \partial \xi_i = \left[1 + \sum_{j=1}^{N} b_j(x)\xi_j^2\right]^{-1/2} b_i(x)\xi_i,$$

it follows from (1.8) that (Q-4) is met.

It remains to show that (Q-5) holds for (1.6). What we have to show is that

(1.9) $$\sum_{i=1}^{N} \int_{\Omega} \int_{0}^{2\pi} A_i\left(x, D_1 v, \ldots, D_N v\right) D_i D_t v = 0 \quad \forall v \in \mathcal{A}$$

where, from (1.8), we see that

(1.10) $$\sum_{i=1}^{N} A_i\left(x, D_1 v, \ldots, D_N v\right) D_i D_t v =$$
$$\sum_{i=1}^{N} \left\{1 + [1 + \sum_{j=1}^{N} b_j(x)|D_j v|^2]^{-1/2}\right\} b_i(x) D_i v D_i D_t v$$

It is easy to see that the right-hand side of (1.10) is

$$\sum_{i=1}^{N} 2^{-1} b_i(x) D_t (D_i v)^2 + D_t [1 + \sum_{j=1}^{N} b_j(x)|D_j v|^2]^{1/2}.$$

Hence it follows from the periodicity of v and $D_i v$ with respect to t that (1.9) holds. Consequently, Qu defined by (1.6) meets (Q-1)–(Q-5) and our first example is complete.

For an example of Qu which meets (Q-1)–(Q-5) but is a higher order elliptic operator, we take m a positive integer and

(1.11) (i) $b_{\alpha\alpha}(x) \in C^0(\overline{\Omega})$ for $|\alpha| = m$;

(ii) $b_{\alpha\alpha}(x) \geq c_1$, $\forall x \in \Omega$, where $c_1 > 0$.

Also, for each α with $|\alpha| = m$, we let

(1.12) $$F_\alpha \colon [0, \infty) \to \mathbb{R}^+$$

be continuous, nondecreasing, and bounded. We then define

$$Qu = \sum_{|\alpha|=m} (-1)^m D^\alpha \left\{ \left[1 + F_\alpha(|D^\alpha u|)\right] b_{\alpha\alpha}(x) D^\alpha u \right\}. \tag{1.13}$$

We consequently see from (1.2) and (1.13) that with $\xi = (\xi_\alpha)_{|\alpha|=m}$

$$A_\alpha(x,\xi) = \left[1 + F_\alpha(|\xi_\alpha|)\right] b_{\alpha\alpha}(x)\xi_\alpha \tag{1.14}$$

Hence, it follows from (1.11), (1.12), and (1.14) that (Q-1) to (Q-3) hold. To see that (Q-4) holds, we observe from (1.11) and (1.14) that

$$\sum_{|\alpha|=m} \left[A_\alpha(x,\xi) - A_\alpha(x,\xi')\right](\xi_\alpha - \xi'_\alpha) \geq c_1 \sum_{|\alpha|=m} (\xi_\alpha - \xi'_\alpha)^2$$
$$+ \sum_{|\alpha|=m} b_{\alpha\alpha}(x)\left[F_\alpha(|\xi_\alpha|)\xi_\alpha - F_\alpha(|\xi'_\alpha|)\xi'_\alpha\right](\xi_\alpha - \xi'_\alpha).$$

Since $F_\alpha(t)$ is a nondecreasing function on $[0,\infty)$, it follows that (Q-4) holds.

It remains to show that (Q-5) holds. In order to do this we set

$$G_\alpha(s) = \int_0^s F_\alpha\left(r^{1/2}\right) dr \quad \text{for } s \geq 0. \tag{1.15}$$

Then it follows that

$$D_t G_\alpha\left(|D^\alpha v|^2\right) = 2F_\alpha\left(|D^\alpha v|\right) D^\alpha v D_t D^\alpha v$$

Hence it follows from (1.14) that

$$2\int_\Omega \left[\int_0^{2\pi} A_\alpha\big(x,\xi(v)\big) D^\alpha D_t v \, dt\right] dx$$
$$= \int_\Omega b_{\alpha\alpha}(x) \left\{\int_0^{2\pi} \left[D_t\left(D^\alpha v\right)^2 + D_t G_\alpha\left(|D^\alpha v|^2\right)\right] dt\right\} dx$$

for $|\alpha| = m$ and $v \in \mathcal{A}$.

It is clear from the periodicity of the v in $\mathcal{A}$ that the integral in braces on the right-hand side of this last equality is zero for each $x \in \Omega$. We conclude that Q does indeed meet (Q-5), and our second example is complete.

2. Time-Periodic Quasi-Linear Reaction-Diffusion Equations

Time-Periodic Quasi-Linear Reaction-Diffusion Equations (joint with Mary Legner), SIAM Journal of Math. Anal, Vol. 26 (1996), pp 135–169.

This paper is concerned with the following system of parabolic partial differential equations:

$$D_t u_i + \rho_i Q_i u_i = \rho_i \lambda_1^* u_i + f_i\big(x,t,\xi_m(u_1),\ldots,\xi_m(u_{N_1})\big) \tag{2.1}$$

for $i = 1,\ldots,N_1$ and for $(x,t) \in \Omega \times (-\pi,\pi)$ where $\Omega \subset \mathbb{R}^N$ is a bounded open set. Here Q_i is a $2m$-th order quasilinear elliptic differential operator, λ_1^* is its common associated first eigenvalue, $D_t u = \partial u/\partial t$, $\xi_m(u) = \left\{D^\alpha u : |\alpha| \leq m\right\}$, and ρ_i is a positive diffusion coefficient.

In the paper, the right-hand side of (2.1) is actually more general, but for expository purposes what we have here will suffice. Systems of the form (2.1) play an important role in applied mathematics and generally are referred to as reaction-diffusion equations. In particular, in mathematical biology when the diffusion coefficients ρ_i are different, they give rise to the Turing mechanism which is important in the theory of pattern formation called morphogenesis (see the interesting article [M1] and further discussion in [M2, pp 375-9]). Also, in predator-prey problems autocatalytic systems of the type (2.1) are important. In this survey, we shall state Theorems 2.1 and 2.2 of the paper and then give a quasilinear autocatalytic system that is covered by Theorem 2.2 and another example which illustrates Theorem 2.1.

To be quite explicit

$$Q_i u = \sum_{1\le|\alpha|\le m} (-1)^{|\alpha|} D^\alpha A^i_\alpha\big(x,\xi_m(u)\big) + a^i_0\big(x,\xi_m(u)\big)u, \tag{2.2}$$

and we see that Q_i is more general than its analogue in (1.2) of §1. Also, in this current section, $\xi_m = \{\xi_\alpha : |\alpha| \le m\}$ and we refer to the dimension of the vector space as $\mathbb{R}^{s_m}$. (Q-1)–(Q-5) now become the following:

(Q-1) $A^i_\alpha(x,\xi_m) : \Omega \times \mathbb{R}^{s_m} \to \mathbb{R}$ satisfies the Carathéodory conditions (i.e., $A^i_\alpha(x,\xi_m)$ is measurable in x $\forall\,\xi_m \in \mathbb{R}^{s_m}$ and continuous in ξ_m for a.e. $x\in\Omega$ for $1\le|\alpha|\le m$).

(Q-2) $\exists\, c_0 > 0$ and $h^\sharp \in L^2(\Omega)$ with $h^\sharp$ nonnegative such that $|A^i_\alpha(x,\xi_m)| \le h^\sharp(x) + c_0|\xi_m|$ $\forall\,\xi_m\in\mathbb{R}^{s_m}$, $1\le|\alpha|\le m$, and a.e. $x\in\Omega$ for $i=1,\ldots,N_1$.

(Q-3) $\sum_{|\alpha|=m}\big[A^i_\alpha(x,\eta_{m-1},\zeta_m) - A^i_\alpha(x,\eta_{m-1},\zeta'_m)\big](\zeta_\alpha - \zeta'_\alpha) > 0$ for a.e. $x\in\Omega$, $\forall\,(\eta_{m-1},\zeta_m),(\eta_{m-1},\zeta'_m)\in\mathbb{R}^{s_m}$ with $\zeta_m\ne\zeta'_m$, where $\xi_m = (\eta_{m-1},\zeta_m)$, and for $i=1,\ldots,N_1$.

(Q-4) $\exists\, c_1 > 0$ such that $\sum_{|\alpha|=m} A^i_\alpha(x,\xi_m)\zeta_\alpha \ge c_1|\zeta_m|^2$ $\forall\,\xi_m\in\mathbb{R}^{s_m}$ and a.e. $x\in\Omega$ for $i=1,\ldots,N_1$.

(Q-5) $a^i_0(x,\xi_m)\colon \Omega\times\mathbb{R}^{s_m}\to\mathbb{R}$ satisfies the Carathéodory conditions and $a^i_0(x,\xi_m) \in L^\infty(\Omega\times\mathbb{R}^{s_m})$.

Next, we designate the Hilbert Space $\tilde H$ and the pre-Hilbert space $\mathcal{A}$ as before §1 and we use the same inner product $\langle u,v\rangle_{\tilde H}$ as in (1.3).

For our quasilinear operator $Q_i u$ in (2.2) we introduce the two-form

$$\mathcal{Q}_i(u,v) = \sum_{1\le|\alpha|\le m}\int_{\tilde\Omega} A^i_\alpha\big(x,\xi_m(u)\big)D^\alpha v + \int_{\tilde\Omega} a_0\big(x,\xi_m(u)\big)uv \tag{2.3}$$

which is well-defined by (Q-1)–(Q-5) $\forall\, u,v\in\tilde H$ and $i=1,\ldots,N_1$. We also suppose in a manner similar to (Q-5) of §1 that

(Q-6) $\mathcal{Q}_i(u,D_t v) = 0$ $\forall\, v\in\mathcal{A}$ and $i=1,\ldots,N_1$

This last key assumption appear to be one of the main ideas which makes things work for quasilinear operators.

Motivated by (1.5) in §1, we define

$$\lambda^*_{1i} = \liminf_{\|u\|_{L^2}\to\infty} \frac{\mathcal{Q}_i(u,u)}{\|u\|^2_{L^2}}, \quad u\in\tilde H. \tag{2.4}$$

It is not difficult on the basis of (Q-1)–(Q-5) to show that λ^*_{1i} is a finite real-valued number. λ^*_{1i} plays the role of the first (or principal) eigenvalue of Q_i. We suppose throughout this paper that there is a common value so that

$$\lambda^*_{1i} = \cdots = \lambda^*_{1N_1} = \lambda^*_1 \tag{2.5}$$

Next we set

$$Lu = \sum_{|\alpha|,|\beta|\le m} (-1)^{|\alpha|} D^\alpha\big[b_{\alpha\beta}(x)D^\beta u\big] \tag{2.6}$$

where $b_{\alpha\beta} \in C^0(\overline{\Omega})$ and satisfy the usual hypotheses that make L a symmetric, uniformly elliptic operator of order $2m$. We then define

$$\lambda_1 = \inf_{u \neq 0} \frac{\mathcal{L}(u,u)}{\|u\|_{L^2}}, \quad u \in \tilde{H} \tag{2.7}$$

where

$$\mathcal{L}(u,v) = \int_{\tilde{\Omega}} \sum_{|\alpha|,|\beta| \leq m} b_{\alpha\beta} D^\alpha u D^\beta v \tag{2.8}$$

for $u, v \in \tilde{H}$. Also, we set

$$\|u\|^2_{W^{m,2,\sim}} = \int_{\tilde{\Omega}} \left|\xi_m(u)\right|^2 \quad \forall u \in \tilde{H}. \tag{2.9}$$

With $0 \leq \kappa < 1$, we say Q_i is $\kappa*$-related to L if the following conditions hold:

$$\lambda^*_{1i} = \lambda_1; \tag{2.10}$$

(2.11) (i)

$$u \in \tilde{H}, \qquad \liminf_{\|u\|_{W^{m,2,\sim}} \to \infty} \frac{\mathcal{Q}_i(u,u) - \mathcal{L}(u,u)}{\|u\|^{1+\kappa}_{W^{m,2,\sim}}} \geq 0,$$

(2.11) (ii) $\exists$ positive constants $c_1^\sharp, c_2^\sharp$ such that

$$\mathcal{Q}_i(u,u) - \mathcal{L}(u,u) \geq -c_1^\sharp \|u\|^{\kappa+1}_{W^{m,2,\sim}} - c_2^\sharp \qquad \forall u \in \tilde{H}.$$

For the f_i given in (2.1), we impose the following conditions:

(f-1) $f_i(x,t,\xi_{m1},\ldots,\xi_{mN_1})\colon \tilde{\Omega} \times \underbrace{\mathbb{R}^{s_m} \times \cdots \times \mathbb{R}^{s_m}}_{N_1} \to \mathbb{R}$ satisfies the Carathéodory conditions;

(f-2) $\exists \kappa$ with $0 \leq \kappa < 1$ and $\exists$ nonnegative constants c_α and a nonnegative function $d(x,t) \in L^2(\tilde{\Omega})$ such that

$$|f_i(x,t,\xi_{m1},\ldots,\xi_{mN_1})| \leq \sum_{|\alpha| \leq m} c_\alpha |\xi_{\alpha i}|^\kappa + d(x,t)$$

$\forall (\xi_{m1},\ldots,\xi_{mN_1}) \in \underbrace{\mathbb{R}^{s_m} \times \cdots \times \mathbb{R}^{s_m}}_{N_1}$ a.e. in $\tilde{\Omega}$ for $i = 1,\ldots,N_1$ where $\xi_{mi} = \left\{\xi_{\alpha i}\right\}_{|\alpha| \leq m}$;

(f-3) Let $S_1' = \{\xi_m \in \mathbb{R}^{s_m} : |\xi_m| = 1 \text{ and } \xi_{(0,\ldots,0)} \neq 0\}$. $\exists \tilde{\Omega}_0 \subset \tilde{\Omega}$ with meas$(\tilde{\Omega} - \tilde{\Omega}_0) = 0$ such that the following uniform limit holds for each $(x,t) \in \tilde{\Omega}_0$

$$\lim_{n \to \infty} \frac{f_i(x,t,\xi_{m1},\ldots,r_n\xi^{(n)}_{mi},\ldots,\xi_{mN_1})}{r_n^\kappa} = h_i(x,t,\xi_{mi})$$

uniformly for $(\xi_{m1},\ldots,\xi_{m(i-1)},\xi_{m(i+1)},\ldots,\xi_{mN_1}) \in \underbrace{\mathbb{R}^{s_m} \times \cdots \times \mathbb{R}^{s_m}}_{N_1 - 1}$

for all sequences $\{r_n\} \subset \mathbb{R}$ and $\xi^{(n)}_{mi} \subset S_1'$ such that $r_n \to \infty$ and $\xi^{(n)}_{mi} \to \xi_{mi} \in S_1'$, where $h_i(x,t,\xi_{mi})$ is a Carathéodory function on $\tilde{\Omega} \times S_1'$ that is uniformly bounded by a function in $L^2(\tilde{\Omega})$.

In this paper, we establish two theorems, the first of which is the following:

THEOREM 2.1. *Let Ω and κ be as above. Suppose that the coefficients of Q_i satisfy* (Q-1)–(Q-6); *Q_i is $\kappa*$-related to L; and f_i satisfies* (f-1)–(f-3). *Also suppose*

(2.12)
$$\int_{\tilde{\Omega}(v)} h_i\left(x,t,\frac{\xi_m(v)}{|\xi_m(v)|}\right)|\xi_m(v)|^{\kappa} v < 0 \text{ for every } v \neq 0 \text{ in } S^{\sharp}_{\lambda_1}$$

for $i = 1,\ldots,N_1$ where h_i is given in (f-3) *and $\tilde{\Omega}(v) = \{(x,t) \in \tilde{\Omega} \,|\, v(x,t) \neq 0\}$. Then $\exists\,(u_1,\ldots,u_{N_1}) \in \tilde{H}^{N_1}$ which is a weak solution to the system* (2.1) *on $\tilde{\Omega}$*

By $S^{\sharp}_{\lambda_1}$ in (2.12) we mean

$$S^{\sharp}_{\lambda_1} = \{v = \sum_{j=1}^{n_1} d_j\phi_j \;:\; d_j \text{ is a real constant, } j = 1,\ldots,n_1\}$$

where n_1 is the multiplicity of λ_1 as an eigenvalue of L and $\{\phi_j\}_{j=1}^{n_1}$ is an orthonormal set of corresponding eigenfunctions.

What we mean by a weak solution of (2.1) is

(2.13)
$$\langle D_t u_i, v\rangle + \rho_i \mathcal{Q}(u_i,v) = \rho_i\lambda_1^*\langle u_i,v\rangle_{L^2} + \big\langle f\big(\cdot,\xi_m(u_1),\ldots,\xi_m(u_{N_1})\big),v\big\rangle_{L^2}$$

$\forall v \in \tilde{H}$ and $i = 1,\ldots,N_1$.

The second theorem established in contrast to the above theorem is a nonresonance result and deals with the following system:

(2.14)
$$D_t u_i + \rho_i Q_i u_i = \rho_i(\lambda_1^* - \delta_i)u_i + f_i\big(x,t,\xi_m(u_1),\ldots,\xi_m(u_{N_1})\big)$$

where $\delta_i > 0$ for $i = 1,\ldots,N_1$.

THEOREM 2.2. *Let Ω and κ be as above. Suppose that the coefficients of Q_i satisfy* (Q-1)–(Q-6), *Q_i is $\kappa*$-related to L, and that f_i satisfies* (f-1) *and* (f-2) *for $i = 1,\ldots,N_1$. Then $\exists\,(u_1,\ldots,u_{N_1}) \in \tilde{H}^{N_1}$ which is a weak solution to the system* (2.14).

In the paper, we show for L given by (2.6) that if furthermore

(2.15)
$$\sum_{|\alpha|=|\beta|=m} b_{\alpha\beta}(\zeta_\beta - \zeta'_\beta)(\zeta_\alpha - \zeta'_\alpha) \geq c'_1 \sum_{|\alpha|=m}(\zeta_\alpha - \zeta'_\alpha)^2 \quad \forall\{\zeta_\alpha\}_{|\alpha|=m} \text{ and } \{\zeta'_\alpha\}_{|\alpha|=m}.$$

where c'_1 is a positive constant then

(2.16)
$$Lu = \sum_{|\alpha|=|\beta|=m}(-1)^{|\alpha|}D^\alpha\left[b_{\alpha\beta}D^\beta u\right]$$
$$+ \sum_{1\leq|\alpha|,|\beta|\leq m-1}(-1)^{|\alpha|}D^\alpha\left[b_{\alpha\beta}D^\beta u\right] + b_{00}(x)u$$

is $\kappa*$-related to itself. So in particular Theorems 1 and 2 hold when the right-hand side of (2.1) and (2.14) are linear, symmetric, uniformly elliptic operators of the form (2.16) satisfying (2.15). The results of both theorems appear to be new even for this linear case. We observe that $(-1)^m\Delta^m u$, where Δ^m is the familiar mth iterated Laplacian, meets (2.15) and (2.16) and hence is $\kappa*$-related to itself.

We now present a second order quasilinear Q which is $\kappa *$-related to L with L being given by (2.20) below. For our Q, we take $m=1$ and Qu to be defined as in (1.6) and (1.7) in §1. Then using our present notation $a_0(x,\xi_1(u))=0$ and

(2.17)
$$A_\alpha(x,\xi_1(u))=\{1+\big[1+\sum_{|\beta|=1} b_{\beta\beta}(x)|\xi_\beta(u)|^2\big]^{-1/2}\}b_{\alpha\alpha}(x)\xi_\alpha(u)$$

where $\xi_\alpha(u)=D^\alpha u$ and

(2.18)
$$b_{\alpha\alpha}(x)\in C^0(\overline{\Omega}) \text{ and } b_{\alpha\alpha}(x)\geq c_1$$
$$\forall\, x\in\overline{\Omega} \text{ and } |\alpha|=1 \text{ where } c_1 \text{ is a positive constant.}$$

Using (2.2) we see that Qu is then given by

(2.19)
$$Qu=-\sum_{|\alpha|=1} D^\alpha A_\alpha(x,\xi_1(u)),$$

with A_α defined by (2.17). As we showed in §1, under slightly different notation, Qu satisfies (Q-1)–(Q-6). We set

(2.20)
$$Lu=-\sum_{|\alpha|=1} D^\alpha\big[b_{\alpha\alpha}(x)D^\alpha u\big]$$

and observe from (2.18) that Lu meets (2.15) and (2.16) so that Lu is a symmetric uniformly elliptic operator which is $\kappa *$-related to itself. What we want to show is that Q is $\kappa *$-related to L. In order to do this, we observe from (2.3), (2.8), (2.17), (2.19), and (2.20) that

(2.21)
$$\mathcal{Q}(u,u)-\mathcal{L}(u,u)=\sum_{|\alpha|=1}\int_{\tilde{\Omega}} b_{\alpha\alpha}(x)\big[1+\sum_{|\beta|=1} b_{\beta\beta}(x)|D^\beta u|^2\big]^{-1/2}|D^\alpha u|^2.$$

It follows from (2.18) that the right-hand side of (2.21) is nonnegative $\forall\, u\in\tilde{H}$. Hence we see that (2.11)(i) and (ii) both hold and that

(2.21′)
$$\mathcal{Q}(u,u)\geq\mathcal{L}(u,u)\quad\forall\, u\in\tilde{H}.$$

To complete the proof that Q is $\kappa *$-related to L it remains to show that (2.10) holds. From (2.4), (2.7), and (2.21′) we see that $\lambda_1^*\geq\lambda_1$. To show the opposite inequality, we recall that it is well-known for L given by (2.20) that λ_1 is a simple eigenvalue and furthermore that $\exists$ a unique $\phi_1>0$ which depends only on x and not on t with $\|\phi_1\|_{L^2}=1$ such that $\mathcal{L}(\phi_1,\phi_1)=\lambda_1$. Taking $u=s\phi_1$ in (2.21) and observing from (2.18) that $\exists\, c_1>0$ such that $b_{\alpha\alpha}(x)\geq c_1$, $\forall\,|\alpha|=1$ and $x\in\overline{\Omega}$, we obtain that

$$\mathcal{Q}(s\phi_1,s\phi_1)\leq\lambda_1 s^2+\sum_{|\alpha|=1}\int_\Omega |b_{\alpha\alpha}(x)|^{1/2}|D^\alpha u|s.$$

Dividing both sides of this inequality by $\|s\phi_1\|_{L^2}^2$ and passing to the limit as $s\to\infty$, we see from (2.4) that $\lambda_1^*\leq\lambda_1$. Hence $\lambda_1^*=\lambda_1$ and Q is indeed $\kappa *$-related to L.

We next give a example of a higher order Q which is $\kappa *$-related to an L. For this, we take Qu to be defined by (1.13) where F_α is defined by (1.12). Then, as we have shown, Q meets (Q-1)–(Q-6). For F_α, we furthermore assume

(2.22) (i)
$$0\leq F_\alpha(s)\leq 1\qquad\forall\, s\in[0,\infty);$$

(2.22) (ii) $$\lim_{s\to\infty} s\big[1 - F_\alpha(s)\big] = 0.$$

Examples meeting (1.12) and (2.22) are $F_\alpha(x) = s/(1+s^2)^{1/2}$, $s^2/(1+s^2)$, $\big[(1+s^2)/(2+s^2)\big]^{1/2}$. We define for m a positive integer

(2.23) $$Lu = 2\sum_{|\alpha|=m}(-1)^{|m|}D^\alpha\big[b_{\alpha\alpha}(x)D^\alpha u\big]$$

where $b_{\alpha\alpha}(x)$ is given by (1.11). It then follows from (1.13) and (2.23) that

(2.24)
$$\mathcal{Q}(u,u) - \mathcal{L}(u,u) = -\sum_{|\alpha|=m}\int_{\tilde\Omega} b_{\alpha\alpha}(x)\big[1 - F_\alpha(|D^\alpha u|)\big]\,|D^\alpha u|^2.$$

Observing from (2.22) that given $\epsilon > 0$, $\exists\ s_0$ such that $s[1-F_\alpha(s)] \le \epsilon$ for $s \ge s_0$, it is easy to see from (2.24) that

$$\limsup_{\|u\|_{W^{m,2,\sim}}\to\infty}\frac{|\mathcal{Q}(u,u)-\mathcal{L}(u,u)|}{\|u\|^{1+\kappa}_{W^{m,2,\sim}}} \le \epsilon$$

and therefore that (2.11)(i) holds. Likewise, it is easy to see from (2.22) that (2.11)(ii) holds. So to show that Q is $\kappa*$-related to L it remains to show that (2.10) holds, i.e., that $\lambda_1^* = \lambda_1$.

Now it follows from (1.13) and (2.24) that

(2.25) $$2^{-1}\mathcal{L}(u,u) \le \mathcal{Q}(u,u) \le \mathcal{L}(u,u).$$

Hence that

(2.26) $$2^{-1}\lambda_1 \le \lambda_1^* \le \lambda_1 \text{ and where } \lambda_1 > 0.$$

To show that $\lambda_1 \le \lambda_1^*$ let $\{u_n\}_{n=1}^\infty$ be a sequence of functions in $L^2(\tilde\Omega)$ such that

(2.27) $$\lim_{n\to\infty}\mathcal{Q}(u_n,u_n)/\|u_n\|^2_{L^2} = \lambda_1^*.$$

Then it follows from this last limit, (2.26) and (1.13) that $\exists\, c_2 > 0$ such that $\sum_{|\alpha|=m}\int_{\tilde\Omega}\big|D^\alpha u_n\big|^2 \le c_2\|u_n\|^2_{L^2}\ \forall n$. But then it follows from this last inequality, (2.22)(ii), and (2.24)

(2.28) $$\lim_{n\to\infty}\frac{\mathcal{Q}(u_n,u_n)-\mathcal{L}(u_n,u_n)}{\|u_n\|^2_{L^2}} = 0.$$

However, from the definition of λ_1, $\mathcal{L}(u_n,u_n) \ge \lambda_1\|u_n\|^2_{L^2}$, and we consequently conclude from (2.27) and (2.28) that $\lambda_1 \le \lambda_1^*$. This coupled with (2.26) gives (2.10) and we conclude that Q given by (1.13) is indeed $\kappa*$-related to L which is given by (2.23).

We next give an example of an autocatalytic reaction-diffusion system which is quasilinear and meets Theorem 2.2. For this system we shall take $m=1$, $N=1$, $\Omega = (0,\pi)$, and $N_1 = 2$. Also, we shall take $\kappa = 0$. We set

$$Q_1(u_1) = -D_1\big[D_1u_1 + D_1u_1\ /\ (1+|D_1u_1|^2)^{1/2}\big]$$
$$Q_2(u_2) = -2^{-1}D_1\big[D_1u_2 + |D_1u_2|^2D_1u_2\ /\ (1+|D_1u_2|^2)\big]$$
$$Lu = -D_1\big(D_1u\big)$$

It follows from our example involving (2.17) and (2.20) with $b_{11} = 1$ that Q_1 is $\kappa*$-related to L. Likewise from our example (1.13) with $b_{11} = 1$ and $F_1(s) =$

$s^2/(1+s^2)$ (so condition (2.22) holds), it follows that Q_2 is also $\kappa*$-related to L. We furthermore observe that $\lambda^*_{11} = \lambda^*_{12} = \lambda_1 = 1$.

Next, we take

$$f_1(x,t,u_1,u_2) = -\frac{\arctan u_2}{1+u_1^2} + b_1(x,t)$$

$$f_2(x,t,u_1,u_2) = \frac{\tanh u_1}{1+u_2^2} + b_2(x,t)$$

where $b_1(x,t)$ and $b_2(x,t)$ are in $C^0\big([0,\pi]\times[0,2\pi],\mathbb{R}\big)$. We observe conditions (f-1) and (f-2) are met with $\kappa = 0$. Also, we set

$$F_1(x,t,u_1,u_2) = \rho_1(1-\delta_1)u_1 + f_1(x,t,u_1,u_2) \tag{2.28$'$}$$

$$F_2(x,t,u_1,u_2) = \rho_2(1-\delta_2)u_2 + f_2(x,t,u_1,u_2),$$

and observe from Theorem 2.2 with $\kappa = 0$ that $\exists\,(u_1,u_2) \in \tilde{H}^2$, which is a weak solution to the reaction-diffusion system

$$\frac{\partial u_1}{\partial t} + \rho_1 Q_1 u_1 = F_1(x,t,u_1,u_2) \tag{2.29}$$

$$\frac{\partial u_2}{\partial t} + \rho_2 Q_2 u_2 = F_2(x,t,u_1,u_2)$$

on the interval $(0,\pi)\times(0,2\pi)$ where ρ_1, ρ_2, δ_1, and δ_2 are positive numbers.

Important in mathematical biology are systems which are autocatalytic of activator-inhibitor type (e.g., see [Be]), i.e., $\frac{\partial F_1}{\partial u_1} > 0$, $\frac{\partial F_2}{\partial u_1} > 0$, and $\frac{\partial F_1}{\partial u_2} < 0$. It is easy to see from $(2.28)'$ that these conditions are met provided $\rho_1 > \pi/(1-\delta_1)$.

To conclude our comments concerning reaction-diffusion equations we give a simple example of a system which meets Theorem 2.1. For this example we take $m = 1$, $0 < \kappa < 1$, $N = 1$, $\Omega = (0,\pi)$, and $N_1 = 2$. Also we take $Lu = -D_1^2 u$ and $Qu = -\{D_1^2 u + D_1[D_1 u/(1+D_1 u^2)^{1/2}]\}$ with $Q_1 u_1 = Q(u_1)$ and $Q_2(u_2) = Q(u_2)$. We also take with $\xi_{1j} = (\eta_{0j},\zeta_{1j})$, $j = 1,2$; so that $\eta_{01}(u_1) = u_1$ and $\zeta_{11}(u_1) = D_1 u_1$. We set for $j = 1,2$ and $k \neq j$,

$$f_j(x,t,\xi_{11},\xi_{12}) = -(\sin x)^4 b_2(t)\,[|\eta_{0j}|^\kappa \arctan\eta_{0j} + |\zeta_{1j}|^\kappa \arctan\zeta_{1j}]$$

$$+ \arctan(\eta_{0k}^2 + \zeta_{1k}^2)/(1+\eta_{0j}^2)^{1/2} + d_j(x,t),$$

where $b_2(t) > 0$ and $b_2(t) \in C^0\big([0,2\pi]\big)$ with $d_j(x,t) \in C^0\big([0,\pi]\times[0,2\pi]\big)$.

We then have that

$$h_j(x,t,\xi_{1j}) = -\tfrac{\pi}{2}(\sin x)^4 b_2(t)\,[|\eta_{0j}|^\kappa \operatorname{sgn}\eta_{0j} + |\zeta_{1j}|^\kappa \operatorname{sgn}\zeta_{1j}]$$

for $|\xi_{1j}| = 1$ and $\eta_{0j} \neq 0$, and it is clear that (f-1), (f-2), and (f-3) hold for $0 < \kappa < 1$. Our system is

$$D_t u_j + Q_j u_j = f_j(x,t,u_1,D_1u_1,u_2,D_2u_2) \quad j = 1,2, \tag{2.30}$$

where now $S^\sharp_{\lambda_1} = \{c\sin x \;:\; c = \text{constant}\}$ and $\lambda_1 = 1$. It is easy to see that the condition (2.12) in Theorem 2.1 holds. So Theorem 2.1 gives that a weak solution $(u_1,u_2) \in \tilde{H}^2$ for the system (2.30) exists.

In the original paper an example for Theorem 2.1 when $\kappa = 0$ is also given.

3. Special Functions and Singular Quasilinear Partial Differential Equations

Special Functions and Singular Quasilinear Partial Differential Equations, SIAM J. Math. Anal., Vol 22 (1991), pp 1411–1429.

Let $\Omega \subset \mathbb{R}^N$, $N \geq 1$, be an open (possibly unbounded) connected set and p and ρ be two positive functions in Ω with $p \in C^0(\overline{\Omega})$ and $\rho \in C^0(\Omega)$. We set

$$Qu = -\sum_{i=1}^{N} D_i\big[p(x)A_i(x,u,Du)\big] + \rho(x)B_0(x,u,Du)u \tag{3.1}$$

where $D_i = \partial/\partial x_i$ and $D = (D_1, \ldots, D_N)$, and we give here a resonance and a nonresonance result with two examples for the equation

$$Qu = \rho f(x,u) + G \tag{3.2}$$

where $G \in \big[H^1_{p,\rho}(\Omega)\big]^*$, the dual of the Hilbert space $H^1_{p,\rho}(\Omega)$ which we now define.

To do this, we suppose that both $\int_\Omega \rho < \infty$ and $\int_\Omega p < \infty$ and set

$$\text{(3.3)}\quad \text{(i)} \qquad C^0_\rho(\Omega) = \big\{u \in C^0(\Omega) \,:\, \int_\Omega u^2\rho < \infty\big\},$$

(3.3) (ii)

$$C^1_{p,\rho}(\Omega) = \big\{u \in C^0(\overline{\Omega}) \cap C^1(\Omega) \,:\, u = 0 \in \partial\Omega \setminus \Gamma_1 \text{ and } \int_\Omega \big[|Du|^2 p + u^2\rho\big] < \infty\big\}$$

where $\Gamma_1 = \{x \in \partial\Omega \,:\, p(x) = 0\}$.

For $u, v \in C^0_\rho(\Omega)$, we set $\langle u, v\rangle = \int_\Omega vu\rho$; for $u, v \in C^1_{p,\rho}(\Omega)$, we set $\langle u, v\rangle_{p,\rho} = \int_\Omega \big[(Du \cdot Du)\rho + uv\rho\big]$. $H^1_{p,\rho}(\Omega)$ will be the Hilbert space that we get by the completing of $C^1_{p,\rho}(\Omega)$ with respect to $\langle u, v\rangle_{p,\rho}$. A similar definition holds for the Hilbert space $L^2(\Omega)$ and $\langle u, v\rangle_\rho$. Also, we shall suppose $\exists\, K > 0$ such that $\|u\|_p \leq K\|u\|_\rho$ $\forall\, u \in L_\rho$ where $\|u\|^2_\rho = \langle u, u\rangle_\rho$. Similarly we define $\|u\|^2_{p,\rho} = \langle u, u\rangle_{p,\rho}$.

For $A_i(x,s,\xi) \colon \Omega \times \mathbb{R} \times \mathbb{R}^N \to \mathbb{R}$ and $B_0(x,s,\xi)$ we assume

(Q-1) A_i and B_0 satisfy the Carathéodory conditions;
(Q-2) $|A_i(x,s,\xi)| \leq h(x) + c_1(s^2 + |\xi|^2)^{1/2}$ a.e. in Ω with $h \in L^2(\Omega)$;
(Q-3) $B_0 \in L^\infty(\Omega \times \mathbb{R} \times \mathbb{R}^N)$;
(Q-4) $\sum_{i=1}^N A_i(x,s,\xi_i)\xi_i \geq c_2|\xi|^2$; for a.e. $x \in \Omega\ \forall\,(s,\xi) \in \mathbb{R} \times \mathbb{R}^N$;
(Q-5) $\sum_{i=1}^N \big[A_i(x,s,\xi) - A_i(x,s,\xi')\big](\xi_i - \xi'_i) > 0$ for a.e. $x \in \Omega$, $\forall\, s \in \mathbb{R}$, and $\forall\, \xi, \xi' \in \mathbb{R}^N$ with $\xi \neq \xi'$.

We introduce the semilinear form

$$\mathcal{Q}(u,v) = \sum_{i=1}^N \int_\Omega \big[pA_i(x,u,Du)D_i v + \rho B_0(x,u,Du)uv\big] \tag{3.4}$$

for $u, v \in H^1_{p,\rho}$ and see from (Q-1)–(Q-3) that $\mathcal{Q}(u,v)$ is well-defined. Also, we set

$$\lambda_1^* = \liminf_{\|u\|_\rho \to \infty} \mathcal{Q}(u,u)/\|u\|^2_\rho, \qquad u \in H^1_{p,\rho}. \tag{3.5}$$

An easy computation using (Q-2)–(Q-4) shows that λ_1^* is finite-valued. With

(f-1) the Carathéodory assumption as in §1, and

(f-2) $\exists\, h_1 \in L^2_\rho$ and $\epsilon_0 > 0$ such that

$$sf(x,s) \le (\lambda_1^* - \epsilon_0)s^2 + h_1(x)|s| \quad \forall\, s \in \mathbb{R} \text{ and a.e. } x \in \Omega$$

and

(f-3) $\exists\, K > 0$ such that

$$|f(x,s)| \le h_1(x) + K|s| \quad \forall\, s \in \mathbb{R} \text{ and a.e. } x \in \Omega,$$

we established the following nonresonance result for p, ρ, Ω, and G as above.

THEOREM 3.1. *Assume* (Q-1)–(Q-5), (f-1)–(f-3), *and* $\mathcal{O}_{p,\rho}(\Omega)$. *Then* $\exists\, u^\sharp \in H^1_{p,\rho}(\Omega)$ *such that*

$$\mathcal{Q}(u^\sharp, v) = \langle f(\cdot, u^\sharp), v\rangle_\rho + G(v) \quad v \in H^1_{p,\rho}(\Omega).$$

By $\mathcal{O}_{p,\rho}(\Omega)$ which is a crucial assumption, we mean

($\mathcal{O}$1) $\exists$ a CONS $\{\psi_n\}_{n=1}^\infty$ in $L^2_\rho(\Omega)$ with $\psi_n \in H^1_{p,\rho}\ \forall\, n$.

($\mathcal{O}$2) $\exists\{\lambda_n\}_{n=1}^\infty$ with $1 \le \lambda_1 < \lambda_2 \le \cdots \le \lambda_m \to \infty$ such that $\langle \psi_n, v\rangle_{p,\rho} = \lambda_n \langle \psi_n, v\rangle_\rho\ \forall\, v \in H^1_{p,\rho}$.

($\mathcal{O}$3) $\psi_1(x) > 0$ a.e. in Ω.

We now give some examples where $\mathcal{O}_{p,\rho}(\Omega)$ is satisfied. For the first example, we take $\Omega = \mathbb{R}^2$ and $p(x_1,x_2) = \rho(x_1,x_2) = e^{-(x_1^2+x_2^2)}$. Then with $H_n(t)$ the familiar Hermite polynomial or order n, we take

$$\phi_{mn}(x_1,x_2) = H_m(x_1)H_n(x_2)/(2^{-(m+n)}m!n!\pi)^{1/2} \tag{3.6}$$

and observe from familiar facts for the Hermite polynomials that $\{\phi_{mn}\}_{m=0,n=0}^{\infty,\infty}$ forms a CONS for $L^2_\rho(\mathbb{R}^2)$.

Consequently, ($\mathcal{O}$1) holds for (3.6). Using the fact that $\left[e^{-t^2}H_n'(t)\right]' = -2ne^{-t^2} H_n(t)$ and that the C^1-functions vanishing outside of a compact set are dense in $C^1_{p,\rho}(\mathbb{R}^2)$, an easy computation shows that $\langle \phi_{mn}, v\rangle_{p,\rho} = \left[2(m+n)+1\right]\langle \phi_{mn}, v\rangle_\rho$ $\forall\, v \in H^1_{p,\rho}$. Consequently, ($\mathcal{O}$2) holds with $\lambda_1 = 1$, $\lambda_2 = 3$, $\lambda_3 = 3$, $\lambda_4 = 5$, etc. Also, it follows from (3.6) that $\psi_1 = \phi_{00} = \pi^{-1/2}$; hence, ($\mathcal{O}$3) holds and $\mathcal{O}_{p,\rho}(\Omega)$ is established for this case.

For our second example, we take $\Omega = (0,1)\times\mathbb{R}^1$ with $p(x_1,x_2) = \rho(x_1,x_2) = x_1e^{-x_2^2}$. In this case, we set

$$\phi_{mn}(x_1,x_2) = J_0(\eta_m x_1)H_n(x_2)/\left[2^{-1}J_1(\eta_m)2^n n!\pi^{1/2}\right]^{1/2} \tag{3.7}$$

for $m = 1,2,\ldots$, $n = 0,1,2,\ldots$, where $J_0(t)$ and $J_1(t)$ are the familiar Bessel functions of the first kind and η_m is the mth positive zero of $J_0(t)$. Also $H_n(t)$ is the Hermite polynomial of order n. As is well-known $\left\{J_0(\eta_m t)/\left[2^{-1}J_1(\eta_m)\right]^{1/2}\right\}_{m=1}^\infty$ is CONS with respect to the weight t on $[0,1]$. Hence it follows that $\{\phi_{mn}\}_{m=1,n=0}^{\infty,\infty}$ is a CONS for $L^2_\rho(\Omega)$. Therefore, ($\mathcal{O}$1) holds for the system (3.7). Observing that the linear space $C^1_{p,\rho,0} = \{v \in C^1_{p,\rho} : \exists a > 0 \text{ s.t. } v(x_1,x_2) = 0 \text{ for } |x_2| > a \text{ and } x_1 \in [0,1]\}$ is dense in the space $C^1_{p,\rho}(\Omega)$, and also that for $v \in C^1_{p,\rho,0}$

$$\int_{-\infty}^{\infty}\int_0^1 e^{-x_2^2}H_n(x_2)x_1D_1J_0(\eta_m x_1)D_1v(x_1,x_2)\,dx_1\,dx_2$$

$$= \eta_m^2\int_{-\infty}^{\infty}\int_0^1 e^{-x_2^2}x_1H_n(x_2)J_0(\eta_m x_1)v(x_1,x_2)\,dx_1\,dx_2,$$

it is easy to see that $\langle \phi_{mn}, v\rangle_{p,\rho} = (\eta_m^2 + 2n + 1)\langle \phi_{mn}, v\rangle_\rho \ \forall v \in H^1_{p,\rho}$. Hence, we obtain that $(\mathcal{O}2)$ holds for the system (3.7) with $\lambda_1 = \eta_1^2+1$, $\lambda_2 = \eta_1^2+3$, etc. Since $J_0[\eta_1 t] > 0$ in $[0, 1)$, we also have $\psi_1(x_1, x_2) = J_0(\eta_1 x_1)\pi^{-1/2} > 0$ in Ω. Therefore, $(\mathcal{O}3)$ holds, and consequently $\mathcal{O}_{p,\rho}(\Omega)$ holds for the system (3.7).

Next, we give an abbreviated version of the resonance theorem and corollary obtained in the paper under review. In order to do this, we set

$$Lu = -\sum_{i=1}^{N} D_i\left[pD_i u\right] + \rho u \tag{3.8}$$

and have that Qu is given by (3.1). Also, we assume $\mathcal{O}_{p,\rho}(\Omega)$ and set $\|u\|^2_{p,\rho} = \langle u, u\rangle_{p,\rho}$. We then say Q is $*$-related to L if

(3.9) (i) $$\lambda_1^* = \lambda_1$$

and

(3.9) (ii) $$\liminf_{\|u\|_{p,\rho}\to\infty} \left[\mathcal{Q}(u,u) - \|u\|^2_{p,\rho}\right]/\|u\|_{p,\rho} \geq 0.$$

We next introduce (f-4) to replace (f-2); so our theorem to be stated will be a resonance (rather than a nonresonance) result.

(f-4) $f(x, s) = \lambda_1 s + g(s)$ where $g \in L^\infty(\mathbb{R}^1) \cap C^0(\mathbb{R}^1)$.

We will also need

$$g_+ < g(t) < g_- \qquad \forall s \in \mathbb{R} \tag{3.10}$$

where $g_+ = \limsup_{s\to\infty} g(s)$ and $g_- = \liminf_{s\to-\infty} g(s)$.

With p, ρ, Ω, and G as before, we then established two results in the paper under review which contains the following

THEOREM 3.2. *Assume* (Q-1)–(Q-5), (f-1), (f-3), (f-4), (3.10), *and* $\mathcal{O}_{p,\rho}(\Omega)$. *Assume also that* Q *is* $*$*-related to* L *where* L *is given by* (3.8). *Suppose furthermore that*

$$\mathcal{Q}(v, \psi_1) = \lambda_1\langle v, \psi\rangle_\rho \quad \forall v \in H^1_{p,\rho}(\Omega) \tag{3.11}$$

Then a necessary and sufficient condition condition that $\exists\, u^\sharp \in H^1_{p,\rho}$ *with the property that*

$$\mathcal{Q}(u^\sharp, v) = \lambda_1\langle v^\sharp, v\rangle_\rho + \langle g(u^\sharp), v\rangle_\rho + G(v) \quad \forall v \in H^1_{p,\rho} \tag{3.12}$$

is that

$$g_+\int_\Omega \psi_1\rho < -G(\psi_1) < g_-\int_\Omega \psi_1\rho. \tag{3.13}$$

The condition (3.13) is called in the literature a "Landesman-Lazer condition." It is clearly a necessary condition for (3.12) to hold. For assume that (3.12) holds. Then taking $v = \psi_1$ in (3.12) and $v = u^\sharp$ in (3.11), we see that $\langle g(u^\sharp), \psi_1\rangle_\rho = -G(\psi_1)$. But then (3.11) follows immediately from (3.10) and the positivity of ψ_1 in $(\mathcal{O}3)$.

We now show that the two special functions situations that we introduced earlier (i.e., (3.6) and (3.7)) give rise to examples which satisfy (Q-1)–(Q-5) as well as (3.9) and (3.11). As a consequence we will therefore have two interesting illustrations of Theorem 3.2.

We return first to the case where $\Omega = \mathbb{R}^2$, $p = \rho = e^{-(x_1^2+x_2^2)}$ and ϕ_{mn} is given by (3.6). With Lu given by (3.8), we take

$$Qu = -\sum_{i=1}^{2} D_i p\{1 + \left[1 + |Du|^2\right]^{-1/2}\} D_i u + \rho u \tag{3.14}$$

and see that $A_i(x, s, \xi) = \xi_i\{1 + \left[1 + |\xi|^2\right]^{-1/2}\}$ and $B_0 = 1$. From this definition of $A_i(x, s, \xi)$, it is clear that (Q-1)–(Q-4) holds. Also, (Q-5) follows from the considerations given in (1.8) in §1 with $b_i(x) = 1$ and $N = 2$. Hence to obtain that Theorem 3.2 holds for Qu given by (3.14), it remains to show that (3.9) and (3.11) hold. Now from (3.6), we see that $\psi_1 = \phi_{00} = (\pi)^{-1/2}$. Consequently, it follows from the definition of $\langle v, \psi_1\rangle_{p,\rho}$ that $\langle v, \psi_1\rangle_{p,\rho} = \langle v, \psi_1\rangle_\rho$ $\forall\, v \in H^1_{p,\rho}$. Therefore, we obtain from ($\mathcal{O}2$) that $\lambda_1 = 1$. Also from (3.4) and (3.14) we see that $\mathcal{Q}(v, \psi_1) = \langle v, \psi_1\rangle_\rho$ $\forall\, v \in H^1_{p,\rho}$. Consequently (3.11) holds. It remains to show that (3.9) obtains, i.e., Q is $*$-related to L. Now from (3.14) and (3.4), $Q(u, u) = \langle u, u\rangle_{p,\rho} + \sum_{i=1}^2 \int_\Omega p(x_1, x_2)|D_i u|^2/[1 + |Du|^2]^{1/2} \geq \lambda_1\langle u, u\rangle_\rho$ because $\lambda_1 = 1$. Consequently, we obtain from (3.5) that $\lambda_1^* \geq \lambda_1$. On the other hand, $\mathcal{Q}(n\psi_1, n\psi_1)/\|n\psi_1\|_\rho^2 = 1$. Therefore we also obtain from (3.5) that $\lambda_1^* \leq 1 = \lambda_1$ which establishes (3.9)(i). As we have just observed $\mathcal{Q}(u, u) - \langle u, u\rangle_{p,\rho} \geq 0$ $\forall\, u \in H^1_{p,\rho}$. Consequently (3.9)(ii) holds. Q is $*$-related to L, and this example is complete.

Next, we deal with a more sophisticated case related to the system (3.7) where $\Omega = (0, 1) \times \mathbb{R}^1$ and $p = \rho = x_1 e^{-x_2^2}$. For this case, we take

$$Qu = Lu + D_2\{p2^{-1}[1 - F(|D_2 u|)]D_2 u\} \tag{3.15}$$

where L is given by (3.8) and $F(s)$ is a nonnegative continuous nondecreasing function on $[0, \infty)$ meeting conditions (2.22)(i) and (ii) in §2. We see from (3.8) and (3.15) that $B_0(x, s, \xi) = 1$ and

$$A_1(x, s, \xi) = \xi_1 \text{ and } A_2(x, s, \xi) = 2^{-1}[1 + F(|\xi_2|)]\xi_2. \tag{3.16}$$

It is clear from the fact that F is nonnegative and nondecreasing that (3.16) implies that (Q-1)–(Q-5) holds for (3.15). Therefore, to show that Theorem 3.2 holds for Q given by (3.15), it remains to show that (3.11) is true and that Q is $*$-related to L.

As we have observed earlier, ($\mathcal{O}2$) holds for the system (3.7) with λ_1 in particular given by $\eta_1^2 + 1$ where η_1 is the first positive zero of J_0 and $\psi_1(x_1, x_2) = \pi^{-1/2}J_0(\eta_1 x_1)$. Consequently $\int_0^1 \int_{-\infty}^{\infty} p2^{-1}[1 - F(|D_2 v|)]D_2 v D_2\psi_1\, dx_2\, dx_1 = 0$ $\forall\, v \in H^1_{p,\rho}$ and we conclude from (3.4), (3.8), (3.15), and ($\mathcal{O}2$) that $\mathcal{Q}(v, \psi_1) = \langle v, \psi_1\rangle_{p,\rho} = (\eta_1^2 + 1)\langle v, \psi_1\rangle_\rho$ $\forall\, v \in H^1_{p,\rho}$. Hence (3.11) is established.

To show Q is $*$-related to L, it remains to show that (3.9) holds. Now from (2.22)(i), we have that $0 \leq F(s) \leq 1$, consequently, we obtain from (3.8) and (3.15) that $\forall\, u \in H^1_{p,\rho}$, $Q(u, u) \leq \langle u, u\rangle_{p,\rho}$. Also, $\mathcal{Q}(u, u) \geq \mathcal{L}(u, u)2^{-1}$ $\forall\, u \in H^1_{p,\rho}$. Using an argument similar to that used in (2.26)–(2.28), we obtain that $\lambda_1 \leq \lambda_1^*$. Also as we have just shown $\mathcal{Q}(n\psi_1, n\psi_1) = \langle n\psi_1, n\psi_1\rangle_{p,\rho} = \lambda_1\|n\psi_1\|_\rho^2$. It therefore follows from (3.5) that $\lambda_1^* \leq \lambda_1$, and we conclude that (3.9)(i) does indeed hold.

It remains to show that (3.9)(ii) holds. From (3.8) and (3.15) we see that to do this we have to show that

$$\lim_{\|u\|_{p,\rho}\to\infty} \int_\Omega p\big[1 - F\big(|D_2 u|\big)\big]\, |D_2 u|^2/\|u\|_{p,\rho} = 0. \tag{3.17}$$

But from (2.22)(ii), we have that $\lim_{s\to\infty} s[1-F(s)] = 0$. Also, from Hölder's inequality, we have that $\int_\Omega p|D_2u| \leq \|u\|_{p,\rho}(\pi/2)^{1/2}$ and (3.17) follows easily from these two facts. We conclude that Q given by (3.15) is $*$-related to L given by (3.8) where $p=\rho=x_1e^{-x_2^2}$, and our final example is complete.

4. Resonance and the Stationary Navier-Stokes Equations

Resonance and the Stationary Navier-Stokes Equations, Proc. Symp. Pure Math., Vol 4 (1986) Part 2.

In this paper, we establish a resonance type existence theorem for periodic solutions of the stationary Navier-Stokes equation. We also will show by an example that in a certain sense the result established is a best possible result. The paper under review was motivated by a previous result of the author [Sh2] plus two others, namely [BN] and [DG].

We let $\mathbb{T}_N = \{x : -\pi \leqq x_j < \pi,\ j=1,\dots,N\}$ designate the N-torus, $N\geq 2$ and $\boldsymbol{\phi}\in\left[C^\infty(\mathbb{T}_N)\right]^N$ mean that $\boldsymbol{\phi}=(\phi_1,\dots,\phi_N)$, where $\phi_j\in C^\infty(\mathbb{R}^N)$ and is periodic of period 2π in each variable $j=1,\dots,N$. $W^{1,2}(\mathbb{T}_N)$ will designate the closure of $C^\infty(\mathbb{T}_N)$ with respect to the usual $W^{1,2}$-norm.

We study the equation

$$(4.1)\qquad \begin{cases} \nu\Delta\mathbf{v}(x) - \big(\mathbf{v}(x)\cdot\nabla\big)\mathbf{v}(x) - \nabla p(x) = \mathbf{g}\big(x,\mathbf{v}(x)\big) - \mathbf{h} \\ \nabla\cdot\mathbf{v}(x) = 0, \end{cases}$$

where ν is a positive constant, $\mathbf{v}$ and $\mathbf{g}$ are vector-valued functions and $\mathbf{h}$ is a vector-valued distribution with components in the dual of $W^{1,2}(\mathbb{T}_N)$ $\left(=\left[W^{1,2}(\mathbb{T}_N)\right]^*\right)$.

In particular, $\mathbf{g}\colon \mathbb{T}_N\times\mathbb{R}^N\to\mathbb{R}^N$ and we also assume that $\mathbf{g}(x,\mathbf{s})$ meets the following four conditions where $x\in\mathbb{T}_N$ and $\mathbf{s}=(s_1,\dots,s_N)\in\mathbb{R}^N$:

(g-1) The jth component of $\mathbf{g}(x,\mathbf{s})$ is of the form $g_j(x,s_j)$, $j=1,\dots,N$.
(g-2) $g_j(x,s_j)$ satisfies the Carathéodory conditions, $j=1,\dots,N$.
(g-3) For each fixed $r>0$, there is a nonnegative function $\zeta_r\in L^1(\mathbb{T}_N)$ such that $|g_j(x,s_j)|\leq\zeta_r(x)$ for $|s_j|\leq r$ and a.e. $x\in\mathbb{T}_N$, $j=1,\dots,N$.
(g-4) $\exists\,\alpha$ and β nonnegative functions $\in L^q(\mathbb{T}_N)$ and $L^1(\mathbb{T}_N)$ respectively, where $q>1$ for $N=2$ and $q=2N/(N+2)$ for $N\geq 3$, such that

$$tg_j(x,t)\geq -\alpha(x)|t| - \beta(x)$$

for $t\in\mathbb{R}$ and a.e. $x\in\mathbb{T}_N$, $j=1,\dots,N$.

For $\mathbf{h}=(h_1,\dots,h_N)$ appearing in (4.1), we shall assume

(h-1) $\mathbf{h}\colon \left[W^{1,2}(\mathbb{T}_N)\right]^N\to\mathbb{R}^N$ where for $\mathbf{u}\in\left[W^{1,2}(\mathbb{T}_N)\right]^N$, the jth component of $\mathbf{h}(\mathbf{u})$ is $h_j(u_j)$ and $h_j\in\left[W^{1,2}(\mathbb{T}_N)\right]^*$, $j=1,\dots,N$.

We say the pair $(\mathbf{v},p)$ is a distribution solution of (4.1) provided $v_j\in W^{1,2}(\mathbb{T}_N)$, $j=1,\dots,N$, $p\in L^1(\mathbb{T}_N)$, the components of $\mathbf{g}\big(x,\mathbf{v}(x)\big)$ are in $L^1(\mathbb{T}_N)$ and

$$(4.2)\qquad \begin{cases} \int_{\mathbb{T}_N}\left[\nu\mathbf{v}\cdot\Delta\boldsymbol{\phi} - \boldsymbol{\phi}\cdot(\mathbf{v}\cdot\nabla)\mathbf{v} + p\nabla\cdot\boldsymbol{\phi}\right]dx \\ = \int_{\mathbb{T}_N}\mathbf{g}\big(x,\mathbf{v}(x)\big)\cdot\boldsymbol{\phi}(x)\,dx - \sum_{j=1}^N h_j(\phi_j), \\ \int_{\mathbb{T}_N}(\nabla\cdot\mathbf{v})\xi\,dx = 0 \end{cases}$$

$\forall\boldsymbol{\phi}\in\left[C^\infty(\mathbb{T}_N)\right]^N$ and $\forall\xi\in C^\infty(\mathbb{T}_N)$.

For $\mathbf{g} = (g_1, \ldots, g_N)$ a vector-valued function satisfying (g-1) and (g-2) above, we set

$$g_{j+}(x) = \liminf_{t\to\infty} g_j(x,t) \text{ and } g_{j-}(x) = \limsup_{t\to-\infty} g_j(x,t). \tag{4.3}$$

Using a Galerkin technique, we establish in this paper the following best possible result.

THEOREM 4.1. *Let $\mathbf{g} = (g_1, \ldots, g_N)$ be a vector-valued function satisfying* (g-1), (g-2), (g-3), *and* (g-4). *Let $\mathbf{h} = (h_1, \ldots, h_N)$ be a vector-valued distribution satisfying* (h-1). *Suppose also that*

$$\int_{\mathbb{T}_N} g_{j-}(x)\,dx < h_j(1) < \int_{\mathbb{T}_N} g_{j+}(x)\,dx \tag{4.4}$$

for $j = 1, \ldots, N$ where g_{j+} and g_{j-} are defined in (4.3). *Then there exists a pair $(\mathbf{v}, p)$ with p, $g_j(x, v_j(x))$, and $g_j(x, v_j(x))v_j(x) \in L^1(\mathbb{T}_N)$ and $v_j \in W^{1,2}(\mathbb{T}_N)$ for $j = 1, \ldots, N$ such that $(\mathbf{v}, p)$ is a distribution solution of* (4.1).

We now show that assumption (4.4) in the above theorem is a best possible assumption, i.e., if it is weakened slightly then the theorem is false.

What we mean by weakening (4.4) slightly is the following:

$$\begin{aligned} &\text{(i)} \quad (4.4) \text{ holds for } j = 2, \ldots, N; \text{ and} \\ &\text{(ii)} \quad \int_{\mathbb{T}_N} g_{1-}(x)\,dx \le h_1(1) < \int_{\mathbb{T}_N} g_{1+}(x)\,dx \end{aligned} \tag{4.5}$$

We see that (4.5)(i) and (ii) together constitute a slight weakening of (4.4) because everything remains the same except that one strict inequality is replaced by $\le$.

We now show that the above theorem is in general false if (4.5)(i) and (ii) replace (4.4) in the hypothesis of the theorem.

To do this we take $h_j = 0$ for $j = 1, \ldots, N$ and

$$\begin{aligned} &\text{(i)} \quad g_j(x, s_j) = s_j \text{ for } j = 2, \ldots, N; \\ &\text{(ii)} \quad g_1(x, s_1) = \begin{cases} s_1 + 1 & \text{for } 0 \le s_1 < \infty \\ (1 + s_1^2)^{-1} & \text{for } -\infty < s_1 \le 0 \end{cases} \end{aligned} \tag{4.6}$$

Then $s_j g_j(x, s_j) \ge -1$ for $j = 1, \ldots, N$ and it is clear that (g-1)–(g-4) is satisfied (i.e., in (g-4) take $\alpha = 0$ and $\beta = 1$) and also that (h-1) holds. Since $g_{j+}(x) = \infty$ and $g_{j-}(x) = -\infty$ and $h_j(1) = 0$ for $j = 2, \ldots, N$, it is also clear that (4.4) holds for $j = 2, \ldots, N$; so (4.5)(i) holds. Furthermore, it is clear that $g_{1+}(x) = \infty$, $g_{1-}(x) = 0$ and $h_1(1) = 0$; hence, (4.5)(ii) is also satisfied.

Now suppose $\exists\, \mathbf{v} \in \left[W^{1,2}(\mathbb{T}_N)\right]^N$ and $p \in L^1(\mathbb{T}_N)$ such that (4.2) holds. In particular, we have that $v_1 \in L^2(\mathbb{T}_N)$, where v_1 is the first component of $\mathbf{v}$. Hence, $\exists\, K > 0$ and $E \subset \mathbb{T}_N$ with $\text{meas}(E) \ge \pi^N$ and $v_1(x) \ge -K\ \forall x \in E$. But then it follows from (4.6)(ii) that $g_1(x, v_1(x)) \ge (1 + K^2)^{-1}$ for $x \in E$. So in particular, we have that

$$\pi^N(1 + K^2)^{-1} \le \int_{\mathbb{T}_N} g_1(x, v_1(x))\,dx. \tag{4.7}$$

Next, we observe that since $v_j \in W^{1,2}(\mathbb{T}_N)$ for $j = 1, \ldots, N$, $\int_{\mathbb{T}_N} v_k \partial v_1/\partial x_k\,dx = -\int_{\mathbb{T}_N} v_1 \partial v_k/\partial x_k\,dx$ for $k = 1, \ldots, N$. (Recall v_k and v_1 are the limits of C^∞, periodic functions using the usual $W^{1,2}$-norm.) Likewise, we see from the second equation in (4.2) that

$$\int_{\mathbb{T}_N} v_1 \left(\sum_{k=1}^{N} \partial v_k / \partial x_k \right) dx = 0.$$

We conclude from these two calculations that

$$\int_{\mathbb{T}_N} \left(\sum_{k=1}^{N} v_k \frac{\partial v_1}{\partial x_k} \right) dx = 0. \tag{4.8}$$

To show that the existence of $\mathbf{v} \in W^{1,2}(\mathbb{T}_N)$ and $p \in L^1(\mathbb{T}_N)$ such that the pair $(\mathbf{v}, p)$ satisfies (4.2) gives rise to a contradiction, take $\boldsymbol{\phi} = (1, 0, 0, \ldots, 0)$ in (4.2). Then it follows from (4.2) that

$$-\int_{\mathbb{T}_N} \left(\sum_{k=1}^{N} v_k \frac{\partial v_1}{\partial x_k} \right) dx = \int_{\mathbb{T}_N} g_1(x, v_1(x))\, dx$$

But by (4.8) the left-hand side of this equality is zero. So, we conclude from (4.7) that $0 \geq \pi^N (1 + K^2)^{-1}$, where K is a positive constant. This is a manifest contradiction. Consequently the hypothesis (4.4) cannot be weakened to (4.5)(i) and (ii) and our example is complete.

5. Removable Isolated Singularities for the Stationary Navier-Stokes Equations

The results obtained in this section are based on theorems proved in the following series of three papers:

1. *Isolated Singularities for Solutions of the Nonlinear Stationary Navier-Stokes Equations*, Trans AMS, Vol 187 (1974), pp. 335–363.
2. *Isolated Singularities in Steady State Fluid Flow*, SIAM Journal Math Anal, Vol 7 (1976), pp. 577–601.
3. *A Counterexample in the Theory of Planar Viscous Incompressible Flow*, J. Diff Eqn., Vol 22 (1976), pp. 164–178.

In this series of papers, we deal with the following system of equations

$$\begin{cases} \nu \Delta \mathbf{v} - (\mathbf{v} \cdot \nabla)\mathbf{v} - \nabla p = 0 \\ \nabla \cdot \mathbf{v} = 0 \end{cases} \tag{5.1}$$

where ν is a positive constant, or the following system

$$\begin{cases} \nu \Delta \mathbf{v} - (\mathbf{v} \cdot \nabla)\mathbf{v} - \nabla p = -\mathbf{f} \\ \nabla \cdot \mathbf{v} = 0 \end{cases} \tag{5.2}$$

For the system (5.1), we shall be interested in stating results of a classical nature. In particular, with $\Omega \subset \mathbb{R}^N$ a bounded open set with $N \geq 2$, we shall say that the pair $(\mathbf{v}, p)$ is a classical solution of (5.1) in Ω if $v_j \in C^2(\Omega)$ and $p \in C^1(\Omega)$ for $j = 1, \ldots, N$ and (5.1) is satisfied at every point in Ω where $\mathbf{v} = (v_1, \ldots, v_N)$.

For the system (5.2), we shall be interested in stating results for distribution solutions. In particular with Ω as above, v_j and f_j locally in $L^2(\Omega)$ and $L^1(\Omega)$ respectively for $j = 1, \ldots, N$ and with p locally in $L^1(\Omega)$, we shall say the pair $(\mathbf{v}, p)$ is a distribution solution of (5.2) with $\mathbf{f} = (f_1, \ldots, f_N)$ provided

$$(5.3)\qquad \begin{cases} \displaystyle\int_\Omega \left[\nu v_j \Delta\phi + \sum_{k=1}^N v_j v_k \frac{\partial \phi}{\partial v_k} + p\frac{\partial \phi}{\partial x_j}\right] = -\int_\Omega \phi f_j \\ j = 1, \dots, N \\ \displaystyle\sum_{k=1}^N \int_\Omega v_k \frac{\partial \phi}{\partial v_k} = 0 \qquad \text{for} \quad \phi \in C_0^\infty(\Omega). \end{cases}$$

We first state some results that we obtained for the case $N = 2$

THEOREM 5.1. *With $N = 2$, let $(\mathbf{v}, p)$ be a classical solution of the system* (5.1) *in the punctured disk $B(0, r_0) \setminus \{0\}$, $r_0 > 0$. Suppose that $v_j(x) = o(|\log|x||)$ as $x \to 0$ for $j = 1, 2$. Then $\mathbf{v}$ and p can be defined at 0 so that $(\mathbf{v}, p)$ is a classical solution of the system* (5.1) *in $B(0, r_0)$.*

In the above theorem, $B(0, r_0)$ represents, as usual, the open N-ball with center 0 and radius r_0. (Of course the N-ball in dimension 2 is usually referred to as a disc.) We notice in particular that in the above theorem no growth conditions were put on the pressure p in a neighborhood of the origin. We do, however, obtain a result which corresponds to the situation for a removable singularity for harmonic functions in the punctured disc, namely the growth condition when v_j is of the form $o(|\log|x||)$. If we put a growth condition on the p in a neighborhood of the origin, we can obtain a better than expected result for the v_j, namely a growth condition of the form $o(|x|^{-1/2})$. In particular the following result holds:

THEOREM 5.2. *With $N = 2$, let $(\mathbf{v}, p)$ be a classical solution of the system* (5.1) *in $B(0, r_0) \setminus \{0\}$, $r_0 > 0$. Suppose that $p(x) = o(|x|^{-1})$ and $v_j(x) = o(|x|^{-1/2})$ as $x \to 0$ for $j = 1, 2$. Then $\mathbf{v}$ and p can be defined at 0 so that $(\mathbf{v}, p)$ is a classical solution of the system* (5.1) *in $B(0, r_0)$.*

Theorem 5.1 was obtained in the SIAM Journal paper listed above and Theorem 5.2 was obtained in the Transactions paper. Now both of the above theorems are in a certain sense best possible. In particular, in (5.1) if $v_j(x) = o(|\log|x||)$ for $j = 1, 2$ is replaced with $v_1(x) = O(|\log|x||)$ and $v_2 = o(|\log|x||)$, then the conclusion of Theorem 5.1 no longer holds.

Also if in Theorem 5.2, $p(x) = o(|x|^{-1})$ is replaced with $p(x) = O(|x|^{-1})$ then the conclusion of Theorem 5.2 no longer holds. Both of these last two stated facts follow from the following theorem, which was obtained in the Journal of Differential Equations article cited above.

THEOREM 5.3. *With $N = 2$, there exists a pair $(\mathbf{v}, p)$ which is a classical solution of the system* (5.1) *in $B(0,1) \setminus \{0\}$ such that $v_2(x)$ and $|x|\,p(x)$ are uniformly bounded in $B(0,1) \setminus \{0\}$ and such that $\lim_{|x|\to 0} v_1(x)/\log|x| = \xi$, where ξ is a finite-valued nonzero constant.*

In particular, we see that $v_1(x)$ in the above theorem cannot be assigned a value at 0 so that it will be continuous in $B(0,1)$. Consequently, the remarks concerning the best possibility of Theorems 5.1 and 5.2 made above are true. Theorem 5.3 itself is a corollary to a theorem in the same paper obtained via double trigonometric series.

With $m = (m_1, m_2)$ designating the integral lattice points in the plane and $(m,x) = m_1x_1 + m_2x_2$, let $u_k^j(x)$ be the function in $L^2(\mathbb{T}_2) \cap C^\infty(\mathbb{R}^2 \setminus \bigcup_m \{2\pi m\})$, whose Fourier coefficients, $\hat{u}_j^k(m)$, are

$$(5.4)\qquad \begin{cases} \hat{u}_j^k(m) = \left[-\delta_j^k |m|^{-2} + m_j m_k |m|^{-4}\right]\nu^{-1} & \text{for} \quad m \neq 0 \\ \hat{u}_j^k(0) = 0 \end{cases}$$

for $j, k = 1, 2$ and δ_j^k is the Kronecker-δ. We then set

$$(5.5)\qquad \mathbf{u}^1(x) = \left(u_1^1(x), u_2^1(x)\right)$$

where u_1^1 and u_2^1 are defined by (5.4).

In the sequel when we say periodic we shall mean periodic of period 2π in each variable. Next we select a function λ with the following properties:

(a) λ is a periodic function in $C^\infty(\mathbb{R}^2)$;
(5.6) (b) $\lambda = 1$ in $B(0,1)$;
(c) $\hat{\lambda}(0) = 0$.

We define

$$(5.7)\qquad \mathbf{f}^\mu = (\mu\mathbf{u}^1 \cdot \nabla)\mu\mathbf{u}^1 + \mu\lambda\boldsymbol{\delta}^1$$

in $\mathbb{R}^2 \setminus \bigcup_m \{2\pi m\}$ where μ is a nonzero constant and $\boldsymbol{\delta}^1 = (1,0)$.

The key to the proof of Theorem 5.3 is to consider the following system

$$(5.8)\qquad \begin{cases} \nu\Delta\mathbf{V} - (\mathbf{V}\cdot\nabla)\mathbf{V} - (\mu\mathbf{u}^1\cdot\nabla)\mathbf{V} - (\mathbf{V}\cdot\nabla)\mu\mathbf{u}^1 - \nabla q = \mathbf{f}^\mu \\ \nabla\cdot\mathbf{V} = 0 \end{cases}$$

We prove the following theorem concerning (5.8) in the Journal of Differential Equations article mentioned at the outset.

THEOREM 5.4. *A positive constant μ_0 exists such that for every μ with $0 < |\mu| < \mu_0$ there is a pair $(\mathbf{V}, q)$, where $\mathbf{V} = (V_1, V_2)$, with the following properties:*

(i) *V_j and q are in $C^\infty(\mathbb{R}^2 \setminus \bigcup_m \{2\pi m\})$ for $j = 1, 2$;*
(ii) *$(\mathbf{V}, q)$ satisfies (5.8) in $\mathbb{R}^2 \setminus \bigcup_m \{2\pi m\}$;*
(iii) *V_j is a periodic function in $C^\alpha(\mathbb{R}^2)$ for $0 < \alpha < 1$ and $j = 1, 2$;*
(iv) *q is a periodic function in $\mathbb{R}^2 \setminus \bigcup_m \{2\pi m\}$ and $q \in L^\beta(\mathbb{T}_2)$ for $1 < \beta < \infty$;*
(v) *$\forall \varepsilon > 0$, $q(x) = O\left(|x|^{-\varepsilon}\right)$ as $x \to 0$.*

To obtain Theorem 5.3 from Theorem 5.4 we proceed as follows: Set $\mathbf{v} = \mathbf{V} + \mu\mathbf{u}^1$ and $p = q + \mu q^1$ where q^1 is the periodic function in $C^\infty(\mathbb{R}^2 \setminus \bigcup_m \{2\pi m\})$ whose Fourier coefficients are

$$(5.9)\qquad \begin{aligned} \hat{q}^1(m) &= im_1|m|^{-2} \quad & m \neq 0 \\ &= 0 & m = 0 \end{aligned}$$

It is well-known (See [Sh3, p.72]) that $|x|q^1(x)$ is uniformly bounded in $B(0,1)\setminus\{0\}$. Consequently, it follows from (v) in Theorem 5.4 that p defined above is such that $|x|p(x)$ is uniformly bounded in $B(0,1)\setminus\{0\}$. Also it follows from [Sh3] that $u_2^1(x)$ defined in (5.4) and (5.5) above is uniformly bounded in $B(0,1)\setminus\{0\}$. Hence we see from (iii) in Theorem 5.4 that v_2 defined above is uniformly bounded in $B(0,1) \setminus \{0\}$. Likewise it follows from (5.4) and (iii) in Theorem 5.4 that $\lim_{x\to 0} v_1(x)/\log|x| = \xi$ a nonzero constant. Hence to complete the proof of the fact that the pair $(\mathbf{v}, p)$ defined above satisfies the conditions in Theorem 5.3 we need only show that $(\mathbf{v}, p)$ satisfies the system (5.1) in $B(0,1) \setminus \{0\}$. Now $\nabla \cdot \mathbf{v} = \nabla \cdot \mathbf{V} + \mu \nabla \cdot \mathbf{u}^1$. From Theorem 5.4 we see that $\nabla \cdot \mathbf{V} = 0$ in $\mathbb{R}^2 \setminus \bigcup_m \{2\pi m\}$. Likewise it follows from (5.4) and (5.5) above and [Sh3] that $\nabla \cdot \mathbf{u}^1 = 0$ in the same region. Hence, in particular, it follows that the second equation in (5.1) is satisfied in $B(0,1) \setminus \{0\}$. To see that the first equation is satisfied in this same region, we observe from (5.4), (5.5), (5.9), and [Sh3] that

$$\nu \Delta \mathbf{u}^1 - \nabla q^1 = -\boldsymbol{\delta}^1 \tag{5.10}$$

where $\boldsymbol{\delta}^1 = (1,0)$. Hence it follows from (5.6)(b) and (5.7) that

$$\mathbf{f}^\mu = (\mu \mathbf{u}^1 \cdot \nabla)\mu \mathbf{u}^1 - \nu\mu\Delta\mathbf{u}^1 + \mu\nabla q^1, \tag{5.11}$$

in $B(0,1) \setminus \{0\}$. But then it follows from (i) and (iii) of Theorem 5.4 and (5.8) and (5.11) that

$$\nu\Delta\mathbf{V} - (\mathbf{V}\cdot\nabla)\mathbf{V} - (\mu\mathbf{u}^1\cdot\nabla)\mathbf{V} - (\mathbf{V}\cdot\nabla)\mu\mathbf{u}^1 = \tag{5.12}$$

$$(\mu\mathbf{u}^1\cdot\nabla)\mu\mathbf{u}^1 - \nu\mu\Delta\mathbf{u}^1 + \mu\nabla q^1 + \nabla q \quad \text{in} \quad B(0,1)\setminus\{0\}$$

But $\mathbf{v} = \mathbf{V} + \mu\mathbf{u}^1$ and $p = q + \mu q^1$, and we conclude from the above equality that

$$\nu\Delta\mathbf{v} - \nabla p = (\mathbf{V}\cdot\nabla)\mathbf{V} + (\mu\mathbf{u}^1\cdot\nabla)\mathbf{V} + (\mathbf{V}\cdot\nabla)\mu\mathbf{u}^1 + (\mu\mathbf{u}^1\cdot\nabla)\mu\mathbf{u}^1$$

in $B(0,1) \setminus \{0\}$. It is easy to see that the right-hand side of this last equality is $(\mathbf{v}\cdot\nabla)\mathbf{v}$ in $B(0,1)\setminus\{0\}$. Hence the pair $(\mathbf{v}, p)$ satisfies the first equation in the system (5.1) in $B(0,1)\setminus\{0\}$ and the proof of our assertion that Theorem 5.3 follows from Theorem 5.4 is complete.

Next we state a theorem covering distribution solutions for the system (5.2) which covers all dimensions $N \geq 2$.

THEOREM 5.5. *Let* $\mathbf{f}$ *be locally in* $L^1(B(0,r_0))$, $r_0 > 0$, *and suppose that*

(i) *there is a* $\beta > 2$ *such that* $\mathbf{v}$ *is locally in* $L^\beta(0, r_0)$;
(ii) p *is locally in* $L^1[B(0,r_0)\setminus\{0\}]$;
(iii) $(\mathbf{v}, p)$ *is a distribution solution of* (5.2) *in* $B(0,r_0)\setminus\{0\}$;
(iv) *for* $N \geqq 3$, $\left\{ r^{-N} \int_{B(0,r)} |\mathbf{v}|^\beta \right\}^{1/\beta} = o\left(r^{-(N-1)/2}\right)$ *as* $r \to 0$;
(v) *for* $N = 2$, $\int_{B(0,r)} |\mathbf{v}| = o\left(r^2 |\log r|\right)$ *as* $r \to 0$.

Then p *is locally in* $L^1[B(0,r_0)]$ *and* $(\mathbf{v}, p)$ *is a distribution solution of* (5.2) *in* $B(0, r_0)$.

For classical solutions of (5.1) in dimension $N \geq 3$, we obtained the following result

THEOREM 5.6. *Let* $N \geq 3$ *and let* $(\mathbf{v}, p)$ *be a classical solution of* (5.1) *in* $B(0,r_0)\setminus\{0\}$, $r_0 > 0$. *Suppose that there is a* $\beta > N$ *such that* $v \in L^\beta[B(0,r_0)]$. *Then* $(\mathbf{v}, p)$ *can be defined at* 0 *so that* $(\mathbf{v}, p)$ *is a classical solution of the system* (5.1) *in* $B(0, r_0)$.

Both of the Theorems were obtained in the SIAM journal paper mentioned at the outset. We observe that if $N = 3$ and $|\mathbf{v}(x)| = o(|x|^{-1})$ as $|x| \to 0$ then $|\mathbf{v}(x)|$ meets the criterion in (i) and (iii) of Theorem 5.5 above because $|x|^{-1} \in L^{2+\varepsilon}[B(0,r_0)]$ for $0 \leq \varepsilon < 1$.

In this same paper, we also obtained best possible removable isolated singularity results for the linear system

$$\begin{cases} \Delta \mathbf{v} = \nabla p \\ \nabla \cdot \mathbf{v} = -cp. \end{cases} \tag{5.13}$$

THEOREM 5.7. *Let* $(\mathbf{v}, p)$ *be a classical solution of the system* (5.13) *in* $B(0,r_0)\setminus\{0\}$, *where* $c \neq -1$ *for* $N \geq 2$ *and* $c \neq -\frac{1}{2}$ *for* $N = 2$. *Suppose that*

(i) $\mathbf{v} \in L^1[B(0,r_0)]$;

(ii) $\lim_{r\to 0} r^{-2} \int_{B(0,r)} |\mathbf{v}| = 0$ *for* $N \geq 3$

$\lim_{r\to 0} |r^2 \log r|^{-1} \int_{B(0,r)} |\mathbf{v}| = 0$ *for* $N = 2$.

Then $(\mathbf{v}, p)$ *can be defined at* 0 *so that* $(\mathbf{v}, p)$ *is a classical solution of the system* (5.13) *in* $B(0,r_0)$.

To see that the above theorem is best possible in dimension $N = 3$, i.e., the conclusion is false if we replace (ii) by

(ii)' $\int_{B(0,r)} |\mathbf{v}| = O(r^2)$ as $r \to 0$,

we set for $|x| \neq 0$ and $c \neq -1$,

$$\begin{aligned} v_1 &= -(1+c)|x|^{-1} + 2^{-1}\partial^2|x|/\partial x_1^2, \\ v_j &= 2^{-1}\partial^2|x|/\partial x_1 \partial x_j, \qquad j = 2,3, \\ p &= \partial |x|^{-1}/\partial x_1. \end{aligned}$$

To see that the result fails for $N \geq 2$ and $c = -1$, we set for $x \neq 0$

$$v_j(x) = x_j|x| \quad \text{for} \quad j = 1,\dots,N \quad \text{and} \quad p(x) = (N+1)|x|$$

Then it is clear that $\Delta \mathbf{v} = \nabla p$ and $\nabla \mathbf{v} = p$ in $B(0,1)\setminus\{0\}$, but $\mathbf{v}$ cannot be defined at 0 so that it is in $C^2(B(0,1))$.

To see that the result also fails for $N = 2$ and $c = -\frac{1}{2}$ we set for $x \neq 0$

$$v_1(x) = x_1^2|x|^{-2}, \quad v_2(x) = x_1x_2|x|^{-2}, \quad p = 2x_1|x|^{-2}.$$

It is clear after some computations that, for $x \neq 0$, $\Delta \mathbf{v} = \nabla p$ and that $\nabla \mathbf{v} = 2^{-1}p$. Also we see that $\mathbf{v}$ cannot be defined at 0 so that it is in $C^0(B(0,1))$.

For further results concerning removable isolated singularities for the systems (5.1), (5.3), and other related systems, we refer the reader to the original papers given at the beginning of this section.

6. Newtonian Capacity and the Navier-Stokes Equations

The results in this section are based on the following two papers:

1. *Positive Newtonian Capacity and Stationary Viscous Incompressible Flow*, Indiana Univ. Math J., Vol **24** (1974), pp. 1–16.

2. *Pressure, Capacity, and the Navier-Stokes Equations*, J. Math. Analysis and Applic., Vol. **65** (1978), pp. 236–246.

We take $N = 3$ throughout this section and let $Z \subset B(0,1)$ be a closed set of Lebesgue measure zero. We set $P(Z) = \{\mu : \mu$ is a nonnegative Borel measure with $\mu(Z) = 1$ and $\mu(\mathbb{R}^3 \setminus Z) = 0\}$. Set $I(\mu) = \int_Z \int_Z |x-y|^{-1} \, d\mu(x) \, d\mu(y)$ for $\mu \in P(Z)$ and define

$$\theta(Z) = \left[\inf\{I(\mu) : \mu \in P(Z)\}\right], \qquad \operatorname{Cap} Z = [\theta(Z)]^{-1} \tag{6.1}$$

In particular, we note that $\operatorname{Cap} Z = 0$ if $I(\mu) = \infty \; \forall \mu \in P(Z)$. The capacity defined by (6.1) corresponds to the usual Newtonian capacity of a conductor in $\mathbb{R}^3$ [GT, p. 27].

We set $\Omega = B(0,1) \setminus Z$ and say that Z is a removable set for the class (S, L^2) with respect to distribution solutions for the system (5.2) if given $\mathbf{f} \in [L^1(\Omega)]^3$, $\mathbf{v} \in S$, $p \in L^2(\Omega)$ and the fact that (5.3) holds with $\Omega = B(0,1) \setminus Z$ implies that

(6.2)

$$\begin{cases} \displaystyle\int_{B(0,1)} \left[\nu v_j \Delta\phi + \sum_{k=1}^{3} v_j v_k \frac{\partial \phi}{\partial x_k} + p \frac{\partial \phi}{\partial x_k}\right] = -\int_{B(0,1)} \phi f_j & j = 1,2,3 \\ \displaystyle\int_{B(0,1)} \left[\sum_{k=1}^{3} v_k \frac{\partial \phi}{\partial x_k}\right] = 0 & \forall\, \phi \in C_0^\infty[B(0,1)]. \end{cases}$$

In the second paper cited above we take the class to be

$$S = \{\mathbf{v} = \mathbf{u} + \mathbf{w} : \mathbf{u} \in \left[L^\infty(B(0,1))\right]^3 \quad \text{and} \quad \mathbf{w} \in \left[W^{1,2}(B(0,1))\right]^3\}$$

and establish the following theorem.

THEOREM 6.1. *A sufficient condition that Z be a removable set for the class (S, L^2) with respect to distribution solutions for the system* (5.2) *is that* $\operatorname{Cap} Z = 0$.

This result, when combined with the techniques of the first named paper above (i.e., the Indiana J. paper), gives the following theorem concerning classical solutions of the system (5.1)

THEOREM 6.2. *A necessary and sufficient condition that Z be a removable set for the class $([L^\infty]^3,\ L^2)$ with respect to classical solutions for the system* (5.1) *is that* $\operatorname{Cap} Z = 0$.

In the Indiana Journal paper cited above, we established the following result:

THEOREM 6.3. *A necessary and sufficient condition that Z be a removable set for the class $([W^{1,2}]^3, L^2)$ with respect to classical solutions for the system* (5.1) *is that* $\operatorname{Cap} Z = 0$.

In particular by the necessary condition in this last cited theorem we mean that if $Z \subset B(0,1)$ is a closed set of 3-dimensional Lebesgue measure zero but with $\operatorname{Cap} Z > 0$, then there exists $\mathbf{v} \in \left[W^{1,2} B(0,1)\right]^3$ and $p \in L^2\big(B(0,1)\big)$ such that $(\mathbf{v}, p)$ is a classical solution of the system (5.1) in $B(0,1) \setminus Z$ but the pair $(\mathbf{v}, p)$ cannot be defined at the points of Z so that $(\mathbf{v}, p)$ is a classical solution of the system (5.1) in $B(0,1)$.

As a good example of the use of multiple Fourier series in dealing with problems in nonlinear differential equations, we give a sketch of the above counterexample. So, in particular, we are given a closed set $Z \subset B(0,1)$ of 3-dimensional Lebesgue measure zero but of positive capacity, i.e., Cap Z defined by (6.1) is positive. We shall assume Z is a removable set for the class $([W^{1,2}]^3, L^2)$ and arrive at a contradiction.

It then follows from well-known facts concerning capacity [We] and [Sh3] that $\operatorname{Cap}(Z) > 0 \Rightarrow \exists\, \mu \in P(Z)$ such that

$$\mu(\mathbb{T}_3 \setminus Z) = 0, \ \mu(Z) = 1, \ \sum_{m \neq 0} |\hat{\mu}(m)|^2 |m|^{-2} < \infty \tag{6.3}$$

where $\hat{\mu}(m) = (2\pi)^{-3} \int_{\mathbb{T}_3} e^{-i(m,x)}\, d\mu(x)$ and we are using the notation of §4 where now $N = 3$. We next take a function $\lambda \in C^\infty(\mathbb{T}_3)$ with the following properties: $\lambda \geq 0$, $\int_{\mathbb{T}_3} \lambda = 1$, $\lambda = 0$ in $B(0, 3/2) \cup [\mathbb{T}_3 \setminus B(0,2)]$. We then set $d\eta = d\mu - \lambda\, dx$ and observe that

$$\sum_{m \neq 0} |\hat{\eta}(m)|^2 \, |m|^{-2} < \infty \quad \text{and} \quad \hat{\eta}(0) = 0. \tag{6.4}$$

Next we say $v_j \in {}_0W^{1,2}(\mathbb{T}_3)$ if $v_j \in W^{1,2}(\mathbb{T}_3)$ and $\hat{v}_j(0) = 0$. We also say $\mathbf{v} \in \mathcal{H}(\mathbb{T}_3)$ if $v_j \in {}_0W^{1,2}$ for $j = 1, 2, 3$ and furthermore $\sum_{j=1}^{3} v_{x_j} = 0$. As an inner product in $\mathcal{H}(\mathbb{T}_3)$ we use

$$[\mathbf{v}, \mathbf{w}] = \sum_{j=1}^{3} \langle v_j, w_j \rangle \quad \text{with} \quad \langle v_j, w_j \rangle = \sum_{k=1}^{3} \int_{\mathbb{T}_3} v_{jx_k} w_{jx_k}. \tag{6.5}$$

We then observe from (5.4) and (6.4) that

$$\mathbf{F} = (F_1, F_2, F_3) \in \mathcal{H}(\mathbb{T}_3) \quad \text{where} \quad \hat{F}_j(m) = \hat{u}_j^1(m)\hat{\eta}(m). \tag{6.6}$$

We also note that $\mathbf{v} \in \mathcal{H}(\mathbb{T}_3)$ implies that $v_j \in L^6(\mathbb{T}_3)$ and consequently that A is a completely continuous mapping of $\mathcal{H}(\mathbb{T}_3)$ into itself (See [Lad, p. 117]) where

$$[A\mathbf{v}, \boldsymbol{\phi}] = (2\pi)^{-3} \sum_{j=1}^{3} \sum_{k=1}^{3} \int_{\mathbb{T}_3} v_j v_k \phi_{jx_k} \quad \forall\, \boldsymbol{\phi} \in \mathcal{H}(\mathbb{T}_3). \tag{6.7}$$

It then follows from the Leray-Schauder Theorem [GT, p. 280] that

$$\exists\, \mathbf{v} \in \mathcal{H}(\mathbb{T}_3) \quad \text{such that} \quad \nu\mathbf{v} = A\mathbf{v} + \mathbf{F}. \tag{6.8}$$

Next, we set

$$g_j = \sum_{k=1}^{3} v_k v_{jx_k} \quad \text{and} \quad V_j(x) = \sum_{k=1}^{3} \int_{\mathbb{T}_3} u_j^k(x - y) g_k(y)\, dy. \tag{6.9}$$

and it follows that $g_j \in L^{3/2}(\mathbb{T}_3)$ and from [GT, p. 159] that

$$\mathbf{V} = (V_1, V_2, V_3) \in \mathcal{H}(\mathbb{T}_3). \tag{6.10}$$

Furthermore, after a computation using (5.4) and (6.9), we see that

$$\nu\, [\mathbf{V}, \boldsymbol{\phi}] = -\sum_{j=1}^{3} (2\pi)^{-3} \int_{\mathbb{T}_3} g_j \phi_j \quad \text{for} \quad \boldsymbol{\phi} \in \mathcal{H}(\mathbb{T}_3). \tag{6.11}$$

Next, we observe that for $\boldsymbol{\phi} \in \mathcal{H}(\mathbb{T}_3)$,

$$-\int_{\mathbb{T}_3} v_k v_{jx_k} \phi_j = \int_{\mathbb{T}_3} v_j \left[v_k \phi_{jx_k} + \phi_j v_{kx_k} \right].$$

Consequently, we obtain from (6.7), (6.9), and (6.11) that $\nu[\mathbf{V},\phi] = [A\mathbf{v},\phi]\ \forall\phi$ in $\mathcal{H}(\mathbb{T}_3)$. But then we obtain from (6.8) that

$$\mathbf{v} = \mathbf{V} + \nu^{-1}\mathbf{F} \tag{6.12}$$

With $q^1(x)$ defined on $\mathbb{T}_3$ by (5.9), we set $P^1(x) = (2\pi)^{-3}\int_{\mathbb{T}_3} q^1(x-y)\,d\eta(y)$ and observe from (6.6) that for $j = 1,2,3$

$$-\nu|m|^2\hat{F}_j(m) - im_j\hat{P}^1(m) = \gamma_j\hat{\eta}(m) \quad \text{where} \quad \gamma = (1,0,0). \tag{6.13}$$

Next, define $q^k(x)$ for $k = 2,3$ by replacing the 1 in (5.9) by k and set

$$\text{(i)}\quad P(x) = (2\pi)^{-3}\sum_{k=1}^{3} q^k(x-y)g_k(y)\,dy \tag{6.14}$$

$$\text{(ii)}\quad p(x) = P(x) + \nu^{-1}P^1(x).$$

A computation using (6.9) shows that for $j = 1,2,3$

$$-\nu|m|^2\hat{v}_j(m) - im_j\hat{p}(m) = \nu^{-1}\gamma_j\hat{\eta}(m) + \hat{g}_j(m) \tag{6.15}$$

Let $B(y,r)$, $r > 0$, be an open 3-ball such that $B(y,r) \subset B(0,3/2)$ and let $\zeta \in C_0^\infty[B(y,r)]$. It follows from the Abel summability of multiple Fourier series (See [Sh3]), (6.9), and (6.15) that for $j = 1,2,3$

$$\int_{B(y,r)} \left[\nu v_j\Delta\zeta + p\zeta_{x_j} + \sum_{k=1}^{3} v_jv_k\zeta_{x_k}\right] = \gamma_j\int_{B(y,r)} \nu^{-1}\zeta\,d\mu(x). \tag{6.16}$$

Using a regularity result from the Transactions paper in §5, and the fact that μ has its support on Z, we obtain from (6.16) that $v_j \in C^2[B(0,3/2)\setminus Z]$ and $p \in C^1[B(0,3/2)\setminus Z]$ and furthermore the pair $(\mathbf{v},p)$ is a classical solution of the system (5.1) in $B(0,3/2)\setminus Z$. Since $\mathbf{v} \in \mathcal{H}(\mathbb{T}_3)$, it follows in particular that $\mathbf{v} \in \left[W^{1,2}[B(0,1)]\right]^3$. Since $g_j \in L^{3/2}(\mathbb{T}_3)$ it follows from (6.14) and [GT, p. 159] that $p \in L^2\big(B(0,1)\big)$. Consequently, if Z were a removable set for the class $([W^{1,2}]^3, L^2)$, it would follow that the pair $(\mathbf{v},p)$ is a classical solution of the system (5.1) in all of $B(0,1)$. Hence we would have that

$$\int_{B(0,1)} \left[\nu v_1\Delta\zeta + p\zeta_{x_1} + \sum_{k=1}^{3} v_1v_k\zeta_{x_k}\right] = 0 \quad \forall\zeta \in C_0^\infty[B(0,1)]. \tag{6.17}$$

Now (6.16) also holds for $B(0,1)$; so we conclude from (6.16) with $j = 1$ and (6.17) that

$$\gamma_1\int_{B(0,1)} \nu^{-1}\zeta\,d\mu(x) = 0 \quad \forall\zeta \in C_0^\infty[B(0,1)]. \tag{6.18}$$

But by (6.13), $\gamma_1 = 1$. It is clear that we can find a nonnegative function $\zeta \in C_0^\infty[B(0,1)]$ which takes the value 1 on Z. Consequently (6.18) is a direct contradiction to the fact that $\mu \in P(Z)$. We conclude that Z is not a removable set for the class $([W^{1,2}]^3, L^2)$, and the proof of our counterexample is complete.

For a more general result on capacity theory and the Navier-Stokes equations see [SW].

7. Capacity and 2nd Order Semilinear Elliptic Supersolutions

The results obtained in this section are based on a paper with same title published in Nonlinear Analysis, Th., Meth., and Appl., vol 9(1985), pp. 605-618.

Let $\Omega \subset \mathbb{R}^n, N \geq 2$, be a bounded open set and let

$$Lu = \sum_{a,j=1}^{N} -\partial[a_{ij}(x)\partial u_j/\partial x_j]/\partial x_i + a(x)u \tag{7.1}$$

be a symmetric uniformly elliptic operator with $a_{ij}(x)$ and $a(x)$ in $L^\infty(\Omega)$.

Let $\mathcal{N}(u)$ be the Nemitsky-operator

$$\mathcal{N}(u) = g(x,u)$$

where $g(x,s) : \Omega \times \mathbb{R} \to \mathbb{R}$ meets the following conditions: (N-1) g meets the usual Caratheodary conditions. (N-2) $\exists g_1 \in L^p(\Omega)$ where $p > 1$ for $N = 2$ and $p = 2N/(N+2)$ for $N \geq 3$ and a nonnegative constant K^* such that

$$|g(x,s)| \leq g_1(x) + K^*|s| \quad \forall s \in \mathbb{R} \quad \text{and a.e. } x \in \Omega \tag{7.2}$$

We study solutions of the inequality

$$Lu + \mathcal{N}u + F \geq f \tag{7.3}$$

where $f \in L^1(\Omega)$ and $F \in W^{-1,2}(\Omega) = [W_0^{1,2}(\Omega)]^* = \text{dual of } W_0^{1,2}(\Omega)$.

Let Ω_1 be an open subset of Ω and suppose that $u \in W^{1,2}(\Omega_1)$. Then u is a solution of inequality (1.3) in Ω_1 means that the following situation prevails:

$$\int_{\Omega_1}\left[\sum_{i,j=1}^{N} a_{ij}\frac{\partial u}{\partial x_i}\frac{\partial \phi}{\partial x_j} + au\phi + g(x,u)\phi\right] + F(\phi) \geq \int_{\Omega_1} f\phi \tag{7.4}$$

$\forall \phi \in C_0^\infty(\Omega_1)$ and $\phi \geq 0$. If u satisfies (7.4) in Ω_1, u is called a supersolution of $Lu + \mathcal{N}u + F = f$ in Ω_1. In this paper, Z always designates a compact subset of Ω. We shall say Z is a removable set for the inequality (7.3) with respect to the class $W^{1,2}(\Omega)$ if $u \in W^{1,2}(\Omega)$ is a supersolution of inequality (7.3) in $\Omega \backslash Z$ implies that u is a supersolution of inequality (7.3) in Ω. The theorem that prevails for this situation is the following:

THEOREM 7.1. *Assume that $L, \mathcal{N}$ and F meet the conditions stated above. Then with respect to the class $W^{1,2}(\Omega)$, a necessary and sufficient condition that Z be a removable set for the inequality (7.3) for every $f \in L^1(\Omega)$ is that Z be of capacity zero.*

By capacity in this paper we mean the classical logaritmic capacity when $N = 2$ and the classical Newtonian capacity when $N \geq 3$. To be specific, we set

$$\Phi(x) = |x|^{2-N} \quad \text{for} \quad N \geq 3 \tag{1}$$

$$= (2\pi)^{-1}\log|x|^{-1} \quad \text{for} \quad N = 2. \tag{2}$$

and define $P(Z)$ as in §6 with $\mathbb{R}^N$ replacing $\mathbb{R}^3$. We then define the energy integral $I(\mu)$ for $\mu \in P(Z)$ to be

$$I(\mu) = \int_Z\int_Z \Phi(x-y)d\mu(x)d\mu(y)$$

$\theta(Z)$ is then defined by (6.1). Cap Z is also defined by (6.1) if $N \geq 3$. For $N = 2$, we define Cap Z $= e^{-\theta(Z)}$. In particular, we observe that Cap $Z = 0$ if $I(\mu) = \infty \forall \mu \in P(Z)$.

The main idea in proving the necessary condition of the above theorem is to use the harmonic analysis that results when the inner product $B(u, v)$ is introduced in the Hilbert space $W_0^{1,2}(\Omega)$ where

$$B(u,v) = \sum_{i,j=1}^{N} \int_\Omega a_{ij} \partial u/\partial x_i \partial v/\partial x_j$$

for $u, v \in W_0^{1,2}(\Omega)$. For the details, we refer the reader to pp. 610-618 of the above cited paper.

8. Orlicz-Sobolev Spaces and the Weak Comparison Principle

The results obtained in this section are based on the paper

N-Functions and the Weak Comparison Principle, J. Diff. Eqns., vol. 36(1980), pp. 257-269.

The dimension of the space will be denoted in this section by $n.A(t)$ is called an N-function if $A(t) = \int_0^t \alpha(s)ds$ for $0 \leq t < \infty$ where (for this paper) α is a continuous strictly increasing function on $[0, \infty)$ with $\alpha(t) > 0$ for $t > 0, \alpha(0) = 0$ and $\lim_{t\to\infty} \alpha(t) = \infty$. We shall also suppose A is twice differentiable on $(0, \infty)$ with $A''(t) > 0$ for $t > 0$. Defining $\alpha^\sim = \alpha^{-1}$ as in [A, p. 229] and setting $A^\sim(t) = \int_0^t \alpha^\sim(s)ds$ we also see that $A^\sim$ is an N-function, called the complementary N-function to A.

We shall deal with the quasilinear differential operator

$$Qu = \nabla \cdot [A'(|Du|)Du/|Du|] + F(x,u) \tag{8.1}$$

where $A'(t) = \alpha(t), Du = (D_1u, \ldots, D_nu), D_iu = \partial u/\partial x_i$, and $\nabla \cdot$ is the divergence operator.

With $\Omega \subset \mathbb{R}^n$ a bounded open set, we establish a comparison principle result along the following lines:

$$Qv \leq Qu \text{ in } u \leq v \text{ on } \partial\Omega, \text{ then } u \leq v \text{ in } \Omega. \tag{8.2}$$

In the sequel, F in (8.1) is such that $F : \Omega \times \mathbb{R} \to \mathbb{R}$. We also suppose that F meets $(F-1) - (F-3)$ where

(F-1) F meets the usual Caratheodory conditions;
(F-2) $F(x, s)$ is nonincreasing in s for fixed $x \in \Omega$;
(F-3) $|F(x,s)| \leq cA(|s|)|s|^{-1} + f(x) \quad \forall s \in \mathbb{R}\backslash\{0\}$ and a.e. $x \in \Omega$

where f is a nonnegative measurable function with $A^\sim(f) \in L^1(\Omega)$. We also assume that A satisfies a Δ_2-condition near infinity (See [A, p. 252]), i.e., that there is a constant k and $t_0 > 0$ such that $A(2t) \leq kA(t)$ for $t_0 \leq t < \infty$. What this means in particular is that

$$\exists c \text{ and } t_0 > 0 \ \ s.t. \ \ A'(t) \leq cA(t)t^{-1} \quad \text{for } t_0 < t < \infty. \tag{8.3}$$

We say $u \in L^A(\Omega)$ if $\int_\Omega A(|u|) < \infty.L^A(\Omega)$ is referred to in the literature as an Orlicz space. We say $u \in W^{1,A}(\Omega)$ (called an Orlicz-Sobolev space) provided first $u \in W_{\text{loc}}^{1,1}(\Omega)$ and second both u and $|Du|$ are in $L^A(\Omega).u \in W_0^{1,A}(\Omega)$ will mean $u \in W^{1,A}(\Omega)$ and $\exists\{\phi_j\}_{j=1}^\infty$ with $\phi_j \in C_0^\infty(\Omega)$ such that $\int_\Omega A(|u - \phi_j|) \to 0$ and $\int_\Omega A(|D(u - \phi_j)|) \to 0$ as $j \to \infty$.

We now interpret (8.2) from a weak point of view. With $u, v \in W^{1,A}(\Omega)$, we say

$$u \leq v \text{ weakly } (A) \text{ on } \partial\Omega \text{ if } (u-v)^+ \in W_0^{1,A}(\Omega) \tag{8.4}$$

where as usual $u^+ = \max(u, 0)$. Next we set

$$\mathcal{Q}(u, v; \phi) = \int_\Omega \{[G(Du) - G(Dv)] \cdot D\phi - [F(x,u) - F(x,v)]\phi\} \tag{8.5}$$

where $G = (G_1, \ldots, G_n)$ and $G_i(x) = A'(|x|)x_i|x|^{-1}$ for $x \neq 0, G_i(0) = 0$. We then say

$$\mathcal{Q}(v) \leq \mathcal{Q}(u) \text{ weakly } (A) \text{ in } \Omega \text{ provided } \mathcal{Q}(u, v; \phi) \leq 0 \tag{8.6}$$

$\forall \phi \in C_0^\infty(\Omega)$ with $\phi \geq 0$. To see that $\mathcal{Q}(u, v; \phi)$ is well-defined for $u, v \in W^{1,A}(\Omega)$ and $\phi \in C_0^\infty(\Omega)$, we need (8.3), Young's inequality [A, p. 229] which says that $st \leq A(t) + A^\sim(s)$ for $0 \leq s, t < \infty$, as well as $\tilde{A}[A(t)t^{-1}] \leq A(t)$. The comparison theorem that we establish for these Orlicz-Sobolev spaces is the following:

THEOREM 8.1. *Let A be an N-function which satisfies a Δ_2-condition near infinity and is twice-differentiable on the open interval $(0, \infty)$ with $A''(t) > 0$ for $0 < t < \infty$. Let Ω be a bounded open set and let $F(x, s)$ meet (F-1) - (F-3). Also, let A meet one of the conditions (8.7) (i), (ii), or (iii) below. Suppose that both u and v are in $W^{1,A}(\Omega)$ and $\mathcal{Q}(v) \leq \mathcal{Q}(u)$ weakly (A) in Ω. Then if $u \leq v$ weakly (A) on $\partial\Omega$, it follows that $u \leq v$ almost everywhere in Ω.*

For a comparison principle for nonlinear elliptic differential equations of a somewhat different nature than presented here, we refer the reader to a result of Serrin, [Se, p.186].

We now state and discuss the additional conditions that we put on A in the above theorem, namely

(8.7) (i) A'' is nonincreasing for $0 < t < \infty$;
(ii) $A''(t)$ is nondecreasing for $0 < t < \infty$;
(iii) $A''(t)$ is continuous for $0 \leq t < \infty$. Furthermore, $\exists t_1 > 0$ s.t. $A''(t)$ is nonincreasing for $t_1 \leq t < \infty$.

Now $A(t) = t^p$ meets (8.7) (i) for $1 < p \leq 2$, but not (8.7) (iii) for $1 < p < 2$. $A(t) = t^p$ meets (8.7) (ii) for $2 < p < \infty$. Also $A(t) = (1+t)^p \log(1+t) - (1+t)^p/p + p^{-1}$ meets (8.7) (i) for $1 \leq p \leq 3/2$ and (8.7) (ii) for $2 \leq p < \infty$; however for $3/2 < p < 2$, A meets neither (8.7) (i) or (8.7) (ii) but it does meet (8.7) (iii). This comes about because $A'''(t) = (1+t)^{p-3}[(p-1)(p-2)\log(1+t) + (2p-3)]$. Hence for $3/2 < p < 2, A'''(t)$ is positive near $t = 0$ but negative for large t. Other $A(t)$ that meet the conditions in the hypothesis of the theorem are, for example, $(1+t)^p[\log(1+t)]^2$ with $1 \leq p < \infty$ and $(1+t)[\log(1+t)]^q$ with $1 \leq q \leq 2$.

9. Pointwise differentiability and the Generalized Cauchy-Riemann Equations

The results obtained in this section are based on the paper

Removable Sets for Pointwise Solutions of the Generalized Cauchy-Riemann Equations, Annals of math., vol. 92(1970), pp., 82-101.

Calderon and Zygmund in their paper [CZ] introduced the notion of a pointwise partial derivative of f in the following manner: Assume that $f \in L^1_{\text{loc}}(\Omega)$ where

$\Omega \subset \mathbb{R}^N, N \geq 2$, is a bounded open set and that $x^* \in \Omega$. Say $f \in t_1^1(x^*)$ if there is a polynomial $P(x)$ of degree 1, i.e., $P(x) = \alpha_0 + \sum_{j=1}^{N} \alpha_j x_j$, s.t.

$$r^{-N} \int_{B(o,r)} |f(x^* + x) - P(x)| dx = o(r) \quad \text{as} \quad r \to 0. \tag{9.1}$$

If $f \in t_1^1(x^*)$, then $D_j f(x^*) = \alpha_j$ is called the pointwise partial derivative of f with respect to x_j at x^*. It is clear that $f \in C^1[B(x^*, r_0)], r_0 > 0$, implies that $f \in t_1^1(x^*)$ and that $D_j f(x^*) = \partial f(x^*)/\partial x_j$ for $j = 1, \ldots, N$. (For a further elucidation of pointwise differentibility see [St, Ch VIII].) An open problem for a number of years was the following quite apparent one: Let $N = 2$ and let u and v be in $t_1^1(x) \forall x \in \Omega$. Suppose $D_1 u(x) = D_2 v(x)$ and $D_2 u(x) = -D_1 v(x) \forall x \in \Omega$, i.e., the Cauchy-Riemann equations hold pointwise $\forall x \in \Omega$. Then is it true that $u + iv$ is analytic function in Ω? This question was answered in the affirmative for the first time in the Annals paper referred to at the beginning of this section. This result was extended to higher dimensions by introducing the notion of the generalized Cauchy-Riemann equations. In particular, we say the vector-valued function $\mathbf{u} = (u_1, \ldots, u_N)$ is a pointwise solution of the generalized Cauchy-Riemann equations at $x^* \in \Omega$ if $u_j \in t_1^1(x^*)$ for $j = 1, \ldots, N$ and

$$\begin{cases} D_j u_k(x^*) = D_k u_j(x^*) \quad for \quad j \neq k, j, k = 1, \ldots, N, \\ \sum_{j=1}^{N} D_j u_j(x^*) = 0 \end{cases} \tag{9.2}$$

For $N = 3$, this says curl $\mathbf{u}(x^*) = 0$ and div $\mathbf{u}(x^*) = 0$.

From now on, it will be understood that if $f \in t_1^1(x^*)$ as in (9.1) then $f(x^*) = \alpha_0$ the constant term in the polynomial. With this fact in mind we say that $Q \subset \Omega$, a Borel set of Lebesgue measure zero, is a removable set for the generalized Cauchy-Riemann equations if given that $\mathbf{u}$ satisfies (i) and (ii) below and (9.2) holds in Ω/Q implies that $\mathbf{u}$ can be defined at the points Q so that $\mathbf{u} \in [C^\infty(\Omega)]^N$ and (9.2) holds at all points in Ω.

(i) $\mathbf{u} \in [t_1^1(x)]^N \forall x \in \Omega/Q$.

(ii) $\int_{B(0,r)} |\mathbf{u}(x + y)| dy = o(r)$ as $r \to o \ \forall x \in Q$.

In the Annals paper above the following result about removable sets for pointwise solutions of the generalized Cauchy-Riemann equations is established.

THEOREM 9.1. *A necessary and sufficient condition that $Q \subset \Omega$, a Borel set of Lebesgue measure zero, be removable for pointwise solutions of the generalized Cauchy-Riemann equations is that Q be a countable set.*

Multiple Fourier series methods are used to establish the above theorem similar to the manner employed in §5 of this survey dealing with removable singularities of the Navier-Stokes equations, but of course the result here appeared first.

10. Generalized Analytic Functions on the N-Torus

The results in this section are based on the paper

Bounded Generalized Analytic Functions on the Torus, Pacific J. Math, vol. 16(1964), pp. 1413-1422.

We use the notation introduced in §5 except now $N \geq 2$. In this section v will always designate a point a unit distance from the origin, i.e. $|v| = 1$. Given $f \in L^1(T_N)$ we say $f \in \mathcal{A}_v^+$ if $(m, v) \leq 0 \Rightarrow \hat{f}(m) = 0$. We set $\mathcal{A}_v = \mathcal{A}_v^+ \cup \mathcal{A}_{-v}^+$ and call $f \in L^1(T_N)$ a generalized analytic function if $\exists v$ such that $f \in \mathcal{A}_v$. We call a generalized analytic function f a strict generalized analytic function if $f \in \mathcal{B}_v$ where $\mathcal{B}_v = \mathcal{B}_v^+ \cup \mathcal{B}_{-v}^+$ and where $f \in \mathcal{B}_v^+$ means $\exists\gamma$ with $0 < \gamma < 1$ such that $(m, v) < \gamma|m| \Rightarrow \hat{f}(m) = 0$. These notions were introduced by Bochner, [Bo]. In the Pacific Journal paper listed above we obtain a result which is true for strict generalized analytic functons but is false for generalized analytic functions, i.e., $\exists$ a generalized analytic function which is not a strict generalized analytic function for which the result fails.

With every unit point $v = (v_1, \dots, v_N)$ there is also associated a one-parameter subgroup of $\mathbb{T}_N$ which we shall call G_v where $G_v = \{x : -\pi \leqq x_j < \pi, x_j \equiv tv_j \mod 2\pi, -\infty < t < \infty\}$. It turns out that G_v is dense in $\mathbb{T}_N$ if and only if v linearly independent with respect to rational coefficients. In either case, the statement that $E \subset G_v$ is of positive linear measure is a well-defined statement.

In the theorem stated below, we work with functions in $L^\infty(\mathbb{T}_N)$. Also, all functions initially defined in $\mathbb{T}_N$ will be understood to be extended to all of $\mathbb{R}^N$ by periodicity of period 2π in each variable. $f(x, s), s > 0$, will designate the Abel partial sum

$$f(x, s) = \sum_m \hat{f}(m)e^{i(m,x)-|m|s}. \tag{10.1}$$

We say f vanishes at x_0 if $\lim_{s\to 0+} f(x_0, s) = 0$. The following theorem is established in the Pacific Journal paper cited above:

THEOREM 10.1. *A necessary and sufficient condition that every $f \in \mathcal{B}_v$ which vanishes in a subset of G_v of positive linear measure be zero almost everywhere on $\mathbb{T}_N$ is that v be linearly independent with respect to rational coefficients.*

It is easy to see that the sufficiency condition of the above theorem is the analog of the familiar F-M. Riesz theorem concerning radial limits of bounded analytic functions in the interior of the unit disc, [Car, p. 50].

In the Pacific Journal paper, it is also shown that given v linearly independent with respect to rational coefficients, there exists an $f \in \mathcal{A}_v$ and also in $L^\infty(\mathbb{T}_N)$ which is nonzero on $E \subset \mathbb{T}_N$ where the N-dimensional Lebesgue measure of E is positive but for which

$$\lim_{s\to 0+} f(x, s) = 0 \quad \forall x \in G_v.$$

Hence, the above theorem is false in general for $\mathcal{A}_v$.

11. Number Theory and Multiple Trigonometric Series

The result in this section is based on the paper

Algebraic Integers and Distributions on the N-Torus, J. Funct. Analysis, vol 13(1973), pp. 138-153.

We use the notation introduced in §4 and 5 and label $C^\infty(\mathbb{T}_N) = \mathcal{D}$; so $\phi \in \mathcal{D}$ is called a test function. $\mathcal{D}'(\mathbb{T}_N)$ will designate the class of real-valued distributions on $\mathbb{T}_N$ having $\mathcal{D}(\mathbb{T}_N)$ as its class of test functions (See [BJS, p. 168] for a precise definition of $\mathcal{D}'(\mathbb{T}_N)$). For $S \in \mathcal{D}'(\mathbb{T}_N)$ and $m = (m_1, \dots, m_N)$ an integral lattice point, we set

$$\hat{S}(m) = (2\pi)^{-N}\{S[\cos(m, x)] - iS[\sin(m, x)]\}. \tag{11.1}$$

In the paper cited at the beginning of this section we deal with various classes of distributions. In this survey article we shall concentrate on only one, namely $\mathcal{A}(\mathbb{T}_N)$ defined as follows:

$\mathcal{A}(\mathbb{T}_N) = \{S \in \mathcal{D}'(\mathbb{T}_N) : |\hat{S}(m)|$ is uniformly bounded and $\lim_{\min(|m_1|,\dots,|m_N|)\to\infty} |\hat{S}(m)| = 0\}$ where $\hat{S}(m)$ is defined by (11.1).

Next, by $C(\zeta)$ with $0 < \zeta < \frac{1}{2}$, we designate the familiar Cantor set with constant ratio of dissection ζ constructed on the interval $[-\pi, \pi)$, i.e., $r \in C(\zeta) \leftrightarrow r = 2\pi(1-\zeta)\sum_1^\infty \epsilon_k \zeta^{k-1} - \pi$ where $\epsilon_k = 0$ or 1 and not all $\epsilon_k = 1$. Following [Z2, p. 148], we say θ is an **S** number if θ is a real algebraic integer which is > 1 with the additional property that all the conjugates of θ (other than θ) have moduli less than 1. Also, we note that if $0 < \zeta_j < \frac{1}{2}, j = 1, \dots, N$, then $C(\zeta_1) \times \dots \times C(\zeta_N)$ is a set contained in $\mathbb{T}_N$ which is closed in the torus topology.

Next, let $G \subset \mathbb{T}_N$ be a set which is open in the torus topology and let $S \in \mathcal{D}'(\mathbb{T}_N)$. By the statement $S = 0$ in G, we then mean the following:

$$S(\phi) = 0 \quad \forall \phi \in \mathcal{D}(\mathbb{T}_N) \text{ with support } \quad \phi \subset G.$$

Given $Z \subset \mathbb{T}_N$ a set closed in the torus topology, we say that Z is a set of uniqueness for the class $\mathcal{A}(\mathbb{T}_N)$ if the following is true:

$$S \in \mathcal{A}(\mathbb{T}_N) \text{ and } S = 0 \text{ in } \mathbb{T}_N \backslash Z \Rightarrow S \equiv 0.$$

The main theorem in the Journal of Functional Analysis paper cited at the beginning of this section is the following:

THEOREM 11.1. *Let $0 < \zeta_j < \frac{1}{2}, j = 1, \dots, N$. A necessary and sufficient condition that $C(\zeta_1) \times \dots \times C(\zeta_N)$ be a set of uniqueness for the class $\mathcal{A}(\mathbb{T}_N)$ is that each ζ_j^{-1} be an* **S** *number for $j = 1, \dots, N$.*

The necessary condition follows easily from the well-known one-dimensional fact (See [Z2, 11.17, p. 152]) that if ζ^{-1} is not an **S** number where $0 < \zeta < \frac{1}{2}$, then there exists a one-dimensional trigonometric series $\sum b_n e^{inr}$ with $|b_n| = o(1)$ as $n \to \infty$, with the property that $\lim_{t\to 0} \sum b_n e^{inr} e^{-n^2 t} = 0$ for $r \in \mathbb{T}_1 \backslash C(\zeta)$. Furthermore, for at least one value of $n, b_n \neq 0$. The main difficulty comes in establishing the sufficiency condition in the above theorem and for that, we refer the reader to the original paper.

We also note in closing this section that $C(\zeta_1) \times \dots \times C(\zeta_N)$ is a good example of an N-dimensional fractal set, and the above theorem gives a number theoretical connection between fractal sets and harmonic analysis.

12. Harmonic Functions and Spherical Harmonics

The results in this section are based on

Topics in Fourier and Geometric Analysis, Memoirs of the AMS, No. 39(1961), Chapter I.

In this section Ω denotes S_{N-1} the $(N-1)$-sphere, in $\mathbb{R}^N$ which is $\partial B(0,1)$. x will designate a point in Ω; so $x = (x_1, \dots, x_N)$ with $|x| = (x_1^2 + \dots + x_N^2)^{\frac{1}{2}} = 1$. $d\Omega$ will designate the natural $(N-1)$ dimensional area element on Ω. Given $f \in L^1(\Omega)$, we set

$$Y_n(x) = \frac{\Gamma(\lambda)(n+\lambda)}{2\pi^{\lambda+1}} \int_\Omega P_n^\lambda[(x,y)] f(y) d\Omega(y)$$

with $\lambda = (N-2)/2$ and P_N^λ the Gegenbauer (ultraspherical) polynomial defined by the equation $(1-2r\cos\theta+r^2)^{-\lambda} = \sum_{n=0}^{\infty} r^n P_n^\lambda(\cos\theta)$. In particular, $P_n^{\frac{1}{2}}$ is the familiar Legendre polynomial. If $f \in C^0(\Omega)$, then it is well-known that $\lim_{r\to 1-} \sum_{n=0}^{\infty} r^n Y_n(x) = f(x)$ uniformly for $x \in \Omega. Y_n(x)$ is called a surface spherical harmonic of order n, and the series $S[f] = \sum_{n=0}^{\infty} Y_n(x)$ when Ω is the 2-sphere is called the Laplace series of f. (See [CH, p. 510]). We set $\|Y_n\| = [\int_\Omega f^2 d\Omega]^{\frac{1}{2}}$.

The following analogue of the Riemann-Lebesgue lemma evidently was obtained on spheres for the first time in the Memoir paper cited above.

THEOREM 12.1. *Let $f \in L^1[\Omega]$ on the $(N-1)$-sphere Ω, $N \geqq 2$, and let $S[f] = \sum_{n=0}^{\infty} Y_n(x)$. Then $\|Y_n\| = o(n^\lambda)$ as $\lambda \to \infty$ where $\lambda = (N-2)/2$.*

In particular on the 2-sphere, for Laplace series, we have $\|Y_n\| = o(n^{\frac{1}{2}})$ as $n \to \infty$. The following uniqueness theorem for Laplace series was obtained in the Memoir paper cited above.

THEOREM 12.2. *Given Ω is the 2-sphere and Y_n is a surface spherical harmonic of order n with $\|Y_n\| = o(n^{\frac{1}{2}})$ as $n \to \infty$. Suppose that $\lim_{r\to 1-} \sum_{n=0}^{\infty} Y_n(x) r^n = f(x)$ $\forall x \in \Omega$ where $f \in L^1(\Omega)$ is a finite-valued function. Then it is true that $\sum_{n=0}^{\infty} Y_n(x)$ is the Laplace series of f, i.e., $Y_n(x) = \frac{(2n+1)}{4\pi} \int_\Omega P_n^{\frac{1}{2}}[(x,y)] f(y) d\Omega(y)$.*

Problems of this nature were worked on previously by Plancherel [P] and Rudin [R]. The above theorem gives rise to the following problem which has been open now for over seventy-five years and is the analogue of the Cantor Theorem in trigonometric series, [Z1, p. 326]:

Problem: Suppose that $\lim_{n\to\infty} \sum_{n=0}^{n} Y_n(x) = f(x)$ $\forall x \in \Omega$ where Ω is the 2-sphere and where $f \in L^1(\Omega)$ is a finite-valued function. Is it true that $\sum_{n=0}^{\infty} Y_n(x)$ is the Laplace series of f?

Theorem 12.2 gives rise to an interesting result concerning function harmonic in the interior of $B(0,1) \subset \mathbb{R}^3$. We define the following class of harmonic functions:

(12.1)

$$\mathcal{S} = \{u \text{ is harmonic in } B(0,1) : \int_\Omega |u(r,x)|^2 d\Omega(x) = o[(1-r)^{-2}] \text{ as } r \to 1-\}$$

We say $Z \subset \Omega$ is a set of uniqueness for the functions in class $\mathcal{S}$ if the following holds:

$u \in \mathcal{S}$ and $\lim_{r\to 1-} u(r,x) = 0$ $\forall x \in \Omega \backslash Z \Rightarrow u \equiv 0$.

The following theorem then holds:

THEOREM 12.3. *Let $Z \subset \Omega$ be a Borel set and let Ω be the 2-sphere and $B(0,1)$ be the unit ball in $\mathbb{R}^3$. Then a necessary and sufficient condition that Z be a set of uniqueness for the class $\mathcal{S}$ is that Z be a countable set.*

The class $\mathcal{S}$ is also best possible for the sufficiency condition in Theorem 12.3 when $B(0,1) \subset \mathbb{R}^3$ because if $\mathcal{S}'$ replaces $\mathcal{S}$ in Theorem 12.3 where $\mathcal{S}'$ is defined by (12.1) with $O[(1-r)^{-2}]$ replacing $o[(1-r)^{-2}]$ then the sufficiency condition is false. To see this, let $u(r,x) = (1-r^2)/[1-2r(x,p)+r^2]^{3/2}$ where p is a fixed point in Ω. Then it is easy that $u(r,x) \in \mathcal{S}'$ and that $\lim_{r\to 1} u(r,x) = 0 \forall x \in \Omega \backslash \{p\}$.

Now it follows from [Sh4, Th 2, p. 640] that if Z is the empty set, the sufficiency condition of Theorem 12.3 is also true when Ω is the one-sphere and $B(0,1) \subset \mathbb{R}^2$. But it is false if Z contains even one point.

We close this survey with the observation that the analogue of Theorem 12.3 when Ω is the $(N-1)$-sphere and $B(0,1)$ is the unit ball in $\mathbb{R}^N, N \geqq 4$, is still an open question.

References

[A] R.A. Adams, *Sobolev Spaces*, Academic Press, New York, 1975.

[Be] E. Beltrami, *Mathematics for Dynamic Modeling*, Academic Press, San Diego, 1987.

[BJS] L. Bers, F. John, and M. Schechter, *Partial Differential Equations*, John Wiley, New York, 1966.

[BN] H. Brezis and L. Nirenberg, *Characterization of the ranges of some nonlinear operators and applications to boundary value problems*, Ann. Scuola Norm. Sup. Pisa **5**(1978), 225-326.

[Bo] S. Bochner, *Boundary values of analytic functions*, Annals of Math, **45** (1944), 708-722.

[Br] F.E. Browder, *Existence theorems in partial differential equations*, in *Proceedings of Symposia in Pure Mathematics* vol. 16, pp. 1-60, Amer. Math. Soc., Providence, 1970.

[Car] C. Caratheodory, *Theory of Functions of a Complex Variable* vol. 2, Chelsea, Newy York, 1954.

[CZ] A. P. Calderon and A. Zygmund, *Local properties of solutions of elliptic partial differential equations*, Studia Math. 20(1961), 171-225.

[CH] R. Courant and D. Hilbert, *Methods of Mathematical Physics*, **Vol. I**, John Wiley, New York, 1966.

[DG] D.G. DeFigueuedo and J.P. Gossez, *Nonlinear partial differential equations near the first eigenvalue*, J. Diff. Eqns. 30(1978), 1-19.

[GT] D. Gilbarg and N.S. Trudinger, *Elliptic Partial Differential Equations of Second Order*, 2nd Edition, Springer-Verlag, Berlin, 1983.

[KS] D. Kinderlehrer and G. Stampachia, *An Introduction to Variational Inequalities and Their Application*, Academic Press, New York, 1980.

[Lad] O.A. Ladyzhenskaya, *The Mathematical Theory of Viscous Incompressible Flow*, Revised 2nd ed., Gordon & Breach, New York, 1969.

[Laz] A.C. Lazer, Book review, *Solvability of nonlinear equations and boundary value problems*, Bull. Amer. Math. Soc., vol. 8 (1983), 482-489.

[LL] J. Leray and J. L. Lions, *Quelques résultats de Visik sur les problèmes elliptiques nonlinéaires par les méthodes de Minty-Browder*, Bull. Soc. Math. France 3(1965), 97-107.

[M1] J. D. Murray, *How the leopard gets its spots*, Scientific American, 258 (1988), 80-87.

[M2] J. D. Murray, *Mathematical Biology*, Springer-Verlag, New York, 1990.

[P] M. Plancherel, *Sur l'unicite du developpment d'une fonction en serie de fonctions spherique*, Bull. Sci. Math. (2) 43(1919), 212-223.

[R] W. Rudin, *Uniqueness for Laplace series*, Trans. Amer. Math. Soc., 68(1950), 287-303.

[S] J. Serrin, *On the strong maximum principle for quasilinear second order differential inequalities*, J. Funct. Analysis 5(1970), 184-193.

[Sh1] V. L. Shapiro, *Quasilinear ellipticity and the 1st eigenvalue*, Comm. Partial Differential Equations 16(1991), 1819-1855.

[Sh2] V. L. Shapiro, *One-sided conditions for the Navier-Stokes equations,* J. Diff. Equations 44(1982), 414-439.

[Sh3] V. L. Shapiro, *Fourier series in several variables*, Bulletin Amer. Math. Soc. 70(1964), 48-93.

[Sh4] V. L. Shapiro, *The uniqueness of functions harmonic in the interior of the unit disc*, Proc. London Math. Soc. 13(1963), 639-652.

[St] E. M. Stein, *Singular Integrals and Differentiability Properties of Functions*, Princeton Univ. Press, Princeton, 1970.

[SW] V. L. Shapiro and G.V. Welland, *Sobolev Spaces, the Navier-Stokes Equations and Capacity,* in Proceedings of symposia in Pure Mathematics, vol. xxxv, pp. 369-375, Amer. Math. Soc., Providence, 1979.

[W] J. Wermer, *Potential Theory,* Lecture Notes in Math. 408, Springer-Verlag, Berlin, 1974.

[Z1] A. Zygmund, *Trigonometric Series* vol. I, Cambridge Univ. Press, Cambridge, 1959.

[Z2] A. Zygmund, *Trigonometric Series* vol. II, Cambridge Univ. Press, Cambridge, 1959.

Department of Mathematics, University of California, Riverside, Riverside, California 92521–0135

E-mail address: shapiro@math.ucr.edu

Contemporary Mathematics
Volume **208**, 1997

A SURVEY OF UNIQUENESS QUESTIONS IN MULTIPLE TRIGONOMETRIC SERIES

J. MARSHALL ASH AND GANG WANG

Dedicated to Victor Shapiro.

ABSTRACT. The issue is uniqueness of representation by multiple trigonometric series. Two basic uniqueness questions, one about series which converge to zero and the other about series which converge to an integrable function, are asked for each of four modes of convergence: unrestricted rectangular convergence, spherical convergence, square convergence, and restricted rectangular convergence. Thus there are eight basic questions in each dimension. In all dimensions, both uniqueness theorems are valid both for unrestricted rectangular convergence and for spherical convergence. All the other questions are still open. The positive work will be surveyed along with related work involving extensions of the Cantor-Lebesgue theorem. In particular, we present a detailed proof of Bourgain's recent important spherical uniqueness theorem.

1. INTRODUCTION

In 1974 Marshall Ash gave a talk entitled "Multiple trigonometric series" in which were listed only 4 substantial uniqueness theorems for convergent multiple trigonometric series [A1]. They were

1 if a double trigonometric series is spherically convergent to zero everywhere, then all its coefficients are zero,
2 if a double trigonometric series is spherically convergent to an everywhere finite, Lebesgue integrable function, then it is the Fourier series of that function,
3 if double trigonometric series is unrestrictedly rectangularly convergent to zero everywhere, then all its coefficients are zero, and
4 if a double trigonometric series is unrestrictedly rectangularly convergent to an everywhere finite, Lebesgue integrable function, then it is the Fourier series of that function.

Not much happened for about 20 years, but in the last few years all of these results have been extended to d dimensions. We will outline how this was done and point out several of the obstacles to further progress. To begin with, we will sketch the very well known one dimensional uniqueness proof given by Cantor.

Date: November 14, 1996.
1991 *Mathematics Subject Classification.* Primary 42B99; Secondary 42B08, 42A63.
Key words and phrases. Uniqueness, square convergence, spherical convergence, restricted rectangular convergence, unrestricted rectangular convergence, Cantor-Lebesgue theorem .
The research of both authors were partially supported by a grant from Faculty and Development Program of the College of Liberal Arts and Sciences, DePaul University.
This paper is in final form and no version of it will be submitted for publication elsewhere.

Theorem 1.1 (Cantor–1870). *Suppose that the one dimensional trigonometric series $\sum c_n e^{inx}$ converges to zero in the sense that at each x, $s_n(x) = \sum_{\nu=-n}^{n} c_\nu e^{i\nu x}$ tends to zero as n tends to ∞. Then all c_n must be zero.*

The first step of the proof is to observe that at each x, $c_{-n}e^{-inx} + c_n e^{inx} = s_n(x) - s_{n-1}(x)$ also tends to 0. The Cantor-Lebesgue theorem asserts that this does not happen because of cancellation between the two terms. Explicitly,

Theorem 1.2 (Cantor-Lebesgue–1870). *If $c_{-n}e^{-inx} + c_n e^{inx} \to 0$ for every $x \in \mathbb{R}$, then $\sqrt{|c_{-n}|^2 + |c_n|^2} \to 0$.*

Remark 1.1. *This was proved by Cantor in 1870. The theorem is so named because Lebesgue proved a stronger version in 1905. For the strongest version we know, see [AKR].*

In other words, our final goal of getting all c_n identically 0 has been partially achieved: at least the c_n tend to 0 as $|n|$ tends to ∞. Next one forms the Riemann function $F(x) = c_0 \frac{x^2}{2} - \sum' \frac{c_n}{n^2} e^{inx}$, where the prime means that the sum is over all nonzero integers. This function has two important properties: 1) it is continuous, and 2) it has a generalized second derivative which is everywhere 0. From these two facts it quickly follows first that F must be linear and consequently that all c_n are 0.

We will see below that all known higher dimensional proofs except one more or less follow this approach of starting with a Cantor-Lebesgue type theorem. Even for the exceptional case of unrestricted rectangular convergence, one of the known proofs [AFR] follows the one dimensional approach. It will also be pointed out that most of the open questions seem totally inaccessible because of a breakdown at this very first step. In most settings the best possible Cantor-Lebesgue extension is known and in some settings the corresponding extension yields much, much less than the one dimensional Cantor-Lebesgue conclusion.

Before leaving the one dimensional setting we will state one further theorem that also serves as a model for higher dimensional generalizations.

Theorem 1.3 (de la Vallée-Poussin–1912). *If a one dimensional trigonometric series T is convergent to an everywhere finite, Lebesgue integrable function f, then it is the Fourier series $S[f]$ of that function.*

Let us consider the relation between Cantor's theorem and de la Vallée-Poussin's theorem. By definition, $S[f] = \sum \hat{f}(n)e^{inx}$, where $\hat{f}(n) = \frac{1}{2\pi}\int_{-\pi}^{\pi} f(x)e^{-inx}dx$. Since the Fourier series of the everywhere zero function has all coefficients equal to zero, it is immediate that Cantor's theorem is a special case of de la Vallée-Poussin's. It is worth a moment's thought to distinguish the questions. After all, Cantor's theorem is called a uniqueness theorem because if T and S are two trigonometric series converging everywhere to the same value, then applying Cantor's theorem to the series $T - S$ shows that all the coefficients of T and S must agree. One might therefore try to deduce de la Vallée-Poussin's theorem from Cantor's theorem as follows. Let T converge to $f \in L^1$. Then $T - S[f]$ converges to 0 everywhere, so by Cantor's theorem, T and $S[f]$ are identical. The fallacy in this argument is that $S[f]$ may very well fail to converge at all. In fact, Kolmogorov has given an example of an L^1 function with an everywhere divergent Fourier series. The proof of de la Vallée-Poussin's theorem requires all the machinery of the original theorem together with additional notions such as majorants and minorants from the theory of Lebesgue integration.

Although we could often save space by only discussing analogues of de la Vallée-Poussin's theorem, it is worth distinguishing the cases because the proofs can be of quite different levels of difficulty, even in the one dimensional case.

We will consider multiple trigonometric series from now on. We begin by introducing some definitions and notation used for the rest of the paper. As a general notation, one dimensional and multiple dimensional indices are both denoted by $m, n, \cdots$. We do not distinguish between them for the sake of notational simplicity. The dimensions of the indices will always be clear from the context.

Let T be a multiple trigonometric series, so that for $x \in \mathbb{T}^d$ we have

$$T = \sum_{m \in \mathbf{Z}^d} c_m e^{i\langle m,x \rangle}.$$

Fix x and set $a_m = a_m(x) = c_m e^{imx}$. There are many ways to add up the terms of $T = \sum a_m$. We will consider 5 summation methods here. All methods will be symmetric, i. e., if a_m is included in a partial sum and if m' differs from m only by some coordinate signs so that $|m'_j| = |m_j|$ for every j, then $a_{m'}$ will also be included in that partial sum.

Definition 1.1. *Let* $|m| = \sqrt{\sum m_j^2}$ *and for each real* $r \geq 0$, *define*

$$T^r = \sum_{|m| \leq r} a_m$$

to be the r*th spherical partial sum of* T. *We say* T *converges spherically to* t *if*

$$\lim_{r \to \infty} T^r = t.$$

For multi-indices m and n, say that $m \geq n$ if for every j, $m_j \geq n_j$, and for any real number r, let $\underline{r}$ denote the multi-index $(r, ..., r)$.

Definition 1.2. *For* $m \geq \underline{0}$, *call*

$$T_m = \sum_{-m \leq n \leq m} a_n$$

to be the m*th rectangular partial sum of* T. *Let* m *be a nonnegative integer. We define*

$$T_{\underline{m}} = \sum_{-\underline{m} \leq n \leq \underline{m}} a_n$$

to be the m*th square partial sum of* T *and say* T *is square convergent to* t *if*

$$\lim_{m \to \infty} T_{\underline{m}} = t.$$

If we define a different norm $\|\cdot\|$ by $\|m\| = \max_{j=1,\ldots,d} \{|m_j|\}$, we see that $T_{\underline{m}}$ can also be expressed as

$$T_{\underline{m}} = \sum_{\|n\| \leq m} a_n.$$

The various methods of summing T can be viewed geometrically. For example, the spherical partial sum T^r are so named because the indices associated with the terms of T appearing in the partial sum are exactly the indices contained within the closed origin-centered sphere of radius r. Spherical summation is a plausible way of adding up all the terms of T because any fixed index m is in all spheres of sufficiently large radius; equivalently, a_m is included in all sufficiently late spherical partial sums. Similarly, if $d = 2$, and $(m_1, m_2) \geq \underline{0}$, the indices n satisfying $-m \leq n \leq m$ are exactly the indices contained within the closed origin-centered rectangle with lower left corner $(-m_1, -m_2)$ and upper right corner (m_1, m_2). In particular, if $m_1 = m_2 = m$, the rectangle is actually a square . (If $\|n\| = m$, then n is on the edge of the square.) Again, as $m \to \infty$, any fixed index n is in all sufficiently late squares; equivalently, a_n is included in all sufficiently late square partial sums. In short, any method which eventually captures all points of the index set $\mathbb{Z}^d$ is a plausible summation method. The three methods yet to be described will also eventually capture the entire index set.

Definition 1.3. *Say that T is unrestrictedly rectangularly convergent to t if*

$$\lim_{\min\{|m_j|\}\to\infty} T_m = t.$$

Next,

Definition 1.4. *T is restrictedly rectangularly convergent to t if for any $e \geq 1$, no matter how large,*

$$\lim_{n\to\infty} \sup \left\{ |T_m - t| : \min\{|m_j|\} \geq n \text{ and all } \left|\frac{m_j}{m_k}\right| \leq e \right\} = 0.$$

Finally,

Definition 1.5. *Say that T is one way iteratively convergent to t if there exists a permutation $\{\sigma(1), \cdots, \sigma(d)\}$ of $\{1, 2, \cdots, d\}$, such that the nested limit,*

$$\lim_{n_{\sigma(d)}\to\infty} \sum_{m_{\sigma(d)}=-n_{\sigma(d)}}^{n_{\sigma(d)}} \left(\lim_{n_{\sigma(d-1)}\to\infty} \sum_{m_{\sigma(d-1)}=-n_{\sigma(d-1)}}^{n_{\sigma(d-1)}} \left(\cdots \left(\lim_{n_{\sigma(1)}\to\infty} \sum_{m_{\sigma(1)}=-n_{\sigma(1)}}^{n_{\sigma(1)}} a_m \right)\right)\right)$$

is equal to t.

Remark 1.2. *If we replace "convergent" in Theorem 1.3 by "one way iteratively convergent", then the multidimensional version of Theorem 1.3 holds by a simple induction argument. Thus, we are only interested in questions where the mode of convergence is not one way iteratively convergence. On the other hand, as we will illustrate below, there are many interesting questions concerning the relationships between one way iterative convergence and other modes of convergence for trigonometric series.*

The rest of the paper is organized as follows. In Section 2 we summarize positive results on multidimensional Cantor-Lebesgue type theorems and give some counter examples. Uniqueness type theorems for multiple trigonometric series under different summation modes will be stated in Section 3. First, we will outline the analogues of Cantor's uniqueness Theorem 1.1 under unrestricted rectangular convergence. Then, we will present Bourgain's recent spherical uniqueness theorem. This will be done by giving a proof which depends on three theorems of Bourgain. Later, in Section 4, the proofs of the three theorems will be given. In Section 5, some open questions are posed.

2. Extensions of the Cantor-Lebesgue theorem

In Cantor's one dimensional proof sketched in Section 1, the first step was to deduce from the strong hypothesis of everywhere convergence to zero, the much weaker consequence that

$$\lim_{n\to\infty} c_n e^{inx} + c_{-n}e^{-inx} = 0, \qquad \text{for all } x. \tag{2.1}$$

As mentioned above, from this Cantor was able to deduce that

$$\lim_{n\to\infty} \sqrt{|c_{-n}|^2 + |c_n|^2} = 0. \tag{2.2}$$

2.1. **Square and restricted rectangular convergence cases.** In the cases of square and restricted rectangular convergence, from the analogue of (2.1) follows much less than the analogue of (2.2). Already in the thesis of Paul Cohen [Coh] there is an example of a double trigonometric series which is convergent off the line $\{(x, y) \in \mathbb{T}^2 : x = 0\}$, but has infinitely many coefficients "almost" satisfying $c_n \sim \|n\|$. This was improved in [AWe] to a double series with much larger coefficients and still convergent off a single line. However, for many years there was still a nagging doubt that maybe the everywhere nature of the hypothesis would rule out such coefficient growth. This hope was finally dashed in [AWa]. In fact, let $\varphi(n)$ be any function which grows more slowly than exponentially in n. This means that for any given $\gamma > 1$ there exists $b = b(\varphi, \gamma) > 0$ such that $|\varphi(n)| \le b\gamma^n$. Note that φ grows more slowly than exponentially in n if and only if $\limsup_{n\to\infty} \varphi(n)^{1/n} \le 1$. For example, $\varphi(n)$ could be $n^{1000000}$ or $e^{n/\log\log n}$. Consider the double trigonometric series

$$T(x, y) = \sum_{n=4}^{\infty}\sum_{m=0}^{\infty} t_{mn} \cos mx \cos ny, \tag{2.3}$$

created by considering the series

$$2\sqrt{\pi}\sum_{n=2}^{\infty} n^{3/2}\varphi(n) \cos^2\left(\frac{x}{2}\right) \sin^{2n-2}\left(\frac{x}{2}\right) \cos ny, \tag{2.4}$$

and using the binomial theorem to expand

$$\cos^2\left(\frac{x}{2}\right)\sin^{2n-2}\left(\frac{x}{2}\right) = \left(\frac{e^{ix/2} + e^{-ix/2}}{2}\right)^2 \left(\frac{e^{ix/2} - e^{-ix/2}}{2i}\right)^{2n-2}.$$

Since this expansion only involves integer frequencies lying between $-n$ and n, it is clear that the index set is supported on the "butterfly-shaped" region $\{(m, n) : -|n| \le m \le |n|\}$, so that the partial sums of the series as expressed in form (2.4) coincide with the square partial sums of the series thought of as a double trigonometric series and expressed in form (2.3). If $|x| < \pi$, so that $a = \sin^2\left(\frac{x}{2}\right)$ is less than 1, then series (2.4) converges by comparison with $\sum n^{3/2}\varphi(n)a^n$, while if $x \in \{-\pi, \pi\}$, then every term of series (2.4) is zero. Thus T is square convergent *everywhere.* On the other hand, some of T's coefficients are huge. For example, a simple calculation shows that

$$t_{0n} = 2\sqrt{\pi}\frac{n^{3/2}}{2n-1}\varphi(n)4^{-n}\binom{2n}{n},$$

and it follows from Stirling's formula, $n! \simeq \sqrt{2\pi n} n^n e^{-n}$, that

$$t_{0n} \simeq \varphi(n) \text{ as } n \text{ tends to} + \infty.$$

Also, T is one-way iteratively convergent everywhere in the sense that

$$\hat{T}(x,y) = \lim_{N\to\infty} \sum_{n=4}^{N} \left\{ \lim_{M\to\infty} \sum_{m=0}^{M} t_{mn} \cos mx \right\} \cos ny$$

exists as a finite real number for each (x, y). This is easy to see, since $t_{mn} = 0$ as soon as $m > n$. The series T also turns out to be restrictedly rectangularly convergent, although the calculations are somewhat technical [AWa]. To summarize,

- T has coefficients growing like $\varphi(n)$, and
- $T$is *everywhere* convergent to an everywhere finite function in the square, restricted rectangular, and one-way iterative senses.

The bad behavior exhibited by these examples is certainly bad enough to destroy any hope of proving uniqueness for square or restrictedly rectangularly convergent trigonometric series by closely following Cantor's proof. On the other hand, perhaps everywhere restricted rectangular or square convergence for trigonometric series implies everywhere one way iterative convergence. If so, Remark 1.2 will immediately lead to a uniqueness proof here. We will come back to this in Section 3.

As bad as these examples are, at least nothing worse can happen. Here are a couple of positive results, which show the examples given by equation (2.4) to be essentially sharp.

Definition 2.1. *Let* $A_0(x) = T_{\underline{0}}(x)$ *and for* $n \geq 1$, *define*

$$A_n(x) = T_{\underline{n}}(x) - T_{\underline{n-1}}(x).$$

Thus, $A_n(x)$ *is the sum of the terms on the edges of the d-dimensional square of size* $2n$.

Theorem 2.1. *Let* $T(x) = \sum_{n\geq 0} A_n(x)$ *be square convergent for all* x *in a full measured subset of* $\mathbb{T}^d$. *Then for any given* $\gamma > 1$, *there exists* $b = b(T,\gamma) > 0$ *such that* $|c_m| \leq b\gamma^{\|m\|}$. *[AWe, p. 408]*

This theorem was proved by Paul Cohen [C1]. We reference [AWe] above since Cohen's (very good and very interesting) thesis, although available in the University of Chicago Library, was never published.

Definition 2.2. *For each* $n \geq \underline{0}$, *define*

$$A_n(x) = \sum_{\substack{|\nu_i|=n_i \\ i=1,\cdots,d}} c_\nu e^{i\langle \nu,x\rangle}.$$

Theorem 2.2. *Let* $T(x) = \sum A_n(x)$ *converge restrictedly rectangularly for all* x *in a full measured subset of* $\mathbb{T}^d$. *Then for any given* $\gamma > 1$, *(1) there exists* $b = b(T,\gamma) > 0$ *such that* $|c_m| \leq b\gamma^{\|m\|}$ *and (2) given any* $E > 1, \{c_n\}$ *tends to 0 as* $\|n\| \to \infty$ *in such a way that every ratio* $|n_i| / |n_j|$ *is bounded by* E *[AWa].*

Again, because of the examples given above, these very weak theorems are essentially the "best possible" Cantor-Lebesgue type theorems for the given methods of convergence.

Remark 2.1. *Recently, Victor Shapiro called our attention to a 1950 paper of Walter Rudin in which an example involving spherical harmonics uses an nth term containing* $\sin^{n-1}\theta\cos\theta$ *and so is asymptotically small at each* θ *while being asymptotically large in another sense* [R]. *It is fascinating to see this identical idea appearing in a different context. Note well and enjoy this idea, because we believe there is still more to be extracted from it.*

2.2. **Unrestricted and spherical convergence cases.** For the methods of unrestricted rectangular convergence and spherical convergence there are much better Cantor-Lebesgue type results. Recall that for $m \in \mathbb{Z}^d$, $A_m(x)$ is defined only for $m \geq \underline{0}$.

Theorem 2.3. *If* $\lim_{\min\{m_i\}\to\infty} A_m(x) = 0$ *for all* x *in a set of positive Lebesgue measure, then* $\lim_{\min\{|m_i|\}\to\infty} c_m = 0$ *and* $\limsup_{\max\{|m_i|\}\to\infty} |c_m| < \infty$ *([AWe], pp. 410–412).*

This means that the coefficients tend to zero "in the corners" and are bounded "along the axes." Again, the result is essentially sharp. In the case of spherical convergence, the situation turns out to be even better.

Theorem 2.4. *If* $d = 2$ *and* $\{c_m\}$ *is a doubly indexed set of complex numbers such that*

$$C_r(x) = \sum_{|m|=r} c_m e^{i\langle m,x\rangle}$$

tends to zero for all x *in a set of positive Lebesgue measure, then*

$$\varepsilon_r = \sqrt{\sum_{|m|=r} |c_m|^2} \text{ tends to } 0 \text{ as } r \to \infty. \tag{2.5}$$

A slightly weaker version of this, requiring that the set $\{x : C_r(x) \to 0\}$ be of full measure was proved by Roger Cooke [Coo]. Shortly thereafter, Antoni Zygmund proved theorem [Z].

If one looks at the proofs of Cooke and Zygmund, it does not seem at all easy to extend them to any dimension higher than 2. But this is exactly what Bernard Connes [Con] did with a very clever geometrical induction argument. Here is his result.

Theorem 2.5 (Connes). *If* $d \geq 3$ *and* $\{c_m\}$ *is a multiply indexed set of complex numbers such that* $C_r(x)$ *tends to zero for all* x *in a non-empty open subset of* $\mathbb{T}^d$, *then* ε_r *tends to zero as* r *tends to infinity.*

Notice that this theorem requires a stronger hypothesis than those of Cooke and Zygmund. However, this will not prove to be a problem for the applications to be made here.

3. Uniqueness theorems

Despite our remarks to the effect that the very first step in proving a uniqueness theorem must be proving a strong Cantor-Lebesgue type theorem, that is not what happened historically. The first important work in the area was done by Victor Shapiro. He assumed that

$$\sigma_r = \sum_{r-1<|m|\leq r} |c_m| = o(r) \text{ as } r \to \infty. \tag{3.1}$$

Let us call conditions (2.5) and (3.1) the Connes condition and the Shapiro condition, respectively. Note that $|m|$ is always the square root of an integer, so letting k always denote an integer, apply first Schwarz's inequality, next the estimate that the number of points of $\mathbb{Z}^d$ in a sphere is asymptotically the volume of that sphere, and finally the fact that in $\mathbb{Z}^d$ there are $O(r^{d-1})$ lattice points of modulus lying between $r-1$ and r (this estimate, due to Landau, can be found in [G]) to get

$$\begin{aligned}\sigma_r &= \sum_{r-1<\sqrt{k}\le r}\left(\sum_{|m|=\sqrt{k}} c_m\right) = \sum_{k=(r-1)^2}^{r^2}\left(\sum_{|m|=\sqrt{k}} c_m\right)\\ &\le \sqrt{\sum_{k=(r-1)^2}^{r^2}\left(\sum_{|m|=\sqrt{k}} 1\right)}\sqrt{\sum_{k=(r-1)^2}^{r^2}\left(\sum_{|m|=\sqrt{k}} |c_m|^2\right)}\\ &= O\left(\sqrt{r^{d-1}}\right)\times\sqrt{\sum_{k=(r-1)^2}^{r^2}\varepsilon_{\sqrt{k}}^2}\\ &= O\left(r^{\frac{d-1}{2}}\right)\times O\left(r^{\frac{1}{2}}\right)\sup_{r-1<\sqrt{k}\le r}\varepsilon_{\sqrt{k}},\end{aligned}$$

Now if the Connes condition (2.5) holds, then $\sup_{r-1<\sqrt{k}\le r}\varepsilon_{\sqrt{k}} = o(1)$ as $r\to\infty$, so that the last calculation shows that in this case $\sigma_r = o(r^{d/2})$. In particular, the Shapiro condition (3.1) is a consequence of the Connes condition (2.5) in dimension 2, but not in any higher dimension. Here is what Shapiro [S] proved.

Theorem 3.1 (Shapiro). *Let the multiple trigonometric series $T = \sum c_m e^{i\langle m,x\rangle}$ be spherically Abel summable to an everywhere finite $L^1(\mathbb{T}^d)$ function and assume that the c_m satisfy condition (3.1), then T is the Fourier series of that function.*

Actually he proved more, but this gives the general idea. The spherical Abel limit of $\sum a_m$ is $\lim_{h\to 0}\sum a_m e^{-|m|h}$.

3.1. **Unrestricted rectangular uniqueness results.** In 1918, a general uniqueness result was announced. It seemed to have a routine inductive proof, so interest in the subject died out. About 1970, Ash and Welland realized that the proof was incorrect. Furthermore, it looked as if some new ideas would be required to get the result. The first result came pretty quickly.

Theorem 3.2 (Ash-Welland). *If a double trigonometric series converges unrestrictedly rectangularly everywhere, then Shapiro's condition (3.1) holds. Also unrestricted rectangular convergence to a function implies circular Abel summability to the same function. Therefore by the theorem of Shapiro, if a double trigonometric series converges to an L^1 function everywhere, then it must be the Fourier series of that function.*

This proof only works in dimension 2. In retrospect, its proof really was as unnatural and lucky as it seemed at the time. For example, unrestricted rectangular convergence of a double series does *not* imply double circular convergence, only double circular Abel summability. Also Shapiro's condition is a consequence of the natural unrestricted rectangular Cantor-Lebesgue theorem only in dimension two.

Twenty years went by without any further progress. Then came two theorems in quick succession that completely settled the problem. Furthermore the first theorem came with two independent and completely different proofs. Here are the theorems.

Theorem 3.3 (Ash–Freiling–Rinne, Tetunashvili). *Let the multiple trigonometric series $T = \sum c_n e^{i\langle n,x\rangle}$ with $x \in \mathbb{T}^d$ be unrestrictedly rectangularly convergent to 0 everywhere. Then all c_n are zero.*

Theorem 3.4 (Freiling–Rinne–Thompson, Tetunashvili). *Let the multiple trigonometric series $T = \sum c_n e^{i\langle n,x\rangle}$ with $x \in \mathbb{T}^d$ be unrestrictedly rectangularly convergent to an everywhere finite $L^1(\mathbb{T}^d)$ function at every x. Then T is the Fourier series of that function.*

We will only discuss the simpler of these two results; namely, we will give the ideas behind both of the proofs of Theorem 3.3.

Tetunashvili's proof is based on a really nice idea as explained in Remark 1.2. His proof opens the door to proving uniqueness theorems without using Cantor-Lebesgue type theorems. But so far, it has not been used to study other modes of convergence.

There are many connections between different modes of convergence and/or summability that do not hold for series of numbers, but which do hold on an everywhere basis. Several of these are mentioned in the paper of Ash and Welland. The one that Tetunashvili found concerns the implication

$$\text{unrestricted rectangular convergence} \implies \text{iterated convergence.}$$

This assertion does not hold for series of numbers. For example, consider the numerical double series $\sum a_{mn}$, where $a_{0n} = (-1)^n$ and $a_{1n} = (-1)^{n+1}$ for $n = 1, 2, \cdots$, and all other $a_{mn} = 0$. Since $T_{(m,n)} = 0$ as soon as $m \geq 1$, this is unrestrictedly rectangularly convergent to 0, but the iterated limit

$$\lim_{m\to\infty} \sum_{|i|\leq m} \left(\lim_{n\to\infty} \sum_{|k|\leq n} a_{ik} \right)$$

obviously doesn't exist. Nevertheless,

Lemma 3.5 (Tetunashvili). *If a multiple trigonometric series converges to 0 unrestrictedly rectangularly at every point of $\mathbb{T}^d$, then it also converges one way iteratively to 0 at every point of $\mathbb{T}^d$.*

Remark 3.1. *Shakro Tetunashvili showed us this related example involving sequences rather than series: if*

$$s_{mn}(x,y) = \left(\frac{m}{m+1} + (-1)^m\right)\frac{\cos mx}{n} + \left(\frac{n}{n+1} + (-1)^n\right)\frac{\cos ny}{m},$$

then $\lim_{\min\{m,n\}\to\infty} s_{mn}(x,y) = 0$ everywhere, but neither iterated limit exists anywhere.

This lemma can be proved by using a theorem of Paul Cohen. Cohen's theorem asserts that if the modulus of a trigonometric polynomial of degree n is bounded by B on a set of positive Lebesgue measure E, then there is a constant $c = c(E)$ such that the supremum norm of that polynomial does not exceed Bc^n. See [Coh] or Lemma 2.1 of [AWe] for this. The first consequence of Cohen's theorem is a theorem of Ash and Welland asserting that unrestricted rectangular convergence at all points of $\mathbb{T}^d$ forces the partial sums to be bounded at each point. (See

Lemma 2.3 of [AWe].) Even though the bound may vary from point to point, look at the example $\sum a_{mn}$, where $a_{0n} = 1$ and $a_{1n} = -1$ for $n = 1, 2, ...$ and all other $a_{mn} = 0$ to see that this is not a trivial result. We will restrict ourselves to the case $d = 2$.

Proof. Assume that the double series $S = \sum c_{mn} e^{i(mx+ny)}$ converges unrestrictedly rectangularly at every point of $\mathbb{T}^2$. It is enough to prove that

$$\begin{aligned} \lim_{N\to\infty} \sum_{n=-N}^{N} c_{mn} e^{imx} &= 0 \text{ for every } x \text{ and } m, \text{ and} \\ \lim_{M\to\infty} \sum_{m=-M}^{M} c_{mn} e^{iny} &= 0 \text{ for every } y \text{ and } n. \end{aligned}$$

Suppose this is not true. We may assume that there is a point y_o and an integer m_o such that

$$\lim_{N\to\infty} \sum_{n=-N}^{N} c_{m_o n} e^{iny_o} = 0 \text{ is false.} \tag{3.2}$$

From now on we will keep $y = y_o$ fixed and think of all functions under consideration as living on the horizontal section $\{(x, y_o) : x \in \mathbb{T}^1\}$. From Lemma 2.3 of [AWe] it follows that there is a function $B(x)$ bounding all the partial sums of S at the point (x, y_o). Hence there is a constant B and a set E of positive one dimensional measure so that all the partial sums of S are uniformly bounded by B on the set $\{(x, y_o) : x \in E\}$. By Paul Cohen's theorem, Lemma 2.1 of [AWe], it follows that for each integer m, there is a constant $B_{|m|}$ so that the partial sums $T_{(|m|,N)}(x, y_o)$ are uniformly bounded by $B_{|m|}$. (Actually, since $\left|T_{(|m|,N)}(x, y_o)\right| = \left|T_{(|m|,N)}(x, y_o) e^{|m|ix}\right|$ and $T_{(|m|,N)}(x, y_o) e^{|m|ix}$ is a trigonometric polynomial of degree $2|m|$, we know that $B_{|m|} = Bc(|E|)^{2|m|}$, but all that matters is that $B_{|m|}$ is independent of both x and N.)

For each integer m, consider the sequence $\{A_{m,N}\}_{N\geq 0}$, where

$$A_{m,N} = \sum_{n=-N}^{N} c_{mn} e^{iny_o}, \ N = 0, 1, 2, ...$$

For each m, we observe that $T_{(|m|,N)}(x, y_o) = \sum_{\mu=-|m|}^{|m|} A_{\mu,N} e^{i\mu x}$, so that Fourier's equation, $A_{m,N} = \frac{1}{2\pi} \int_{\mathbb{T}^1} T_{(|m|,N)}(x, y_o) e^{-imx} dx$ leads to the estimate

$$|A_{m,N}| \leq \sup \left|T_{(|m|,N)}(x, y_o)\right| \leq B_{|m|}. \tag{3.3}$$

Now let $\mathbb{Z}' = \{i_1, i_2, ...\}$ be a well-ordering of the integers with m_o coming first. For example, if $m_o = -2$, then $\mathbb{Z}' = \{-2, 0, -1, 1, 2, -3, 3, -4, 4, ...\}$. We now define inductively a number λ_{i_k} for each $k = 1, 2, ...$ The sequence $\{A_{m_o,N}\}_{N\geq 0}$ is bounded by inequality (3.3), but not convergent to zero by assumption (3.2), so there is a sequence of nonnegative integers $\{N_k^1\}_{k\geq 0}$ and a number $\lambda_{m_o} = \lambda_{i_1}$ so that

$$\lim_{k\to\infty} A_{m_o,N_k^1} = \lim_{k\to\infty} A_{i_1,N_k^1} = \lambda_{m_o}$$

where

$$\lambda_{m_o} \neq 0. \tag{3.4}$$

Now the sequence $\{A_{i_2,N_k^1}\}_{k\geq 0}$ is a subsequence of the bounded sequence $\{A_{i_2,N}\}_{N\geq 0}$ (see inequality (3.3) again) and hence itself a bounded sequence, so there is a subsequence $\{N_k^2\}_{k\geq 0}$ of $\{N_k^1\}_{k\geq 0}$ and a number λ_{i_2} so that

$$\lim_{k\to\infty} A_{i_2,N_k^2} = \lambda_{i_2}.$$

Once more, the sequence $\{A_{i_3,N_k^2}\}_{k\geq 0}$ is a bounded sequence, so there is a subsequence $\{N_k^3\}_{k\geq 0}$ of $\{N_k^2\}_{k\geq 0}$ and a number λ_{i_3} so that

$$\lim_{k\to\infty} A_{i_3,N_k^3} = \lambda_{i_3}.$$

Continue inductively. This procedure produces a number λ_m for each integer m. To finish the lemma, it is enough to prove that

$$\lim_{M\to\infty} \sum_{m=-M}^{M} \lambda_m e^{imx} = 0 \text{ for every } x, \tag{3.5}$$

because, by Cantor's one dimensional uniqueness theorem, this implies that every λ_m, and in particular λ_{m_o} is zero, contrary to condition (3.4). But to prove (3.5), it suffices to fix x and $\varepsilon > 0$ and to then prove that there is an integer M_o so that

$$M \geq M_o \text{ implies } \left| \sum_{m=-M}^{M} \lambda_m e^{imx} \right| < \varepsilon.$$

The series S converges unrestrictedly rectangularly to 0 everywhere; in particular, there is an integer $M_o = M_o(x, y_o, \varepsilon)$ such that $\left|T_{(M,N)}(x, y_o)\right| < \frac{\varepsilon}{2}$ whenever M and N exceed M_o. Fix any $M \geq M_o$, and note that $T_{(M,N)}(x, y_o) = \sum_{m=-M}^{M} A_{m,N} e^{imx}$ so that

$$\left| \sum_{m=-M}^{M} A_{m,N} e^{imx} \right| < \frac{\varepsilon}{2}.$$

Finally, pick $p = p_M$ so large that $\{0, -1, 1, -2, 2, ..., -M, M\} \subset \{i_1, i_2, ..., i_p\}$ and in this inequality let N tend to infinity through values belonging to the subsequence $\{N_k^p\}_{k\geq 0}$. Since $A_{m,N_k^p} \to \lambda_m$ as $k \to \infty$ for each integer $m \in [-M, M]$, we get

$$\left| \sum_{m=-M}^{M} \lambda_m e^{imx} \right| \leq \frac{\varepsilon}{2} < \varepsilon.$$

This completes the proof. □

Now Theorem 3.3 is almost immediate since a uniqueness theorem for uniqueness for iterated convergence can be proved by simple mathematical induction. Thus Tetunashvili's very clever contribution was to build an unexpected relationship between unrestricted rectangular convergence and one way iterative convergence.

Remark 3.2. *It is not hard to see that the above argument also leads to the following lemma:*

Lemma 3.6. *If a multiple trigonometric series converges everywhere to an everywhere finite $f(x) \in L^1(\mathbb{T}^d)$, then it also converges one way iteratively to $f(x)$ everywhere.*

Thus, Tetunashvili's proof can be easily adopted to prove Theorem 3.4.

The Ash-Freiling-Rinne approach, on the other hand, required a whole new technology. It will prove useful in the long run only if the new techniques turn out to have further applications. We will give an idea of how these techniques work by using them to reprove Theorem 1.1, Cantor's original one-dimensional uniqueness theorem. Suppose that

$$\sum c_n e^{inx} = 0 \text{ for all } x. \tag{3.6}$$

Assume that the standard proof has been started, i.e., that the Cantor-Lebesgue result $|c_n| + |c_{-n}| \to 0$ has been established, that the Riemann function $F(x) = c_0 \frac{x^2}{2} - \sum' \frac{c_n}{n^2} e^{inx}$ (where the prime on the summation sign denotes the omission of the $n = 0$ term) has been shown to be a continuous function, that the identity

$$\frac{F(x+h) - 2F(x) + F(x-h)}{h^2} = c_0 + \sum{}' c_n e^{inx} \left(\frac{\sin n\frac{h}{2}}{n\frac{h}{2}} \right)^2 \tag{3.7}$$

has been established, and that the consistency of Riemann summability,

$$\lim_{h \to 0} c_0 + \sum{}' c_n e^{inx} \left(\frac{\sin n\frac{h}{2}}{n\frac{h}{2}} \right)^2 = 0 \tag{3.8}$$

has been shown. The next step is to prove this lemma.

Lemma 3.7. *If $F(x)$ is continuous and if $\lim_{h\to 0} \frac{F(x+h)-2F(x)+F(x-h)}{h^2} = 0$ for all x, then $F(x + 2\pi) - 2F(x) + F(x - 2\pi) = 0$ for all x.*

Use this lemma and set $h = 2\pi$ in identity (3.7) to see that $c_0 = 0$. The first new idea is to finish the proof by desymmetrizing this conclusion: it also follows that all the other c's are zero. To see this, fix an integer j. Our assumption (3.6) is that

$$\lim_{N \to \infty} \sum_{n=-N}^{N} c_n e^{inx} = 0 \text{ for all } x.$$

Now consider the shifted series $\sum c_{n+j} e^{inx}$. If this also converges to zero everywhere, then c_j also is zero and, since j was arbitrary, the proof is complete. But if $j > 0$, then for N large

$$\sum_{n=-N}^{N} c_{n+j} e^{inx} = - \sum_{k=-N}^{-N+j-1} c_k e^{i(k-j)x} + e^{-ijx} \left(\sum_{k=-N}^{N} c_k e^{ikx} \right) + \sum_{k=N+1}^{N+j} c_{k-j} e^{i(k-j)x}.$$

As N tends to infinity, the first and third sums have fixed length and hence tend to zero by the Cantor-Lebesgue result. That the middle sum tends to zero also is exactly our assumption. The case of $j < 0$ is done in a similar way. It only remains to prove Lemma 3.7. The usual proof uses the maximum principle and is very quick. To find a proof that would be independent of dimension a new technique which avoids the maximum principle was found. We will only sketch it here, but the details can be found in [A2].

Proof of Lemma 3.7. Fix x_0 and $\varepsilon_0 > 0$. It is enough to prove

$$\frac{|F(x_0 + 2\pi) - 2F(x_0) + F(x_0 - 2\pi)|}{(2\pi)^2} < \varepsilon_0. \tag{3.9}$$

We will associate to F a function $\mathcal{F}$ of 2 real variables. First, identify the point $x \in \mathbb{R}$ with the line having slope -1 and passing through the point (x, x). Then let $\mathcal{F}$ have the value $F(x)$ at all points of this line. In other words, for every real $k, \mathcal{F}(x+k, x-k) = F(x)$, or, equivalently, $\mathcal{F}(x,y) = F(\frac{x+y}{2})$. Thus we first think of the domain of F as being the line $y = x$, and then create $\mathcal{F}$ from F in the real-valued case by sweeping the graph of F through three-dimensional space parallel to the vector $(-1,1,0)$. (When F is complex-valued, this last visualization has to be applied to its real and imaginary parts separately.) Whenever we want to move from $\mathcal{F}$ to its associated $\mathbb{R}$-domained function F, we will use the terms *shadow* and *project.* The shadow of a point (x, y) is the point $\frac{x+y}{2}$ or, equivalently, (x, y) projects onto the point $\frac{x+y}{2}$, since (x, y) is orthogonally projected onto the point $(\frac{x+y}{2}, \frac{x+y}{2})$ of the one-dimensional subspace $\{(x, y) : x = y\}$ and then the point $(\frac{x+y}{2}, \frac{x+y}{2})$ of that subspace is identified with $\frac{x+y}{2}$. The point of all this is that if S is a square $[a, a+p] \times [b, b+p]$, and if Δ_S is a function defined on squares by the formula $\Delta_S = \mathcal{F}(a,b) - \mathcal{F}(a, b+p) - \mathcal{F}(a+p, b) + \mathcal{F}(a+p, b+p)$, then Δ_S also has the representation $F(x+h) - 2F(x) + F(x-h)$ where $x = \frac{a+b+p}{2}$ and $h = \frac{p}{2}$. Also $|S|$, the area of S is obviously $p^2 = 4h^2$, so we have the correspondence

$$4\frac{\Delta_S}{|S|} = \frac{F(x+h) - 2F(x) + F(x-h)}{h^2}.$$

It is not hard to see that the set function Δ is additive in the sense that if a square S (all squares and rectangles to be discussed here will have their sides parallel to the coordinate axes) is partitioned into nonoverlapping subsquares $\{S_j\}$, then $\Delta_S = \sum_j \Delta_{S_j}$. This formula allows us to expand a symmetric second difference with a large h into a sum of symmetric second differences, each of which has a small h. This is the essential idea behind the entire proof. In other words, the proof has been translated into the geometric problem of decomposing the square $[x_0, x_0 + 2\pi] \times [x_0, x_0 + 2\pi]$ into appropriate subsquares. Actually, the situation is slightly more complicated than this. It turns out that we will need to use rectangles of bounded eccentricity for our partition elements. We will arbitrarily restrict ourselves to rectangles of eccentricity bounded by 2. Again, for such a rectangle R, define $\Delta_R = \mathcal{F}(A) - \mathcal{F}(B) - \mathcal{F}(D) + \mathcal{F}(C)$ where A and C are the lower left and upper right corners of R, and B and D are the upper left and lower right corners of R. The function Δ is still additive, i.e., if S is partitioned into nonoverlapping rectangles $\{R_j\}$, then $\Delta_S = \sum_j \Delta_{R_j}$ still holds. Furthermore, if R is the rectangle $[a, a+p] \times [b, b+q]$, then the usual projection from $\mathcal{F}$ to F yields $\Delta_R = F(x-h) - F(x-k) - F(x+k) + F(x+h)$, where $x = \frac{a+b}{2} + \frac{p+q}{4}, h = \frac{p+q}{4}$, and $k = \frac{|p-q|}{4}$. It is easy to see that the condition of eccentricity bounded by 2 converts into the condition $\frac{k}{h} \leq \frac{1}{3}$. The area of R is $pq = \frac{h^2-k^2}{4}$, so by analogy with the case of the square, we are naturally led to considering

$$4\frac{\Delta_R}{|R|} = \frac{F(x-h) - F(x-k) - F(x+k) + F(x+h)}{h^2 - k^2}.$$

The limit of the right side as h and k go to 0 in such a way that $0 \leq \frac{k}{h} \leq \frac{1}{3}$ defines a generalized second derivative that was studied by Riemann. From the identity

$$\begin{aligned} 4\frac{\Delta_R}{|R|} = \frac{F(x-h) - 2F(x) + F(x+h)}{h^2} \frac{h^2}{h^2 - k^2} \\ + \frac{F(x-k) - 2F(x) + F(x+k)}{k^2} \frac{k^2}{h^2 - k^2}, \end{aligned}$$

and equation (3.7)–(3.8), it follows that

$$\lim_{\substack{h\to 0 \\ 0\le \frac{k}{h}\le \frac{1}{3}}} 4\frac{\Delta_R}{|R|} = 0, \qquad \text{for each } x. \tag{3.10}$$

If we were lucky enough to know that this limit was uniform in x, then the proof would be easy: just partition the square $I = [x_0, x_0 + 2\pi] \times [x_0, x_0 + 2\pi]$ very finely into tiny rectangles and then use the additivity of Δ to reduce the original difference quotient into a lot of small difference quotients, each with a small value. In fact, this almost works, but since the rate at which the limit is reached in equation (3.10) varies from point to point, the following more delicate argument is needed.

The strategy of the proof is to partition the square I into rectangles $\{R_j\} \bigcup \{R'_k\}$ with the property that for each R_j, $4\frac{|\Delta_{R_j}|}{|R_j|} < \frac{\varepsilon_0}{2}$, while the R'_k have the property that $\sum \left|\Delta_{R'_k}\right| < \frac{|I|}{4} \cdot \frac{\varepsilon_0}{2}$. This will be sufficient to complete the proof, for Δ and $|\cdot|$ are both additive, so that we have

$$\frac{F(x_0 + 2\pi) - 2F(x_0) + F(x_0 - 2\pi)}{(2\pi)^2} = 4\frac{\Delta_I}{|I|} = 4\frac{\sum_j \Delta_{R_j}}{|I|} + 4\frac{\sum_k \Delta_{R'_k}}{|I|} = A + B,$$

where, since $\{R_j\}$ is part of a partition of I,

$$|A| \le \frac{1}{|I|}\sum_j 4\frac{|\Delta_{R_j}|}{|R_j|}|R_j| \le \frac{\varepsilon_0}{2}\cdot\frac{1}{|I|}\sum_j |R_j| \le \frac{\varepsilon_0}{2},$$

and

$$|B| \le \frac{4}{|I|}\cdot\frac{|I|}{4}\cdot\frac{\varepsilon_0}{2} = \frac{\varepsilon_0}{2}.$$

It is reasonable to believe that such a partitioning can be accomplished in view of equation (3.10). Carrying out the actual partitioning also uses the fact that there holds

$$\lim_{h\to 0} F(x + 2h) - 2F(x + h) + F(x) = 0, \quad \text{for each } x,$$

and is, unfortunately, quite technical. See [A2] for the details.

3.2. **Spherical uniqueness results.** Here are the two results in this area.

Theorem 3.8 (Bourgain). *Let a multiple trigonometric series $T = \sum c_m e^{i\langle m,x\rangle}$ with $x \in \mathbb{T}^d$ be spherically convergent to zero at every x. Then all c_m are zero.*

Theorem 3.9 (Ash-Wang). *Let a multiple trigonometric series $T = \sum c_m e^{i\langle m,x\rangle}$ with $x \in \mathbb{T}^d$ be spherically convergent to an everywhere finite $f(x)$ for all $x \in \mathbb{T}^d$ and $f(x) \in L^1(\mathbb{T}^d)$. Then c_m are the Fourier coefficients of $f(x)$.*

If $d = 2$, both of the theorems are due to Shapiro [S] and Cooke [Coo]. For Theorem 3.8, Bourgain needs to present some major new ideas at one point of the argument. Theorem 3.9 builds a little further on the foundation laid by Shapiro and Bourgain. The proof of Theorem 3.9 will appear elsewhere.

We feel that Bourgain's quite brilliant argument is a major breakthough in this area. Since the proof given in Bourgain's paper is quite brief, we will devote the reminder of this paper to presenting the details of his proof.

Proof. Let T be spherically convergent to 0 everywhere. First notice that the theorems of Cooke (if $d = 2$) or Connes (if $d \geq 3$) apply. Thus, as in the one dimensional case, we begin with the much weaker consequence that the coefficients at least tend to zero in the sense that Connes condition (2.5) holds. But, as we calculated just above the statement of Shapiro's theorem, if $d = 2$, then Connes condition (2.5) implies Shapiro's condition (3.1). This fact, together with the consistency of Abel summability, implies that when $d = 2$, the hypothesis of Shapiro's theorem is fulfilled, so that Bourgain's theorem is an immediate corollary of Shapiro's. Hence we may as well assume that $d \geq 3$.

It is sufficient to show a special case of Theorem 3.8 given below.

Theorem 3.10. *If for any $x \in \mathbb{T}^d$,*

$$\lim_{r\to\infty} \sum_{0<|m|\leq r} c_m e^{i\langle x,m\rangle} = 0$$

Then $c_m = 0$ for all $m \in \mathbb{Z}^d \setminus \{0\}$.

In fact, for all $x \in \mathbb{T}^d$, if

$$\lim_{r\to\infty} \sum_{|m|\leq r} c_m e^{i\langle x,m\rangle} = 0, \tag{3.11}$$

then replace x by $2x$ in (3.11), we get

$$\lim_{r\to\infty} \sum_{|m|\leq r} c_m e^{i\langle 2x,m\rangle} = \lim_{r\to\infty} \sum_{|m|\leq r} c_m e^{i\langle x,2m\rangle} = 0.$$

So, taking the difference of the above two limits, we have

$$\lim_{r\to\infty} \sum_{0<|m|\leq r} \tilde{c}_m e^{i\langle x,m\rangle} = 0, \tag{3.12}$$

where $\tilde{c}_m = c_m$, if some of the element in m are odd; and $\tilde{c}_m = c_m - c_{m/2}$, if all of the elements in m are even. Thus, by Theorem 3.10 and (3.12), we have $c_m = 0$ for all $m \in \mathfrak{I} = \{$ some of the element in m are odd $\}$, and $c_m = c_{m/2}, m \notin \mathfrak{I}, m \neq 0$. Since for any $m \neq 0$, $m = 2^k \tilde{m}$ for some $\tilde{m} \in \mathfrak{I}$ and $k \geq 0$, we have $c_m = 0$ for all $m \neq 0$ and consequently $c_0 = 0$ from (3.11).

We now prove Theorem 3.10. Define a Riemann function $F(x)$:

$$F(x) = -\sum_{m\neq 0} \frac{c_m}{|m|^2} e^{i\langle x,m\rangle}. \tag{3.13}$$

Since

$$\lim_{r\to\infty} \sum_{0<|m|\leq r} c_m e^{i\langle x,m\rangle} = 0 \text{ for all } x \in \mathbb{T}^d,$$

the series defining $F(x)$ is convergent everywhere spherically in $\mathbb{T}^d$ by simply observing that

$$F(x) = -\sum_{k\geq 1} \frac{1}{k} \sum_{|m|^2=k} c_m e^{i\langle x,m\rangle} = -\sum_{k\geq 1} \left(\sum_{|m|^2\leq k} c_m e^{i\langle x,m\rangle} \right) (\frac{1}{k} - \frac{1}{k+1}).$$

Moreover, based on Theorem 2.5, $F(x)$ is in $L^2(\mathbb{T}^d)$ since

$$\left\| \sum_{m\neq 0} \frac{c_m}{|m|^2} e^{i\langle x,m\rangle} \right\|_2^2 = \sum_{k\geq 1} \frac{1}{k^2} \sum_{|m|^2=k} |c_m|^2 \leq \sum_{k\geq 1} \frac{1}{k^2} \varepsilon^2_{\sqrt{k}}. \tag{3.14}$$

Thus, the series is also the Fourier series of $F(x)$.

The goal is to show that $F(x)$ is harmonic in $\mathbb{R}^d$. Once this is true, then by periodicity, $F(x)$ is bounded in $\mathbb{R}^d$. Thus, $F(x)$ is a constant K because it is bounded and harmonic. Compare the Fourier coefficients of function $g(x) = K$ with $F(x)$, we get $c_m = 0$ for all $m \neq 0$. This will complete the proof.

In order to show that $F(x)$ is harmonic, the first step is to introduce the generalized Laplacian for L^1 functions. Let $B(x,\rho)$ be an open ball in $\mathbb{R}^d$ centered at $x \in \mathbb{T}$ with radius $\rho > 0$ and $m(B(x,\rho))$ be the volume of $B(x,\rho)$. Then $m(B(x,\rho)) = v_d\rho^d$, where v_d is the volume of unit ball in $\mathbb{R}$. For an L^1 function $F(x)$, designate $A_\rho F(x) = \frac{1}{v_d\rho^d}\int_{B(x,\rho)} F(y)\,dy$ to be the average of F over $B(x,\rho)$. Let $I(x) = \frac{1}{v_d} I_{B(0,1)}(x)$ be the normalized indicator function of the unit ball. Then the Fourier transform $\hat{I}(m)$ of $I(x)$ satisfies the following

$$\lim_{\rho\to 0} \frac{\hat{I}(\rho m) - 1}{\rho^2|m|^2} = -\frac{1}{2}\int_{B(0,1)} x_1^2 I(x)\,dx = \mu_d < 0, \tag{3.15}$$

and for $|m| = 1$

$$\int_0^\infty \left| \partial_r \left[\frac{\hat{I}(\rho r m) - 1}{\rho^2 r^2} \right] \right| dr = \int_0^\infty \left| \partial_r \left[\frac{\hat{I}(rm) - 1}{r^2} \right] \right| dr < c. \tag{3.16}$$

Note that the constant c in (3.16) is independent of ρ. The above two equations can be verified easily by calculus.

We now define the generalized Laplacian operator for a L^1 function $f(x)$ to be

$$\tilde{\Delta} f(x) = \lim_{\rho\to 0} \frac{A_\rho f(x) - f(x)}{\rho^2}$$

if such a limit exists. For $F(x)$ defined by (3.13), by (3.15) and (3.16),

(3.17)

$$\begin{aligned} \frac{A_\rho F(x) - F(x)}{\rho^2} &= -\sum_{m\neq 0} \frac{c_m}{|m|^2} \frac{\hat{I}(\rho m) - 1}{\rho^2} e^{i\langle x,m\rangle} \\ &= -\sum_{k\geq 1} \frac{\hat{I}(\rho\sqrt{k}) - 1}{\rho^2 k} \sum_{|m|^2=k} c_m e^{i\langle x,m\rangle} \\ &= -\sum_{k\geq 1} \left(\sum_{|m|^2\leq k} c_m e^{i\langle x,m\rangle} \right) \left(\frac{\hat{I}(\rho\sqrt{k}) - 1}{\rho^2 k} - \frac{\hat{I}(\rho\sqrt{k+1}) - 1}{\rho^2(k+1)} \right) \\ &\longrightarrow 0 \text{ as } \rho \longrightarrow 0 \end{aligned}$$

we have for all $x \in \mathbb{T}^d$. Thus, we have for all $x \in \mathbb{R}^d$

$$\tilde{\Delta} F(x) = 0. \tag{3.18}$$

A theorem of Rado [Ra] and (3.18) imply that $F(x)$ is harmonic once it is continuous everywhere in $\mathbb{R}^d$. By periodicity, it is enough to show that $F(x)$ is continuous everywhere in $\mathbb{T}^d$. Let

$$Z_\varepsilon = \left\{ x \in \mathbb{T}^d : \sup_{\substack{|x-x_i|<\delta \\ x_i \in \mathbb{T}^d, i=1,2}} |F(x_1) - F(x_2)| > \varepsilon \text{ for all } \delta > 0 \right\}. \tag{3.19}$$

It is easy to see that Z_ε is a closed set. The discontinuity set of $F(x)$ in $\mathbb{T}^d$ is

$$Z = \bigcup_{\varepsilon>0} Z_\varepsilon.$$

We will show that $Z = \emptyset$.

Assume $Z \neq \emptyset$. By assumption, $\sum\limits_{|m|\leq r} c_m e^{i\langle x,m\rangle}$ is uniformly bounded in r at each point x of $\mathbb{T}^d$, thus

$$\bigcup_{n\geq 1} \bigcap_{k\geq 1} \left\{ x \in \overline{Z} : \Big| \sum_{|m|^2\leq k} c_m e^{i\langle x,m\rangle} \Big| \leq n \right\} = \overline{Z}.$$

By Baire's category theorem, for some $n \geq 1$,

$$\bigcap_{k\geq 1} \left\{ x \in \overline{Z} : \left| \sum_{|m|^2\leq k} c_m e^{i\langle x,m\rangle} \right| \leq n \right\}$$

has a nonempty interior relative to $\overline{Z}$. Namely, there exists an open ball $B(p, \rho_0), p \in Z$ such that

$$\max_k \max_{x\in B(p,\rho_0)\cap\overline{Z}} \left| \sum_{|m|^2\leq k} c_m e^{i\langle x,m\rangle} \right| \leq n < \infty. \tag{3.20}$$

Consequently,

$$F(x) = -\sum_{k\geq 1} \frac{1}{k} \sum_{|m|^2=k} c_m e^{i\langle x,m\rangle} = -\sum_{k\geq 1} (\frac{1}{k} - \frac{1}{k+1}) \sum_{|m|^2\leq k} c_m e^{i\langle x,m\rangle}$$

is convergent uniformly on $B(p, \rho_0)\cap\overline{Z}$. Thus, when restricted to $B(p, \rho_0)\cap Z$, $F(x)$ is continuous. In addition, the arguments given in (3.17) combined with (3.20) show that

$$|F(x) - A_\rho F(x)| \leq C\rho^2 \tag{3.21}$$

uniformly on $B(p, \rho_0) \cap \overline{Z}$.

We need the following three theorems to show that this will lead to a contradiction.

The first theorem generalizes a result of Rado [Ra].

Theorem 3.11. *Let F be a bounded function defined on ball $\overline{B}(p, r)$ and Z be the set of discontinuity of F in $\overline{B}(p, r)$. Suppose for each $x \in B(p, r)$,*

$$\tilde{\Delta} F(x) = 0. \tag{3.22}$$

If for all $x \in B(p, r) \setminus \overline{Z}$, the harmonic measure

$$\omega(B(p, r) \setminus \overline{Z}, \partial Z, x) = 0, \tag{3.23}$$

then F is harmonic on $B(p, r)$ provided that F is continuous when restricted to $\overline{Z}$.

For an open set G, the harmonic measure of $F \subset \partial G$ related to G at $x \in G$, $\omega(G, F, x)$, is closely related to Brownian motion. Let $\{X_t\}_{t\geq 0}$ be the standard Brownian motion in $\mathbb{R}^d$ and $(\Omega, \mathfrak{F}, P^x)$ the be corresponding probability space such that $P^x(X_0 = x) = 1$. Denote T be the stopping time of X_t hitting ∂G: $T = \inf\{t \geq 0 : X_t \in \partial G\}$. We use the convention that $\inf \emptyset = 0$. Then the harmonic measure on ∂G at $x \in G$ is the distribution of X_T under P^x. That is, for $F \subset \partial G$,

$$\omega(G, F, x) = P^x(X_T \in F) = E^x I_F(X_T),$$

where E^x is the expectation operator under P^x. If $f(x)$ is a bounded measurable function defined on ∂G, then

$$H_f(x) = E^x f(X_T) = \int_{\partial G} f(z)\, P^x(X_T \in dz) = \int_{\partial G} f(z)\, \omega(G, dx, z) \tag{3.24}$$

is a harmonic function in G. Moreover, if f is continuous on ∂G and ∂G satisfies the exterior cone condition, for example, then H_f is continuous in $\overline{G}$ and converges to f at every point of ∂G. See [H], for example, for details.

An important property regarding harmonic measure is summarized in the following lemma. It follows almost immediately from the continuity of Brownian motion paths.

Lemma 3.12. *Let G be an open set and let the closed set F be contained in $\overline{G}$. Suppose G_k is a sequence of nested open sets containing F such that $F = \bigcap_{k\geq 1} \overline{G}_k$, then for $x \in G \setminus \overline{G}_1$,*

$$\omega(G \setminus \overline{G}_k, \partial(G \cap G_k), x) \downarrow \omega(G \setminus F, \partial F, x) \quad \text{as } k \longrightarrow \infty.$$

The second theorem, which is Bourgain's key contribution, is the following inequality.

Theorem 3.13. *Let $F, B(p, \rho_0)$ and Z be the objects described in the discussion preceding Theorem 3.11. If $p_1 \in Z, B(p_1, \rho_1) \subset B(p, \frac{1}{2}\rho_0)$ and $p_2 \in B(p_1, \frac{1}{2}\rho_1)$, then,*

$$\begin{aligned} &|F(p_1) - F(p_2)| \\ &\leq c\Big(\Big[|F(p_1)| + \rho_1^{-\frac{3}{4}(d-1)}\Big]\big[1 - \omega(B(p_1, \rho_1) \setminus \overline{Z}, \partial(Z \cap B(p_1, \rho_1)), p_2)\big]^{\frac{1}{4}} \\ &\quad + \sup_{q \in B(p_1, 2\rho_1) \cap \overline{Z}} |F(p_1) - F(q)|\Big). \end{aligned} \tag{3.25}$$

The last theorem is concerned with general harmonic measure again.

Theorem 3.14. *Let $B(p_0, r)$ be a ball in $\mathbb{R}^d$ and F a closed set such that $B(p_0, r) \cap F \neq \emptyset$. Suppose for some $x \in B(p_0, r) \setminus F$,*

$$\omega(B(p_0, r) \setminus \overline{F}, \partial(B(p_0, r) \cap F), x) > 0.$$

Then there exists $p_1 \in B(p_0, r) \cap F$, such that

$$\inf_{\delta_1 > 0} \liminf_{\delta_2 \to 0} \inf_{x \in B(p_1, \delta_2)} \omega(B(p_1, \delta_1) \setminus F, \partial(B(p_1, \delta_1) \cap F), x) = 1.$$

We will prove these three theorems in Section 4. Now we finish the proof of Theorem 3.10.

If for all $\varepsilon > 0$,

$$\omega(B(p, \rho_0) \setminus Z_\varepsilon, \partial(B(p, \rho_0) \cap Z_\varepsilon), x) = 0$$

for all $x \in B(p, \rho_0) \setminus Z_\varepsilon$, then

$$\omega(B(p,\rho_0) \setminus \overline{Z}, \partial(B(p,\rho_0) \cap Z_\varepsilon), x) \le \omega(B(p,\rho_0) \setminus Z_\varepsilon, \partial(B(p,\rho_0) \cap Z_\varepsilon), x) = 0,$$

for all $x \in B(p,\rho_0) \setminus \overline{Z}$ by the maximum principle. Thus,

$$\omega(B(p,\rho_0) \setminus \overline{Z}, \partial(B(p,\rho_0) \cap Z_\varepsilon), x) = 0$$

for all $x \in B(p,\rho_0) \setminus \overline{Z}$ and all $\varepsilon > 0$. Since $Z = \bigcup_{\varepsilon>0} Z_\varepsilon$, we have

$$\omega(B(p,\rho_0) \setminus \overline{Z}, \partial(B(p,\rho_0) \cap Z), x) = 0.$$

Let $p_1 = p$ and $\rho_1 = \frac{1}{2}\rho_0$ in (3.25) to see that F is bounded on $B(p, \frac{\rho_0}{4})$. Since it is also continuous when restricted to $B(p,\rho_0) \cap \overline{Z}$, (3.18) and the above imply that the hypotheses of Theorem 3.11 hold. Thus, F is harmonic and hence continuous on $B(p, \frac{\rho_0}{4})$, which is a contradiction.

Therefore, for some $\varepsilon > 0$, we must have

$$\omega(B(p,\rho_0) \setminus Z_\varepsilon, \partial(B(p,\rho_0) \cap Z_\varepsilon), x) > 0$$

for some $x \in B(p,\rho_0) \setminus Z_\varepsilon$. By Theorem 3.14, there exists $p_1 \in Z_\varepsilon \cap B(p, \frac{\rho_0}{8})$ such that

$$\inf_{\delta_1>0} \liminf_{\delta_2\to 0} \inf_{x\in B(p_1,\delta_2)} \omega(B(p_1,\delta_1) \setminus Z_\varepsilon, \partial(B(p_1,\delta_1) \cap Z_\varepsilon), x) = 1. \tag{3.26}$$

Because F restricted to $\overline{Z} \cap B(p, \frac{\rho_0}{8})$ is continuous, there exists $0 < \delta_1$ such that

$$|F(y) - F(p_1)| \le \frac{\varepsilon}{10} \text{ for all } y \in B(p_1, 2\delta_1) \cap \overline{Z}. \tag{3.27}$$

Let $\eta > 0$ be any positive number. By (3.26), there exists $0 < \delta_2 = \delta_2(\eta)$ such that

$$\omega(B(p_1,\delta_1) \setminus \overline{Z}, \partial(B(p_1,\delta_1) \cap Z), y) > 1 - \eta \text{ for all } y \in B(p_1,\delta_2). \tag{3.28}$$

Since $p_1 \in Z_\varepsilon$, by definition, there exists $p_2 \in B(p_1, \delta_2)$, such that

$$|F(p_1) - F(p_2)| \ge \frac{\varepsilon}{2}.$$

Applying Theorem 3.13 with $\rho_1 = \delta_1$, by (3.27)-(3.28) and the above inequality, we have

$$\frac{\varepsilon}{2} \le |F(p_1) - F(p_2)| \le [|F(p_1)| + \delta_1^{-\frac{3}{4}(d-1)}]\eta^{1/4} + \frac{\varepsilon}{10}.$$

This is also seen to be a contradiction by letting $\eta \to 0$. Thus we will have completed the proof when we have proved the above three theorems. □

4. Proofs of Theorems 3.11, 3.13, and 3.14

4.1. **Proof of Theorem 3.11.** Denote the average of F on the surface of $B(x,\rho)$ by

$$S_\rho F(x) = \frac{1}{\sigma_d \rho^{d-1}} \int_{\partial B(x,\rho)} F(z)\, d\sigma(z),$$

where σ is the surface measure, and $\sigma_d = dv_d$ is the surface area of unit ball in $\mathbb{R}$. Then,

$$A_\rho F(x) = \frac{d}{\rho^d} \int_0^\rho S_\beta F(x) \beta^{d-1}\, d\beta. \tag{4.1}$$

For any $\eta > 0$ and $x \in B(p,r)$, by (3.22) there exists $0 < \rho' = \rho'_{x,\eta} < r - |x-p|$ such that for any $\rho < \rho'$, $|F(x) - A_\rho F(x)| \le \eta\rho^2$. Thus, by (4.1), for any $\rho < \rho'$, we have

$$\int_0^\rho [F(x) - S_\beta F(x) - a\eta\beta^2]\beta^{d-1}\, d\beta \le 0$$

where $a = \frac{d+2}{d}$. So there exists $\beta = \beta_{x,\eta,\rho} \le \rho$, such that $F(x) - S_\beta F(x) \le a\eta\beta^2$. Consequently, for any $\eta > 0$ and $x \in B(p,r)$, there exists a sequence of $\{\beta_n\} = \{\beta_{x,\eta}(n)\}$ of positive numbers decreasing to 0 such that

$$F(x) - S_{\beta_n} F(x) \le a\eta\beta_n^2 \quad \text{for all } n \ge 1. \tag{4.2}$$

Let $B(q,\rho_1) \subset B(p,r)$. We will show below that for $y \in B(q,r_1) \setminus \overline{Z}$

$$\begin{aligned} F(y) &= \int_{\partial B(q,r_1)} F(z)\omega(B(q,r_1),dz,y) \\ &= \frac{1}{\sigma_d r_1}\int_{\partial B(q,r_1)} \frac{r_1^2 - |q-y|^2}{|z-y|^d} F(z)\, d\sigma(z). \end{aligned} \tag{4.3}$$

Assume (4.3) holds. For $q \in Z$, by (4.2), there exists a decreasing sequence $r_1(n)$ of positive numbers going to 0 such that for each n

$$F(q) - \frac{1}{\sigma_d r_1^{d-1}(n)}\int_{\partial B(q,r_1(n))} F(z)\, d\sigma(z) \le r_1^2(n). \tag{4.4}$$

So combining (4.3) and (4.4), by the mean value theorem and the boundedness of F,

(4.5)

$$\begin{aligned} F(q) - F(y) \le & \frac{1}{\sigma_d r_1(n)}\int_{|z-q|=r_1(n)} \left| \frac{1}{r_1^{d-2}(n)} - \frac{r_1^2(n) - |q-y|^2}{|z-y|^d} \right| |F(z)| d\sigma(z) \\ & + r_1^2(n) \\ \le & c\frac{|q-y|}{r_1(n)} + r_1^2(n) \end{aligned}$$

if $|y-q| < \frac{1}{2}r_1(n)$. Let $y \to q$ and then let $n \to \infty$ to see that (4.5) implies that

$$\liminf_{\substack{y\to q \\ y \in B(p,r)\setminus\overline{Z}}} F(y) \ge F(q). \tag{4.6}$$

Note that if F satisfies the hypothesis of Theorem 3.11, then so does $-F$. So inequality (4.6) holds for $-F$, whence

$$\liminf_{\substack{y\to q \\ y \in B(p,r)\setminus\overline{Z}}} F(y) \le F(q).$$

This implies that

$$\lim_{\substack{y\to q \\ y \in B(p,r)\setminus\overline{Z}}} F(y) = F(q).$$

Since F restricted to $\overline{Z}$ is continuous, F is continuous at q. Hence, $Z = \emptyset$. Thus, we have representation (4.3) for all $y \in B(p,r)$. Consequently, F is harmonic.

To show (4.3), it is enough to show that for $y \in B(q, r_1) \setminus \overline{Z}$,

$$F(y) \leq \int_{\partial B(q,r_1)} F(z)\omega(B(q,r_1), dz, y), \tag{4.7}$$

because if (4.7) holds for F then it also holds for $-F$ since they both satisfies the assumption of the Theorem 3.11.

Let $\{X_t\}_{t\geq 0}$ be the standard Brownian motion starting from $y \in B(q, r_1) \setminus \overline{Z}$ in probability space $(\Omega, \mathfrak{F}, P^y)$. Define

$$T = \inf\{t \geq 0 : X_t \in \partial B(q, r_1)\}$$

to be the exit time of X_t from $B(q, r_1)$. Then (4.7) is equivalent to

$$F(y) \leq E^y[F(X_T)]. \tag{4.8}$$

Our next goal is to show that for any stopping time $S \leq T$,

$$P^y(X_S \in \overline{Z}) = 0. \tag{4.9}$$

Let R be the first hitting time of X_t with $\partial(B(q, r_1) \setminus \overline{Z})$: $R = \inf\{t \geq 0 : X_t \in \partial(B(q, r_1) \setminus \overline{Z})\}$. Then $R \leq T$. We first show a special case of (4.9):

$$P^y(X_T \in \overline{Z}) = 0. \tag{4.10}$$

Note by (3.23) and the maximum principle, for $y \in B(q, r_1) \setminus \overline{Z}$,

$$\begin{aligned} 0 &= \omega(B(p,r) \setminus \overline{Z}, \partial Z, y) \\ &\geq \omega(B(q,r_1) \setminus \overline{Z}, \partial(B(q,r_1) \cap Z), y) \\ &= P^y(X_R \in \overline{Z}). \end{aligned} \tag{4.11}$$

Furthermore, since $\{R < T\} \subset \{X_R \in \partial Z\}$, $P^y(R < T) \leq P^y(X_R \in \partial Z) = 0$. Thus, by (4.11) we have

$$\begin{aligned} P^y(X_T \in \overline{Z}) &= P^y(X_T \in \overline{Z}, R = T) + P^y(X_T \in \overline{Z}, R < T) \\ &\leq P^y(X_R \in \overline{Z}) + P^y(R < T) \\ &= 0. \end{aligned}$$

We now show (4.9) for a general stopping time S. By Lemma 3.12, for any $\tau > 0$, there exists an open set G such that $\overline{Z} \subset G$ and

$$\omega(B(p,r) \setminus \overline{G}, \partial(B(p,r) \cap G), y) < \tau. \tag{4.12}$$

Define a function u on $\overline{B}(p, r)$ as follows:

$$u(x) = \begin{cases} \omega(B(p,r) \setminus \overline{G}, \partial(B(p,r) \cap G), x), & \text{on } x \in B(p,r) \setminus \overline{G}, \\ 1, & \text{on } \overline{B(p,r) \cap G}, \\ 0, & \text{on } \partial B(p,r) \setminus \partial(B(p,r) \cap G). \end{cases}$$

Then u is superharmonic on $B(p, r)$. Let $\{\tilde{r}_n\}$ be an increasing sequence going up to r and $y \in B(p, \tilde{r}_1)$. Denote T_n to be the exit time of X_t from $B(p, \tilde{r}_n)$. Clearly T_n is increasing and convergent to T. Since Brownian motion is continuous, we have

$$\{X_S \in \overline{Z}, S < T\} \subset \bigcup_{n \geq 1} \{X_{S \wedge T_n} \in \overline{Z}, S < T\},$$

where $S \wedge T_n = \min\{S, T_n\}$. So (4.9) is implied by

$$P^y(X_{S\wedge T_n} \in \overline{Z}) = 0, \quad \text{for each } n, \tag{4.13}$$

since by (4.12)

$$\begin{aligned} P^y(X_S \in \overline{Z}) &= P^y(X_S \in \overline{Z}, S = T) + P^y(X_S \in \overline{Z}, S < T) \\ &\leq P^y(X_T \in \overline{Z}) + P^y(X_S \in \overline{Z}, S < T) \\ &\leq \sum_{n=1}^{\infty} P^y(X_{S\wedge T_n} \in \overline{Z}, S < T). \end{aligned}$$

For a superharmonic function u and for each $n \geq 1$, there exists a sequence of increasing superharmonic functions $\{u_j\}$ such that $u_j \in C^2$ and

$$\lim_{j\to\infty} u_j = u \quad \text{on } \overline{B}(p, \tilde{r}_n) \tag{4.14}$$

(see, for example, Theorem 4.20 of [H]). Applying Itô's formula to $u_j(X_{S\wedge T_n})$, we have $E^y[u_j(X_{S\wedge T_n})] \leq u_j(y)$. Let j go to infinity and apply (4.14) to see that $E^y[u(X_{S\wedge T_n})] \leq u(y)$. Consequently

$$\tau \geq u(y) \geq E^y[u(X_{S\wedge T_n})] \geq P^y(X_{S\wedge T_n} \in \overline{B(p,r) \cap G}) \geq P^y(X_{S\wedge T_n} \in \overline{Z}).$$

Letting $\tau \to 0$ proves (4.13).

As a consequence of (4.9), since F is continuous on $\overline{B}(p,r) \setminus Z$, we have almost everywhere with respect to the probability measure P^y

$$\lim_{n\to\infty} F(X_{S_n}) = F(X_{S_\infty}) \text{ if } S_n \uparrow S_\infty \leq T. \tag{4.15}$$

Let $\eta > 0$. Consider the family of stopping times

$$\mathfrak{S} = \{S \leq T : F(y) - E^y F(X_S) \leq a\eta E^y|y - X_S|^2\}.$$

Define $S_0 = \inf\{t \geq 0 : |X_t - y| \geq \beta_{y,\eta}\}$. Then X_{S_0} is uniformly distributed on $\partial B(y, \beta_{y,\eta})$. So by (4.2), we have $F(y) - E^y F(X_{S_0}) \leq a\eta E^y|y - X_{S_0}|^2$. Thus $S_0 \in \mathfrak{S}$ and $\mathfrak{S}$ is not empty.

For a sequence of increasing stopping times S_n in $\mathfrak{S}$, let $S_\infty = \lim_{n\geq 1} S_n$. Then by (4.15) and the boundedness of F on $\overline{B}(p,r)$, we have $F(y) - E^y F(X_{S_\infty}) \leq a\eta E^y|y - X_{S_\infty}|^2$. So $S_\infty \in \mathfrak{S}$. Thus by Zorn's lemma, there exists $S^* \in \mathfrak{S}$ such that S^* is a maximal element of $\mathfrak{S}$. We show $S^* = T$. If $S^* < T$ with positive P^y probability, then define

$$S_1^* = \inf\{T \geq t \geq S^* : |X_t - X_{S^*}| \geq \beta_{X_{S^*},\eta}\},$$

where for $x \in \partial B(p,r)$, define $\beta_{x,\eta} = 0$. Then, clearly, $S_1^* \geq S^*$ with strict inequality on $\{S^* < T\}$. On the other hand, conditional on X_{S^*}, $X_{S_1^*}$ is uniformly distributed on the surface of $B(X_{S^*}, \beta_{X_{S^*},\eta})$, if $S^* < S_1^*$. So by (4.2)

$$\begin{aligned} F(X_{S^*}) - E^{X_{S^*}} F(X_{S_1^*}) &= F(X_{S^*}) - S_{\beta_{X_{S^*},\eta}} F(X_{S^*}) \\ &\leq a\eta|\beta_{X_{S^*},\eta}|^2 \\ &= a\eta E^{X_{S^*}}|X_{S^*} - X_{S_1^*}|^2. \end{aligned}$$

Hence, by the above, the strong Markovian property, orthogonality between $X_{S_1^*} - X_{S^*}$ and $X_{S^*} - y$, and $S^* \in \mathfrak{S}$, we have

$$\begin{aligned} F(y) - E^y F(X_{S_1^*}) &= F(y) - E^y F(X_{S^*}) + E^y[F(X_{S^*}) - F(X_{S_1^*})] \\ &\le a\eta E^y |y - X_{S^*}|^2 + a\eta E^y[E^{X_{S^*}} |X_{S^*} - X_{S_1^*}|^2, S^* < S_1^*] \\ &= a\eta E^y |y - X_{S_1^*}|^2. \end{aligned}$$

Thus $S_1^* \in \mathfrak{S}$. This contradicts the maximality of S^* in $\mathfrak{S}$ since $S_1^* \ge S^*$ and $S_1^* \ne S^*$. Thus we must have $T \in \mathfrak{S}$. So

$$\begin{aligned} F(y) - \int_{\partial B(q,r_1)} F(z)\, \omega(B(q,r_1), dz, y) &= F(y) - E^y F(X_T) \\ &\le a\eta E^y |X_T - y|^2 \\ &\le 4a\eta r_1^2. \end{aligned}$$

Letting $\eta \to 0$ shows that this implies (4.8). The proof is complete.

4.2. **Proof of Theorem 3.13.** To better analyze $F(x)$ on $B(p, \rho_0)$, we decompose it as follows. For $k \ge 1$, define

$$F_k(x) = -\sum_{2^{k-1} \le |m| < 2^k} \frac{c_m}{|m|^2} e^{i\langle x,m \rangle}. \tag{4.16}$$

Then

$$F(x) = \sum_{k \ge 1} F_k(x). \tag{4.17}$$

Our first lemma is for general multiple trigonometric polynomials and it is well known.

Lemma 4.1. *Let $x \in \mathbb{T}^d$, let $\mathfrak{R}$ be the family of all rectangles containing x and having sides parallel to the coordinate axes, and let*

$$Mh(x) = \sup_{R \in \mathfrak{R}} \frac{1}{|R|} \int_R |h(z)|\, dz$$

be the maximal function of $h(x) = \sum_{|m| \le n} c_m e^{i\langle x,m \rangle}$, where $\{c_m\}_{m \in \mathbb{Z}}$ is a sequence of complex numbers. Then if $|x - y| \le \frac{c}{n}$, there exists a constant C, depending on c only, such that $|h(x)| \le CMh(y)$.

Remark 4.1. *The operator M differs from the standard strong maximal operator M_S in that we do not demand that the chosen rectangles be centered at x. However, since any rectangle R containing x can be contained in a rectangle centered at x and having volume no more than $2^d |R|$, it is easy to see that $Mh(x) \le 2^d M_S h(x)$.*

Proof. Let $t \in [-\pi, \pi)$ and consider the delayed arithmetic means

$$\tau_n(t) = 2\,(2K_{2n-1}(t) - K_{n-1}(t)) = \frac{1}{n} \frac{\sin^2(nt) - \sin^2(nt/2)}{\sin^2 \frac{t}{2}}$$

where $K_n(t)$ is the Fejer kernel. Then the Fourier coefficients of τ_n satisfy $\hat{\tau}(m) = 1$ whenever $|m| \le n$ and it is easy to show that $|\tau_n(t)| \le \psi_n(t) = \frac{100n}{1+(nt)^2}$. Let $m \in \mathbb{Z}^d$. If $|m| \le n$, then every m_j satisfies $|m_j| \le n$, whence $h(x) = (2\pi)^{-d} \int_{[-\pi,\pi)^d} h(z)\tau_n(x - z)dz$, where $\tau_n(x) = \tau_n(x_1)\tau_n(x_2) \cdots \tau_n(x_d)$.

Hence $|h(x)| \leq (2\pi)^{-d} \int_{[-\pi,\pi)^d} |h(z)|\, \psi_n(x-z)dz$, where $\psi_n(x) = \psi_n(x_1)\cdots\psi_n(x_d)$. Decompose the integral dyadically by writing $[-\pi,\pi)^d = \left(\bigcup_{k_1\geq 0} I_{k_1}(y_1)\right) \times \cdots \times \left(\bigcup_{k_d\geq 0} I_{k_d}(y_d)\right)$, where $I_k(t) = \{s \in [-\pi,\pi) : a_{k-1} \leq \frac{n}{c}|s-t| < a_k\}$ where $a_{-1} = 0$ and $a_k = 2^k$ for $k \geq 0$. Then make the estimates

$$\int_{I_{k_1}(y_1)\times\cdots\times I_{k_d}(y_d)} |h(z)|\, \psi_n(x-z)dz \leq \psi_n(b_{k_1}\frac{c}{n})\cdots\psi_n(b_{k_d}\frac{c}{n}) \times$$

$$|I_{k_1}(y_1)|\cdots|I_{k_d}(y_d)|\, Mh(y),$$

where $b_0 = 0$ and $b_k = a_{k-1} - 1$ if $k \geq 1$. Add everything up to complete the proof. □

The next lemma is for the special multiple trigonometric polynomials defined at the beginning of the proof of Theorem 3.13.

Lemma 4.2. *Let $0 < r < 2$ and let $B(p_1, \rho_1)$ be any open ball contained in $\frac{1}{2}\mathbb{T}^d$ and let*

$$F_k(x) = -\sum_{2^{k-1}\leq|m|<2^k} \frac{c_m}{|m|^2} e^{i\langle x,m\rangle}$$

be a multiple trigonometric polynomial. If

$$\sup_r \varepsilon_r^2 = \sup_r \sum_{|m|=r} |c_m|^2 \leq M < \infty, \tag{4.18}$$

then

$$\int_{\partial B(p_1,\rho_1)} |F_k(x)|^r\, d\sigma(x) \leq c(M+1)2^{-\frac{rk}{2}}$$

for a constant c, independent of k, r, M, p_1 and ρ_1.

As a corollary of Lemma 4.2, if $F(x) = \sum_{k\geq 1} F_k(x)$, using Minkowski's inequality, when $1 \leq r < 2$, we get

$$\begin{aligned}\left(\int_{\partial B(p_1,\rho_1)} |F(x)|^r\, d\sigma(x)\right)^{1/r} &\leq \sum_{k\geq 1}\left(\int_{\partial B(p_1,\rho_1)} |F_k(x)|^r\, d\sigma(x)\right)^{1/r}\\ &\leq c(M+1)\sum_{k\geq 1}\frac{1}{(\sqrt{2})^k}\\ &< \infty.\end{aligned}$$

When $0 < r < 1$, we have

$$\int_{\partial B(p_1,\rho_1)} |F(x)|^r\, d\sigma(x) \leq \sum_{k\geq 1}\int_{\partial B(p_1,\rho_1)} |F_k(x)|^r\, d\sigma(x) < \infty.$$

Thus we have shown

Corollary 4.3. *Let $0 < r < 2$ and $B(p_1, \rho_1)$ be any open ball contained in $\frac{1}{2}\mathbb{T}^d$. For a sequence of complex numbers $\{c_m\}_{m\in\mathbb{T}^d}$ satisfying*

$$\sup_R \varepsilon_R^2 = \sup_R \sum_{|m|=R} |c_m|^2 \leq M < \infty, \tag{4.19}$$

the L^2 function defined by

$$F(x) = -\sum_{|m|\neq 0} \frac{c_m}{|m|^2} e^{i\langle x,m\rangle} \tag{4.20}$$

satisfies

$$\int_{\partial B(p_1,\rho_1)} |F(x)|^r \, d\sigma(x) < c(M+1) < \infty,$$

where c is independent of M, p_1 and ρ_1.

Proof. Let $E = E_{k,\gamma}$ be a set of 2^{-k}-separated points $x \in \partial B(p_1, \rho_1)$ satisfying $|F_k(x)| > \gamma$. Use $|E|$ to denote the cardinality of E. Then by Lemma 4.1 and the boundedness of the strong maximal operator on L^2 (coupled with the remark above relating the operator M to the strong maximal operator), we have

$$\begin{aligned}
|E|\gamma^2 &\leq \sum_{x\in E} |F_k(x)|^2 \leq c v_d 2^{kd} \sum_{x\in E} \int_{B(x,2^{-k})} |MF_k(y)|^2 \, dy \\
&\leq c v_d 2^{kd} \int_{\mathbb{T}^d} |MF_k(y)|^2 \, dy \leq c 2^{kd} \int_{\mathbb{T}^d} |F_k(y)|^2 \, dy \\
&\leq c 2^{kd} \sum_{2^{2k-2}\leq n < 2^{2k}} \frac{1}{n^2} \varepsilon_{\sqrt{n}}^2 \leq c 2^{k(d-2)} \delta_k^2
\end{aligned}$$

where

$$\delta_k^2 = 2^{-2k} \sum_{2^{2k-2}\leq n<2^{2k}} \varepsilon_{\sqrt{n}}^2 \leq M \text{ for all } k \geq 1. \tag{4.21}$$

Thus,

$$|E| \leq c\gamma^{-2} 2^{k(d-2)} M. \tag{4.22}$$

Consequently,

$$\sigma\{|F_k(x)| > \gamma\} \leq c 2^{-k(d-1)} |E| \leq c\gamma^{-2} 2^{-k} M. \tag{4.23}$$

Thus, use the fact that $\sigma\{|F_k(x)| > \gamma\} \leq \sigma\{\partial B(p_1, \rho_1)\} \leq c$ for any $\gamma > 0$ and the above, we have,

$$\begin{aligned}
\int_{\partial B(p_1,\rho_1)} |F_k|^r d\sigma &= r \int_0^\infty \gamma^{r-1} \sigma\{|F_k(x)| > \gamma\} \, d\gamma \\
&= r \int_0^{2^{-k/2}} \gamma^{r-1} \sigma\{|F_k(x)| > \gamma\} \, d\gamma \\
&\quad + r \int_{2^{-k/2}}^\infty \gamma^{r-1} \sigma\{|F_k(x)| > \gamma\} \, d\gamma \\
&\leq c(M+1) 2^{-\frac{r}{2}k}.
\end{aligned}$$

This finishes the proof. □

The next lemma describes the "continuity" property of the average function $A_\rho F$ under the condition (2.5). Shapiro showed $A_\rho F(x)$ is equi-continuous in ρ as a function of x under the his condition (3.1). (See Lemma 5 of [S].) Such a strong condition may not hold under (2.5), but we have the following result.

Lemma 4.4. *Let F, Z, p_1, p_2 be given as in Theorem 3.13. For $\kappa \ll \tau$, where the quantity $\tau = dist(p_2, \overline{Z}) \leq \frac{1}{2}\rho_1$, define*

$$G_\kappa = \{x \in B(p_1, \rho_1) : dist(x, \overline{Z}) < \kappa\}.$$

For each $x \in \partial G_\kappa$, let $\tilde{x} \in \overline{Z}$ be such that $|x - \tilde{x}| = \kappa$. Then

$$\lim_{\kappa \to 0} \int_{\partial(G_\kappa \cap B(p_1,\rho_1))} |A_\kappa F(x) - A_\kappa F(\tilde{x})| \omega(B(p_1, \rho_1) \setminus \overline{G}_\kappa, dx, p_2) = 0.$$

Proof. It is sufficient to show that

$$\lim_{\kappa \to 0} \int_0^\infty \omega(B(p_1, \rho_1) \setminus \overline{G}_\kappa, S_\alpha, p_2)\, d\alpha = 0, \tag{4.24}$$

where

$$S_\alpha = \{x \in \partial(G_\kappa \cap B(p_1, \rho_1)) : |A_\kappa F(x) - A_\kappa F(\tilde{x})| > \alpha\}. \tag{4.25}$$

For a sequence of positive numbers $\{\alpha_k\}_{k \geq 1}$ to be chosen later such that $\alpha \geq \sum_{k \geq 1} \alpha_k$, we denote

$$S_{\kappa,\alpha_k} = \{x \in \partial(G_\kappa \cap B(p_1, \rho_1)) : |A_\kappa F_k(x) - A_\kappa F_k(\tilde{x})| > \alpha_k\}. \tag{4.26}$$

If $x \in S_{\kappa,\alpha_k}$, by the mean value theorem, there exists $\overline{x} \in B(p_1, 2\rho_1)$, such that $|\nabla A_\kappa F_k(\overline{x})| > \kappa^{-1}\alpha_k$. Therefore it will be useful to also consider the sets

$$S'_{\kappa,\alpha_k} = \{\overline{x} \in B(p_1, 2\rho_1) : |\nabla A_\kappa F_k(\overline{x})| > \frac{\alpha_k}{\kappa}\}. \tag{4.27}$$

Using contour integration (see page 158 of [SW]), we have

$$\hat{I}_d(m) = O(|m|^{-\frac{d+1}{2}}) \text{ as } |m| \longrightarrow \infty.$$

Thus,

$$\left\|\nabla A_\kappa F_k\right\|_2^2 \leq c\delta_k^2 \sup_{2^{k-1} \leq |m| < 2^k} |\hat{I}(\kappa|m|)|^2 \leq \begin{cases} c\delta_k^2(\kappa 2^k)^{-(d+1)}, & \kappa 2^k > 1 \\ c\delta_k^2, & \kappa 2^k \leq 1 \end{cases}$$

where δ_k is defined in (4.21). Hence, using (4.27) and applying Lemma 4.1 and the maximum function inequality, an argument similar to the one leading to (4.22) shows that the number of radius 2^{-k} balls covering S'_{κ,α_k}, $|E_k|$, is bounded by

$$|E_k| \leq \begin{cases} c\alpha_k^{-2}\delta_k^2\kappa^{-d+1}2^{-k}, & \kappa 2^k > 1, \\ c\alpha_k^{-2}\kappa^2 2^{kd}\delta_k^2, & \kappa 2^k \leq 1. \end{cases} \tag{4.28}$$

Let $\kappa < \sqrt{\kappa} \ll \tau$. Consider the following three cases.

Case (i): $\kappa^{-1} \leq 2^k$

Enlarge the radii of the balls from 2^{-k} to 2κ. Then by (4.28), there is a covering of S_{κ,α_k} by at most $c\alpha_k^{-2}\delta_k^2\kappa^{-d+1}2^{-k}$ many balls of radius 2κ. Hence

$$\begin{aligned} \omega(B(p_1, \rho_1) \setminus \overline{G}_\kappa, S_{\kappa,\alpha_k}, p_2) &\leq c\alpha_k^{-2}\delta_k^2\kappa^{-d+1}2^{-k}\left(\frac{2\kappa}{\frac{1}{2}\tau}\right)^{d-2} \\ &< \frac{c}{\tau^{d-2}}(\alpha_k^{-2}\delta_k^2(\kappa 2^k)^{-1}). \end{aligned} \tag{4.29}$$

To see why the first inequality follows, note if $B(p_3, \rho_3) \subset B(p_1, \rho_1)$ and $x \in B(p_1, \rho_1) \setminus (\overline{B}(p_3, \rho_3) \cap \overline{G}_\kappa)$, then

$$(4.30)\qquad \begin{aligned} \omega(B(p_1,\rho_1) \setminus \overline{G}_\kappa, &\partial(B(p_3,\rho_3) \cap G_\kappa), x) \\ &\le \omega(B(p_1,\rho_1) \setminus \overline{B}(p_3,\rho_3), \partial B(p_3,\rho_3), x) \\ &\le \frac{\rho_3^{d-2}}{|x-p_3|^{d-2}} \end{aligned}$$

by the maximum principle for harmonic functions. Now use the fact that the covering ball of S_{κ,α_k} centered at p_3 of radius κ is at least $\tau - 2\kappa > \frac{1}{2}\tau$ away from p_2 since $\operatorname{dist}(p_2, \overline{Z}) = \tau \gg \kappa$, and $\rho_3 = 2\kappa$, the first inequality of (4.29) follows easily from (4.30).

Case (ii): $\kappa^{-1/2} \le 2^k < \kappa^{-1}$

By (4.28) and (4.30) with $\rho_3 = 2^{-k}$, we have

$$(4.31)\qquad \begin{aligned} \omega(B(p_1,\rho_1) \setminus \overline{G}_\kappa, S_{\kappa,\alpha_k}, p_2) &\le c\alpha_k^{-2}\delta_k^2\kappa^2 2^{kd}(\frac{1}{2^k\tau})^{d-2} \\ &= \frac{c}{\tau^{d-2}}(\alpha_k^{-2}\delta_k^2(\kappa 2^k)^2). \end{aligned}$$

Case (iii): $2^k < \kappa^{-1/2}$

In this case, $\sqrt{\kappa} < 2^{-k}$. Using Lemma 4.1, the maximal function inequality and (4.27), it is easy to see that there are at most $c\alpha_k^{-2}\kappa^2\delta_k^2\kappa^{-d/2}$ balls of radius $2\sqrt{\kappa}$ covering S_{κ,α_k}. Thus, by (4.30) with $\rho_3 = 2\sqrt{\kappa}$

$$(4.32)\qquad \begin{aligned} \omega(B(p_1,\rho_1) \setminus \overline{G}_\kappa, S_{\kappa,\alpha_k}, p_2) &\le c\alpha_k^{-2}\delta_k^2\kappa^2\kappa^{-d/2}\left(\frac{2\sqrt{\kappa}}{\tau}\right)^{d-2} \\ &= \frac{c}{\tau^{d-2}}\left(\alpha_k^{-2}\delta_k^2\kappa\right). \end{aligned}$$

Combining (4.29), (4.31) and (4.32), it follows from $S_\alpha \subset \bigcup_{k\ge1} S_{\kappa,\alpha_k}$ that

$$(4.33)\qquad \begin{aligned} \omega(B(p_1,\rho_1) \setminus \overline{G}_\kappa, S_\alpha, p_2) \le \frac{c}{\tau^{d-2}}\Bigg\{ &\sum_{2^k \ge \kappa^{-1}} \alpha_k^{-2}\delta_k^2(2^k\kappa)^{-1} \\ &+ \sum_{\kappa^{-1/2} \le 2^k < \kappa^{-1}} \alpha_k^{-2}\delta_k^2(\kappa 2^k)^2 + \sum_{2^k < \kappa^{-1/2}} \alpha_k^{-2}\delta_k^2\kappa \Bigg\}. \end{aligned}$$

Now let

$$\alpha_k = \begin{cases} c\alpha[\log(1+2^k\kappa)]^{-2}, & \text{for } 2^k \ge \kappa^{-1} \\ c\alpha[\log(1+2^{-k}\kappa^{-1})]^{-2}, & \text{for } 2^k < \kappa^{-1} \end{cases}$$

The positive constant c is so chosen that $\sum_{k\ge1} \alpha_k \le 2c\alpha \sum_{n\ge0}(\log(1+2^n))^{-2} \le \alpha$. It is easy to see that c can be an absolute constant independent of κ.

Let k_0 be the smallest integer such that $2^{k_0} \geq \kappa^{-1}$. Then

$$\begin{aligned}
\sum_{2^k \geq \kappa^{-1}} \alpha_k^{-2} \delta_k^2 (2^k \kappa)^{-1} &= \sum_{k \geq k_0} \alpha_k^{-2} \delta_k^2 (2^k \kappa)^{-1} \\
&\leq c\alpha^{-2} \max_{2^k \geq \kappa^{-1}} \delta_k^2 \sum_{n \geq 0} \frac{(\log(1+2^{n+1}))^4}{2^n} \\
&\leq c\alpha^{-2} \max_{2^k \geq \kappa^{-1}} \delta_k^2.
\end{aligned} \tag{4.34}$$

Similarly

$$\begin{aligned}
\sum_{\kappa^{-1/2} \leq 2^k < \kappa^{-1}} \alpha_k^{-2} \delta_k^2 (\kappa 2^k)^2 &\leq c\alpha^{-2} \max_{2^k \geq \kappa^{-1/2}} \delta_k^2 \sum_{n \geq 0} \frac{(\log(1+2^n))^4}{4^n} \\
&\leq c\alpha^{-2} \max_{2^k \geq \kappa^{-1/2}} \delta_k^2.
\end{aligned} \tag{4.35}$$

Finally,

$$\sum_{2^k < \kappa^{-1/2}} \alpha_k^{-2} \delta_k^2 (\kappa/\kappa_0)^2 \leq c\alpha^{-2} \kappa (\log \frac{1}{\kappa})^5 \tag{4.36}$$

since there are at most $\log \frac{1}{\kappa}$ many terms in the sum and each summand is bounded by $\log(1+\kappa^{-1/2})$. Thus, from (4.34)-(4.36), we have

$$\omega(B(p_1,\rho_1) \setminus \overline{G}_\kappa, S_\alpha, p_2) \leq \frac{c}{\tau^{d-2}} \alpha^{-2} \left\{ \max_{2^k > \kappa^{-1/2}} \delta_k^2 + \kappa (\log \frac{1}{\kappa})^5 \right\} \tag{4.37}$$

Consequently, for any given τ, by choosing

$$\beta_\kappa = \left\{ \max_{2^k > \kappa^{-1/2}} \delta_k^2 + \kappa (\log \frac{1}{\kappa})^5 \right\}^{1/2}$$

we have,

$$\begin{aligned}
\int_{\partial G_\kappa \cap B(p_1,\rho_1)} & |A_\kappa F(x) - A_\kappa F(\tilde{x})| \omega(B(p_1,\rho_1) \setminus \overline{G}_\kappa, dx, p_2) \\
&= \int_0^\infty \omega(B(p_1,\rho_1) \setminus \overline{G}_\kappa, S_\alpha, p_2)\, d\alpha \\
&\leq \int_0^{\beta_\kappa} d\alpha \\
&\quad + \frac{c}{\tau^{d-2}} \left\{ \max_{2^k > \kappa^{-1/2}} \delta_k^2 + \kappa (\log \frac{1}{\kappa})^5 \right\} \int_{\beta_\kappa}^\infty \alpha^{-2}\, d\alpha \\
&= \left(\frac{c}{\tau^{d-2}} + 1 \right) \left\{ \max_{2^k > \kappa^{-1/2}} \delta_k^2 + \kappa (\log \frac{1}{\kappa})^5 \right\}^{1/2} \\
&\to 0 \text{ as } \kappa \to 0
\end{aligned}$$

since $\delta_k \to 0$ as $k \to \infty$. □

The last lemma of this section is an analogue of Lemma 4.4 for the function F.

Lemma 4.5. *Let p_1, p_2, ρ_1, and G_κ be as given in Lemma 4.4. Then*

$$\int_{\partial B(p_1,\rho_1)} |F(p_1) - F(x)|\omega(B(p_1, \rho_1 \setminus \overline{G}_\kappa, dx, p_2)$$
$$\leq c\left[|F(p_1) + \rho_1^{-\frac{3}{4}(d-1)}\right][1 - \omega(B(p_1,\rho_1)\setminus\overline{Z}, \partial(Z\cap B(p_1,\rho_1)), p_2)]^{\frac{1}{4}}.$$

Proof. Denote

$$I_1 = \int_{\partial B(p_1,\rho_1)} |F(p_1) - F(x)|\omega(B(p_1, \rho_1 \setminus \overline{G}_\kappa, dx, p_2).$$

Let T be the hitting time of the Brownian motion X_t starting at p_2 and exiting through $\partial B(p_1, \rho_1)$, and T_κ be the hitting time starting at $p_2 \in B(p_1, \rho_1)\setminus G_\kappa$ and exiting through $\partial(B(p_1, \rho_1)\setminus G_\kappa)$. Then $T = T\circ\theta_{T_\kappa}$, where θ is a transformation on X_t:

$$\theta_t(X_s) = X_{s+t}.$$

Using the strong Markovian property of Brownian motion, for $S \subset \partial B(p_1, \rho_1) \cap \partial(B(p_1, \rho_1)\setminus \overline{G}_\kappa)$, we have

$$\begin{aligned}\omega(B(p_1,\rho_1), S, p_2) &= P^{p_2}(X_T \in S)\\ &= E^{p_2}(P^{X_{T_\kappa}}(X_T\in S))\\ &= \int_{\partial(B(p_1,\rho_1)\setminus G_\kappa)} P^x(X_T\in S)\,\omega(B(p_1,\rho_1)\setminus\overline{G}_\kappa, dx, p_2)\\ &\geq \int_S P^x(X_T\in S)\,\omega(B(p_1,\rho_1)\setminus\overline{G}_\kappa, dx, p_2)\\ &= \omega(B(p_1,\rho_1)\setminus\overline{G}_\kappa, S, p_2).\end{aligned}$$

The last equality holds because $P^x(X_T \in S) = 1$ when $x \in S$. On the other hand, if $S\cap\partial(B(p_1, \rho_1)\setminus\overline{G}_\kappa) = \emptyset$, then

$$\omega(B(p_1,\rho_1)\setminus\overline{G}_\kappa, S, p_2) = P^{p_2}(X_{T_\kappa}\in S) = 0.$$

Combining these two cases, we have for $S \subset \partial B(p_1, \rho_1)$,

$$\omega(B(p_1,\rho_1), S, p_2) \geq \omega(B(p_1,\rho_1)\setminus\overline{G}_\kappa, S, p_2). \tag{4.38}$$

Thus, the harmonic measure $\omega(\cdot) = \omega(B(p_1, \rho_1)\setminus\overline{G}_\kappa, \cdot, p_2)$ is absolutely continuous with respect to $\omega_1(\cdot) = \omega(B(p_1, \rho_1), \cdot, p_2)$ when restricted to $\partial B(p_1, \rho_1)$, and the Radon-Nikodym derivative $\dfrac{d\omega}{d\omega_1}$ is bounded by 1. But the harmonic measure $\omega_1(\cdot) = \omega(B(p_1, \rho_1), \cdot, p_2)$ is absolutely continuous with respect to the surface measure σ of $\partial B(p_1, \rho_1)$ and the Radon-Nikodym derivative is the Poisson kernel:

$$\omega(B(p_1,\rho_1), S, p_2) = \frac{1}{\sigma_d\rho_1}\int_S \frac{\rho_1^2 - |p_2-p_1|^2}{|x-p_2|^d}\,d\sigma(x).$$

Since $|p_1 - p_2| \leq \frac{1}{2}\rho_1$, the Poisson kernel

$$\frac{1}{\sigma_d\rho_1}\frac{\rho_1^2-|p_2-p_1|^2}{|x-p_2|^d} \leq c\rho_1^{-(d-1)} \text{ if } x\in\partial B(p_1,\rho_1),$$

so when restricted to $\partial B(p_1,\rho_1)$, the harmonic measure $\omega(\cdot) = \omega(B(p_1,\rho_1)\setminus \overline{G}_\kappa, \cdot, p_2)$ is absolutely continuous with respect to the surface measure $\sigma(\cdot)$. Moreover the Radon-Nikodym derivative satisfies

$$\frac{d\omega}{d\sigma} = \frac{d\omega}{d\omega_1}\frac{d\omega_1}{d\sigma} \le c\rho_1^{-(d-1)}. \tag{4.39}$$

By Corollary 4.3 with $r = \frac{4}{3}$, (4.39), and Hölder's inequality, we have

$$\begin{aligned}
|I_1| &\le |F(p_1)|\omega(B(p_1,\rho_1)\setminus\overline{G}_\kappa, \partial B(p_1,\rho_1), p_2) + \|F\|_{L^r(d\sigma)}\left\|\frac{d\omega}{d\sigma}\right\|_{L^{r'}(d\sigma)} \\
&\le |F(p_1)|\omega(B(p_1,\rho_1)\setminus\overline{G}_\kappa, \partial B(p_1,\rho_1), p_2) + c\left\|\frac{d\omega}{d\sigma}\right\|_{L^4(d\sigma)} \\
&\le |F(p_1)|\omega(B(p_1,\rho_1)\setminus\overline{G}_\kappa, \partial B(p_1,\rho_1), p_2)^{\frac14} \\
&\quad + c\left(\int_{\partial B(p_1,\rho_1)} \left|\frac{d\omega}{d\sigma}\right|^3 d\omega\right)^{\frac14} \\
&\le |F(p_1)|\omega(B(p_1,\rho_1)\setminus\overline{G}_\kappa, \partial B(p_1,\rho_1), p_2)^{\frac14} \\
&\quad + c\rho_1^{-\frac34(d-1)}\omega(B(p_1,\rho_1)\setminus\overline{G}_\kappa, \partial B(p_1,\rho_1), p_2)^{\frac14} \\
&\le c\left[|F(p_1) + \rho_1^{-\frac34(d-1)}\right][1-\omega(B(p_1,\rho_1)\setminus\overline{Z}, \partial(Z\cap B(p_1,\rho_1)), p_2)]^{\frac14}
\end{aligned}$$

since

$$\begin{aligned}
&\omega(B(p_1,\rho_1)\setminus\overline{G}_\kappa, \partial B(p_1,\rho_1), p_2) \\
&\qquad = 1-\omega(B(p_1,\rho_1)\setminus\overline{G}_\kappa, \partial(B(p_1,\rho_1)\cap G_\kappa), p_2) \\
&\qquad \le 1-\omega(B(p_1,\rho_1)\setminus\overline{Z}, \partial(B(p_1,\rho_1)\cap Z), p_2).
\end{aligned}$$

The last inequality follows from Lemma 3.12. □

We now prove Theorem 3.13. Let $p_1, \rho_1, p_2, \tau, G_\kappa$ be as given in Lemma 4.4. It is clear that $\overline{Z}\cap B(p_1,\rho_1)\subset G_\kappa$. By Theorem 3.11, F is harmonic on $B(p_1,\rho_1)\setminus\overline{Z}$. Thus, $F(p_1)-F(x)$ is harmonic on $B(p_1,\rho_1)\setminus\overline{G}_\kappa$. Consequently, we have

$$F(p_1)-F(p_2) = \int_{\partial(B(p_1,\rho_1)\setminus\overline{G}_\kappa)} [F(p_1)-F(x)]\,\omega(B(p_1,\rho_1)\setminus\overline{G}_\kappa, dx, p_2).$$

Therefore

$$\begin{aligned}
|F(p_1)-F(p_2)| &\le \int_{\partial B(p_1,\rho_1)} |F(p_1)-F(x)|\,\omega(B(p_1,\rho_1)\setminus\overline{G}_\kappa, dx, p_2) \\
&\quad + \int_{\partial(B(p_1,\rho_1)\cap G_\kappa)} |F(p_1)-F(x)|\,\omega(B(p_1,\rho_1)\setminus\overline{G}_\kappa, dx, p_2) \\
&= I_1 + I_2.
\end{aligned}$$

Applying Lemma 4.5 to I_1, we get the first half of (3.25). To estimate I_2, for each $x\in\partial G_\kappa$, let $\tilde{x} = \tilde{x}(x)\in\overline{Z}$ be such that $|\tilde{x}-x| = \kappa$. Thus, for $x\in\partial(G_\kappa\cap B(p_1,\rho_1))$, by the triangle inequality, $|\tilde{x} - p_1| < |\tilde{x} - x| + |x - p_1| \le \kappa + \rho_1 < 2\rho_1$ and

$$|F(p_1) - F(x)| \le \sup_{q \in \overline{Z} \cap B(p_1, 2\rho_1)} |F(p_1) - F(q)| + |F(x) - F(\tilde{x})|. \tag{4.40}$$

Since $x \in \partial G_\kappa$, $B(x, \kappa) \cap \overline{Z} = \emptyset$. Thus, by harmonicity

$$F(x) = A_\kappa F(x). \tag{4.41}$$

From (3.21),

$$|F(\tilde{x}) - A_\kappa F(\tilde{x})| \le C\kappa^2 \tag{4.42}$$

uniformly for all $\tilde{x} \in \overline{Z} \cap B(p_1, 2\rho_1)$. Hence, by (4.41) and (4.42),

$$|F(x) - F(\tilde{x})| \le |A_\kappa F(x) - A_\kappa F(\tilde{x})| + C\kappa^2.$$

Apply Lemma 4.4 and (4.40) and then let $\kappa \to 0$ to see that

$$|I_2| \le \sup_{q \in \overline{Z} \cap B(p_1, 2\rho_1)} |F(p_1) - F(q)|.$$

This completes the proof.

4.3. **The Proof of Theorem 3.14.** We need the following lemma.

Lemma 4.6. *Let B be an open ball and let F be a closed set such that $F \subset B$. If $\omega(B \setminus F, \partial F, x) > 0$ for some $x \in B \setminus F$, then for any $\eta > 0$, there exists $p \in F$ and $\tilde{\delta} > 0$, such that*

$$\omega(B \setminus F, \partial F, z) > 1 - \eta \text{ for all } z \in B(p, \tilde{\delta}).$$

Proof. By Lemma 3.12, there exists an open set G such that $F \subset \overline{G} \subset B$ and such that for a positive number η_1 to be chosen later,

$$\omega(B \setminus \overline{G}, \partial G, x) < (1 + \eta_1)\omega(B \setminus F, \partial F, x). \tag{4.43}$$

Let X_t be the standard Brownian motion starting from x in probability space $(\Omega, \mathfrak{F}, P^x)$. Define hitting times

$$T = \inf\{t \ge 0 : X_t \in \partial B \cup \partial G\}, \text{ and}$$
$$S = \inf\{t \ge 0 : X_t \in \partial B \cup \partial F\}.$$

Then $T \le S$ and $S = S \circ \theta_T$. Thus by the strong Markovian property, we have

$$\begin{aligned} \omega(B \setminus F, \partial F, x) &= P^x(X_S \in \partial F) \\ &= E^x[P^{X_T}(X_S \in \partial F)] \\ &= \int_{\partial G} \omega(B \setminus F, \partial F, z)\, \omega(B \setminus \overline{G}, dz, x) \\ &\le \sup_{z \in \partial G} \omega(B \setminus F, \partial F, z)\omega(B \setminus \overline{G}, \partial G, x). \end{aligned} \tag{4.44}$$

Combining (4.43) and (4.44), we have $\sup_{z \in \partial G} \omega(B \setminus F, \partial F, z) > 1 - \eta_1$. So there exists $y \in \partial G$ such that

$$\omega(B \setminus F, \partial F, y) > 1 - \eta_1. \tag{4.45}$$

Let $\tau = \operatorname{dist}(y, F)$. Note that $1 - \omega(B \setminus F, F, x)$ is a positive harmonic function on $B(y, \tau)$. For $0 < \delta < 1$ to be specified later, using Harnack's inequality for positive harmonic functions, we have for all $z \in B(y, \delta\tau)$,

$$\frac{1 - \omega(B \setminus F, \partial F, z)}{1 - \omega(B \setminus F, \partial F, y)} \le c_\delta$$

where c_δ is a constant depending on δ only. Thus, from (4.45), we have

$$\omega(B \setminus F, \partial F, z) > 1 - c_\delta \eta_1 \text{ if } z \in B(y, \delta\tau). \tag{4.46}$$

Moreover, since $\overline{G} \subset B$, there exists $C_1 > 1$ such that $B(y, C_1\tau) \subset B$. Thus, by the maximum principle for harmonic functions,

$$\begin{aligned} \omega(B \setminus \overline{B}(y, \delta\tau), \partial B(y, \delta\tau), x) &\ge \omega(B(y, C_1\tau) \setminus \overline{B}(y, \delta\tau), \partial B(y, \delta\tau), x) \\ &= \left(\frac{(C_1\tau)^{d-2}}{|x - y|^{d-2}} - 1\right) \frac{\delta^{d-2}}{C_1^{d-2} - \delta^{d-2}} \end{aligned} \tag{4.47}$$

for $x \in B(y, C_1\tau) \setminus \overline{B}(y, \delta\tau)$. Consequently, for any c such that $0 < \delta < 1 < c < C_1$ and $x \in B(y, c\tau) \setminus \overline{B}(y, \delta\tau)$, by (4.47) we have

$$\omega(B \setminus \overline{B}(y, \delta\tau), \partial B(y, \delta\tau), x) \ge \frac{(C_1/c)^{d-2} - 1}{(C_1/\delta)^{d-2} - 1}. \tag{4.48}$$

Now choose $0 < \delta < 1$ and $C_1 > c > 1$ with both δ and c close enough to 1 to force

$$\frac{(C_1/c)^{d-2} - 1}{(C_1/\delta)^{d-2} - 1} > 1 - \frac{\eta}{2}. \tag{4.49}$$

Next choose $\eta_1 = \eta_1(\delta)$ such that

$$1 - c_\delta \eta_1 > 1 - \frac{\eta}{2}. \tag{4.50}$$

Denote $F_1 = F \cup \overline{B}(y, \delta\tau)$. Then $\partial F_1 = \partial F \cup \partial B(y, \delta\tau)$ and $\partial F \cap \partial B(y, \delta\tau) = \emptyset$. Let $z \in B(y, c\tau) \setminus \overline{B}(y, \delta\tau)$. Combining the strong Markovian property, (4.46), and (4.48)–(4.50) with an argument similar to the one which led to (4.44); we have

$$\begin{aligned} \omega(B \setminus F, \partial F, z) &= \int_{\partial F_1} \omega(B \setminus F, \partial F, x)\omega(B \setminus F_1, dx, z) \\ &= \int_{\partial F} \omega(B \setminus F, \partial F, x)\omega(B \setminus F_1, dx, z) \\ &\quad + \int_{\partial B(y,\delta\tau)} \omega(B \setminus F, \partial F, x)\omega(B \setminus F_1, dx, z) \\ &\ge \omega(B \setminus F_1, \partial F, z) + (1 - c_\delta\eta_1)\omega(B \setminus F_1, \partial B(y, \delta\tau), z) \\ &\ge (1 - c_\delta\eta_1)\omega(B \setminus F_1, \partial F_1, z) \\ &\ge (1 - c_\delta\eta_1)\omega(B \setminus \overline{B}(y, \delta\tau), \partial B(y, \delta\tau), z) \\ &\ge 1 - \eta. \end{aligned} \tag{4.51}$$

The penultimate inequality is from the maximum principle for harmonic functions since $B(y, \delta\tau) \subset F_1$. Let $p \in F$ be such that $|p - y| = \tau = \operatorname{dist}(y, F)$. Then $B(p, \tilde{\delta}) \subset B(y, c\tau) \setminus \overline{B}(y, \delta\tau)$ for any $\tilde{\delta} < \min\{(1 - \delta)\tau, (c - 1)\tau\}$. Consequently, from (4.51),

$$\omega(B \setminus F, \partial F, z) \ge 1 - \eta \text{ for } z \in B(p, \tilde{\delta}). \tag{4.52}$$

This proves the lemma. □

We now prove Theorem 3.14

We need to find $p_1 \in B(p_0,r)\cap F$, such that for any $1>\eta>0$ and $\delta_1>0$, there exists $0<\delta_2=\delta_2(\eta)<\delta_1$, such that

$$\omega(B(p_1,\delta_1)\setminus F,\partial(B(p_1,\delta_1)\cap F),x)>1-\eta \tag{4.53}$$

for all $x\in B(p_1,\delta_2)$.

Note by the maximum principle for harmonic functions, for fixed x and δ_2, the left hand side of (4.53) is an increasing function of δ_1. Thus (4.53) would be implied by the following: There exists $p_1\in B(p_0,r)\cap F$ and a sequence of positive numbers $\delta_n>\tilde{\delta}_n$ decreasing to 0, such that

$$\omega(B(p_1,\delta_n)\setminus F,\partial(B(p_1,\delta_n)\cap F),x)>1-\frac{1}{n} \tag{4.54}$$

for all $x\in B(p_1,\tilde{\delta}_n)$.

To prove (4.54), note that $\omega(B(p_0,r)\setminus F,\partial(B(p_0,r)\cap F),x)>0$ implies for any $\varepsilon>0$, there exists a ball $B_\varepsilon=B(q,\varepsilon r')$ such that $B(q,r')\subset B(p_0,r)$ and such that $\omega(B(p_0,r)\setminus F,\partial(B_\varepsilon\cap F),x)>0$. Choose $0<\varepsilon_1\le\frac{1}{9}$ such that $(9\varepsilon_1)^{d-2}<1/4$. Let $F_1=\overline{B_{\varepsilon_1}}\cap F$ and denote $B(p_0,r)$ by B_0. Since

$$\omega(B(p_0,r)\setminus F_1,\partial F_1,x)\ge\omega(B(p_0,r)\setminus F,\partial F_1,x)>0,$$

we may apply Lemma 4.6 to see that there exists $q_1\in F_1$, and $\tilde{\delta}_1>0$ such that $B(q_1,\tilde{\delta}_1)\subset overline B(q_1,2\varepsilon_1 r')\subset B(q_1,\frac{2}{9}r')\subset B_0$ and

$$\omega(B_0\setminus F_1,\partial F_1,z)>3/4 \tag{4.55}$$

for all $z\in B(q_1,\tilde{\delta}_1)$. If $z\in B(q_1,\tilde{\delta}_1)$, by the strong Markovian property and (4.55),

$$\begin{aligned}3/4 &\le \omega(B_0\setminus F_1,\partial F_1,z)\\ &= \int_{\partial B(q_1,\frac{2}{9}r')\cup\partial F_1}\omega(B_0\setminus F_1,\partial F_1,x)\omega(B(q_1,\frac{2}{9}r')\setminus F_1,dx,z)\\ &\le \sup_{x\in\partial B(q_1,\frac{2}{9}r')}\omega(B_0\setminus F_1,\partial F_1,x)+\omega(B(q_1,\frac{2}{9}r')\setminus F_1,\partial F_1,z)\\ &\le 1/4+\omega(B(q_1,\frac{2}{9}r')\setminus F_1,\partial F_1,z),\end{aligned} \tag{4.56}$$

where the last inequality holds because when $x\in\partial B(q_1,\frac{2}{9}r')$

$$\begin{aligned}\omega(B_0\setminus F_1,\partial F_1,x) &\le \omega(B_0\setminus\overline{B}(q_1,2\varepsilon_1 r'),\partial B(q_1,2\varepsilon_1 r'),x)\\ &\le \frac{(2\varepsilon_1 r')^{d-2}}{|x-q_1|^{d-2}}\\ &\le (9\varepsilon_1)^{d-2}\\ &< \frac{1}{4}\end{aligned}$$

by (4.30) and the choice of ε_1. So

$$\omega(B(q_1,\frac{2}{9}r')\setminus F_1,\partial F_1,z)>1/2 \text{ for } z\in B(q_1,\tilde{\delta}_1). \tag{4.57}$$

By the maximum principle again,

$$(4.58)\quad \omega(B(q_1,\frac{2}{9}r')\setminus(\overline{B}(q_1,2\varepsilon_1 r')\cap F),\partial(B(q_1,2\varepsilon_1 r')\cap F),z) \geq \omega(B(q_1,\frac{2}{9}r')\setminus F_1,\partial F_1,z)\geq 1/2$$

for $z\in B(q_1,\tilde{\delta}_1)$. Thus, we have shown that there exists $q_1\in B(q,r')\cap F$, $\gamma_1=\frac{2}{9}r'>\eta_1=2\varepsilon_1 r'>\tilde{\eta}_1=\tilde{\delta}_1>0$ such that

$$\omega(B(q_1,\gamma_1)\setminus(\overline{B}(q_1,\eta_1)\cap F),\partial(B(q_1,\eta_1)\cap F),z)\geq 1/2 \text{ for all } z\in B(q_1,\tilde{\eta}_1).$$

Since $(9\varepsilon_1)^{d-2}<1/4$ we may choose $\tilde{\eta}_1$ so small that $\left(\frac{\eta_1+\tilde{\eta}_1}{\gamma_1-\tilde{\eta}_1}\right)^{d-2}\leq\frac{1}{2}$ holds.

Repeating the above process with $B(p_0,r)$ replaced by $B(q_1,\tilde{\eta}_1)$, we can find $q_2\in B(q_1,\tilde{\eta}_1)\cap F$, $\frac{2}{9}\tilde{\eta}_1>\gamma_2>\eta_2>\tilde{\eta}_2>0$ such that

$$\omega(B(q_2,\gamma_2)\setminus(\overline{B}(q_2,\eta_2)\cap F),\partial(B(q_2,\eta_2)\cap F),z)\geq 3/4 \text{ for all } z\in B(q_2,\tilde{\eta}_2)$$

and $\left(\frac{\eta_2+\tilde{\eta}_2}{\gamma_2-\tilde{\eta}_2}\right)^{d-2}\leq\frac{1}{4}$.

In general, for any $n\geq 1$, there exists $q_n\in B(q_{n-1},\tilde{\eta}_{n-1})\cap F$, $\frac{2}{9}\tilde{\eta}_{n-1}>\gamma_n>\eta_n>\tilde{\eta}_n>0$, such that

$$(4.59)\qquad \omega(B(q_n,\gamma_n)\setminus(\overline{B}(q_n,\eta_n)\cap F),\partial(B(q_n,\eta_n)\cap F),z)>1-\frac{1}{2n}$$

for all $z\in B(q_n,\tilde{\eta}_n)$ and

$$(4.60)\qquad \left(\frac{\eta_n+\tilde{\eta}_n}{\gamma_n-\tilde{\eta}_n}\right)^{d-2}\leq\frac{1}{2n}.$$

It is clear that $\{\eta_n\}$ and $\{\tilde{\eta}_n\}$ are decreasing to 0.

Let $p_1=\bigcap_{n\geq1}B(q_n,\eta_n)\cap F=\bigcap_{n\geq1}B(q_n,\tilde{\eta}_n)\cap F$. For any $n\geq 1$, since $p_1\in B(q_n,\tilde{\eta}_n)\subset B(q_n,\eta_n)$, there exists $\tilde{\delta}_n<\tilde{\eta}_n$ such that $B(p_1,\tilde{\delta}_n)\subset B(q_n,\tilde{\eta}_n)$. Let $\tilde{\gamma}_n=\tilde{\eta}_n+\eta_n$. From (4.60) there follows $\tilde{\gamma}_n<\gamma_n-\tilde{\eta}_n$. So $B(q_n,\eta_n)\subset B(p_1,\tilde{\gamma}_n)\subset B(q_n,\gamma_n)$. Thus by the maximum principle for harmonic functions and (4.59), we have

$$(4.61)\qquad \begin{aligned}&\omega(B(q_n,\gamma_n)\setminus(\overline{B}(p_1,\tilde{\gamma}_n)\cap F),\partial(B(p_1,\tilde{\gamma}_n)\cap F),z)\\ &\quad\geq\omega(B(q_n,\gamma_n)\setminus(\overline{B}(q_n,\eta_n)\cap F),\partial(B(q_n,\eta_n)\cap F),z)\\ &\quad\geq 1-\frac{1}{2n}\qquad\text{for all } z\in B(p_1,\tilde{\delta}_n).\end{aligned}$$

Let $B=B(q_n,\gamma_n), F_1=\overline{B}(p_1,\tilde{\gamma}_n)\cap F$, and replace q_1 with p_1 and $\frac{2}{9}r'$ with $\delta_n=\gamma_n-\tilde{\eta}_n$ in (4.56). Then (4.61), (4.30) and (4.60) imply,

$$\begin{aligned}1-\frac{1}{2n} &\leq \omega(B\setminus F_1,\partial F_1,z)\\ &= \int_{\partial B(p_1,\delta_n)\cup\partial F_1} \omega(B\setminus F_1,\partial F_1,x)\omega(B(p_1,\delta_n)\setminus F_1,dx,z)\\ &= \int_{\partial B(p_1,\delta_n)} \omega(B\setminus F_1,\partial F_1,x)\omega(B(p_1,\delta_n)\setminus F_1,dx,z)\\ &\quad+\int_{\partial F_1} \omega(B\setminus F_1,\partial F_1,x)\omega(B(p_1,\delta_n)\setminus F_1,dx,z)\\ &\leq \sup_{x\in\partial B(p_1,\delta_n)} \omega(B\setminus F_1,\partial F_1,x)+\omega(B(p_1,\delta_n)\setminus F_1,\partial F_1,z)\\ &\leq \frac{1}{2n}+\omega(B(p_1,\delta_n)\setminus F_1,\partial F_1,z),\end{aligned} \tag{4.62}$$

for all $z \in B(p_1, \tilde{\delta}_n)$. Thus, by (4.62),

$$\omega(B(p_1,\delta_n)\setminus F_1,\partial F_1,z) \geq \frac{1}{n} \text{ for all } z \in B(p_1,\tilde{\delta}_n).$$

By the maximum principle again, since $F_1 = B(p_1, \tilde{\gamma}_n) \cap F \subset B(p_1, \delta_n)$, we finally have

$$\begin{aligned}\omega(B(p_1,\delta_n)\setminus F,\partial(B(p_1,\delta_n)\cap F),z) &\geq \omega(B(p_1,\delta_n)\setminus F_1,\partial F_1,z)\\ &\geq 1-\frac{1}{n} \text{ for all } z\in B(p_1,\tilde{\delta}_n).\end{aligned}$$

This proves the theorem since $\delta_n > \tilde{\delta}_n > \delta_{n+1}$ is decreasing to 0.

5. Open Questions

We end this paper by posing some open questions. We have seen all the known uniqueness proofs either taking the direct approach of extending Cantor's one dimensional proof to a truly multi-dimensional one, or taking the indirect way of relating another mode of convergence to one way iterative convergence and then using induction. Let us denote the different modes of convergence defined in Section 1 by the following abbreviation:

URC: unrestricted rectangular convergence.
RC: restricted rectangular convergence.
SC: square convergence.
IC: one way iterative convergence.

It is clear we have the following diagram for general complex series:

$$\textbf{URC} \longrightarrow \textbf{RC} \longrightarrow \textbf{SC}.$$

For multiple trigonometric series, we can say more because of Lemma 3.6. Namely we have

$$\begin{array}{ccccc}\textbf{URC} & \longrightarrow & \textbf{RC} & \longrightarrow & \textbf{SC}\\ \downarrow & & & & \\ \textbf{IC}. & & & & \end{array}$$

Our first question is to see if we can generalize this diagram, or equivalently, Lemma 3.6 further.

Open question 1. *Can we replace "unrestricted rectangular convergence" in Lemma 3.6 by "restricted rectangular convergence" or, even better, by "square convergence"?*

If the answer is yes, then some or all of the following open questions are solved positively. If the answer is no, then the following four open questions seem to require a new method of proof.

Open question 2. *If a multiple trigonometric series is square convergent to zero everywhere, must all of its coefficients be zero?*

Open question 3. *If a multiple trigonometric series is square convergent to an everywhere finite $L^1(T^d)$ function, $d \geq 2$, must its coefficients be the Fourier coefficients of that function?*

Open question 4. *If a multiple trigonometric series is restrictedly rectangularly convergent to zero everywhere, must all of its coefficients be zero?*

Open question 5. *If a multiple trigonometric series is restrictedly rectangularly convergent to an everywhere finite $L^1(T^d)$ function, $d \geq 2$, must its coefficients be the Fourier coefficients of that function?*

References

[A1] J. M. Ash, *Multiple trigonometric series,* Studies in Harmonic Analysis, J. M. Ash, ed., Math. Assoc. of America Studies in Mathematics, **13**(1976), pp. 76–96.

[A2] J. M. Ash, *A new, harder proof that continuous functions with Schwarz derivative 0 are lines,* Fourier Analysis, Analytical and Geometric Aspects, eds. W.O. Bray, P. S. Milojević, and Č. V. Stanojević, Marcel Dekker, New York, 1994, 35–46.

[AFR] J. M. Ash, C. Freiling, and D. Rinne, *Uniqueness of rectangularly convergent trigonometric series,* Ann. of Math. **137**(1993), 145–166.

[AKR] J. M. Ash, R. P. Kaufman, and E. Rieders, *The Cantor-Lebesgue property,* Israel J. of Math., **84**(1993), 179–191.

[AWa] J. M. Ash and G. Wang, *One and two dimensional Cantor-Lebesgue type theorems*, Trans. Amer. Math. Soc. to appear.

[AW] J. M. Ash and G. Wang, *Spherical uniqueness of multiple trigonometric series for L^1 functions*, preprint.

[AWe] J. M. Ash and G. V. Welland, *Convergence, uniqueness, and summability of multiple trigonometric series*, Trans. Amer. Math. Soc. **163**(1972), 401–436. MR **45** # 9057.

[B] J. Bourgain, *Spherical summation and uniqueness of multiple trigonometric series,* Internat. Math. Res. Notices (1996), 93-107.

[Coh] P. J. Cohen, *Topics in the theory of uniqueness of trigonometrical series,* Thesis, University of Chicago, Chicago, IL., 1958.

[Con] Bernard Connes, *Sur les coefficients des séries trigonométriques convergentes sphériquement, C. R. Acad. Sci. Paris*, Sér. A, **283**(1976), 159–161. MR **54** #10975.

[Coo] R. Cooke, *A Cantor-Lebesgue theorem in two dimensions,* Proc. Amer. Math. Soc., **30**(1971), 547–550. MR **43** #7847.

[Ge] H. Geringer, *Trigonometrische Doppelreihen*, Monatsh. Math. Phys. **29**(1918), 65–144.

[FRT] C. Freiling, D. Rinne, and B. S. Thomson, *A Riemann-type integral based on the second symmetric derivative*, J. of the London Math. Soc., to appear.

[G] V. Guillemin, *Some classical theorems in spectral theory revisited*, Seminar on Singularities of Solutions of Linear Partial Differential Equations, ed. L. Hörmander, Annals of Math. Studies 91, Princeton University Press, (1979), 219–259.

[H] L. Helms, *Introduction to Potential Theory,* John Wiley &Sons, New York, 1969.

[Ra] T. Rado, *Subharmonic Functions,* Chelsea, New York, 1949.

[R] W. Rudin. *Uniqueness theory for Laplace series,* Trans. Maer. Math. Soc., **68**(1950), 287–303.

[S] V. L. Shapiro, *Uniqueness of multiple trigonometric series,* Ann. of Math. **66**(1957), 467–480. MR **19**, 854; 1432.

[SW] E. M. Stein and G. Weiss, *Introduction to Fourier Analysis on Euclidean Spaces,* Princeton Univ. Press, Princeton, 1971.

[T] S. Tetunashvili, *On some multiple function series and the solution of the uniqueness problem for Pringsheim convergence of multiple trigonometric series,* Mat. Sb. **182**(1991), 1158–1176 (Russian), Math. USSR Sbornik **73**(1992), 517–534 (English).

[Zy] A. Zygmund, *Trigonometric Series*, Vol. 1, 2nd rev. ed. Cambridge Univ. Press, New York, 1959.

[Z] A. Zygmund, *A Cantor-Lebesgue theorem for double trigonometric series*, Studia Math., **43**(1972), 173–178. MR **47** #711.

Department of Mathematics, DePaul University, Chicago IL 60614
E-mail address, J. Marshall Ash: `mash@math.depaul.edu`
E-mail address, Gang Wang: `gwang@condor.depaul.edu`

Contemporary Mathematics
Volume **208**, 1997

A NEW LOOK AT SOME OLD TRIGONOMETRIC EXPANSIONS

RICHARD ASKEY

In honor of Victor Shapiro on his 70th Birthday.

ABSTRACT. A new look is taken at two classical trigonometric series: $\sum_{n=1}^{\infty}\frac{\sin n\theta}{n}$ and $\sum_{n=0}^{\infty}\frac{\sin(n+\frac{1}{2})\theta}{n+\frac{1}{2}}$. Related sums of some orthogonal polynomials are also considered.

1. INTRODUCTION.

Many of us have had blind spots which made it hard or impossible to solve a problem that after it was solved seemed very simple. One of my blind spots was revealed when Malcolm Brown and Mourad Ismail solved a problem which had defeated me for almost twenty years. See [5] for their solution. A formal version is given below. The problem was to find the inverse of a divided difference operator that had arisen in the study of the most general example of a set of orthogonal polynomials which have properties analogous to these of the classical orthogonal polynomials.

The classical orthogonal polynomials are usually described as those orthogonal with respect to the beta, gamma and normal distributions. These are named after Jacobi, Laguerre and Hermite respectively. The most general of these, Jacobi polynomials, can be given in a number of ways. They have a representation as a hypergeometric series

$$P_n^{(\alpha,\beta)}(x)=\frac{(\alpha+1)_n}{n!}\,{}_2F_1\left(\begin{matrix}-n,n+\alpha+\beta+1\\ \alpha+1\end{matrix};\frac{1-x}{2}\right). \tag{1.1}$$

The shifted factorial $(a)_n$ is defined by

$$(a)_n=a(a+1)\cdots(a+n-1)=\Gamma(n+a)/\Gamma(a) \tag{1.2}$$

and the hypergeometric series is

$${}_pF_q\left(\begin{matrix}a_1,\cdots,a_p\\ b_1,\cdots,b_q\end{matrix};t\right)=\sum_{n=0}^{\infty}\frac{(a_1)_n\cdots(a_p)_n}{(b_1)_n\cdots(b_q)_n}\frac{t^n}{n!}. \tag{1.3}$$

1991 *Mathematics Subject Classification.* Primary 26A36, Secondary 33C25, 33D25, 39A70, 33D10, 42A16.

Key words and phrases. Divided difference operator, theta functions, Brown-Ismail operator.

Supported in part by NSF Grant DMS-9622816

Another way of representing Jacobi polynomials is by a differentiation formula of Rodrigues type:

$$(1.4)\qquad (1-x)^\alpha(1+x)^\beta P_n^{(\alpha,\beta)}(x) = \frac{(-1)^n}{2^n n!}\frac{d^n}{dx^n}\left[(1-x)^{n+\alpha}(1+x)^{n+\beta}\right].$$

Their orthogonality is

$$(1.5)\qquad \begin{aligned}\int_{-1}^{1} P_n^{(\alpha,\beta)}(x)P_m^{(\alpha,\beta)}(x)(1-x)^\alpha(1+x)^\beta dx &= 0, \quad m \neq n,\\ &= \frac{2^{\alpha+\beta+1}\Gamma(n+\alpha+1)\Gamma(n+\beta+1)}{(2n+\alpha+\beta+1)\Gamma(n+\alpha+\beta+1)\Gamma(n+1)} = h_n^{\alpha,\beta}, m = n,\end{aligned}$$

when $\alpha, \beta > -1$.

Another useful fact about Jacobi polynomials is that

$$(1.6)\qquad \frac{d}{dx}P_n^{(\alpha,\beta)}(x) = \frac{(n+\alpha+\beta+1)}{2}P_{n-1}^{(\alpha,\beta)}(x).$$

Formula (1.6) follows immediately from the representation (1.1) as a hypergeometric series. See Szegö [20] for these and further results involving Jacobi polynomials.

Either (1.4) or (1.6) can be used to characterize the classical orthogonal polynomials of Jacobi, Laguerre and Hermite. For (1.6), Hahn [9] proved the following theorem.

Theorem A. If $\{p_n(x)\}_{n=0}^\infty$ is a set of polynomials which are orthogonal with respect to a positive measure on the real line and $\{p'_{n+1}(x)\}_{n=0}^\infty = \{q_n(x)\}_{n=0}^\infty$ are also orthogonal, then up to a change of scale and normalization, $\{p_n(x)\} = \{P_n^{(\alpha,\beta)}(x)\}$or the limiting cases of Laguerre and Hermite polynomials.

A very nice survey of characterization theorems for different classes of orthogonal polynomials was written by Waleed Al-Salam [1].

Fortunately, these classical polynomials are not the only sets of orthogonal polynomials with properties like (1.4) and (1.6). What needs to be done to find other sets is to replace the derivative with other operators.

The first set of examples was found by Chebychev [6], [7], but are known as Hahn polynomials because earlier appearances were not noticed. These polynomials can be given as

$$(1.7)\qquad Q_n(x;\alpha,\beta,N) = {}_3F_2\left(\begin{matrix}-n, n+\alpha+\beta+1, -x\\ \alpha+n, -N\end{matrix};1\right)$$

$x, n = 0, 1, \cdots, N$. They are orthogonal on $x = 0, 1\cdots, N$ with respect to

$$\binom{x+\alpha}{x}\binom{N-x+\beta}{N-x} = \frac{(\alpha+1)_x(\beta+1)_{N-x}}{x!(N-x)!}.$$

The substitute for the derivative is the finite difference operator

$$(1.8)\qquad \Delta f(x) = f(x+1) - f(x).$$

A routine calculation gives

$$\Delta Q_n(x;\alpha,\beta,N) = \frac{-n(n+\alpha+\beta+1)}{(\alpha+1)N} Q_{n-1}(x;\alpha+1,\beta+1,N-1). \tag{1.9}$$

Hahn found a q-extension of these polynomials [10].

The most general set of polynomials which Andrews and I called classical orthogonal polynomials is given by a basic hypergeometric series. See [3]. Hypergeometric series are series $\sum a_n$ with a_{n+1}/a_n a rational function of n. Basic hypergeometric series have a_{n+1}/a_n a rational function of q^n for a fixed number q. The most general of these neoclassical polynomials (to use a term suggested by David and Gregory Chudnovsky) is

$$w_n(x;a,b,c,d|q) = \sum_{k=0}^{n} \frac{(q^{-n};q)_k (q^{n-1}abcd;q)_k (ae^{i\theta};q)_k (ae^{-i\theta};q)_k q^k}{(ab;q)_k (ac;q)_k (ad;q)_k (q;q)_k} \tag{1.10}$$

where $x = \cos\theta$ and

$$\begin{aligned} (a;q)_n &= \prod_{j=0}^{n-1} (1-aq^j), \quad n = 1,2,\cdots, \\ (a;q)_0 &= 1. \end{aligned} \tag{1.11}$$

The extension of the derivative to this setting is

$$D_q f(x) = \frac{\delta_q h(e^{i\theta})}{\delta_q \cos\theta} \tag{1.12}$$

with $x = \cos\theta$, $f(x) = f(\cos\theta) = h(e^{i\theta})$, and

$$\delta_q h(e^{i\theta}) = h(q^{1/2}e^{i\theta}) - h(q^{-1/2}e^{i\theta}). \tag{1.13}$$

A simple calculation shows that

$$\begin{aligned} D_q \prod_{j=0}^{k-1} (1-2axq^j + a^2q^{2j}) &= \frac{\delta_q (ae^{i\theta};q)_k (ae^{-i\theta};q)_k}{\delta_q (e^{i\theta}+e^{-i\theta})/2} \\ &= -2a\frac{(1-q^k)}{(1-q)} (aq^{1/2}e^{i\theta};q)_{k-1} (aq^{1/2}e^{-i\theta};q)_{k-1} \\ &= -2a\frac{(1-q^k)}{(1-q)} \prod_{j=0}^{k-2} (1-2axq^{j+1/2} + a^2q^{2j+1}). \end{aligned}$$

For the polynomials $w_n(x)$ defined in (1.10)

$$D_q w_n(x;a,b,c,d|q) = \frac{2aq^{1-n}(1-q^n)(1-q^{n-1}abcd)}{(1-ab)(1-ac)(1-ad)(1-q)}. \tag{1.14}$$

$$w_{n-1}(x; aq^{1/2}bq^{1/2}, cq^{1/2}, dq^{1/2}|q).$$

See Askey and Wilson [4] for these results and Gasper and Rahman [8] for many more results about q-series.

There are two reasons to be interested in the operator (1.12). It operates on the most general of the neoclassical orthogonal polynomials. In addition, there is a structural reason which was found by Alphonse Magnus, [15].

Consider the divided difference operator

$$Df(x) = \frac{f(y_2(x)) - f(y_1(x))}{y_2(x) - y_1(x)}. \tag{1.15}$$

If this operator gives a polynomial of degree $n-1$ when $f(x)$ is a polynomial of degree n, then from $f(x) = x^2$ and $f(x) = x^3$,

$$y_1(x) + y_2(x) = ax + b \tag{1.16}$$

and

$$[y_1(x)]^2 + y_1(x)y_2(x) + [y_2(x)]^2 = \alpha x^2 + \beta x + \gamma.$$

These give

$$y_1(x)y_2(x) = cx^2 + dx + f, \tag{1.17}$$

for constants, $a, b, c, d, f, \alpha, \beta, \gamma$.

Removing $y_2(x)$ gives

$$Ay^2 + Bxy + Cx^2 + Dy + Ex + F = 0 \tag{1.18}$$

for constants A, B, C, D, E, F with $y_1(x) = y$ here. A similar equation exists for $y_2(x)$. Magnus considered the different cases which arise from the solutions of (1.18). Essentially, they are the operator D_q defined in (1.12), a hypergeometric limit of this:

$$D_1 f(x^2) = \frac{f((x+\frac{i}{2})^2) - f((x-\frac{i}{2})^2)}{(x+\frac{i}{2})^2 - (x-\frac{i}{2})^2}, \tag{1.19}$$

the finite difference operator (1.8), and its q-extension

$$\Delta_q f(x) = \frac{f(qx) - f(x)}{(q-1)x}, \tag{1.20}$$

after rescaling is done. The derivative also arises as a limit from each of these operators after they are rescaled.

2. THE INVERSE OPERATOR TO THE q-DIVIDED DIFFERENCE OPERATOR.

The fundamental theorem of calculus is a connection between differentiation and integration. It is natural to look for similar relations for the operators defined above. For the difference operator Δ defined in (1.8), the inverse operator is just a sum.

For, if

$$S(x) = \sum_{k=0}^{x-1} f(k), \tag{2.1}$$

then

$$\Delta S(x) = f(x) \tag{2.2}$$

and

$$\sum_{k=0}^{x-1} \Delta f(k) = \sum_{k=0}^{x-1} [f(k+1) - f(k)] = f(x) - f(0). \tag{2.3}$$

The inverse of $\Delta_q f(x)$ defined in (1.20) is

$$\int_0^x f(t) d_q t = x(1-q) \sum_{k=0}^{\infty} f(xq^k) q^k \tag{2.4}$$

This integral has traditionally been called the Jackson integral, but it was introduced a number of years earlier by Thomae. Since the idea of using point masses on the terms of a geometric progression to approximate a measure is due to Fermat, I call this integration with respect to Fermat measure. See Toeplitz [21].

It is easy to see that

$$\Delta_q \int_0^x f(t) d_q t = f(x).$$

Also

$$\begin{aligned}\int_0^x \Delta_q f(t) d_q t &= \int_0^x \frac{f(t) - f(tq)}{t(1-q)} d_q t \\ &= x(1-q) \sum_{n=0}^{\infty} \frac{\left[f(xq^n) - f(xq^{n+1})\right] q^n}{xq^n(1-q)} \\ &= \sum_{n=0}^{\infty} [f(xq^n) - f(xq^{n+1})] = f(x) - f(0).\end{aligned}$$

However, the operator (1.12) defeated me, and I did not know how to find its inverse. One evening, Mourad Ismail called to say that this inverse operator could be found by applying (1.12) to a series expansion in terms of a special case of the polynomials defined in (1.10). When $a = 1, b = q^{1/2}, c = -1, d = -q^{1/2}$, the polynomials in (1.10) are just multiples of the Chebychev polynomials $T_n(x) =$

$\cos n\theta$ when $x = \cos\theta$. When $a = -c = q, b = -d = q^{1/2}$, these polynomials are multiples of the Chebychev polynomials $U_n(\cos\theta) = \sin(n+1)\theta/\sin\theta$.

If $T_n(x) = T_n(\cos\theta) = \cos n\theta$, then

$$D_q T_n(x) = q^{(1-n)/2}\frac{(1-q^n)}{(1-q)}\frac{\sin n\theta}{\sin\theta} = q^{(1-n)/2}\frac{(1-q^n)}{(1-q)}U_{n-1}(x). \tag{2.5}$$

To find the inverse operator to D_q, consider the cosine series

$$f(\cos\theta) = \sum_{k=1}^{\infty} a_k \cos k\theta, \tag{2.6}$$

where we have assumed that

$$\int_0^{\pi} f(\cos\theta)d\theta = 0$$

for simplicity. Also assume that the coefficients decrease rapidly enough so that the series below converges uniformly. Then

$$D_q f(x) = \sum_{k=1}^{\infty} a_k q^{(1-k)/2}\frac{(1-q^k)}{(1-q)}\frac{\sin k\theta}{\sin\theta} = g(x). \tag{2.7}$$

Define I_q by

$$\begin{aligned} I_q g(x) = f(x) &= \sum_{k=1}^{\infty}\frac{2}{\pi}\int_0^{\pi} g(\cos\varphi)q^{(k-1)/2}\frac{(1-q)}{(1-q^k)}\frac{\sin k\varphi}{\sin\varphi}\cos k\theta\sin^2\varphi d\varphi \\ &= \frac{1}{\pi}\int_0^{\pi} g(\cos\varphi)\sum_{k=1}^{\infty} q^{(k-1)/2}\frac{(1-q)}{(1-q^k)}[\sin k(\varphi+\theta) + \sin k(\varphi-\theta)]\sin\varphi d\varphi \\ &= \frac{1}{\pi}\int_{-\pi}^{\pi} g(\cos\varphi)K(\theta+\varphi)\sin\varphi d\varphi \end{aligned} \tag{2.8}$$

with

$$K(\varphi) = \sum_{K=1}^{\infty}\frac{(1-q)}{(1-q^k)}q^{(k-1)/2}\sin k\varphi. \tag{2.9}$$

The series defining $K(\varphi)$ can be represented using a theta function, but the formula is not widely known, so a derivation will be given.

$$\begin{aligned} \sum_{k=1}^{\infty}\frac{q^k}{1-q^{2k}}e^{ik\theta} &= \sum_{k=1}^{\infty} q^k e^{ik\theta}\sum_{j=0}^{\infty} q^{2jk} \\ &= \sum_{j=0}^{\infty}\sum_{k=1}^{\infty} q^{(2j+1)k}e^{ik\theta} = \sum_{j=0}^{\infty}\frac{q^{2j+1}e^{i\theta}}{1-q^{2j+1}e^{i\theta}} \\ &= -\frac{1}{i}\frac{d}{d\theta}\sum_{j=0}^{\infty}\log(1-q^{2j+1}e^{i\theta}) \\ &= i\frac{d}{d\theta}\log\prod_{j=0}^{\infty}(1-q^{2j+1}e^{i\theta}). \end{aligned} \tag{2.10}$$

Then

$$\begin{aligned}\sum_{k=1}^{\infty}\frac{q^{k/2}\sin k\theta}{1-q^k} &= \frac{d}{d\theta}\log\prod_{j=0}^{\infty}(1-q^{j+1/2}e^{i\theta})(1-q^{j+1/2}e^{-i\theta}) \\ &= \frac{d}{d\theta}\log\prod_{j=0}^{\infty}(1-q^{j+1/2}e^{i\theta})(1-q^{j+1/2}e^{-i\theta})(1-q^{j+1}).\end{aligned} \tag{2.11}$$

But

$$\prod_{j=0}^{\infty}(1-q^{j+1})(1-q^{j+1/2}e^{i\theta})(1-q^{j+1/2}e^{-i\theta}) = \sum_{j=-\infty}^{\infty} q^{j^2/2}e^{ij\theta}$$

by the triple product identity [8], so $K(\varphi)$ is a logarithmic derivative of a theta function.

A detailed treatment of this result including the identification of a space of functions on which these operators act to give the identity operator is given by Brown and Ismail [5]. Ismail and Zhang [13] and Ismail, Rahman and Zhang [12] give further information on the integral transform with this kernel.

3. Fourier series.

Once this argument was outlined for me, it was natural to look for other settings where a similar argument could be used. Some arise in the elementary parts of Fourier analysis.

Assume $f(\theta)$ is a sufficiently smooth function defined for $0 \le \theta \le \pi$ with $f(0) = f(\pi) = 0$. Consider the expansion

$$f(\theta) = \sum_{n=1}^{\infty} a_n \sin n\theta. \tag{3.1}$$

Then,

$$f'(\theta) = \sum_{n=1}^{\infty} na_n \cos n\theta = g(\theta). \tag{3.2}$$

Integration gives

$$Ig(\theta) = \sum_{n=1}^{\infty} a_n \sin n\theta \tag{3.3}$$

with

$$na_n = \frac{2}{\pi}\int_0^{\pi} g(\varphi)\cos n\varphi d\varphi \tag{3.4}$$

Then

$$\begin{aligned}Ig(\theta) &= \frac{2}{\pi}\int_0^{\pi} g(\varphi)\sum_{n=1}^{\infty}\frac{\sin n\theta\cos n\varphi}{n}d\varphi \\ &= \frac{1}{\pi}\int_0^{\pi} g(\varphi)\sum_{n=1}^{\infty}\left[\frac{\sin n(\theta+\varphi)}{n}+\frac{\sin n(\theta-\varphi)}{n}\right]d\varphi.\end{aligned} \tag{3.5}$$

It is well known and easy to show that

$$\sum_{n=1}^{\infty}\frac{\sin n\theta}{n}=\frac{\pi-\theta}{2},\quad 0<\theta<2\pi,$$
$$=\frac{-\pi-\theta}{2},\quad -2\pi<\theta<0. \tag{3.6}$$

Then (3.5) becomes

$$\begin{aligned}Ig(\theta)&=\frac{1}{\pi}\int_0^\pi g(\varphi)\left[\frac{\pi-\theta-\varphi}{2}\right]d\varphi\\&+\frac{1}{\pi}\int_0^\theta g(\varphi)\left[\frac{\pi-\theta+\varphi}{2}\right]d\varphi+\frac{1}{\pi}\int_\theta^\pi g(\varphi)\left[\frac{-\pi-\theta+\varphi}{2}\right]d\varphi\\&=\int_0^\theta g(\varphi)d\varphi-\frac{\theta}{2}\int_0^\pi g(\varphi)d\varphi.\end{aligned}\tag{3.7}$$

However, the constant term in (3.2) is missing, so

$$\int_0^\pi g(\varphi)d\varphi=0.$$

Thus

$$Ig(\theta)=\int_0^\theta g(\varphi)d\varphi \tag{3.8}$$

as it should.

Another example for Fourier series is

$$f(\theta)=\sum_{n=0}^{\infty}a_n\sin(n+\frac{1}{2})\theta. \tag{3.9}$$

A similar argument follows.

$$g(\theta)=f'(\theta)=\sum_{n=0}^{\infty}(n+\frac{1}{2})a_n\cos(n+\frac{1}{2})\theta$$

and

$$(n+\frac{1}{2})a_n=\frac{2}{\pi}\int_0^\pi g(\varphi)\cos(n+\frac{1}{2})\varphi d\varphi.$$

Thus

$$\begin{aligned}Ig(\theta)&=\sum_{n=0}^{\infty}a_n\sin(n+\frac{1}{2})\theta\\&=\frac{2}{\pi}\int_0^\pi g(\varphi)\sum_{n=0}^{\infty}\frac{\sin(n+\frac{1}{2})\theta\cos(n+\frac{1}{2})\varphi}{n+\frac{1}{2}}d\varphi\\&=\frac{1}{\pi}\int_0^\pi g(\varphi)\left[\sum_{n=0}^{\infty}\frac{\sin(n+\frac{1}{2})(\theta+\varphi)}{n+\frac{1}{2}}+\frac{\sin(n+\frac{1}{2})(\theta-\varphi)}{n+\frac{1}{2}}\right]d\varphi.\end{aligned}\tag{3.10}$$

The relevant series expansion is

$$\sum_{n=0}^{\infty} \frac{\sin(n+\frac{1}{2})\theta}{n+\frac{1}{2}} = \begin{matrix} \pi/2, & 0<\theta<2\pi, \\ -\pi/2, & -2\pi<\theta<0. \end{matrix} \tag{3.11}$$

This gives

$$\begin{aligned} Ig(\theta) &= \frac{1}{\pi}\int_0^{\pi} \frac{\pi}{2} g(\varphi)d\varphi + \frac{1}{\pi}\int_0^{\theta} \frac{\pi}{2} g(\varphi)d\varphi - \frac{1}{\pi}\int_{\theta}^{\pi} \frac{\pi}{2} g(\varphi)d\varphi \\ &= \int_0^{\theta} g(\varphi)d\varphi. \end{aligned} \tag{3.12}$$

4. Jacobi expansions.

It is natural to look for similar bilinear series for Jacobi polynomials. Consider the characteristic function of the interval $x<t<1$,

$$\chi_x(t) = \begin{matrix} 1, & x<t<1, \\ 0, & -1<t<x. \end{matrix} \tag{4.1}$$

Then

$$\chi_x(t) = \sum_{n=0}^{\infty} a_n(x) P_n^{(\alpha,\beta)}(t)$$

with

$$\begin{aligned} h_n^{\alpha,\beta} a_n(x) &= \int_{-1}^{1} \chi_x(t) P_n^{(\alpha,\beta)}(t)(1-t)^{\alpha}(1+t)^{\beta} dt \\ &= \int_x^1 P_n^{(\alpha,\beta)}(t)(1-t)^{\alpha}(1+t)^{\beta} dt \\ &= \frac{(-1)^n}{2^n n!}\int_x^1 \frac{d^n}{dt^n}(1-t)^{n+\alpha}(1+t)^{n+\beta} dt \\ &= \frac{(-1)^{n+1}}{2^n n!}\frac{d^{n-1}}{dx^{n-1}}(1-x)^{n+\alpha}(1+x)^{n+\beta} \\ &= \frac{1}{2n} P_{n-1}^{(\alpha+1,\beta+1)}(x), \qquad n=1,2,\cdots. \end{aligned}$$

$$h_0^{\alpha,\beta} a_0 = \int_x^1 (1-t)^{\alpha}(1+t)^{\beta} dt,$$

These give

$$\chi_x(t) = \frac{1}{h_0^{\alpha,\beta}}\int_x^1 (1-t)^{\alpha}(1+t)^{\beta} dt + \sum_{n=1}^{\infty} \frac{P_n^{(\alpha,\beta)}(t) P_{n-1}^{(\alpha+1,\beta+1)}(x)}{2n h_n^{\alpha,\beta}}, \tag{4.2}$$

with $\chi_x(t)$ defined in (4.1) and $h_n^{\alpha,\beta}$ in (1.5).

5. Olinde Rodrigues.

A couple of years ago, Victor Shapiro gave an analysis seminar talk in Madison. Among other things, he told us something about the French mathematician Olinde Rodrigues. Those who know something about special functions, and Legendre polynomials in particular, know Rodrigues's name from the formula (1.4). What is not well known is that Rodrigues's paper, which contains (1.4) for $\alpha = \beta = 0$, and a bit more, was lost for many decades. Heine [11, pages 20-21] gives a detailed early history of this formula. After Hermite's discovery of Rodrigues's thesis [16], and Heine's historical account, Rodrigues's name replaced those of Ivory and Jacobi, which had been used earlier.

There are other important results of Rodrigues which escaped notice for a long time. A description of one is included in [2]. Another seems to have escaped all notice until about 1970.

Stern [19] posed the problem of finding the total number of inversions of the permutations of $1, 2, \cdots, n$. The total number of permutations is $n!$, and these can be paired up symmetrically. Since the two extreme cases have 0 and $n(n-1)/2$ inversions, it is easy to see that the total number of inversions is $n! \cdot n(n-1)/4$. In addition to the pairing, a simple induction also works, adding "$n+1$" in any of the $n+1$ places before, between, or after the numbers $1, 2, \cdots, n$.

What Rodrigues [17] did was to introduce a more general problem and solve it. He found the generating function for these inversions. For example, the six permutations of $\{1, 2, 3\}$ and the number of inversions are

1 2 3	0
1 3 2	1
2 1 3	1
2 3 1	2
3 1 2	2
3 2 1	3

The generating function is

$$1 + 2q + 2q^2 + q^3 = (1+q)(1+q+q^2) = 3!_q.$$

Rodrigues showed that the generating function for the inversions of the permutations of $\{1, 2, \cdots, n\}$ is

$$n!_q = 1(1+q)(1+q+q^2)\cdots(1+q+\cdots+q^{n-1}). \tag{5.1}$$

This is easily shown by induction.

The point of mentioning this here is that Rodrigues's result has been extended by MacMahon to count the inversions of a multiset. For example, the number of ways of putting k 0's and $n-k$ 1's into n spots in $\binom{n}{k}$. This also can be interpreted as the number of walks from $(0,0)$ to $(k, n-k)$ where steps are to the right and up. MacMahon showed that

$$\binom{n}{k}_q = \frac{n!_q}{k!_q(n-k)!_q}$$

is the generating function for the inversions of these arrangements.

For example, the arrangements of two 0's and two 1's and the number of inversions are

$$\begin{array}{cccccc} 0&0&1&1&&0\\ 0&1&0&1&&1\\ 0&1&1&0&&2\\ 1&0&0&1&&2\\ 1&0&1&0&&3\\ 1&1&0&0&&4 \end{array},$$

The generating function is

$$\begin{aligned} 1+q+2q^2+q^3+q^4 &= (1+q^2)(1+q+q^2) = \frac{(1-q^3)(1-q^4)}{(1-q)(1-q^2)} \\ &= \frac{4!_q}{2!_q 2!_q} \end{aligned}$$

This can also be interpreted via lattice points, now counting with respect to the area under the paths. This is illustrated by

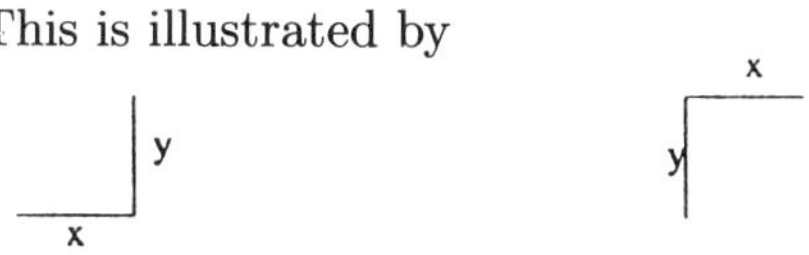

with the first having no area, the second one. Schützenberger [18] observed that MacMahon's generating function can be stated as the binomial theorem with a noncommutative multiplication.

$$(5.2) \qquad (x+y)^n = \sum_{k=0}^{n} \binom{n}{k}_q x^{n-k} y^k$$

when $yx = qxy, xq = qx, yq = qy$.

This noncommutative multiplication can be used to define a twisted version of $SU(2)$, called $SU_q(2)$. It is well known that Jacobi polynomials arise in the representation theory of $SU(2)$. It has recently been discovered that q-versions of Jacobi polynomials, including many special cases of (1.10), arise in the representation theory of $SU_q(2)$. See [22].

This is a fitting twist, for the first serious treatment of basic hypergeometric series in a book was Heine's in [11]. This book primarily dealt with spherical harmonics, which at that time were only represented as hypergeometric functions. However, Heine had done some early work on basic hypergeometric functions, and did not want it to be lost. As a result, he included some on it, but in small type to suggest that it was not directly related to the rest of his book. Now we know that it is closely related. The connection with differential equations seems to be gone, but there are interesting and important $q-$ difference equations instead.

Since Victor Shapiro appreciates both special functions and some of the work of Rodrigues, I hope he will catch a mild case of the $q-$ disease and see how these beautiful functions occur. We could use his insights as this subject continues to develop. In particular, separation of variables can be used when the operator (1.12) arises. See [14].

REFERENCES

[1] W. Al-Salam, *Characterization theorems for orthogonal polynomials*, Orthogonal Polynomials: Theory and Practice (P. Nevai, ed.), Kluwer, Dordrecht, 1990 pages 1–24.
[2] Simon L. Altman, *Hamilton, Rodrigues and the Quaternion Scandal*, Mathematics Magazine **62** (1989), 291-308.
[3] G. E. Andrews and R. Askey, *Classical orthogonal polynomials*, Polynômes Orthogonaux et Applications - (Bar-Le-Duc 1984) (C. Brezinski and others, ed), Lecture Notes in Mathematics, vol. 1171, Springer, Berlin-New York, 1985.
[4] R. Askey and J. Wilson, *Some basic hypergeometric polynomials that generalize Jacobi polynomials*, Mem. Amer. Math. Soc., Number 319, Amer. Math. Soc., Providence, RI, 1985.
[5] M. Brown and M. Ismail, *A right inverse of the Askey-Wilson operator*, PAMS **123** (1995), 2071–2079.
[6] P. L. Chebychev (P. L. Tchebychef), *Sur les fractions continues*, J. Math. Pures Appl., series II, 3 (1858), 289-323. Translation of Russian paper published in 1855; reprinted in Oeuvres de P. L. Tchebychef, tome I, eds. A. Markoff and N. Sonin, St. Petersburg, 1899; reprinted, Chelsea, NY, 1961, 201-230.
[7] P. L. Chebychev (P. L. Tchebychef), *Sur l' interpolation des valeus equidistantes*, in Oeuvers de P. L. Tchebychev, tome II, eds. A. Markoff and N. Sonin, St. Petersburg, 1899; reprinted, Chelsea, NY 1961, 219-242.
[8] G. Gasper and M. Rahman, *Basic Hypergeometric Series*, Cambridge Univ. Press, 1990.
[9] W. Hahn, *Über die Jacobischen Polynome und zwei verwandte Polynomklasse*, Math. Zeit. **39** (1935), 634–638.
[10] W. Hahn, *Über Orthoogonalpolynome, die q-Differentengleichungen genügen*, Math. Nach. **2** (1949), 4–34.
[11] E. Heine, *Handbuch der Kugelfunctionen*, vol. I, 2nd edition, Berlin, 1878.
[12] M. Ismail, M. Rahman and R. Zhang, *Diagonalization of certain integral operators, II*, J. Comp. Appl. Math. **68** (1996), 163-196.
[13] M. Ismail and R. Zhang, *Diagonalization of certain integral operators*, Adv. Math. **109** (1994), 1–33.
[14] V. B. Kuznetzov and E. K. Sklyanin, *Separation of variables and integral relations for special functions*, preprint, October, 1996.
[15] A. Magnus, *Associated Askey-Wilson polynomials as Laguerre-Hahn orthogonal polynomials*, Orthogonal Polynomials and Their Applications (Segovia, 1986), ed. M. Alfaro and others, Lecture Notes in Math. 1329, Springer, Berlin, New York, 1988, 261-378.
[16] O. Rodrigues, *Mémoire sur l' attraction des sphéroïdes*, Correspondence sur l'École Polytechnique **3** (1816), 361-385.
[17] O. Rodrigues, *Note sur les Inversions, ou dérangements produits dans les permutations*, Jour. de Mathématiques **4** (1939), 236–240.
[18] M. Schützenberger, *Une interprétation de certaines solutions de l'equation fonctionnelle* $F(x+y) = F(x)F(y)$, C. R. Acad. Sci. Paris **236** (1953), 352-353.
[19] M. Stern, *Aufgaben*, Jour. reine und angew. Math. **17** (1837), 100.
[20] G. Szegő, *Orthogonal Polynomials,*, 4th edition, Amer. Math. Soc. Colloq. Pub., vol 23, Providence, RI, 1975.
[21] O. Toeplitz, *The Calculus - A Genetic Approach*, Univ. of Chicago Press, Chicago, 1963.
[22] N. Ja Vilenkin and A. U. Klimyk, *Representations of Lie Groups and Special Functions*, volumes 1,2,3, Kluwer, Dordrecht, 1991-1992.

DEPARTMENT OF MATHEMATICS, UNIVERSITY OF WISCONSIN-MADISON, MADISON, WI 53706, U.S.A.
E-mail address: askey@math.wisc.edu

Contemporary Mathematics
Volume **208**, 1997

Analysis Results and Problems Related to Lattice Points on Surfaces

Jean Bourgain

ABSTRACT. Various problems in Harmonic Analysis and PDE have close relations to distribution of lattice points on quadratic surfaces and the behaviour of the associated exponential sums. This paper gives an overview of several research directions in this context, mainly describing some of the author's recent work. Topics discussed here include:

- Problems on the behaviour of eigenfunctions for the flat torus;
- Uniqueness of multiple trigonometric series under spherical summation;
- Periodic Strichartz' inequalities and the Cauchy problem for the nonlinear Schrödinger equation with periodic boundary conditions;
- The role of separation of lattice points on the singular surface in recent developments of the Kolmogorov-Arnold-Moser persistency theory for periodic and quasi-periodic solutions of nonlinear PDEs.

1. Introduction

The aim of this short exposé is to describe a few issues in Analysis depending on the structure of the set of lattice points on certain surfaces in $\mathbb{R}^d$ (or nearby that surface). The topics discussed below are chosen mainly according to the author's own research interests and earlier work. We will not always be very systematic and the discussion will often not be selfcontained or precise.

The surfaces considered will be quadratic with defining equations

$$S = [\Phi(x) = 0] \tag{1.1}$$

where $\Phi(x)$ is a quadratic polynomial. The following examples will be of particular interest to us

$$\phi(x) = x_1^2 + \cdots + x_d^2 - a \qquad \text{(sphere)} \tag{1.2}$$

$$\phi(x) = x_1^2 + \cdots + x_{d-1}^2 + x_d \qquad \text{(paraboloid)} \tag{1.3}$$

$$\phi(x) = x_1^2 + \cdots + x_{d-1}^2 - \lambda x_d^2 \qquad \text{(cone).} \tag{1.4}$$

Specific questions addressed here are for instance

(*i*) *Moment inequalities for trigonometric polynomials with frequencies in S*

1991 *Mathematics Subject Classification.* Primary: 42B05, 35J60; Secondary: 35J10.
Key words and phrases. Moment Inequalities, Eigenfunctions, Laplacian, Torus, Uniqueness problem for spherical Summation, KAM theory, Schrödinger equation.
This paper is in final form and no version of it will be submitted for publication elsewhere.

Thus we consider trigonometric polynomials of the form

$$P(x) = \sum_{n\in S}{}' a_n\, e^{in.x} \tag{1.5}$$

as functions on $\mathbb{T}^d$ and try to estimate $\|P\|_p = \|P(x)\|_{L^p(\mathbb{T}^d)}$ for various p. In particular, in the cases (1.2), (1.3) we are interested in the values of $p > 2$ for which there is a uniform inequality valid

$$\|P\|_p \leq K.\|P\|_2 = K.\Big(\sum_{n\in S} |a_n|^2\Big)^{1/2} \tag{1.6}$$

for some constant K (bounding thus the so-called "Λ_p-constant" of the set $\{n \in \mathbb{Z}^d | n \in S\}$).

This problem, especially in larger dimension, is mainly open. In the case of (1.3), the resulting inequalities are periodic versions of R. Strichartz's inequalities for the linear Schrödinger equation

$$iu_t + \Delta u = 0. \tag{1.7}$$

They play an important role in the theory of the Cauchy problem in the nonlinear case (NLS), for instance the equation

$$iu_t + \Delta u \pm u|u|^{p-2} = 0 \tag{1.8}$$

with Hamiltonian

$$H(\phi) = \frac{1}{2}\int |\nabla\phi|^2 \mp \frac{1}{p}\int |\phi|^p. \tag{1.9}$$

(ii) Distribution of the lattice points on S or nearby S

With Φ as in (1.2)-(1.4), consider the set of lattice points

$$\Omega = \{x \in \mathbb{Z}^d | \quad |\Phi(x)| < 1\}. \tag{1.10}$$

In cases (1.2), (1.3) and in (1.4) for typical $\lambda \in \mathbb{R}$, (1.10) has a structure of "separated clusters", in the sense that for any given number K, there is a partition

$$\Omega = \bigcup_\alpha \Omega_\alpha \tag{1.11}$$

satisfying

$$\text{dist}\,(\Omega_\alpha.\Omega_\beta) > K \quad \text{for} \quad \alpha \neq \beta \tag{1.12}$$

$$\text{diam}\ \Omega_\alpha < K^C \quad \text{for all} \quad \alpha \tag{1.13}$$

where the constant C in (1.19) depends only on the dimension d.

In the case of the sphere, the previous statement is part of the work of B. Connes [Co] proving that any sequence of eigenfunctions of Δ of $\mathbb{T}^d$ converging pointwise to zero also converges to zero in L^2-norm. This last fact is a key step in the solution of the uniqueness problem for spherical summation of trigonometric series in dimension ≥ 3, which was recently completed in [Bo1].

The partition result for lattice points on a paraboloid is an observation due to A. Granville and T. Spencer [G-S] motivated by Green's function estimates for lattice Schrödinger operators. We will discuss here its relevance in a closely related context namely the KAM theory of NL Schrödinger equations (for instance with periodic boundary conditions) on the existence of time periodic and quasi-periodic solutions. In the same context, the case of the cone (1.4) appears naturally in the corresponding problems for the wave equation. The "KAM-theory for PDEs", is a relatively recent line of research and we will indicate later on in this exposé some of the methods, results and problems. The type of problems mentioned above are

not restricted to quadratic surfaces. For instance certain investigations regarding Korteweg de Vries equation

$$y_t + \partial_x^3 y + y\partial_x y = 0 \tag{1.14}$$

and generalizations are intimately related to properties of the lattice point set

$$\{(n, n^3) | n \in \mathbb{Z}\}. \tag{1.15}$$

We will not discuss this here and refer the reader for this to some of the references listed at the end (see [Gi]).

2. Eigenfunction bounds for the Laplacian on the torus

Let Δ denote the Laplacian on the d-torus $\mathbb{T}^d = \mathbb{R}^d/\mathbb{Z}^d$. An eigenfunction ψ of Δ with eigenvalue $-\lambda$, thus $\Delta\psi + \lambda\psi = 0$, may be written as

$$\psi(x) = \sum_{|n|^2=\lambda} \widehat{\psi}(n)\, e^{in.x}. \tag{2.1}$$

In [Bo2], estimates are given on the quantities ($p > 2$)

$$M_{d,p}(\lambda) = M_p(\lambda) = \sup_{\Delta\psi+\lambda\psi=0} \frac{\|\psi\|_p}{\|\psi\|_2}. \tag{2.2}$$

According to a result of Cooke [C], often attributed to Zygmund [Zy2], there is in $d = 2$ a uniform bound

$$M_{2,4}(\lambda) < C. \tag{2.3}$$

This fact results from a simple direct computation of the L^4-norm and the observation that the equation in n_1, n_2

$$n_1 + n_2 = n, \qquad n,\, n_1, n_2 \in \mathbb{Z}^2 \text{ and } |n_1|^2 = \lambda = |n_2|^2 \tag{2.4}$$

has only 2 solutions at most, for given $n \neq 0$.

Denoting $\mathcal{S}_{d,\lambda}$ the set of lattice points on the sphere in $\mathbb{R}^d$ of radius $\sqrt{\lambda}$, it is a classical fact on representing integers by sums of squares that

$$\#\mathcal{S}_{d,\lambda} \ll \lambda^{\frac{d}{2}-1+\varepsilon} \qquad (d \geq 2) \tag{2.5}$$

(for $d \geq 5$, the ε may be dropped).

It follows from (2.5) that

$$M_{d,\infty}(\lambda) \ll \lambda^{\frac{d-2}{4}+\varepsilon}. \tag{2.6}$$

It seems reasonable to conjecture that in dimension d

$$\begin{cases} M_p(\lambda) < C_p \ \text{ for } \ p < \frac{2d}{d-2} & (2.7) \\ M_p(\lambda) \ll \lambda^\varepsilon \ \text{ for } \ p = \frac{2d}{d-2} & (2.8) \\ M_p(\lambda) \ll \lambda^{\frac{d-2}{4}-\frac{d}{2p}+\varepsilon} \ \text{ for } \ \frac{2d}{d-2} \leq p \leq \infty & (2.9) \end{cases}$$

((2.9) is obtained interpolating between (2.6) and (2.8)).

This conjecture is open in any dimension. Using arguments of an arithmetical nature, one may show that $M_{3,4}(\lambda) \ll \lambda^\varepsilon$ (bounding the number of lattice points on an ellipse of size $\sqrt{\lambda}$). On the other hand, based on the circle method, the following partial verification of (2.9) was obtained in [Bo2].

Proposition 2.10. *Let $d \geq 4$ and $p \geq \frac{2(d+1)}{d-3}$. Then (2.9) holds, thus*

$$M_{d,p}(\lambda) \ll \lambda^{\frac{d-2}{4} - \frac{d}{2p} + \varepsilon}.$$

Recall at this point the continuous counterpart of the problem considered here, which is the Tomas-Stein restriction theorem for the sphere, see [St] for instance.

Proposition 2.11. *Denote $S_{d-1} = \{x \in \mathbb{R}^d \,|\, |x| = 1\}$ with surface measure $\sigma = \sigma_{d-1}$. Assume μ a measure on S_{d-1} which is absolutely continuous wrt σ and such that $\frac{d\mu}{d\sigma} \in L^2(\sigma)$. Then the Fourier transform*

$$\widehat{\mu}(\xi) = \int_{S_{d-1}} e^{i\langle x, \xi\rangle} \mu(dx) \tag{2.12}$$

is in $L^p(\mathbb{R}^d)$ for

$$p \geq \frac{2(d+1)}{d-1} \tag{2.13}$$

and there is an inequality

$$\|\widehat{\mu}\|_{L^p(\mathbb{R}^d)} \leq c \left\|\frac{d\mu}{d\sigma}\right\|_{L^2(\sigma)} \tag{2.14}$$

in the range (2.13) for p.

The result is optimal.

Observe that the exponent (2.13) differs from the exponent in (2.8) for the lattice point case.

The proof of Prop. 2.10 combines the circle method and the interpolation arguments underlying Prop. 2.11. If ψ is given by (2.1), one has

$$\psi = \psi * K \tag{2.15}$$

where

(2.16)

$$K(x) \equiv \sum_{n \in S_\lambda} e^{i\pi i n.x} = \int_{\mathbb{T}} \Pi_{j=1}^d \left[\sum_{k=1}^N e^{2\pi i(kx_j + k^2 t)}\right] . e^{-2\pi i\lambda t} dt \quad (N \sim \sqrt{\lambda}).$$

One then performs a standard "major-case" type decomposition of $\mathbb{T}$, defining for dyadic values of $Q < N$ neighborhoods of rational numbers

(2.17)

$$V_{Q,s} = \{t \in \mathbb{T} \,|\, |t - \frac{a}{q}| \sim \frac{1}{N.2^s} \text{ for some } 1 \leq a < q,\ (a,q) = 1,\ Q \leq q < 2Q\}$$

and write essentially

$$K = \sum_{Q,s} \Lambda_{Q,s} \tag{2.18}$$

where

$$\Lambda_{Q,s} = \int_{V_{Q,s}} \Pi_{j=1}^d \left[\sum_{k=1}^N e^{2\pi i(kx_j + k^2 t)}\right] . e^{-2\pi i\lambda t} dt. \tag{2.19}$$

Estimates on $\|*\Lambda_{Q,s}\|_{L^{p'}\to L^p}$ are then obtained interpolating between $\|*\Lambda_{Q,s}\|_{L^2\to L^2}$ and $\| * \Lambda_{Q,s}\|_{L^1\to L^\infty}$ (as in the Tomas-Stein argument). Observe that

$$\| * \Lambda_{Q,s}\|_{L^1\to L^\infty} = \|\Lambda_{Q,s}\|_\infty. \tag{2.20}$$

The main point next is thus to bound (2.19) uniformly in x, which is done by processing the Gauss sums the usual way. See [Bo2] for details.

The case $d = 4$ in Prop. 2.10 uses Kloosterman's refinement of the circle method, consisting in exploiting cancelations for fixed modulus q, when summing over the $1 \le a \le q$, through a Kloosterman (for d even) or Salié sum (for d odd). For $d = 3$, no nontrivial result is gotten. To achieve this would require a so-called "double-Kloosterman refinement", where also cancelation in the summation over different moduli q are exploited. This problem is essentially equivalent to a circle method solution to the Linnik problem on representing integers by a sum of three squares, as Kloosterman did for four squares (cf. [Sar]).

Coming back to the eigenfunctions for $\mathbb{T}^d$, it is clear that any statement of the form (2.7) for some $p > 2$ implies the uniform integrability of the set of squares of L^2-normalized eigenfunctions

$$\bigcup_\lambda \{|\psi|^2 | \Delta\psi + \lambda\psi = 0,\ \|\psi\|_2 = 1\}. \tag{2.21}$$

Thus it would in particular follow that if $\{\psi_j\}$ is a sequence of eigenfunctions satisfying

$$\lim_{j\to\infty} \psi_j(x) = 0 \text{ for almost all } x \in \mathbb{T}^d \tag{2.22}$$

then

$$\lim \|\psi_j\|_2 = 0.$$

This fact is only known in $d = 2$, as a consequence of (2.3).

The following weaker result however was obtained by B. Connes [Co] in arbitrary dimension.

Proposition 2.23. *Let $\{\psi_j\}$ be a sequence of eigenfunctions of Δ on $\mathbb{T}^d$ such that*

$$\lim_{j\to\infty} \psi_j(x) = 0 \textit{ at each point } x \in \mathbb{T}^d. \tag{2.24}$$

Then

$$\lim \|\psi_j\|_2 = 0. \tag{2.25}$$

If (2.24) holds, then by Baire's theorem, $\{\psi_j\}$ is uniformly bounded on some open subset G of $\mathbb{T}^d$ and hence there is a smooth function $0 \le \tau \le 1$, $\tau \ne 0$ such that

$$\lim_{j\to\infty} \int_{\mathbb{T}^d} |\psi_j|^2(x)\,\tau(x)dx = 0. \tag{2.26}$$

To deduce (2.25) from (2.26), we may then use the following property related to the distribution of the lattice points on spheres in separated clusters, as mentioned in the introduction.

Lemma 2.27. *Let $\{\psi_j\}$ be a sequence of eigenfunctions (with eigenvalues tending to infinity) normalized in L^2. Then for all fixed $\varepsilon > 0$, one may extract a subsequence, which we still denote $\{\psi_j\}$, such that*

$$\psi_j = \psi_j' + \psi_j'' \tag{2.28}$$

with

$$\|\psi_j''\|_2 < \varepsilon \tag{2.29}$$

and where ψ_j', ψ_j'' may be written in the form

$$\psi_j' = \sum_\alpha \Big[\sum_{n\in\Omega_{j,\alpha}'} \widehat{\psi}_j(n)e^{in.x}\Big],\ \psi_j'' = \sum_{n\in\Omega_j''} \widehat{\psi}_j(n)e^{in.x} \tag{2.30}$$

where

$$\sup_{j,\alpha} \mathit{diam}\ \Omega_{j,\alpha}' < \infty \tag{2.31}$$

and

$$\lim_{j\to\infty} \min_{\alpha\neq\beta} \mathit{dist}\,(\Omega_{j,\alpha}', \Omega_{j,\beta}') = \infty \text{ and } \lim_{j\to\infty} \min_\alpha \mathit{dist}\,(\Omega_{j,\alpha}', \Omega_j'') = \infty. \tag{2.32}$$

From (2.32) one gets clearly

$$\begin{aligned} 0 = \lim \textstyle\int |\varphi_j|^2 &= \lim \left(\textstyle\int |\varphi_j'|^2\tau + \int |\varphi_j'|^2\tau\right) \\ &= \lim \left\{ \textstyle\sum_\alpha \int \Big| \sum_{n\in\Omega_{j,\alpha}'} \widehat{\varphi}_j(n)e^{in.x}\Big|^2 \tau + \int |\psi_j''|^2 \right\} \end{aligned}$$

so that by (2.26)

$$\lim_{j\to\infty} \sum_\alpha \int \Big| \sum_{n\in\Omega_{j,\alpha}'} \widehat{\varphi}_j(n)e^{in.x}\Big|^2 \tau = 0. \tag{2.33}$$

By (2.13) however, there is some constant $\alpha > 0$, depending on $\sup_{j,\alpha}$ diam $\Omega_{j,\alpha}'$ and on τ, such that

$$\int \Big| \sum_{n\in\Omega_{j,\alpha}'} \widehat{\varphi}_j(n)e^{in.x}\Big|^2 \tau \geq a \sum_{n\in\Omega_{j,\alpha}'} |\widehat{vp}_j(n)|^2. \tag{2.34}$$

Thus the left side of (2.33) is at least $a\|\varphi_j'\|_2^2 > a(1-\varepsilon)$ by (2.29) (let $\varepsilon = \frac{1}{2}$ say), a contradiction. This proves Prop. 2.23.

Another application of Lemma 2.27 is a result on quantum limits for the flat torus, appearing in D. Jacobson's forthcoming paper [Jac].

Call "quantum limit" any probability measure μ on $\mathbb{T}^d$ obtained as ω^*-limit

$$\mu = \lim_{j\to\infty} |\psi_j|^2$$

for a sequence $\{\psi_j\}$ of eigenfunctions of the Laplacian, normalized in L^2.

Proposition 2.35 *[Jac] Quantum limits for the flat torus in any dimension are absolutely continuous.*

Applying the lemma to the sequence $\{\psi_j\}$ with any fixed $\varepsilon > 0$. Thus again by (2.29), (2.32)

$$\mu = \lim_j \sum_\alpha \Big| \sum_{n\in\Omega_{j,\alpha}} \widehat{\psi}_j(n)\, e^{in.x}\Big|^2 + 0(\varepsilon) \tag{2.36}$$

where, by (2.31), the first term in (2.36) is obviously a trigonometric polynomial. This is a soft statement with no estimates.

For $d = 3$, quantum limits are in $L^4(\mathbb{T}^3)$, see [Jac].

We conclude this section with some comments on a recent work [Bl, Bo] on the behaviour of the error term for the number of lattice points inside a shifted ball. Fix $x \in \mathbb{R}^d$ and define for $R \to \infty$

$$N_x(R) = \#\{n \in \mathbb{Z}^d \,|\, |x-n| \leq R\} \tag{2.37}$$

and the appropriately normalized error term

$$F_x(R) = \frac{N_x(R) - R^d \mathrm{Vol}_d B(1)}{R^{\frac{d-1}{2}}}. \tag{2.38}$$

When $d = 2$, it was proved in [B-C-D-L] that for any fixed $x \in \mathbb{R}^d$, the random variable $F_x(R)$ considered for $R \in [0, T]$ has a limit distribution ν_x when $T \to \infty$, which density $P_x(\lambda)$ is an entire function of λ decaying, for real λ, faster than $\exp(-|\lambda|^{4-\varepsilon})$. In [Bl, Bo] the problem is studied for $d \geq 3$. Here one needs to make an assumption on the center x, which we assume to satisfy the diophantine condition

(2.39)

$$\left(\Pi_{j=1}^d |k_j|\right)^{1+\varepsilon} \left\| \sum_{j=1}^d k_j x_j \right\| > C_\varepsilon \text{ for all } k = (k_j)_{j=1,\cdots,d} \in \mathbb{Z}^d \backslash \{0\},\ \varepsilon > 0.$$

If x satisfies (2.39), again the limit distribution ν_x exists and has a tail distribution $\nu_x[|\lambda'| > \lambda] < \exp(-\lambda^{4-\varepsilon})$ for $\lambda \to \infty$. A question unanswered in [Bl, Bo] is whether ν_x is absolutely continuous (this is the case for randomly chosen center x and the density here is a smooth function). To settle this problem would require better estimates on the eigenfunctions

$$r_\lambda(x) = \sum_{|n|^2 = \lambda} e^{2\pi i n.x} \tag{2.40}$$

assuming x satisfies (2.39). The role of (2.40) results from the following calculation. Write by Poisson's summation formula

$$F_x(R) \sim \sum_{n \neq 0} |n|^{-\frac{d+1}{2}} \cos(2\pi|n|R - \phi) e^{2\pi i n.x} \tag{2.41}$$

which with the notation (2.40) we rewrite as

$$\sum_{m \text{ square free}} f_m(R\sqrt{m}) \tag{2.42}$$

with periodic functions $f_m(t)$ given by

$$f_m(t) = m^{-\frac{d+1}{4}} \sum_{k=1}^{\infty} k^{-\frac{d+1}{2}} r_{k^2 m}(x) \cos(2\pi k t - \phi). \tag{2.43}$$

The summands in (2.42) behave like independent random variables, due to the fact that the numbers $\{\sqrt{m} | m \in \mathbb{Z}_+,\ m \text{ square free}\}$ are rationally independent. To substantiate this requires to establish appropriate convergence properties for the series depending in particular on estimates for $r_\lambda(x)$. We prove that if x satisfies (2.39), then

$$|r_\lambda(x)| \ll \lambda^{\frac{d-1}{4}} + \varepsilon \tag{2.44}$$

while in fact $|r_\lambda(x)| \ll \lambda^{\frac{d-2}{4}+\varepsilon}$ should hold (by (2.5), this is the mean estimate). Observe that (2.44) is trivial for $d = 3$. Improving (2.44) would again result from a "double-Kloosterman" improvement to the circle method approach we mentioned earlier. Recent work on the Linnik problem permits us to establish strong equidistribution results for lattice points on the sphere $[n \in \mathbb{Z}^3 \,|\, |n|^2 = \lambda]$, $\lambda \to \infty$ (see [Sar]) but do not seem to yield nontrivial bounds on $r_\lambda(x)$ assuming x diophantine.

3. Uniqueness problem for spherical summation

Theorem 3.1 *Let $\sum_{n\in\mathbb{Z}^d} a_n\, e^{in.x}$ be a trigonometric series which converges spherically to 0 at any point x of the d-torus $\mathbb{T}^d$, i.e.*

$$\sum_{|n|<R} q_n\, e^{ix.n} \xrightarrow[R\to\infty]{} 0 \text{ for all } x \in \mathbb{T}^d. \tag{3.2}$$

Then $a_n = 0$ for all $n \in \mathbb{Z}^d$.

The case $d = 1$ is Cantor's uniqueness theorem, the case $d = 2$ results from subsequent work of V. Shapiro [S] and R. Cooke [C] and finally the general case $d \geq 3$ was settled in the author's paper [Bo1]. The general idea of the proof is as follows.

It follows in particular from (3.2) that

$$\psi_R(x) = \sum_{|n|=R} a_n\, e^{ix.n} \xrightarrow[R\to\infty]{} 0 \text{ for all } x \in \mathbb{T}^d. \tag{3.3}$$

Since $\{\psi_R\}$ are eigenfunctions of the Laplacian on the d-torus, B. Connes' result (Prop. 2.23) implies that

$$\|\psi_R\|_2^2 = \sum_{|n|=R} |a_n|^2 \xrightarrow[R\to\infty]{} 0. \tag{3.4}$$

Next one considers Riemann's function

$$F(x) = \sum_{n\neq 0} \frac{a_n}{|n|^2}\, e^{in.x} \tag{3.5}$$

(we assume $a_0 = 0$, which is always possible replacing $F(x)$ by $F(x) - F(kx)$, $k \neq 1$). It follows from (3.3) and (3.4) that (3.5) converges pointwise to F and $F \in L^2(\mathbb{T}^d)$. Moreover, one has the following facts

(*i*) As a consequence of (3.4), the set of discontinuities of F has Hausdorff-dimension at most $d - 2$.

(*ii*) As a consequence of (3.3), F is formally harmonic

$$\text{“}\Delta\text{”} F = 0. \tag{3.6}$$

If F were continuous, then fact (*ii*) would imply that F is constant, hence $F = 0$. The key point is that since the harmonic measure of a ρ-ball in $\mathbb{R}^d$

$$\omega(\rho\text{-ball}) \sim \rho^{d-2} \tag{3.7}$$

$d - 2$ is the critical dimension of removability of the discontinuities (for bounded harmonic functions). At this stage, fact (*i*) comes into play and in fact the more precise statement

$$\sum_{R<|n|<2R} |\widehat{F}(n)|^2 \leq \frac{\varepsilon_R}{R^2} \tag{3.8}$$

where

$$\lim_{R\to\infty} \varepsilon_R = 0. \tag{3.9}$$

The proof is completed by harmonic measure analysis and is basically a play-off between facts (i) and (ii).

4. Periodic Strichartz inequalities and applications to NLS

In the theory of the Cauchy problem for the time dependent Schrödinger equation, an important role is played by Strichartz's inequality for the linear Schrödinger equation $iu_t + \Delta u = 0$. The solution of the IVP

$$\begin{cases} iu_t + \Delta u = 0 \\ u(0) = \phi \end{cases} \tag{4.1}$$

with space variable $x \in \mathbb{R}^d$ is explicitly given by an oscillatory integral

$$u(x,t) = e^{it\Delta}\phi = \int_{\mathbb{R}^d} \widehat{\phi}(\xi)\, e^{i(x\xi + |\xi|^2 t)} d\xi. \tag{4.2}$$

Thus (4.2) may be seen as the Fourier transform of a density carried by the paraboloid $(\xi, |\xi|^2)$. One has the following basic inequality closely related to Prop. 2.11 for the sphere.

Theorem 4.3 (R. Strichartz [Str]).

If $\phi \in L^2(\mathbb{R}^d)$, then $e^{it\Delta}\phi \in L^p(\mathbb{R}^{d+1})$ for $p = \frac{2(d+2)}{d}$ and

$$\|e^{it\Delta}\phi\|_p \leq C_d \|\phi\|_2. \tag{4.4}$$

In the case of periodic boundary conditions, i.e. $\mathbb{T}^d$ replaces $\mathbb{R}^d$, (4.2) is replaced by the exponential sum

$$\sum_{n\in\mathbb{Z}^d} \widehat{\phi}(n)\, e^{i(n.x + |n|^2 t)} \tag{4.5}$$

and the problem is to find the periodic analogue of Theorem 4.3. This problem is far from understood, especially in dimension $d \geq 3$. Partial results are obtained by either direct calculation (when p is an even integer) or applying more analytical techniques and the circle method, much in the spirit of what was discussed in section 2 for the sphere. It turns out that in the periodic case, the exact analogue of Prop. 4.3 fails, as we will see below. The proper conjecture here is the following

$$\begin{cases} \|e^{it\Delta}\phi\|_{L^p(\mathbb{T}^{d+1})} \leq C_{p,d} \|\phi\|_{L^2(\mathbb{T}^d)} \text{ if } p < \frac{2(d+2)}{d} & (4.6) \\ \|P_N(e^{it\Delta}\phi)\|_{L^p(\mathbb{T}^{d+1})} \ll N^{\varepsilon} \|\phi\|_{L^2(\mathbb{T}^d)} \text{ if } p = \frac{2(d+2)}{d},\ \varepsilon > 0 & (4.7) \\ \|P_N(e^{it\Delta}\phi)\|_{L^p(\mathbb{T}^{d+1})} \leq C_{p,d} N^{(\frac{d}{2} - \frac{d+2}{p})} \|\phi\|_2 \text{ for } \frac{2(d+2)}{d} < p \leq \infty & (4.8) \end{cases}$$

where $P_N\psi = \sum_{|n|\leq N} \widehat{\psi}(n) e^{in.x}$ restricts the Fourier series to modes $|n| < N$.

For $d = 1$, (4.6) may be verified for $p = 4$, thus

$$\Big\| \sum_{n\in\mathbb{Z}} a_n\, e^{i(nx + n^2 t)} \Big\|_{L^4(\mathbb{T}^2)} \leq C\Big(\sum |a_n|^2\Big)^{1/2} \tag{4.9}$$

using a direct calculation in the spirit of the Cooke-Zygmund theorem for the circle. Also by arithmetical means, one verifies (4.7), thus

$$\Big\| \sum_{|n|<N} a_n\, e^{i(n.x+n^2 t)} \Big\|_{L^6(\mathbb{T}^2)} \ll N^\varepsilon \Big(\sum |a_n|^2 \Big)^{1/2}. \tag{4.10}$$

Here one needs to consider the number of solutions in $\mathbb{Z}^3$ of the system of equations

$$\begin{cases} n_1 + n_2 + n_3 = a \\ n_1^2 + n_2^2 + n_3^2 = b \end{cases} \tag{4.11}$$

with $|n_i| < N$. This clearly amounts to counting lattice points on ellipses $X^2 + 3Y^2 = A$ in $\mathbb{Z}^2$, leading to a bound by the divisor function $\exp \frac{\log N}{\log \log N}$.

As mentioned before, the inequality (4.10) does not hold if N^ε is replaced by a constant. For instance, one verifies that

$$\Big\| \sum_{n=0}^{N} e^{i(nx+n^2 t)} \Big\|_{L^6(\mathbb{T}^2)} \sim (\log N)^{1/6} \sqrt{N}. \tag{4.12}$$

It seems an interesting question is to determine the size of the factor in (4.10). A logarithmic divergence factor instead of the divisor function would have significant PDE applications.

For $\underline{d=2}$, (4.7) may be again verified by lattice point counting. Thus

$$\Big\| \sum_{n\in\mathbb{Z}^2, |n|<N} a_n\, e^{i(n.x+|n|^2 t)} \Big\|_{L^4(\mathbb{T}^3)} \ll N^\varepsilon \Big(\sum |a_n|^2 \Big)^{1/2}. \tag{4.13}$$

No result of the type (4.6) seems to be known for $d = 2$. It may be worth here to make the comparison with W. Rudin's well-known problem ([Ru]).

(4.14)

Is the set of the squares $\{n^2 | n \in \mathbb{Z}\}$ a Λ_p-set for some $p > 2$

(this fails for $p \geq 4$)?

For $\underline{d=3}$, (4.13) becomes

$$\Big\| \sum_{n\in\mathbb{Z}^3, |n|<N} a_n\, e^{i(n.x+|n|^2 t)} \Big\|_{L^4(\mathbb{T}^4)} \ll N^{\frac{1}{4}+\varepsilon} \Big(\sum |a_n|^2 \Big)^{1/2} \tag{4.15}$$

and one gets similarly the L^4-inequality for $d > 3$.

On the other hand, for $d \geq 4$, better results are gotten by the Hardy-Littlewood type analysis outlined in section 2. One has in particular the following

Proposition 4.16

(i) Inequality (4.8) holds if $p > \frac{2(d+4)}{d}$

(ii) Define

$$F(x,t) = \sum_{|n|\leq N} a_n\, e^{i(n.x+|n|^2 t)}, \ \sum |a_n|^2 \leq 1. \tag{4.17}$$

Then one has the distributional inequality

(4.18)

$$mes\{(x,t) \in \mathbb{T}^{d+1} \Big| |F(x,t)| > \lambda\} \ll N^\varepsilon \lambda^{-\frac{2(d+2)}{d}} \text{ for } \lambda > N^{\frac{d}{4}}.$$

Thus (4.18) is a partial verification of (4.7). Establishing the distributional inequality for small values of λ is usually harder, as appears for instance also in the theory of Dirichlet sums (see [Mo] for instance).

There are a certain number of applications of the preceding inequalities to the Cauchy problem for NLS with periodic boundary conditions. We consider wellposedness of the IVP

$$\begin{cases} iu_t + \Delta u + f'(|u|^2)u = 0 \\ u(0) = \phi \end{cases} \tag{4.19}$$

which is rewritten as the integral equation

$$u(t) = e^{it\Delta}\phi + i\int_0^t e^{i(t-\tau)\Delta}\,[f'(|u|^2)u](\tau)d\tau. \tag{4.20}$$

The approach followed is similar in the periodic case to what is done on the line or $\mathbb{R}^d$. Thus we solve (4.20) (locally in time) as a fixpoint problem, estimating the nonlinear term using the Strichartz inequalities for $e^{it\Delta}$. Because for periodic boundary conditions there is only a partial knowledge of those, the results here are not as complete. We list some in the next

Theorem 4.21

(1) $\underline{d=1}$ *(i) the equation* $iu_t + \Delta u \pm u|u|^2 = 0$ *(1D cubic NLS) is globally wellposed for data* $\phi \in H^s(\mathbb{T})$, $s \geq 0$.

 (ii) *The equation* $iu_t+\Delta u\pm u|u|^{p-2} = 0$, $p > 4$, *is locally wellposed for data* $\phi \in H^s(\mathbb{T})$, $s > 0$ *(provided compatible smoothness of nonlinearity* $u|u|^{p-2}$ *for large* s*). There is also global wellposedness for data* $\phi \in H^s(\mathbb{T})$, $s \geq 1$, *provided no blowup of the solution (for* $p \geq 6$ *in the focusing case* $iu_t + \Delta u + |u|u|^{p-2} = 0$*).*

(2) $\underline{d=2}$ *(i) The equation* $iu_t + \Delta u \pm u|u|^2 = 0$ *(2D cubic NLS) is locally wellposed for data* $\phi \in H^s(\mathbb{T}^2)$, $s > 0$.

 (ii) *Equations* $iu_t + \Delta u \pm u|u|^{p-2} = 0$ *are globally wellposed for all* p *and data* $\phi \in H^s(\mathbb{T})$, $s \geq 1$, *if no blowup occurs in* H^1*-norm (modulo the same comment on the smoothness of the nonlinearity as above).*

(3) $\underline{d=3}$ *There is global wellposedness of the cubic NLS* $iu_t + \Delta u \pm u|u|^2 = 0$ *for data* $\phi \in H^s(\mathbb{T}^3)$, $s \geq 1$, *provided no blowup of the solution in* H^1.

(4) $\underline{d=4}$ *Consider an NLS of the form* $iu_t+\Delta u+uf'(|u|^2) = 0$ *with* $f(t)$ *smooth and* $f(t) = 0(t^{3/2})$ *for* $t \to \infty$. *Then there is global wellposedness for data* $\phi \in H^s(\mathbb{T}^4)$, $s \geq 2$ *with sufficiently small* L^2*-norm.*

In the case (1), (2), (3), (4) global solutions $u(x,t)$ *for data* $\phi \in H^s$, $s > 1$ *will satisfy* $u(t) \in H^s$ *for all time (we assume again the nonlinearity* $u|u|^{p-2}$ *sufficiently smooth) and in cases discussed in (1), (2), (3), there is a derivative estimate of the form*

$$\|u(t)\|_{H^s} < (1+|t|)^{A.s} \text{ for } |t| \to \infty \tag{4.22}$$

(thus polynomial in time).

Remark The case of the $1D$ NLS $iu_t + \Delta u \pm u|u|^2 = 0$ with cubic nonlinearity is special because it is integrable (the Zakharov-Shabat theory). In this situation the Cauchy problem may be solved by other methods (inverse spectral techniques).

Most of the results discussed here may be found in [Bo3, Bo4]. A survey presentation appears in [Gi].

5. KAM problems for Hamiltonian PDEs

The classical KAM theory (Kolmogorov, Arnold, Moser) refers to the persistency of invariant tori of Hamiltonian systems

$$\begin{cases} \dot{I}_j = -\varepsilon \frac{\partial H_1}{\partial \varphi_j} \\ \dot{\varphi}_j = \frac{\partial H_0}{\partial I_j} + \varepsilon \frac{\partial H_1}{\partial I_j} \end{cases} \qquad (j = 1, \dots, J) \tag{5.1}$$

(written in action-angle variables (I, φ)) with Hamiltonian $H(I, \varphi) = H_0(I) + \varepsilon H_1(I, \varphi)$ obtained as perturbation of the integrable Hamiltonian $H_0(I)$.

Rather recently, there has been considerable interest in developing this theory in the context of an infinite dimensional phase space ($J = \infty$) and in particular to apply it to Hamiltonian PDEs. Our aim here is to establish the existence of solutions that are

- periodic (1 dim tori)
- quasi-periodic (finite dim tori)
- almost-periodic (infinite dim tori)

in time. We will only discuss briefly here the first two cases, $n\ell$ periodic and quasi-periodic solutions. Thus we study finite dimensional tori in infinite dimensional phase space and the main difficulty compared with the classical setting is the appearance of additional resonance problems.

Before going on, we briefly mention some model problems. The basic model would be that of a Hamiltonian perturbation of a parameter dependent linear equation, for instance the NLS

$$iu_t - \Delta u + V(x)u + \varepsilon \frac{\partial H}{\partial \overline{u}}(u, \overline{u}, x) = 0 \tag{5.2}$$

where $iu_t - \Delta u + V(x)u = 0$ is the linear equation with part of the spectrum of $-\Delta + V$ considered as parameters, letting the (real periodic) potential V range in an appropriate manifold say. The linear equation has quasi-periodic solutions of the form

$$u_0(x,t) = \sum_{j=1}^{b} a_j \varphi_j(x)\, e^{i\lambda_j t} \tag{5.3}$$

where $\{\lambda_1, \dots, \lambda_b\}$ are the distinguished eigenvalues of the Sturm-Liouville operators $-\Delta + V(x)$ (with specified boundary conditions, for instance Dirichlet, Neumann or periodic) which are considered as parameters in the problem and $\{\varphi_1, \dots, \varphi_b\}$ are the corresponding eigenfunctions

$$(-\Delta + V)\varphi_j = \lambda_j \varphi_j. \tag{5.4}$$

The problem is then to determine which of the quasi-periodic solutions (5.3) will persist for the perturbed equation. Persistency means that there are perturbed frequencies

$$\lambda_j' = \lambda_j + 0(\varepsilon) \qquad (j = 1, \dots, b) \tag{5.5}$$

and a quasi-periodic solution $u_\varepsilon = u_\varepsilon(x,t)$ with those frequencies, of the form

$$u_\varepsilon(x,t) = \sum_{\substack{m \in \mathbb{Z}^b \\ n \in \mathbb{Z}^d}} \widehat{u}_\varepsilon(m,n)\, e^{i\langle m, \lambda' \rangle t} \varphi_n(x) \tag{5.6}$$

with

(5.7) $$\widehat{u}_\varepsilon(e_j, j) = a_j \text{ for } j = 1, \dots, b \ (e_j = j^{th} \text{ unit vector or } \mathbb{Z}^b)$$

and the remainder

(5.8) $$\{\widehat{u}_\varepsilon(m,n)\big|_{(m,n)\notin S}\}; \ S = \{(e_j, j)\big|_{j=1,\dots,b}\}$$

small for $\varepsilon \to 0$ in an appropriate function space.

The function space considered here would be typically the real analytic or Gevrey classes, although many of the results probably carry over in a setting of finitely many derivatives (as in the classical KAM theory).

The specific model (5.2) is well understood in $1D$ (1 space dimension $d = 1$). A parameter dependent linear equation may sometimes be avoided by extracting the parameters from the nonlinearity itself, using normal form techniques and amplitude-frequency modulation. This procedure is applied successfully in the construction of quasi-periodic solutions of NLS of the form

(5.9) $$iu_t + u_{xx} + mu + |u|^2 u + (\text{higher order}) = 0$$

($1D$, S. Kuksin and J. Pöschel [K-P] assuming Dirichlet boundary conditions and by the author under periodic boundary conditions) and in $2D$, see [Bo7]

(5.10) $$iu_t + \Delta u + mu + |u|^2 u + (\text{higher order}) = 0$$

for certain frequency configurations.

Finally, there is also the model of perturbation of an integrable equation, for instance a perturbed KdV equation

(5.11) $$u_t + u_{xxx} + uu_x + \varepsilon \partial_x f(u) = 0$$

where the unperturbed frequency λ is given by $\nabla H_0(I_0)$, $H_0(I)$ being the (integrable Hamiltonian) corresponding to the (integrable) KdV equation $u_t + u_{xxx} + uu_x = 0$. Roughly speaking, the amplitude-frequency modulation amounts here to the usual twist property. See [K2] for the analysis of (5.11).

To solve these problems, there are presently two approaches. First, one may rework the classical KAM technology in this setting, the main idea being to remove the perturbation by a sequence of canonical transformations of the phase space. This approach is very Hamiltonian and the specific question considered gets quickly immersed in the general abstract setting. This theory is mainly due to S. Kuksin and an exposition of it may be found in his book [K1]. Its range of applicability is mainly restricted to $1D$ equations. This is due to certain non-resonance conditions required for this method to work (at least in this setting), ruling out multiplicities or near-multiplicities in the normal frequencies of the system (present in the higher dimensional case and in fact also in $1D$, considering periodic boundary conditions). Recall at this point for instance that the periodic spectrum of the $1D$ Sturm-Liouville operator

(5.13) $$-\frac{d^2}{dx^2} + V(x)$$

is a sequence $\lambda_0 < \lambda_1 \leq \lambda_2 < \lambda_3 \leq \lambda_4 < \dots < \lambda_{2i-1} \leq \lambda_{2i} < \dots$ where

(5.14) $$\lambda_{2i-1}, \lambda_{2i} = \pi^2 i^2 + 0(1)$$

(5.15) $$|\lambda_{2i} - \lambda_{2i-1}| \to 0 \text{ when } i \to \infty$$

the decay of $\lambda_{2i} - \lambda_{2i-1}$ being roughly comparable to that of the Fourier transform $\widehat{V}(i)$. The distinguished set of b modes corresponding to the b-dimensional invariant terms under consideration yields a subset of b-frequencies $\lambda^1, \lambda^2, \dots, \lambda^b \subset \{\lambda_i\}$ and the other frequencies will be referred to as normal frequencies $\{\mu_n\}$, appearing in nearby pairs when $n \to \infty$.

(This problem disappears considering Dirichlet boundary conditions, since then the frequencies $\lambda_{2i-1} \leq \mu_i \leq \lambda_{2i}$ are separated as $\pi^2 i^2$).

In higher dimension, one gets typically (considering NLS modules for instance)

$$\mu_n \sim |n|^2 \tag{5.16}$$

and there is the problem of many lattice points lying on a sphere of given radius. The second method to approach the persistency problem originates from a paper of W. Craig and E. Wayne [C-W] where it is applied to the model of a $1D$ wave equation with periodic boundary conditions of the form

$$y_{tt} - y_{xx} + \rho y + y^3 + \text{(higher order)} = 0 \tag{5.17}$$

in the construction of time periodic solutions. The method of [C-W] consists in a Lyapounov-Schmidt decomposition of the problem and a direct application of a Newton iteration scheme to construct approximate solutions of the equation. This technique avoids certain resonance problems that seem an artifax of the KAM approach discussed before. Presently, one may apply it in order to extend Kuksin's work in $1D$ to the setting of periodic boundary conditions (including quasi-periodic solutions). In fact, it yields a new approach to the classical KAM theory which is much less Hamiltonian.

In space dimension $d \geq 2$, many problems remain. The understanding of time periodic solutions (1 dim tori) both for NLS and NLW equations is pretty satisfactory in any dimension (a brief sketch of the argument will be given below). In $2D$, quasi-periodic solutions may be constructed for the NLS model, cf. (5.10), but significant difficulties due to resonances appear if one considers for instance $2D$ NLW equations. At this stage, the full range of applicability of this approach is not yet understood and is an active research area.

A summary of the Lyapounov-Schmidt method is given next. One expresses the equation as an infinite system passing to Fourier transform (using the eigenfunction basis in space and exponentials $e^{i\langle m,\lambda'\rangle t}$ in time). This system is split in a finite "resonant" piece (the Q-equations) and an infinite "non-resonant" part (the P-equations) which however involves small divisor problems. The general idea is to solve the P-equations (using a rapidly convergent scheme, such as Newton's method) and then substitute the solution in the finite system of Q-equations to determine the new frequencies $\lambda_1, \dots, \lambda_b$. The analytically "hard" part is the solution of the P-equations and is essentially non-Hamiltonian. Newton's method avoids the difficulties of divergent series appearing in a standard ε-expansion (only recently resolved in Eliassons' work [Gal]) because its double-exponentially fast convergence makes it less sensitive to the small-divisor phenomena. On the other hand, at each step of the iteration, one needs to deal with the inverse of a non-diagonal linear operator that appears by linearizing the equations. This linearization is singular and the role of separation of "singular sites" is very important for its analysis. As we will see, this separation stems from the distribution of lattice points on surfaces in separated clusters as described in the Introduction. To obtain the desired estimates requires certain restrictions in the parameter set, the parameter being for instance the unperturbed frequency vector $\lambda = (\lambda_1, \dots, \lambda_b)$. Here the conditions do not usually have an explicit diophantine nature and the result of the procedure is persistency of the solution (5,3) for λ taken in a (Cantor-like) set of positive measure, depending on ε in (5.2), and with complement of measure $\to 0$ when $\varepsilon \to 0$.

Technical details involved in carrying out the preceding are often long and complicated. A rather systematic exposition may be found in [Bo5]. The case of time periodic solutions is considerably simpler and already permits to explain many of the ideas, in particular the role of the separation of resonant sites when inverting the linearized operator.

In the remainder of this section we describe briefly the construction of time periodic solutions for NLW equations of the form

$$y_{tt} - \Delta y + \rho y + F(x,y) = 0, \quad \text{as} \quad y \to 0 \tag{5.18}$$

in dimension $d \geq 2$ (with periodic boundary condition). (The case $d = 1$ appears in [C-W]). We let $F(x,y) = y^3$ as model example.

The discussion will also show the flexibility of the [C-W] method to deal with certain resonant situations.

Replacing y by δy, (5.18) becomes

$$y_{tt} - \Delta y + \rho y + \delta^2 y^3 = 0 \tag{5.19}$$

appearing thus as perturbation of the linear equation

$$y_{tt} - \Delta y + \rho y = 0. \tag{5.20}$$

Fix $m_0 \in \mathbb{Z}^d \backslash \{0\}$ and consider (time periodic) solutions of (5.20), say

$$y_0(x,t) = p \cos(\langle m_0, x\rangle + \lambda_0 t) \text{ with } \lambda_0 = (|m_0|^2 + \rho)^{1/2}. \tag{5.21}$$

The amplitude p will be the parameter in this problem. We are looking for a perturbed solution y of (5.19) of the form

$$y(x,t) = \sum_{m \in \mathbb{Z}^d, n \in \mathbb{Z}_+} \widehat{y}(m,n) \cos(\langle m, x\rangle + n\lambda t) \tag{5.22}$$

where

$$\widehat{y}(m_0, 1) = p \tag{5.23}$$

$$|\lambda - \lambda_0| < 0(\delta) \tag{5.24}$$

$$\sum_{(m,n) \neq (m_0,1)} |\widehat{y}(m,n)| \exp(|m| + |n|)^c < 0(\delta).^* \tag{5.25}$$

We define the (extended) resonant set of Fourier modes

$$S_1 = \{(m, \pm 1) | m \in \mathbb{Z}^d, |m| = |m_0|\}. \tag{5.26}$$

With representation (5.22), equation (5.19) becomes

$$(-n^2\lambda^2 + \mu_m^2)\widehat{y}(m,n) + \delta^2 \widehat{y^3}(m,n) = 0 \tag{5.27}$$

where

$$\mu_m = (|m|^2 + \rho)^{1/2}. \tag{5.28}$$

---

*We let here $0 < c < 1$ be a fixed constant. One could probably proceed with real analytic norms (as in [C-W]) instead of a Gevrey norm as well.

Rewrite y as

(5.29)
$$y(x,t) = \sum_{\substack{m\in\mathbb{Z}^d\\ |m|=|m_0|}} p_m \cos(\langle m,x\rangle + \lambda t) + \sum_{(m,n)\notin S_1} \widehat{y}(m,n)\cos(\langle m,x\rangle + n\lambda t).$$

The P-equations (resp. Q-equations) are obtained restricting (5.27) to frequencies $(m,n)\notin S_1$ (resp. frequencies $(m,n)\in S_1$).

The strategy is as follows. Fix $\lambda=\lambda_0+0(\delta^2)$, $p_0=p_{m_0}=0(1)$ and $(p_m)_{\substack{|m|=|m_0|\\ m\neq m_0}} = 0(\delta^2)$.

Using the P-equation, we solve in $\widehat{y}|_{S_1^c}$. This will require some arithmetical assumptions on ρ in (5.19) and also the restriction of (λ,p) to a Cantor type set, constructed along the Newton iteration procedure. The function $\widehat{y}|_{S^c}$ will however be smoothly defined on the entire (λ,p)-parameter set. Next, substitute $\widehat{y}|_{S^c}$ in the remaining Q-equations

$$(-\lambda^2+|m_0|^2+\rho)p_m + \delta^2\widehat{y^3}(m,1)=0 \qquad (|m|=|m_0|) \tag{5.30}$$

leading to equations in $(p_m)_{|m|=|m_0|}$ and λ. Parametrizing in p_0, one obtains a solution of the form

$$p_m = 0(\delta^2) \text{ for } |m|=|m_0|, m\neq m_0 \tag{5.31}$$

$$\lambda^2 = |m_0|^2+\rho+\frac{3}{4}p_0^2\delta^2+0(\delta^4) \tag{5.32}$$

and an appropriate restriction of p_0 permits to fulfill the conditions on (λ,p) imposed in order to solve the P-equations. This is mainly possible because of the presence of the "twist" term $\frac{3}{4}p_0^2\delta^2$ in (5.32) providing amplitude-frequency modulation. As a conclusion, one gets the following result (see [Bo6]).

Proposition 5.33 *Let $\rho\in\mathbb{R}$ satisfy a condition of the form*

$$\Big|\sum_{j=0}^{r} a_j\rho^j\Big| > \Big(\sum|a_j|\Big)^{-C} \quad \textit{for all } \{a_j\}\in\mathbb{Z}^{r+1}\backslash\{0\} \tag{5.34}$$

and $r=r(d)$ sufficiently large. Consider the periodic wave equation in dimension d

$$y_{tt}-\Delta y+\rho y+y^3+F(x,y)=0. \tag{5.35}$$

with $F(x,y)=0(y^4)$ and periodic in x. Fix $m_0\in\mathbb{Z}^d\backslash\{0\}$. There is a Cantor set $\mathcal{C}$ of positive measure in an interval $[0,\delta]$ (which relative measure $\to 1$ for $\delta\to 0$) and for $p_0\in\mathcal{C}$ a solution of (5.36) of the form

$$y = p_0\cos(\langle m_0,x\rangle+\lambda t)+0(p_0^3) \tag{5.36}$$

where

$$\lambda^2=\lambda(p_0)^2=|m_0|^2+\rho+\frac{3}{4}p_0^2+0(p_0^4). \tag{5.37}$$

When applying Newton's method, one is led to consider the linearized operator corresponding to (5.27). This operator T has the form $T=D+T'$ where D is diagonal

$$D(m,n) = -n^2\lambda^2+\mu_m^2 \tag{5.38}$$

and T' is given by

$$T' = 3\delta^2 S_{y^2}. \tag{5.39}$$

Here, we denote S_ϕ the Toeplitz-operator

$$S_\phi\big((m_1,n_1),(m_2,n_2)\big) = \widehat{\phi}(m_1 - m_2, n_1 - n_2) \tag{5.40}$$

which is the lattice representation of the multiplication operator by ϕ. Thus T' depends on the approximative solution obtained at a given stage. Its main properties are fast off-diagonal decay as a consequence of (5.25) and stationarity in the sense that the matrix elements only depend on the difference of the sites (m_1, n_1) and (m_2, n_2). In order to perform the Newton iteration step, one needs to bound the inverse of a restriction $T_M = T|_{\substack{|m|<M\\|n|<M}}$ in a reasonable way say

$$\|T_M^{-1}\| < M^C \tag{5.41}$$

and establish off-diagonal decay estimates

$$|T_M^{-1}(x,y)| < e^{-|x-y|^c} \text{ for } |x-y| > M^{\frac{1}{2}}. \tag{5.42}$$

It follows in particular from (5.34) that ρ satisfies a diophantine condition

$$|a\rho + b| > c|a|^{-C} \quad (a \in \mathbb{Z}\backslash\{0\},\, b \in \mathbb{Z}) \tag{5.43}$$

and hence for $|m|, |n| < M$, $|m| \neq |m_0|$

(5.44)

$$|-n^2\lambda^2 + \mu_m^2| > |-n^2\lambda_0^2 + \mu_m^2| - n^2|\lambda^2 - \lambda_0^2| > M^{-C} - M^2\delta^2 > \frac{1}{2}M^{-C}$$

provided M is not too large compared with δ. This fact permits us to control the inverse operator T_M^{-1} for the initial steps in the iteration (first $\sim \log\frac{1}{\delta}$ steps). After that, further restrictions on (λ, p) need to be imposed.

Coming back to the definition of $T = D + T'$ with diagonal elements $D(m,n) = \lambda^2 n^2 - |m|^2$, the "singular sites" is the set of lattice elements

$$\Omega = \{(m,n) \in \mathbb{Z}^{d+1} \big|\, |\lambda^2 n^2 - |m|^2| < 1\} \tag{5.45}$$

and its part Ω_M contained in a box B_M of size M. If for instance we require λ to fulfill a condition (5.34), thus

$$\Big|\sum_{j=0}^{10d} a_j \lambda^j\Big| > \Big(\sum |a_j|\Big)^{-C} \quad \text{for all } \{a_j\} \in \mathbb{Z}^{10d+1}\backslash\{0\} \tag{5.46}$$

then the lattice points in Ω appearing nearby the cone (1.4) appear in separated clusters as described in the Introduction. More precisely, we write

$$\Omega_M = \bigcup_\alpha \Omega_\alpha \tag{5.47}$$

where

$$\text{diam } \Omega_\alpha < M^\delta \tag{5.48}$$

and

$$\text{dist }(\Omega_\alpha, \Omega_\beta) > M^{\delta'} \text{ for } \alpha \neq \beta \tag{5.49}$$

for some small $\delta = \delta(d) > 0$, $\delta' = \delta'(d) > 0$.

In the case of NLS, the cone in (5.45)

$$\lambda^2 z^2 = |\xi|^2 \tag{5.50}$$

is replaced by a paraboloid

$$\lambda z = |\xi|^2 \tag{5.51}$$

and the same partition result in clusters hold, without the assumption (5.46), due to the fact that the paraboloid has positive curvature.

Clearly assumption (5.34) on ρ will permit us to fulfill (5.46) for $|a_j| < \delta^{-C}$, since recalling (5.32)

$$\lambda = [|m_0|^2 + \rho + \frac{3}{4}p_0^2\delta^2 + 0(p_0^4\delta^4)]^{1/2}. \tag{5.52}$$

For larger values of $\{a_j\}$, (5.52) permits then to satisfy (5.46) by appropriate restriction of p_0.

It remains to explain the role of the partition (5.47) in controlling the inverse T_M^{-1}. The main ingredient here is a simple lemma going back to the work of J. Fröhlich and T. Spencer on Green's function estimates for lattice Schrödinger operators; see also J. Pöschel's paper [Pos] for the most general and precise formulation.

For our purpose, we state it in the following form

Lemma 5.53 *Let $T = D + T'$ be as above (in particular $\|T'\| = 0(\delta^2) = 0(1)$ by (5.39)). Let $B = B_M = M$-ball in $\mathbb{Z}^{d+1}$ and $\Omega_M = \Omega \cap B_M$, Ω given by (5.45), the set of singular sites in B_M. Thus*

$$|D(x)| \geq 1 \text{ for } x \in B_M \backslash \Omega_M. \tag{5.54}$$

Let further $\delta_1 > \delta_2 > \delta_3 > 0$ be some small constants and assume there is a partition

$$\Omega_M = \bigcup \Omega_\alpha \tag{5.55}$$

satisfying

$$\text{diam } \Omega_\alpha < M^{\delta_1} \tag{5.56}$$

$$\text{dist}(\Omega_\alpha, \Omega_\beta) > M^{\delta_2} \text{ for } \alpha \neq \beta \tag{5.57}$$

and

$$\|(T|_{\tilde{\Omega}_\alpha})^{-1}\| < M^{C_1} \text{ for all } \alpha \tag{5.58}$$

where $\tilde{\Omega}_\alpha$ denote a M^{δ_3}-neighborhood of Ω_α.

Then, recalling the off-diagonal decay property of T

$$|T'(x,y)| < e^{-|x-y|^c} \tag{5.59}$$

one gets

$$\|(T|_{B_M})^{-1}\| < M^{C_2} \tag{5.60}$$

and off-diagonal decay

$$|T_M^{-1}(x,y)| < e^{-\frac{1}{10}|x-y|^c} \text{ for } |x-y| > M^{10\varepsilon_1}. \tag{5.61}$$

In the present case, the partition (5.55) is produced by (5.47). This partition clearly depends on an initial choice of λ, which will be sufficiently restricted however so that we may assume the set Ω_M and the partition fixed.

In order to apply Lemma 5.53, it remains to fulfill conditions (5.58). The numbers of "islands" Ω_α is clearly at most M^{d+1}. Realizing (5.58) may easily be done, invoking first order eigenvalue variation (the operators are selfadjoint), the eigenvalues E of $T|_{\tilde{\Omega}_\alpha}$ being considered as functions of λ, p and a further restriction in the (λ, p) parameter set. Observe here that if $E_\Omega(\lambda, p)$ is an eigenvalue of $T|_\Omega$ where Ω is at distance $> K$ from 0 and containing a singular site, then clearly

$$\frac{\partial E_\Omega(\lambda,p)}{\partial p_0} = \frac{\partial E}{\partial \lambda}\frac{\partial \lambda}{\partial p_0} + \frac{\partial E}{\partial p}\frac{\partial p}{\partial p_0} \tag{5.62}$$

where

$$\left|\frac{\partial E}{\partial \lambda}\right| > K^2,\ \frac{\partial \lambda}{\partial p_0} = 0(\delta^2),\ \left|\frac{\partial E}{\partial p}\right| < 0(\delta^2),\ \frac{\partial p}{\partial p_0} = 0(1) \tag{5.63}$$

invoking (5.31), (5.32), (5,38), (5.39).

More details of the preceding may be found in [C-W], [Bo6] and [Bo5].

We conclude this section by observing that in the quasi-periodic case $(b > 1)$, the diagonal of the linearization is given by

$$-\langle n, \lambda\rangle^2 + \mu_m^2. \tag{5.64}$$

Even considering only finitely many frequencies μ_m (or simply ignoring μ_m), a more complicated structure of singular sites appears here. Their pattern has clearly a self-similar structure and the control of the inverse operator is done along a multi-scale analysis much in the spirit of [F-S-W]. See again [Bo5] for details.

6. Appendix

We outline the proof of the following extension of B. Connes' result

Proposition 6.1 *Let (ψ_j) be a sequence of solutions of the linear Schrödinger equation*

$$i\dot{\psi}_j + \Delta\psi_j = 0 \tag{6.2}$$

with periodic boundary conditions, such that

$$\lim_{j\to\infty} \psi_j(x) = 0 \text{ for every } x \in \mathbb{T}^d. \tag{6.3}$$

Then

$$\lim \|\psi_j\|_2 = 0. \tag{6.4}$$

(This statement is easily seen to imply Connes' result).

As part of the argument, we will prove the cluster separation of lattice points on the paraboloid. Let

$$P = \{(n, |n|^2) | n \in \mathbb{Z}^d\} \tag{6.5}$$

and denote $\Gamma : \mathbb{Z}^d \longrightarrow P$ the map $\Gamma(n) = (n, |n|^2)$.

Assume (6.4) does not hold, hence let $\|\psi_j\|_2 = 1$ after normalization.

As in Connes' argument, there is by Baire's theorem a smooth function $0 \le \tau \le 1$, $\tau \ne 0$ such that

$$\|\psi_j\|_2 = 1,\ \lim \int |\psi_j|^2 \tau = 0. \tag{6.6}$$

If ψ is a solution to (6.2), then

$$\sup \widehat{\psi} \subset P \tag{6.7}$$

and

$$\psi(x,t)=\sum_n a_n\, e^{i(n.x+|n|^2t)}. \tag{6.8}$$

Denote

$$\sigma(\psi)=\{n\in\mathbb{Z}^d \big| a_n\neq 0\} \tag{6.9}$$

and

$$\dim[\widehat{\psi}]=\dim[n_1-n_2 \big| n_1,n_2\in\sigma(\psi)]. \tag{6.10}$$

Assume

$$\dim[\widehat{\psi}_j]\leq d_1\leq d. \tag{6.11}$$

The main idea of the argument consists then in replacing the sequence $\{\psi_j\}$ by a sequence $\{\psi_j'\}$ of solutions of (6.2) for which (6.6) still holds and such that

$$\dim[\widehat{\psi}_j']\leq d_1-1. \tag{6.12}$$

The difficulty here compared with the sphere is that for lower dimensional sections of P, the curvature is not necessarily uniformly small, although the section is unbounded (if paraboloid) or are ellipsoids of diameter $\to\infty$.

We will use the following additional fact which is a general statement for strictly convex surfaces

Lemma 6.13 *Let $G\subset\mathbb{T}^{d+1}$ be a set of positive measure and ψ satisfying (6.2), $\|\psi\|_2\leq 1$. Then for all $\xi\in P$*

$$\log|\widehat{\psi}(\xi)|\leq\int_G\log|\psi(x)|+1. \tag{6.14}$$

Proof Using strict convexity of P, one may find $\nu\in\mathbb{Z}^{d+1}$ such that

$$\nu.\xi>\max_{\xi'\in P,\,\xi'\neq\xi}\nu.\xi' \tag{6.15}$$

Consider for fixed $x\in\mathbb{T}^{d+1}$, the function F_x on $\mathbb{T}$ given by

(6.16)
$$F_x(\theta)=\psi(x+\nu,\theta)=\widehat{\psi}(\xi)e^{i\xi.x}\,e^{i(\xi.\nu)\theta}+\sum_{\xi'\neq\xi}\widehat{\psi}(\xi')e^{i\xi'.x}e^{i(\xi'.\nu)\theta}.$$

It follows then from (6.15) and Jensen's inequality that

$$\log|\widehat{\psi}(\xi)|=\log|\widehat{F}_x(\xi.\nu)|\leq\int_{\mathbb{T}}\log|F_x(\theta)|d\theta \tag{6.17}$$

and integrating in x yields

(6.18)
$$\log|\widehat{\psi}(\xi)|\leq\int_{\mathbb{T}}\log|\psi(x)|\leq\int_G\log|\psi|+(1-|G|)\log\left[\frac{\int_{G^c}|\psi|}{1-|G|}\right]$$

and hence (6.14).

As an immediate consequence of (6.14) and the assumption (6.6), we get

$$\lim_{j\to\infty}\|\widehat{\psi}_j\|_\infty=0. \tag{6.19}$$

Lemma 6.20 *Let $\mathcal{V}\subset\mathbb{Z}^d$, $n_1,n_2\in\mathcal{V}$ and R sufficiently large such that*

$$|n_1-n_2|>R^{d+2}. \tag{6.21}$$

Then

$$\dim[n - n_1 | n \in \mathcal{V}, |\Gamma(n) - \Gamma(n_1)| < |\Gamma(n_1) - \Gamma(n_2)| + 10R] > \tag{6.22}$$
$$\min\{\dim[n - n_1 | n \in \mathcal{V}, |\Gamma(n) - \Gamma(n_1)| < R], \dim[n - n_2 | n \in \mathcal{V}, |\Gamma(n) - \Gamma(n_2)| < R]\}$$

Proof If (6.22) fails, then clearly the linear spaces

$$[n - n_1 || n \in \mathcal{V}, |\Gamma(n) - \Gamma(n_1)| < |\Gamma(n_1) - \Gamma(n_2)| + 10R]$$
$$= [n - n_1 | n \in \mathcal{V}, |\Gamma(n) - |\Gamma(n_1)| < R] = [n - n_2 | n \in \mathcal{V}, |\Gamma(n) - \Gamma(n_2)| < R] = H \tag{6.23}$$

coincide.

Let $k = \dim H$ and $H = [\Delta_1 n_1, \ldots, \Delta_k n_1]$ with for $\ell = 1, \cdots, k$

$$\Delta_\ell n_1 \in \mathbb{Z}^d, \ |\Gamma(n_1 + \Delta_\ell n_1) - \Gamma(n_1)| < R. \tag{6.24}$$

It follows from (6.24) that

$$\begin{cases} |\Delta_\ell n_1| < R & (6.25) \\ \textit{and} \\ \big| |n_1 + \Delta_\ell n_1|^2 - |n_1|^2 \big| < R & (6.26) \end{cases}$$

hence

$$|\langle n_1, \Delta_\ell n_1 \rangle| < R^2 \qquad (\ell = 1, \cdots, k). \tag{6.27}$$

Denote P_H the orthogonal projection on H. Write

$$P_H n_1 = \sum_{\ell=1}^{k} c_\ell (\Delta_\ell n_1). \tag{6.28}$$

Then by (6.27), (6.28)

$$R^2 > \left| \sum_{\ell'=1}^{k} c_{\ell'} \langle \Delta_{\ell'} n_1, \Delta_\ell n_1 \rangle \right| \text{ for } \ell = 1, \ldots, k. \tag{6.29}$$

Since the matrix

$$A = (\langle \Delta_\ell n_1, \Delta_{\ell'} n_1 \rangle)_{\ell, \ell' = 1, \ldots, k} \tag{6.30}$$

has integer coefficients and is non-singular,

$$|\det A| \geq 1. \tag{6.31}$$

$$|c_\ell| \lesssim R^{2k} \qquad (1 \leq \ell \leq k) \tag{6.32}$$

and thus by (6.27), (6.28), (6.32)

$$|P_H n_1|^2 \leq \sum_{\ell=1}^{k} c_\ell |\langle P_H n_1, \Delta_\ell n_1 \rangle| \lesssim R^{2(k+1)}$$
$$|P_H n_1| \lesssim R^{k+1} \tag{6.33}$$

Similarly, one gets

$$|P_H n_2| \lesssim R^{k+1} \tag{6.34}$$

hence

$$|P_H(n_1 - n_2)| \lesssim R^{k+1}. \tag{6.35}$$

By (6.23), $n_1 - n_2 \in H$ and (6.35) contradicts the assumption (6.21).

As a consequence of (6.20) for the lattice points of P in the partitioning property in separated clusters, we obtain the following

Lemma 6.36 *Given any large number K, there is a partition*

$$\mathbb{Z}^d = \bigcup \Omega_\alpha \tag{6.37}$$

satisfying

$$\textit{diam } \Gamma(\Omega_\alpha) < K^C \tag{6.38}$$

and

$$\textit{dist}\,\big(\Gamma(\Omega_\alpha), \Gamma(\Omega_\beta)\big) > K \textit{ for } \alpha \neq p \tag{6.39}$$

where $C = C_d$ is a fixed constant.

Proof The construction of such partition is clear provided we show that there are no chains of distinct lattice points

$$n_1, n_2, \dots, n_J \qquad J > K^C \tag{6.40}$$

with

$$|\Gamma(n_j) - \Gamma(n_{j+1})| < K \qquad (1 \le j < J). \tag{6.41}$$

Assume otherwise. We may moreover assume (passing to some subchain) that

$$d_1 \equiv \dim[n_j - n_{j'} \mid 1 \le j, j' \le J] = \dim[n_j - n_{j'} | j, j' \in I] \tag{6.42}$$

where I is any interval in $\{1, 2, \dots, J\}$ of size

$$|I| \ge J_1 \tag{6.43}$$

(J_1 to be specified). Clearly, by (6.41)

$$|\Gamma(n_j) - \Gamma(n_{j'})| \le K J_1 \text{ if } |j - j'| \le J_1 \tag{6.44}$$

and since the $\{n_j | j = 1, \dots, J\}$ are distinct

$$\text{diam } \{n_j | j = 1, \dots, J\} > J^{1/d}. \tag{6.45}$$

Let in lemma 6.20

$$\mathcal{V} = \{n_j | j = 1, \dots, J\} \tag{6.46}$$

and choose according to (6.45) elements $n^1, n^2 \in \mathcal{V}$ such that

$$|n^1 - n^2| > J^{1/d}. \tag{6.47}$$

Let K, J_1 satisfy

$$(K J_1)^{d+2} < J^{1/d}. \tag{6.48}$$

Taking $R = K.J_1$, (6.22) clearly contradicts (6.42), (6.43). This proves lemma 6.36.

Coming back to the sequence ψ_j satisfying (6.6), (6.11), fix a sequence $K_j \nearrow \infty$. For each j, Lemma 6.36 allows to write

$$\sigma(\psi_j) = \bigcup \Omega_{j,\alpha} \tag{6.49}$$

where

$$\text{diam } \Gamma(\Omega_{j,\alpha}) < K_j^C \text{ and dist}_{\alpha \neq \beta}\big(\Gamma(\Omega_{j,\alpha}), \Gamma(\Omega_{j,\beta})\big) > K_j. \tag{6.50}$$

Write then

(6.51)
$$\psi_j = \sum_\alpha \psi_{j,\alpha}$$

where

(6.52)
$$\psi_{j,\alpha} = \sum_{n\in\Omega_{j,\alpha}} \widehat{\psi}_j(n)\, e^{i(n.x+|n|^2t)}.$$

By (6.50) and since $K_j \nearrow \infty$, it follows that

(6.53)
$$\sum_{\alpha\neq\beta} \psi_{j,\alpha}.\overline{\psi_{j,\beta}} \xrightarrow{j\to\infty} 0(\text{weak}^*).$$

(Recall that $\|\psi_j\|_2 = 1$) and thus

(6.54)
$$(\omega^*)\lim|\psi_j|^2 = (\omega^*)\lim\sum_\alpha \|\psi_{j,\alpha}\|_2^2|\psi'_{j,\alpha}|^2$$

denoting

(6.55)
$$\psi'_{j,\alpha} = \frac{\psi_{j,\alpha}}{\|\psi_{j,\alpha}\|_2}.$$

Consequently, there is a sequence $\{\psi'_j = \psi'_{j,\alpha_j}\}$ still satisfying (6.6).

Our next aim (as in Connes' argument) is to perform the previous construction with lowering of dimension, i.e. (6.12) holds

(6.56)
$$\dim[\widehat{\psi'_j}] \leq d_1 - 1.$$

Define

(6.57)
$$R_j = \min\{R | \dim[n-a | n\in\sigma(\psi_j), |\Gamma(n)-\Gamma(a)| < R] = \dim[\widehat{\psi}_j] = d_1, \quad \text{for some } a\in\sigma(\psi_j)\}$$

If $\lim R_j = \infty$, it suffices to take in the previous construction K_J such that $K_j^C < \frac{1}{2}R_j$, since then diam $\Gamma(\Omega_{j,\alpha}) < \frac{1}{2}R_j$, hence

(6.58)
$$\dim\sigma(\psi_{j,\alpha}) = \dim[\Omega_{j,\alpha}] < d_1.$$

Assume otherwise $R_j \leq R_0$ bounded (passing to some subsequence).

Thus for each j, there is $a_j \in \sigma(\psi_j)$ such that

(6.59)
$$\dim[n-n_j | n\in\sigma(\psi_j),\ |\Gamma(n)-\Gamma(a_j)| < R_0] = d_1.$$

Obviously, invoking (6.19), one has for any fixed number R'

(6.60)
$$\sum_{n\in B(a_j,R')\cap\mathbb{Z}^d} \left|\widehat{\psi}_j(n,|n|^2)\right| \lesssim (R')^d\|\widehat{\psi}_j\|_\infty \xrightarrow{j\to\infty} 0.$$

Passing thus to a subsequence, one may assume

(6.61)
$$\|\psi_j - \eta_j\|_2 \to 0$$

where

(6.62)
$$\eta_j = \sum_{n\in\sigma(\psi_j)\backslash B(a_j,j)} \widehat{\psi}_j(n,|n|^2)e^{i(n.x+|n|^2t)}$$

and thus also

$$\int |\eta_j|^2 \tau \to 0. \tag{6.63}$$

We now apply the previous partitioning construction to the sequence $\{\eta_j\}$ instead of $\{\psi_j\}$, letting the sequence $\{K_j\}$ satisfy

$$K_j^{C(d+2)} < j. \tag{6.64}$$

The sets $\Omega_{j,\alpha} \subset \mathbb{Z}^d \backslash B(a_j, j)$, diam $\Omega_{j,\alpha} < K_j^C$ necessarily satisfy

$$\text{diam } [\Omega_{j,\alpha}] < d_1. \tag{6.65}$$

This follows from lemma 6.20. Let $\mathcal{V} = \sigma(\eta_j), n_1 = a_j$ and $n_2 \in \Omega_{j,\alpha}$. Hence

$$|n_1 - n_2| > j > K_j^{C(d+2)}. \tag{6.66}$$

Taking (6.59) into account, (6.65) follows from (6.22), applying Lemma 6.20 with $R = K_j^C$.

This permits us to produce a sequence $\{\psi_j'\}$ of normalized solutions of (6.2) such that (6.6) and $\dim[\widehat{\psi_j'}] \leq d_1 - 1$ holds.

The proof of Prop. 1 is then easily completed.

Remark If $\{\psi_j\}$ is a normalized sequence of solutions of (6.2), then weak*-limits of $\{|\psi_j|^2\}$ are absolutely continuous (similarly to Prop. 2.36). One has indeed in the preceding that $|\eta_j|^2 \in \bar{c}\{|\psi_{j,\alpha}'|^2 | \alpha\}$ (the norm-closed convex hull), where $\dim[\widehat{\psi_j'}] \leq d_1 - 1$. Denote $\mathcal{S}_{d_1} = \{|\psi|^2 | \psi$ is a solution of (2), $\dim[\widehat{\psi}] \leq d_1$ and $\|\psi\|_2 \leq 1\}$ and K_{d_1} the ω^*-closure of $\mathcal{S}_{d_1}$. Thus $\{|\eta_j|^2\} \subset \bar{c}(\mathcal{S}_{d_1-1}) \subset \bar{c}(K_{d_1-1})$. By hypothesis say, K_{d_1-1} consists of absolutely continuous measures and hence is norm-separable in $M(\mathbb{T}^{d+1})$. Hence $\bar{c}(K_{d_1-1})$ is weak*-compact and hence (weak*) $\lim |\psi_j|^2 =$ (weak*) $\lim |\eta_j|^2 \in \bar{c}(K_{d_1-1})$, i.e. $K_{d_1} \subset \bar{c}(K_{d_1-1})$.

References

[Bo1] J. Bourgain, *Spherical summation and uniqueness of multiple trigonometric series*, IMRN, 1996, N3, 93-107.

[Bo2] J. Bourgain, *Eigenfunction bounds for the Laplacian on the n-torus*, IMRN, Vol. 3 (1996), 61-66.

[Bo3] J. Bourgain, *Restriction phenomena for lattice subsets and applications to nonlinear evolution equations I, Schrödinger equations*, Geometric and Functional Analysis (GAFA), Vol. 3, N2 (1993), 107-156.

[Bo4] J. Bourgain, *On the growth in time of higher Sobolev norms of smooth solutions of Hamiltonian PDE*, IMRN, 1996, N6, 277-304.

[Bo5] J. Bourgain, *Nonlinear Schrödinger Equations*, Park City Lectures 1995.

[Bo6] J. Bourgain, *Construction of periodic solutions of nonlinear wave equations in higher dimensions*, GAFA, Vol. 5, N4 (1995), 629-639.

[Bo7] J. Bourgain, *Quasi-periodic solutions of Hamiltonian perturbations of 2D linear Schrödinger equations*, preprint IHES (1994).

[Bl-Bo] P. Bleher, J. Bourgain, *Distribution of the error term for the number of lattice points inside a shifted ball, in Analytic Number Theory: Proc. of a Conference in 'Honor of H. Halberstam' 1996 Birkhauser, 141-153.*

[B-C-D-L] P. Bleher, Z. Cheng, F. Dyson, J. Lebowitz, *Distribution of the error term of the number of lattice points inside a shifted circle*, CMP 154 (1994), 433-469.

[C] R. Cooke, *A Cantor-Lebesgue theorem in two dimensions*, Proc. AMS 30 (1971), 547-550.

[Co] B. Connes, *Sur les coefficients des séries trigonométriques convergentes sphériquement*, C. R. Acad. Sci. Paris, Sér. A 283 (1976), 159-161.

[C-W] W. Craig, G. Wayne, *Newton's method and periodic solutions of nonlinear wave equations*, Comm. Pure and Appl. Math, 46 (1993), 1409-1501.

[F-S-W] J. Fröhlich, T. Spencer, R. Wittwer, *Localization for a class of one dimensional quasi-periodic Schrödinger operators*, CMP 132, N1 (1990), 5-25.

[Gal] G. Galavotti, *Invariant tori: a field theoretic point of view on Eliasson's work*, in Advances in Dynamical Systems and Quantum Physics, 117-132, Ed. R. Figari, World Scientific 1995.

[Gi] J. Ginibre, *Bourbaki Sem*, March 1996, (Asterisque, Soc. Math de France).

[G-S] A. Granville, T. Spencer, private communication.

[Jac] D. Jacobson, *Quantum limits of flat tori*, preprint to appear in Annals of Math.

[K1] S. Kuksin, *Nearby Integrable Infinite Dimensional Hamiltonian Systems*, Springer LNM 1556 (1993).

[K2] S. Kuksin, *Perturbation theory of conditionally periodic solutions of infinite dimensional Hamiltonian systems and its application to the Korteweg de Vries Equation*, Math. Sb. (NS), 136 (178):3 (1988), Math. USSR Sb., 64, 397-413 (1989).

[K-P] S. Kuksin, J. Pöschel, *Invariant Cantor manifolds of quasi-periodic oscillations for a nonlinear Schrödinger equation*, Annals, Vol. 1, 143, N1, 149-179 (1996).

[Mo] H. Montgomery, *Topics in multiplicative number theory*, Springer LNM 227 (1971).

[Pos] J. Pöschel, *On Fröhlich-Spencer estimates on Green's function*, Manuscripta Math. 70, 27-37.

[Ru] W. Rudin, *Trigonometric series with gaps*, J. Math. Mech. 9 (1960), 203-227.

[Sar] P. Sarnak, *Some applications of modular forms*, Cambridge Tracts in Math. 99 (1990).

[S] V. L. Shapiro, *Uniqueness of multiple trigonometric series*, Annals of Math. (2) 66 (1957), 467-480.

[St] E. Stein, *Oscillatory integrals in Fourier Analysis*, in Beijing lectures in Harmonic Analysis, Ann. of Math. Stud. 112, Princeton UP, 1986, 307-355.

[Str] R. Strichartz, *Restrictions of Fourier transforms to quadratic surfaces and decay of solutions of wave equations*, Duke Math. J. 44 (1977), 705-714.

[Zy1] A. Zygmund, *Trigonometric Series*, Cambridge UP (1968).

[Zy2] A. Zygmund, *A Cantor-Lebesgue theorem for double trignometric series*, Studia Math., 43(1972), 173-178.

IAS AND UNIVERSITY OF ILLINOIS

Contemporary Mathematics
Volume **208**, 1997

A SEMILINEAR WAVE EQUATION WITH DERIVATIVE OF NONLINEARITY CONTAINING MULTIPLE EIGENVALUES OF INFINITE MULTIPLICITY

JOSE F. CAICEDO[1] AND ALFONSO CASTRO[2]

Dedicated to Professor Victor Shapiro on the occasion of his 70th. birthday.

ABSTRACT. We give a sufficient condition for a semilinear wave equation to have a time periodic solution with space variable in $[0, \pi]$ and zero Dirichlet boundary condition. The period is an irrational multiple of π such that, under these boundary conditions, all the eigenvalues of the wave operator have infinite multiplicity. We allow the range of the derivative of the nonlinearity to include multiple eigenvalues.

1. INTRODUCTION.

Guided by second order boundary value problems such as

$$u'' + g(x, u) = 0 \qquad x \in [0, \pi], \quad u(0) = u(\pi) = 0, \tag{1.1}$$

a great deal of information has been discovered about semilinear equations of the form

$$Lu + Nu = 0, \tag{1.2}$$

where L is a linear operator with discrete spectrum. When L is defined by an elliptic operator, generally, the spectrum consists of eigenvalues of finite multiplicity. When L is defined by the wave operator

$$\partial_{tt} - \partial_{xx} \equiv \Box$$

acting on functions $u : [0, \pi] \times \mathbb{R} \to \mathbb{R}$ that satisfy

$$u(0, t) = u(\pi, t) = 0, \quad u(x, t) = u(x, t + T) \quad \text{for all} \quad (x, t) \in [0, \pi] \times \mathbb{R} \tag{1.3}$$

the spectrum of L is discrete if $T = \sqrt{d}\pi$ with d a positive integer. When d is the square of a positive integer only 0 is an eigenvalue of infinite multiplicity. This case

---

1991 *Mathematics Subject Classification.* Primary 35L20, 35L70: Secondary 47H17.

Key words and phrases. Wave equation, spectrum, multiple eigenvalues, infinite multiplicity, a priori estimates.

[1]This author thanks the Department of Mathematics of the University of North Texas for their kind hospitality and support.

[2] Partially supported by NSF Grant DMS 9215027.

The authors thank the referee for a very careful reading of the original manuscript.

has been extensively studied when the *numerical range of* N' omits 0 (See [2], [6], [8]). Studies in which 0 is allowed in the numerical range of N' can be found in ([3], [5], [6], [9]). When d is not the square of a positive integer all the eigenvalues have infinity multiplicity. This case is considered in [7] under the assumption that the numerical range of N' does not include any eigenvalue. Here we present a sufficient condition for the boundary value problem

$$u_{tt} - u_{xx} = h(u) + p(x,t) \qquad (x,t) \in [0,\pi] \times \mathbb{R}, \tag{1.4}$$

$$u(0,t) = u(\pi,t) = 0, \quad u(x,t) = u(x,t+\sqrt{2}\pi) \quad (x,t) \in [0,\pi] \times \mathbb{R}. \tag{1.5}$$

to have a solution. Our results are motivated by the developments in [3]. Our main result deals with the case where the numerical range of N' contains several eigenvalues of infinite multiplicity. We assume $h : \mathbb{R} \to \mathbb{R}$ to be a function of class C^1 such that

$$\lim_{|u|\to\infty} h'(u) = \lambda \in \mathbb{R} - \sigma(\Box), \tag{1.6}$$

where $\sigma(\Box) \equiv \{k^2 - 2j^2 \, : k = 1, \dots \quad j = 0, 1 \dots\}$, the set of eigenvalues of $\Box$ subject to (1.5). For the sake of simplicity in the calculations we assume, without loss of generality, that

$$\lambda > 0 \quad \text{and} \quad g(u) \equiv h(u) - \lambda u \quad \text{is bounded .} \tag{1.7}$$

Let m be a positive integer and $0 < a < 1$, such that

$$-m + a \leq h'(s) \leq m - a, \quad \text{for all} \quad s \in \mathbb{R}. \tag{1.8}$$

Let L^2 be the space of functions $v : [0,\pi] \times \mathbb{R} \to \mathbb{R}$ that are $\sqrt{2}\pi$-periodic in the second variable, and are square integrable on $[0,\pi] \times [0,\sqrt{2}\pi] \equiv \Omega$. Let H^1 be the Sobolev space of elements $u \in L^2$ with first order partial derivatives in L^2 and trace on $\{0,\pi\} \times \mathbb{R}$ equal to 0 (See [1]). The norm in L^2 is denoted by $\| \ \|$, and the norm in H^1 is given by $\|u\|_1^2 = \int_\Omega u_t^2 + u_x^2$. We denote by $\| \ \|_\infty$ the norm in $L^\infty([0,\pi] \times \mathbb{R}) = L^\infty$.

For each $\rho = k^2 - 2j^2 \in \sigma(\Box)$, we will denote by α_{kj}, β_{kj} the eigenfunctions associated to ρ defined by

$$\alpha_{kj}(x,t) = \sin(kx)\sin(jt\sqrt{2}), \ \beta_{kj}(x,t) = \sin(kx)\cos(jt\sqrt{2}), \ (x,t) \in [0,\pi] \times \mathbb{R}.$$

For each positive integer $M \geq 2$ we denote by X_M the closed subspace of L^2 generated by the functions $\alpha_{kj}, \beta_{k,j}$, where $k^2 - 2j^2 \in (-M, M)$, and by Y_M its orthogonal complement. That is :

$$\begin{cases} X_M = \langle \{\alpha_{kj}, \beta_{kj} : k^2 - 2j^2 \in (-M,M)\} \rangle, \\ Y_M = \langle \{\alpha_{kj}, \beta_{kj} : k^2 - 2j^2 \notin (-M,M)\} \rangle = X_M^{\perp}. \end{cases} \tag{1.9}$$

Also we let P_M, Q_M denote the orthogonal projections of L^2 onto X_M, Y_M respectively.

The purpose of this paper is to establish a sufficient condition for the solvability of the problem (1.4)-(1.5). We prove the following result.

Theorem 1.2. *Let m, λ be as in (1.7)-(1.8). Let $p(x,t) = cq(x,t) + r(x,t)$ with $r \in X_m$, $q \in Y_m$, and $c \in \mathbb{R}$. We assume that*

$$r = \sum_{\substack{k=1,2,\dots \\ j=0,1,\dots}} a_{kj}\alpha_{kj} + b_{kj}\beta_{kj} \quad \textit{and} \sum_{\substack{k=1,2,\dots \\ j=0,1,\dots}} \frac{(a_{kj}^2 + b_{kj}^2)(k^2 + 2j^2)}{(k^2 - 2j^2 - \lambda)^2} < \infty \tag{1.10}$$

Let φ be the solution to the problem:

$$\varphi_{tt} - \varphi_{xx} - \lambda\varphi = q(x,t), \quad (x,t) \in [0,\pi] \times [0, \sqrt{2}\pi] \tag{1.11}$$

$$\varphi(0,t) = \varphi(\pi,t) = 0, \quad \textit{and} \quad \varphi(x,t) = \varphi(x, t + \sqrt{2}\pi). \tag{1.12}$$

Let μ denote the one-dimensional Lebesgue measure. If $\varphi \in H^1$ and

$$\mu(\{x \in [0,\pi] \ : |\varphi(x, \zeta \pm x)| \le \delta\}) \to 0 \quad as \quad \delta \to 0, \tag{1.13}$$

uniformly in ζ, then there exists a constant $c(r,q)$ such that if $c \ge c(r,q)$ the equation (1.4) – (1.5) *has a weak solution $u \in H^1 \cap L^\infty([0,\pi] \times \mathbb{R})$. That is u satisfies*

$$\int_\Omega \partial_x u \partial_x w - \partial_t u \partial_t w = \int_\Omega (h(u) + p) w$$

for all $w \in H^1$.

The reader can easily verify that Theorem 1.2 holds when $\sqrt{2}$, is replaced by $\sqrt{d}$ and d is an integer not in the set $\{1, 4, 9, \dots, n^2, \dots\}$. Also it should be noted that condition (1.13) holds for $q's$ in a dense set.

2. Preliminaries and Lyapunov-Schmidt Reduction

The spectrum of the wave operator $\Box$ is a proper subset of the set of integers $\mathbb{Z}$. Each eigenvalue has infinite multiplicity. That is, if $\rho = k^2 - 2j^2$ then there are infinitely many pairs of integers k_n, j_n that satisfy $k_n - 2j_n^2 = \rho$. For example, $-1 \in \sigma(\Box)$, and $-1 = k_n^2 - 2j_n^2$, for each pair of integers (k_n, j_n) satisfying the recurrence relation:

$$1 = k_0 = j_0, \quad k_{n+1} = 3k_n + 4j_n, \quad j_{n+1} = 3j_n + 2k_n$$

For each $f \in L^2$ with $f = \sum_{\substack{k=1,2,\dots \\ j=0,1,\dots}} (a_{kj}\alpha_{kj} + b_{kj}\beta_{kj})$ we see that

$$\hat{f} = \sum_{\substack{k=1,\dots \\ j=0,1,\dots}} \frac{a_{kj}\alpha_{kj} + b_{kj}\beta_{kj}}{k^2 - 2j^2}$$

satisfies

$$\int_\Omega \hat{f}(\Box\psi) = \int_\Omega f\psi$$

for each function φ of class C^2 satisfying (1.5). That is $\hat{f}$ is a weak solution to the problem $\Box\hat{f} = f$, (1.5). Thus we define

$$\Box^{-1} f \equiv \hat{f}. \tag{2.1}$$

Thus for $u \in Y_m$ (See (1.9)) we have

$$\|\Box^{-1} u\| \leq \frac{1}{m}\|u\| \tag{2.2}$$

and

$$\|\Box^{-1} u\|_1 \leq \frac{1}{m}\|u\|_1 \tag{2.3}$$

Let $p(x,t)$ and φ be as in Theorem 1.2. We note that u satisfies (1.4) if and only if $w = u - c\varphi$ satisfies

$$\begin{aligned} w_{tt} - w_{xx} &= u_{tt} - u_{xx} - c(\varphi_{tt} - \varphi_{xx}) \\ &= \lambda u + g(u) + cq + r - c(\lambda\varphi + q) \\ &= \lambda(u - c\varphi) + g(u) + r = \lambda w + g(w + c\varphi) + r. \end{aligned} \tag{2.4}$$

Because $\lambda \notin \sigma(\Box)$, for $u \in L^2$ with

$$u = \sum_{\substack{k=1,2,\ldots \\ j=0,1,\ldots}} a_{kj}\alpha_{kj} + b_{kj}\beta_{kj} \quad ,$$

then

$$A(u) = \sum_{\substack{k=1,2\ldots \\ j=0,1\ldots}} \frac{a_{kj}\alpha_{kj} + b_{kj}\beta_{kj}}{k^2 - 2j^2 - \lambda} \tag{2.5}$$

defines a continuous linear transformation from L^2 into L^2 as well as from H^1 into H^1. Let $\Gamma = \max\{|(k^2 - 2j^2 - \lambda)|^{-1}; \quad k^2 - 2j^2 \in \sigma(\Box)\} > 0$ we have

$$\|A(u)\| \leq \Gamma\|u\| \quad \text{for all} \quad u \in L^2, \quad \text{and} \quad \|A(u)\|_1 \leq \Gamma\|u\|_1 \tag{2.6}$$

If $M > \lambda$ then for all $u \in Y_M \cap H^1$ we have

$$\|A(u)\| \leq \frac{1}{M-\lambda}\|u\| \quad \text{and} \quad \|A(u)\|_1 \leq \frac{1}{M-\lambda}\|u\|_1. \tag{2.7}$$

Lemma 2.1. *There exists a continuous function $\psi : X_m \to Y_m$ such that $z = \psi(v)$ if and only if*

$$z = AQ_m\big(g(v+z+c\varphi)\big) \tag{2.8}$$

Proof. First we note that $z \in Y_m$ satisfies (2.8) if and only if

$$z = \square^{-1}\Big(Q_m(\lambda z + g(c\varphi + v + z))\Big). \tag{2.9}$$

Let $F : X_m \times Y_m \to Y_m$ be the function defined by $F(u,v) = \square^{-1}Q_m(\lambda z + g(v + z + c\varphi))$. It is clear that F is a continuous function. We next show that F is a contraction in the second variable. Indeed, for a fixed $v \in X_m$ and $z_1, z_2 \in Y_m$ we have

$$\begin{aligned}\|F(v,z_2) - F(v,z_1)\| &= \|\square^{-1}Q_m(g(z_2+v+c\varphi) - g(z_1+v+c\varphi) + \lambda(z_2 - z_1))\| \\ &\le \frac{1}{m}\|Q_m(\lambda(z_2-z_1) + g(z_2+v+c\varphi) - g(z_1+v+c\varphi)\| \\ &\le \frac{1}{m}\|g(z_2+v+c\varphi) - g(z_1+v+c\varphi) + \lambda(z_2-z_1)\| \\ &= \frac{1}{m}\|h(z_2+v+c\varphi) - h(z_1+v+c\varphi)\| \\ &\le \frac{m-a}{m}\|z_2 - z_1\|.\end{aligned} \tag{2.10}$$

Where we have used (2.2), and the fact that $\|h(u_1) - h(u_2)\| \le (m-a)\|u_1 - u_2\|$ for all $u_1, u_2 \in L^2$ (See (1.8)). Since $\frac{m-a}{m} < 1$ by the contraction mapping principle (See [4]) it follows that such a ψ exists, which proves the lemma. ∎

Lemma 2.2. *Let v and z be as in Lemma 2.1. If $v \in H^1$ then $z \in H^1$*

Proof. For l positive integer we define Y^l to be the subspace of L^2 generated by $\{\alpha_{kj}, \beta_{kj} \in Y_m;\ k < l,\ j < l\}$. We let Π_l we denote the orthogonal projection onto Y^l. Let $F_l : X_m \times Y^l \to Y^l$ be defined by $F_l(v,z) = \Pi_l(F(v,z))$, where F is the map defined in the proof of the Lemma 2.1. Arguing as in (2.10) we see that for each positive integer l there exists a unique $z_l \in Y^l$ such that $z_l = \Pi_l(F(v, z_l))$. Since Y^l's are finite dimensional the z_l's are of class C^∞. As in (2.8)-(2.9) we see that $z_l = \square^{-1}\Pi_l\Big(Q_m(\lambda z_l + g(c\varphi + v + z_l))\Big)$. Since $(z_l)_t \in Y^l$, we have

$$\begin{aligned}\int_\Omega (z_l)_t(z_l)_t &= \int_\Omega \Pi_l(\square^{-1}Q_m(\lambda z_l + g(c\varphi + z_l + v)))_t(z_l)_t \\ &= \int_\Omega (\Pi_l(\square^{-1}Q_m(h(c\varphi + z_l + v) - \lambda c\varphi)_t))(z_l)_t \\ &\le \frac{m-a}{m}(c\|\varphi_t\| + \|v_t\| + \|(z_l)_t\|)\|(z_l)_t\| + \frac{c\lambda}{m}\|\varphi_t\|\|(z_l)_t\|.\end{aligned}$$

Since $v, \varphi \in H^1$, $\|(z_l)_t\| \le \frac{1}{a}\Big(c(m-a+\lambda)\|\varphi_t\| + (m-a)\|v_t\|\Big) < \infty$. Also, because $\square z_l = \Pi_l\Big(Q_m(\lambda z_l + g(c\varphi + v + z_l))\Big)$, $\int_\Omega (z_l)_x(z_l)_x - (z_l)_t(z_l)_t = \int_\Omega(\lambda z_l + g(c\varphi + v +$

$z_l))z_l \leq \lambda\|z_l\|^2 + |g|_\infty\|z_l\||\Omega|^{1/2}$. Since $\|z_l\| \leq \Gamma|g|_\infty|\Omega|^{1/2}$, we see that $\{\|(z_l)_x\|$ is bounded. Thus $\{z_l\}$ is bounded in H^1. By Rellich-Kondrachov's Theorem (See [1, Theorem 6.2]) we may assume, without loss of generality, that $\{z_l\}$ converges weakly in H^1 and strongly in L^2 to $\widetilde{z} \in H^1$. Therefore, since $v \in H^1$, $\{AQ_m g(c\varphi + v + z_l)\}$ converges to $\{AQ_m g(c\varphi + v + \widetilde{z})\}$ and $\{\|AQ_m g(c\varphi + v + z_l)\|_1\}$ is bounded. Hence $(I - \Pi_l)(AQ_m g(c\varphi + v + z_l))$ converges to zero as l tends to infinity. Therefore $z_l = \Pi^l\Big(AQ_m\big(g(v + c\varphi + z_l)\big)\Big) \to AQ_m\big(g(v + c\varphi + \widetilde{z})\big) = \widetilde{z}$ as $i \to \infty$. Thus $\widetilde{z} = AQ_m g(c\varphi + v + \widetilde{z})$. By Lemma 2.1, $z = \widetilde{z} \in H^1$, which proves the lemma.

3 . Linear Estimates.

In this section we prove several regularity properties of the wave operator subject to the boundary condition (1.5). We make extensive use of Fourier series properties for elements of L^2. First we note that

$$\int_\Omega (P_M u)_t v_t = \int_\Omega u_t v_t, \qquad \text{for all} \quad u \in H^1,\ v \in H^1 \cap X^M, \tag{3.1}$$

and

$$\int_0^{\pi\sqrt{2}} v^2(x,s)\,ds \leq (\pi/2)\int_\Omega (v_z)^2(z,t)\,dzdt, \quad \text{for each} \quad x \in [0,\pi],\ v \in H^1. \tag{3.2}$$

For k, l positive integers we define:

$$I_{1kl} = \int_0^{\pi\sqrt{2}} \Big(\sqrt{l+k^2}\cos\big(t\sqrt{l+k^2}\big) - k\cos(kt)\Big)^2 dt,$$

$$I_{2kl} = \int_0^{\pi\sqrt{2}} \Big(\sqrt{l+k^2}\sin\big(t\sqrt{l+k^2}\big) - k\sin(kt)\Big)^2 dt.$$

If $l \leq k^2$ then we also define

$$I_{3kl} = \int_0^{\pi\sqrt{2}} \Big(\sqrt{k^2-l}\cos\big(t\sqrt{k^2-l}\big) - k\cos(kt)\Big)^2 dt,$$

$$I_{4kl} = \int_0^{\pi\sqrt{2}} \Big(\sqrt{k^2-l}\sin\big(t\sqrt{k^2-l}\big) - k\sin(kt)\Big)^2 dt.$$

Next we estimate these integrals assuming that $k^2 + l = 2j^2$ for some non-negative integer j if $i = 1, 2$, and $k^2 - l = 2j^2$ for some non-negative integer j if $i = 3, 4$.

Lemma 3.1. *Let k, l be as above. There is a constant $a > 0$ such that $I_{ikl} \le a(k + l^2)$ for $i = 1, 2, 3, 4$.*

Proof. We carry out in detail the proof for $i = 1$ only. The other three cases follow similarly. For the sake of simplicity in the notation we write $r = \sqrt{l + k^2}$. Expanding the integrand in I_{1kl} we have

$$I_{1kl} \le \frac{(2k^2 + l)\pi\sqrt{2}}{2} + \frac{k}{4} - kr \int_0^{\pi\sqrt{2}} \Big[\cos(t(r + k)) + \cos(t(r - k))\Big] dt \le (2k^2 + l)\frac{\pi\sqrt{2}}{2} + \frac{k}{4} - kr\Big(\frac{\sin(\pi\sqrt{2}(r + k))}{k + r} + \frac{\sin(\pi\sqrt{2}(r - k))}{r - k}\Big).$$

Since $\sin(x) \ge x - x^3$, for all $x \ge 0$, we have

$$I_{1kl} \le \frac{\pi\sqrt{2}(2k^2 + l)}{2} + \frac{k}{4} - kr\Big(\frac{-1}{r} + \pi\sqrt{2} - (\pi\sqrt{2})^3[r - k]^2\Big)$$

Since $[r - k]^2 = \left[\frac{l}{r+k}\right]^2 = \frac{l^2}{l+2k^2+2kr}$, and $\frac{kr}{l+2k^2+2kr} \le \frac{1}{4}$, we conclude

$$\begin{aligned} I_{1kl} &\le \frac{\pi\sqrt{2}(2k^2 + l)}{2} + \frac{5k}{4} - (k\pi\sqrt{2})r + (\pi\sqrt{2})^3 kr[r - k]^2 \\ &\le \frac{5k}{4} + \frac{l\pi\sqrt{2}}{2} + \frac{(\pi\sqrt{2})^3 l^2}{4}. \end{aligned}$$

Thus

$$I_{1kl} \le \frac{5k}{4} + \frac{\sqrt{2}(\pi^3 + \pi)l^2}{2}. \tag{3.3}$$

Hence $I_{1kl} \le a_1(k + l^2)$, where $a_1 > 0$ is independent of k, l. Similarly, there exist constants a_2, a_3, a_4 such that $I_{ikl} \le a_i(k + l^2)$. Taking $a_5 = max(a_1, a_2, a_3, a_4)$, the lemma is proven. For the cases $i = 3, 4$ one uses that if $k^2 - 2j^2 = l$ then

$$k - \sqrt{k^2 - l} = \frac{l}{\sqrt{k^2 - l} + k} \le \frac{l}{k}.$$

■

For $u \in L^2$ with $u = \sum_{k,j} a_{kj}\alpha_{kj} + b_{kj}\beta_{kj}$, and $\alpha \in (0, 1]$ we let

$$\|u\|_\alpha^2 = \sum_{k,j} (k^2 + 2j^2)^\alpha (a_{kj}^2 + b_{kj}^2) \tag{3.4}$$

Lemma 3.2. *For each $v \in H^1 \cap X_M$ there exists a function $w(x,t) = f(x+t) + f(x-t) \in H^1((0,\pi) \times [0, \pi\sqrt{2}])$ such that*

$$\|v_t - w_t\| \leq D\left(\|v\|_{\frac{1}{2}} + M^{\frac{3}{2}}\|v\|\right),$$

with D independent of v.

Proof. By the definition of X_M we have

$$v = \sum_{|k^2-2j^2|\leq M} a_{kj}\alpha_{kj} + b_{kj}\beta_{kj}.$$

We let

$$w = \sum_{|k^2-2j^2|\leq M} a_{kj}\sin(kx)\sin(kt) + b_{kj}\sin(kx)\cos(kt).$$

Since by Lemma 3.1 $\int_0^\pi (sin(kx))^2 dx I_{1kl} \leq a_5\pi(k+l^2)$ for $i = 1,2,3,4$, k,l positive integers we conclude that

$$\|(v-w)_t\|^2 \leq 2a_5\pi\left(\sum_{|k^2-2j^2|\leq M-1} (a_{kj}^2 + b_{kj}^2)(l^2+k).\right)$$

Now since $\sum_{l=1}^{M-1} l^2 \leq M^3$, then

$$\|v_t - w_t\| \leq (2a_5\pi)^{1/2}\left(M^{\frac{3}{2}}\|v\| + \|v\|_{\frac{1}{2}}\right).$$

This proves the Lemma.

Lemma 3.3. *Let A be as in* (2.5)*. If $u \in Y_M \cap H^1$, then*

$$\|A(u)\|_\infty \leq \sqrt{\pi}(M-\lambda)^{-\frac{3}{4}}\|u\|_1.$$

Proof. First we observe that for each positive integer s

$$\sum_{j=1}^{\infty} \frac{1}{4j^2+s} \leq \int_0^\infty \frac{dx}{4x^2+s} = \frac{\pi}{4\sqrt{s}}.$$

Therefore

$$\sum_{s\geq M}\sum_{j=1}^{\infty} \frac{1}{(s-\lambda)^2(4j^2+s)} \leq \sum_{s=M}^{\infty} \frac{\pi}{4(s-\lambda)^2\sqrt{s}} =$$

$$\frac{\pi}{4}\sum_{s\geq M}^{\infty}(s-\lambda)^{-2}s^{-\frac{1}{2}} \leq \frac{\pi}{4}\left(\frac{1}{M^{\frac{1}{2}}(M-\lambda)^2} + \int_M^\infty \frac{1}{(x-\lambda)^2 x^{-\frac{5}{2}}}dx\right)$$

$$\leq \frac{\pi}{2}(M-\lambda)^{-\frac{3}{2}}$$

Now for $u \in H^1 \cap Y_M$ writing $u = \sum_{|k^2-2j^2|\geq M} a_{kj}\alpha_{kj} + b_{kj}\beta_{kj}$ we have

$$
\begin{aligned}
|A(u(x,t))| &= \left| \sum_{|k^2-2j^2|\geq M} \frac{a_{kj}\alpha_{kj}(x,t) + b_{kj}\beta_{kj}(x,t)}{k^2-2j^2-\lambda} \right| \\
&\leq \left[\sum_{|k^2-2j^2|\geq M} (k^2-2j^2-\lambda)^{-2}(k^2+2j^2)^{-1} \right]^{1/2} \\
&\times \left[\left(\sum_{|k^2-2j^2|\geq M} a_{kj}^2(k^2+2j^2) \right)^{1/2} + \left(\sum_{|k^2-2j^2|\geq M} b_{kj}^2(k^2+2j^2) \right)^{1/2} \right] \\
&\leq \sqrt{\frac{\pi}{2}}(M-\lambda)^{-\frac{3}{4}} \left[2 \sum_{|k^-2j^2|\geq M} (a_{kj}^2 + b_{kj}^2)(k^2+2j^2) \right]^{\frac{1}{2}}
\end{aligned}
$$

which proves the Lemma. ■

For future reference we note that if $w \in L^2([0,\pi]\times[0,2\pi])$ and

$$
w(x,t) = \sum_{k=1}^{\infty} a_k \sin(kt)\sin(kx) + b_k \sin(kx)\cos(kt), \tag{3.5}
$$

then $\int_0^{2\pi}\int_0^{\pi} w^2 = 2\int_0^{\pi}\int_0^{\pi} w^2$. Also using elementary trigonometric identities one can see that if w is as in (3.5) then there exist 2π-periodic functions $a, b : \mathbb{R} \to \mathbb{R}$ such that

$$
w(x,t) = a(x+t) + b(t-x).
$$

Lemma 3.4. *Given $\tau > 0$ there exist $\delta_1 > 0$, $\delta_2 > 0$ such that if $\|u\|_\infty < \delta_1|c|$, and $|c| > \delta_2$ then:*

$$
I \equiv \int_\Omega \left[g'(c\varphi + u)\right]^2 w^2 dxdt \leq \tau^2\|w\|^2, \tag{3.6}
$$

for any $w(x,t) = a(x+t) + b(-x+t)$, with a, b locally integrable and 2π-periodic.

Proof. Let $\epsilon_1 > 0$ be such that $\epsilon_1(|g'|_\infty)^2 + \epsilon_1^2\pi < \frac{\tau^2}{32}$. By hypotesis (1.6) there exists s_1 such that if $|s| \geq s_1$ then $|g'(s)| < \epsilon_1$. By (1.13) there exists $\delta > 0$ such that $\mu(\{x; |\varphi(x,\xi+x)| < \delta\}) < \epsilon_1$, uniformly in ξ.
Now if $|\varphi(x,\xi+x)| \geq \delta$, $\|u\|_\infty \leq \frac{|c|\delta}{2} \equiv |c|\delta_1$, and $|c| \geq \frac{2s_1}{\delta} \equiv \delta_2$, we have:

$$
|c\varphi(x,x+\xi) + u(x,x+\xi)| \geq |c|\delta - \frac{|c|\delta}{2} = \frac{|c|\delta}{2} \geq s_1. \tag{3.7}
$$

Letting $\eta = x$, $\xi = t - x$ we see that

$$I_1 \equiv \int_\Omega \left[g'(c\varphi + u)(x,t))\right]^2 b^2(t-x)dxdt$$

(3.8)
$$\leq \int_0^\pi \int_{x-\pi}^{x+\sqrt{2}\pi} [g'(c\varphi + u)(x,t)]^2 b^2(t-x)dtdx$$

$$= \int_{-\pi}^{\sqrt{2}\pi} b^2(\xi) \int_0^\pi [g'(c\varphi+u)(\eta, \eta+\xi)]^2 d\eta d\xi.$$

For $\xi \in [(\sqrt{2}-1)\pi, \sqrt{2}\pi]$ let

$$W(\xi) = \int_0^\pi [g'(c\varphi+u)(\eta, \xi+\eta)]^2 d\eta$$

For $\xi \in R$ and $\delta > 0$ let $\gamma_{\xi,\delta} = \{x \in [0,\pi]; |\varphi(x, \xi+x)| < \delta\}$. Thus if $|c| \geq \delta_2$ and $\|u\|_\infty \leq |c|\delta_1$ then

$$W(\xi) = \int_{\gamma_{\xi,\delta}} [g'(c\varphi+u)(\eta, \xi+\eta)]^2 d\eta + \int_{[0,\pi]-\gamma_{\xi,\delta}} [g'(c\varphi+u)(\eta, \eta+\xi)]^2 d\eta$$

$$\leq \epsilon_1 |g'|_\infty^2 + \epsilon_1^2 \pi < \frac{\tau^2}{32}.$$

Therefore

(3.9)
$$I_1 \leq \int_{-\pi}^{\sqrt{2}\pi} b^2(\xi) W(\xi) d\xi \leq \frac{\tau^2}{32} \int_{-\pi}^{\sqrt{2}\pi} b^2(\xi) d\xi \leq \frac{\tau^2}{32} \int_0^{4\pi} b^2(\xi) d\xi.$$

By (3.8) and (3.9) we have

$$I_1 \leq \frac{\tau^2}{32} \int_{-\pi}^{\sqrt{2}\pi} b^2(\xi) d\xi \leq \frac{\tau^2}{32} \int_{-\pi}^{2\pi} b^2(\xi) d\xi \leq \frac{\tau^2}{16} \int_0^{2\pi} b^2(\xi) d\xi \leq \frac{\tau^2}{8} \int_0^\pi b^2(\xi) d\xi,$$

where we have used that

$$\int_0^\pi \int_0^{2\pi} b^2(t-x) dtdx = 2 \int_0^\pi b^2(\xi) \left[\int_0^\pi d\eta\right] d\xi = 2\pi \int_0^\pi b^2(\xi) d\xi.$$

In conclusion

(3.10)
$$I_1 \leq \frac{\tau^2}{8} \int_0^\pi \int_0^{2\pi} b^2(t-x) dxdt \leq \frac{\tau^2}{4} \|b\|^2.$$

Similarly
(3.11)

$$I_2 \equiv \int_\Omega \left[g'(c\varphi+u)(x,t))\right]^2 a^2(t+x) dxdt \leq \frac{\tau^2}{8} \int_0^\pi \int_0^{2\pi} a^2(x+t) dxdt \leq \frac{\tau^2}{4} \|a\|^2.$$

From (3.10) and (3.11) the lemma follows. ∎

Double checking the proof of Lemma 3.4 one sees that by taking $w = 1$ and ϵ_1 such that $\epsilon_1(|g'|_\infty)^4 + \epsilon_1^4 \pi < \frac{\tau^4}{32}$ then one can estimate $\|g^2(c\varphi + u)\|$. More precisely one has the following result.

Corollary 3.5. *Given $\tau > 0$ there exists $\delta_1 > 0, \delta_2 > 0$ such that if $\|u\|_\infty < \delta_1|c|$, and $|c| > \delta_2$ then :*

$$\Big(\int_\Omega g'(c\varphi + u)^4 dxdt\Big)^{\frac{1}{4}} \le \tau^4.$$

Lemma 3.6. *If $u \in X_M$ then*

$$\|u\|_\infty \le (32/15)^{1/2}(M-1)^{\frac{3}{8}}\|u\|_{\frac{3}{4}}.$$

Proof. Let $u \in X_M$, that is $u = \sum_{|k^2-2j^2|<M} a_{kj}\alpha_{kj} + b_{kj}\beta_{kj}$. Therefore

$$\|u\|_\infty \le \Big(\sum_{|k^2-2j^2|<M} (a_{kj}^2 + b_{kj}^2)(k^2+2j^2)^{\frac{3}{4}}\Big)^{\frac{1}{2}}\Big(\sum_{|k^2-2j^2|<M}(k^2+2j^2)^{-\frac{3}{4}}\Big)^{\frac{1}{2}}. \tag{3.12}$$

Let us see that the latter series converges. Since $\int_0^\infty (2x^2+l)^{-\frac{3}{4}}dx \le \frac{8}{5}l^{-\frac{1}{4}}$ for each positive integer, we have

$$\sum_{l=1}^{M-1}\sum_{\substack{k=1,\ldots\\ k^2-2j^2=-l}}(k^2+2j^2)^{-\frac{3}{4}} \le \sum_{l=1}^{M-1}\int_0^\infty (2x^2+l)^{-\frac{3}{4}}dx \le \sum_{l=1}^{M-1}\frac{8}{5}l^{-\frac{1}{4}}$$
$$\le \frac{8}{5}\Big[(M-1)^{\frac{3}{4}} - 1\Big]\frac{4}{3} \le \frac{32}{15}(M-1)^{-\frac{3}{4}}. \tag{3.13}$$

Also, for each positive integer l, $\sum_{k^2-2j^2=l}(k^2+2j^2)^{-\frac{3}{4}} = \sum_{j=0}^\infty (4j^2+l)^{-\frac{3}{4}} \le \int_0^\infty (4x^2+l)^{-\frac{3}{4}}dx \le \frac{6}{5}l^{-\frac{1}{4}}$. Thus

$$\sum_{l=1}^{M-1}\sum_{\substack{k^2-2j^2=l\\ j=0,\ldots}}(k^2+2j^2)^{-\frac{3}{4}} \le \sum_{l=1}^{M-1}\frac{6}{5}l^{-\frac{1}{4}} \le \int_0^\infty x^{-1/4}dx \le \frac{32}{15}(M-1)^{\frac{3}{4}} \tag{3.14}$$

By (3.13) and (3.14) we see that $\sum_{|k^2-2j^2|<M}(k^2+2j^2)^{-\frac{3}{4}} \le \frac{32}{15}(M-1)^{\frac{3}{4}}$. Combining this with (3.12) the lemma is proven.

4. Iterative process and proof of the Theorem 1.2.

Let $p, q, r, \lambda, \varphi$ be as in Theorem 1.2. For the sake of simplicity in the calculations we will assume, without loss of generality, that g **is bounded,** $g(0) = 0$**, and that** c **and** λ **are positive.**

We let $v_0 = 0$ and $z_1 \in Y_m$ be the only solution to

$$z_1 = AQ_m(g(0 + z_1 + c\varphi)) \tag{4.1}$$

(See Lemma 2.1). Inductively we define the sequences $v_0,\ v_1, \ldots$ and $z_1,\ z_2, \ldots$ by

$$v_n = AP_m(r + g(z_n + v_{n-1} + c\varphi)) \quad \text{and} \quad z_{n+1} = AQ_m(g(v_n + z_{n+1} + c\varphi)). \tag{4.2}$$

For future reference (See (1.8)) we note that for each $u \in H^1$

$$\|g(u)\|_1 \leq |g'|_\infty \|u\|_1 \quad \text{and} \quad \|h(u)\|_1 \leq (m-a)\|u\|_1. \tag{4.3}$$

We let $K = \Gamma |g'|_\infty \|\varphi\|_1 + 1$. Let δ_1, δ_2 be as in Lemma 3.4 for

$$\tau \leq \min\left\{1, \frac{1}{16\Gamma^2}, \frac{1}{160\Gamma}, \frac{a}{16m\Gamma}\right\}. \tag{4.4}$$

In the first place we establish:

Lemma 4.1. *Given $\epsilon > 0$ there exists $c_1(\epsilon) = c_1$ such that if $c > c_1$ then, for each positive integer n,*

$$\|z_{n+1} + v_n\|_1 \leq 3Kc \quad \textit{and} \quad \|z_{n+1} + v_n\|_\infty \leq c \min\{\delta_1, \epsilon\}. \tag{4.5}$$

Proof. Let us see that (4.5) holds for $n = 0$. By (2.3), (2.9), (4.1) and (4.3)

$$\begin{aligned}
\|z_1\|_1 = \|F(0, z_1)\|_1 &= \|\Box^{-1}\Big(Q_m(g(c\varphi + z_1) + \lambda z_1)\Big)\|_1 \\
&\leq \frac{1}{m}\|\lambda z_1 + g(c\varphi + z_1)\|_1 \\
&\leq \frac{1}{m}\|\lambda(z_1 + c\varphi) + g(c\varphi + z_1)\|_1 + \frac{c\lambda}{m}\|\varphi\|_1 \\
&\leq \frac{1}{m}\|h(z_1 + c\varphi)\|_1 + \frac{c\lambda}{m}\|\varphi\|_1 \\
&\leq \frac{m-a}{m}(c\|\varphi\|_1 + \|z_1\|_1) + \frac{\lambda\|\varphi\|_1}{m}c.
\end{aligned}$$

Therefore

$$\|z_1\|_1 \leq k_1 c \quad \text{with} \quad k_1 = \frac{m(\lambda+1)\|\varphi\|_1}{a} \tag{4.6}$$

We let $k_2 = |g'|_\infty(\|\varphi\|_1 + k_1)$. Now we choose M to be a positive integer such that

$$\begin{cases} M & \geq \lambda + (1 + 8\Gamma|g'|_\infty)(8\Gamma k_2 + (1 + |g'|_\infty)(k_5 + \|\varphi\|_1 + 3K)) \\ M & \geq \lambda + ((2k_6 + \|\varphi\|_1 + k_1)\max\{\frac{8}{\delta_1}, \frac{8}{\epsilon}\})^{4/3}, \end{cases} \tag{4.7}$$

where $k_5 \equiv \frac{(m-a+\lambda)(\|\varphi\|_1 + 3K)}{a}$ and $k_6 \equiv |g'|_\infty \sqrt{\pi}(M-\lambda)^{-\frac{3}{4}}(\|\varphi\|_1 + k_5 + 1.5K)$. Let $z_1 = x_1 + y_1$ with $x_1 \in X_M$, $y_1 \in Y_M$. By (2.7) and (4.3)

$$\begin{aligned}
\|y_1\|_1 = \|AQ_M(g(c\varphi + z_1))\|_1 &\leq \frac{1}{M-\lambda}\|g(c\varphi + z_1)\|_1 \\
&\leq \frac{|g'|_\infty}{(M-\lambda)}(c\|\varphi\|_1 + \|z_1\|_1) \\
&\leq \frac{|g'|_\infty}{(M-\lambda)}(c\|\varphi\|_1 + k_1 c) \\
&\leq \frac{|g'|_\infty(\|\varphi\| + k_1)}{M-\lambda}c \\
&= \frac{k_2}{M-\lambda}c.
\end{aligned} \tag{4.8}$$

Now by Lemma 3.3 and (4.6) we have

$$\begin{aligned}\|y_1\|_\infty &= \|AQ_M(g(c\varphi + z_1)\|_\infty \le \sqrt{\pi}(M-\lambda)^{-\frac{3}{4}}\|Q_M(g(z_1+c\varphi))\|_1 \\ &\le \sqrt{\pi}(M-\lambda)^{-\frac{3}{4}}\|g(c\varphi+z_1)\|_1 \\ &\le \sqrt{\pi}|g'|_\infty(M-\lambda)^{-\frac{3}{4}}(c\|\varphi\|_1+\|z_1\|_1) \\ &\le \sqrt{\pi}|g'|_\infty(M-\lambda)^{-\frac{3}{4}}(\|\varphi\|_1+k_1)c,\end{aligned} \tag{4.9}$$

for $M \ge 2$.

By interpolation properties of Sobolev spaces (See [1] and (3.4)) there exists $\alpha_1 \in (0,1)$, and $a_1 > 0$, such that $\|u\|_{\frac{3}{4}} \le a_1\|u\|^{\alpha_1}\|u\|_1^{1-\alpha_1}$, for each $u \in H^1$. Since $\|z_1\| = \|AQ_m(g(z_1+c\varphi))\| \le \Gamma\|g(z_1+c\varphi)\| \le \Gamma|g|_\infty\pi 2^{\frac{1}{4}}$ (See (2.6)), we have

$$\begin{aligned}\|x_1\|_{\frac{3}{4}} \le \|z_1\|_{\frac{3}{4}} &\le a_1\|z_1\|^{\alpha_1}\|z_1\|_1^{1-\alpha_1} \\ &\le a_1(\Gamma|g|_\infty\pi 2^{1/4})^{\alpha_1}(k_1c)^{1-\alpha_1} \equiv k_3c^{1-\alpha_1}.\end{aligned}$$

Hence, by Lemma 3.6

$$\|x_1\|_\infty \le k_4(M-1)^{\frac{3}{8}}c^{1-\alpha_1}, \tag{4.10}$$

where $k_4 = 2k_3$. From (4.6), (4.9) and (4.10) we obtain

$$\begin{aligned}\|z_1\|_\infty &\le \|x_1\|_\infty + \|y_1\|_\infty \\ &\le k_4(M-1)^{\frac{3}{8}}c^{1-\alpha_1} + 2\sqrt{\pi}(\|\varphi\|_1+k_1)(M-\lambda)^{-\frac{3}{4}}c \\ &\le \big(k_4(M-1)^{\frac{3}{8}}c^{-\alpha_1} + 4(\|\varphi\|_1+k_1)(M-\lambda)^{-\frac{3}{4}}\big)c.\end{aligned} \tag{4.11}$$

Thus if

$$c > \max\{\delta_2, \Big[\frac{2k_4(M-1)^{\frac{3}{8}}}{\epsilon}\Big]^{\frac{1}{\alpha_1}}, \Big[\frac{2k_4(M-1)^{\frac{3}{8}}}{\delta_1}\Big]^{\frac{1}{\alpha_1}}\}, \tag{4.12}$$

the by (4.7) we have

$$\|z_1\|_\infty < \min\{\epsilon, \delta_1\}c. \tag{4.13}$$

Next we estimate $\|(x_1)_t\|$. For the sake of simplicity in the notations we write $x = x_1,\ y = y_1$. Since A is a selfadjoint operator in L^2, and $x_t \in X_M$ we have

$$\begin{aligned}\int_\Omega x_tx_t &= \int_\Omega g'(c\varphi+x+y)(c\varphi_t+x_t+y_t)Ax_t \\ &\le c\Gamma\|x_t\||g'|_\infty(\|\varphi_t\| + \frac{k_2}{M-\lambda}) + \int_\Omega g'(c\varphi+x+y)x_tAx_t,\end{aligned} \tag{4.14}$$

(See (4.8)). Let now $w : \Omega \to \mathbb{R}$ be such that $\|x_t - w_t\| \le D(\|x\|_{\frac{1}{2}} + M^{\frac{3}{2}}\|x\|)$ (See Lemma 3.2). Thus, since $c > \delta_2$ by (4.4), Lemma 3.4 and (4.13)-(4.14)

$$\begin{aligned}
\int_\Omega g'(c\varphi + x + y)(x_t - w_t + w_t)Ax_t &\le \Big(\int_\Omega (g'(c\varphi + x + y)w_t)^2\Big)^{\frac{1}{2}}\Gamma\|x_t\| \\
&\quad + \Gamma|g'|_\infty\|x_t\|\|x_t - w_t\| \\
&\le \Gamma(\tau\|w_t\| + |g'|_\infty\|x_t - w_t\|)\|x_t\| \\
&\le \frac{\|x_t\|^2}{4} + \|x_t - w_t\|\|x_t\|(\tau + |g'|_\infty)\Gamma \\
&\le \frac{\|x_t\|^2}{4} + D(\|x\|_{\frac{1}{2}} + M^{\frac{3}{2}}\|x\|)\Gamma(\tau + |g'|_\infty)\|x_t\|.
\end{aligned} \tag{4.15}$$

By (4.1) $\|x\| \le \|z\| \le \Gamma|g|_\infty|\Omega|^{\frac{1}{2}}$ and, by (4.6), $\|x\|_1 \le k_1 c$. Also by interpolation properties of Sobolev spaces we have $\|x\|_{\frac{1}{2}} \le \|x\|^{\frac{1}{2}}\|x\|_1^{\frac{1}{2}}$ (See (3.4)). Combining these with (4.13)-(4.15) we conclude:

$$\begin{aligned}
\|x_t\| \le \frac{4}{3}\Gamma c\Big(&|g'|_\infty(\|\varphi_t\| + \frac{k_2}{M - \lambda}) \\
&+ (\tau + |g'|_\infty)(D(\Gamma^{1/2}|g|_\infty^{\frac{1}{2}}|\Omega|^{1/4}k_1^{1/2}c^{-\frac{1}{2}} + M^{\frac{3}{2}}\Gamma|g|_\infty|\Omega|^{1/2}c^{-1})\Big).
\end{aligned} \tag{4.16}$$

Therefore if

$$\Gamma(\tau + |g'|_\infty)D(\Gamma^{1/2}|g|_\infty^{\frac{1}{2}}|\Omega|^{1/4}k_1^{1/2}c^{-\frac{1}{2}} + M^{\frac{3}{2}}\Gamma|g|_\infty|\Omega|^{1/2}c^{-1}) < \frac{1}{8}, \tag{4.17}$$

then by (4.7)

$$\|x_t\| \le \frac{4}{3}c(\Gamma|g'|_\infty\|\varphi_t\| + \frac{1}{4}). \tag{4.18}$$

Now we estimate $\|x_x\|$. Multiplying z by $\Box x$ and integrating by parts (See (2.8), (2.9) and (4.1)) we have

$$\int_\Omega x_x^2 - x_t^2 = \int_\Omega g(c\varphi + x + y)x + \lambda x^2.$$

Using this, (4.18), and $\|z\| \le \Gamma|g|_\infty|\Omega|^{1/2}$ we have

$$\|x_x\| \le \frac{4}{3}c(\Gamma|g'|_\infty\|\varphi_t\| + \frac{1}{4}) + |g|_\infty\{|\Omega|\Gamma(\Gamma + \lambda)\}^{1/2} \le \frac{4}{3}c(\Gamma|g'|_\infty\|\varphi_t\| + 1) \tag{4.19}$$

if

$$c > |g|_\infty\{|\Omega|\Gamma(\Gamma + \lambda)\}^{1/2} \tag{4.20}$$

Hence (See (4.7), (4.8), (4.18) and (4.19))

$$\begin{aligned}\|z_1\|_1 &\le \|x_1\|_1 + \|y_1\|_1 \le \|x_x\| + \|x_t\| + \|y_1\|_1 \\ &\le \frac{4}{3}c(\Gamma 2|g'|_\infty\|\varphi_t\| + \frac{5}{4} + \frac{3k_2}{4(M-\lambda)}) \\ &\le \frac{8}{3}c(\Gamma|g'|_\infty\|\varphi_t\| + 1) \\ &\le (8/3)cK.\end{aligned} \tag{4.21}$$

Thus (4.5) holds for $n = 0$ provided that c satisfies (4.12), (4.17) and (4.20). This completes the proof of (4.5) for $n = 0$.

Now we suppose that (4.5) is true for $n \ge 0$. Hence

$$\|g(v_n + z_{n+1} + c\varphi)\|_1 \le |g'|_\infty(\|\varphi\|_1 + 3K)c.$$

Let $x_{n+1} = P_M(z_{n+1})$ and $y_{n+1} = Q_M(z_{n+1})$. Thus by (2.6) and (2.7)

$$\begin{aligned}\|y_{n+1}\|_1 &= \|AQ_M(g(v_n + z_{n+1} + c\varphi))\|_1 \\ &\le \frac{1}{M-\lambda}\|Q_M(g(v_n + z_{n+1} + c\varphi))\|_1 \\ &\le \frac{|g'|_\infty}{M-\lambda}(\|\varphi\|_1 + 3K)c.\end{aligned} \tag{4.22}$$

Differentiating v_{n+1} with respect to t, multiplying by $(v_{n+1})_t$ and integrating over Ω (See (4.2)) we obtain:

$$\int_\Omega (v_{n+1})_t^2 = \int_\Omega (g(z_{n+1} + v_n + c\varphi))_t A(v_{n+1})_t + Ar_t(v_{n+1})_t \tag{4.23}$$

$$\le \int_\Omega g'(c\varphi + z_{n+1} + v_n)(c(\varphi)_t + (z_{n+1})_t + (v_n)_t)A(v_{n+1})_t + \|Ar_t\|\|(v_{n+1})_t\|.$$

Let $u_n = v_n + x_{n+1}$. Let $w \in H^1$ (See Lemma 3.2) be such that $\|(u_n)_t - w_t\| \le D(\|u_n\|_{\frac{1}{2}} + M^{\frac{3}{2}}\|u_n\|)$. Thus

$$\int_\Omega g'(c\varphi + z_{n+1} + v_n)((u_n)_t - w_t + w_t)A(v_{n+1})_t \tag{4.24}$$

$$\le \Gamma\|(v_{n+1})_t\|\left(|g'|_\infty(D(\|u_n\|_{\frac{1}{2}} + M^{\frac{3}{2}}\|u_n\|) + (\int_\Omega (g'(c\varphi + z_{n+1} + v_n)^2(w_t)^2))^{1/2}\right).$$

Also

$$\int_\Omega g'(c\varphi + z_{n+1} + v_n)(y_{n+1})_t A(v_{n+1})_t \le \Gamma|g'|_\infty\|(y_{n+1})_t\|\|(v_{n+1})_t\|. \tag{4.25}$$

Combining (4.23),(4.25) and Lemma 3.4 we have

$$
\begin{aligned}
\|(v_{n+1})_t\| &\le \|Ar_t\| + \Gamma\Big(|g'|_\infty(c\|\varphi_t\| + \|(y_{n+1})_t\| \\
&\quad + D(\|u_n\|_{\frac{1}{2}} + M^{\frac{3}{2}}\|u_n\|)) + \tau\|w_t\|\Big) \\
&\le \|Ar_t\| + \Gamma|g'|_\infty\Big(\|y_{n+1}\|_1 + c\|\varphi_t\| + D(M^{\frac{3}{2}}\|u_n\| + \|u_n\|_{\frac{1}{2}})\Big) \\
&\quad + \Gamma\tau(\|(u_n - w)_t\| + \|(u_n)_t\|).
\end{aligned} \tag{4.26}
$$

By induction hypothesis $\|u_n\|_1 \le \|z_{n+1}\|_1 + \|v_n\|_1 \le 6Kc$. This, (4.22) and (4.26) and the choice of w given by Lemma 3.2 imply

$$
\begin{aligned}
\|(v_{n+1})_t\| &\le \|Ar_t\| + \Gamma|g'|_\infty c(\|(\varphi)_t\| + \frac{|g'|_\infty(\|\varphi\|_1 + 3K)}{M-\lambda}) \\
&\quad + D\Gamma(|g'|_\infty + \tau)(\|u_n\|_{\frac{1}{2}} + M^{\frac{3}{2}}\|u_n\|) + \Gamma\tau 6Kc.
\end{aligned} \tag{4.27}
$$

By the definition of z_{n+1} we have $\|x_{n+1}\| \le \|z_{n+1}\| \le \Gamma|g|_\infty|\Omega|^{1/2}$. Also $\|v_n\| \le \|Ar\| + \Gamma|g|_\infty|\Omega|^{1/2}$. Thus $\|u_n\|_{1/2} \le (6Kc)^{1/2}(\|Ar\| + 2\Gamma|g|_\infty|\Omega|^{1/2})^{1/2}$, by interpolation properties of Sobolev spaces. Replacing this in (4.27) we conclude

$$
\begin{aligned}
\|(v_{n+1})_t\| &\le c\Big(\|Ar_t\|c^{-1} + \Gamma|g'|_\infty(\|\varphi_t\| + \frac{|g'|_\infty(\|\varphi\|_1 + 3K)}{M-\lambda}) \\
&\quad + D(\Gamma(|g'|_\infty + \tau)((8K)^{\frac{1}{2}}c^{-\frac{1}{2}}(\|Ar_t\| + 2\Gamma|g|_\infty||\Omega|^{\frac{1}{2}})^{\frac{1}{2}} \\
&\quad + M^{\frac{3}{2}}(\|Ar_t\| + 2\Gamma|g|_\infty|\Omega|^{\frac{1}{2}})c^{-1} + 6K\Gamma\tau\Big)
\end{aligned}
$$

Now choosing c such that

$$
\begin{aligned}
\frac{K}{8} &\ge D\Gamma(|g'|_\infty + \tau)\Bigg((6K)^{1/2}c^{-1/2}(\|Ar\| + 2\Gamma|g|_\infty|\Omega|^{1/2})^{1/2} \\
&\quad + M^{\frac{3}{2}}(\|Ar\| + 2\Gamma|g|_\infty|\Omega|^{1/2})c^{-1}\Bigg) + \|Ar_t\|c^{-1},
\end{aligned} \tag{4.28}
$$

we have (See (4.4) and (4.7))

$$
\|(v_{n+1})_t\| \le 1.3Kc. \tag{4.29}
$$

Using Fourier series one sees that if $u, v \in X_m$, then $\int_\Omega Au\Box v \le m\Gamma\|u\|\|v\|$. Thus, multiplying (4.2) by $\Box v_{n+1}$, and integrating by parts we have

$$
\int_\Omega (v_{n+1})_x^2 \le \int_\Omega (v_{n+1})_t^2 + m\Gamma(\|r\| + |g|_\infty|\Omega|^{1/2})^2. \tag{4.30}
$$

Thus for

$$
c \ge \frac{5m\Gamma(\|r\| + |g|_\infty|\Omega|^{1/2})^2}{K} \tag{4.31}
$$

we have (See (4.29)) $\|(v_{n+1})_x\| \le 1.5Kc$. and thus

$$\|v_{n+1}\|_1 \le 2.8Kc \tag{4.32}$$

Let $x_{n+2} = P_M(z_{n+2})$ and $y_{n+2} = Q_M(z_{n+2})$. Imitating the arguments in (4.9) and using (4.32) one sees that

$$\|y_{n+2}\|_\infty \le \sqrt{\pi}(M-1)^{-\frac{3}{4}}|g'|_\infty(c\|\varphi\|_1 + \|z_{n+2}\|_1 + 2.8Kc). \tag{4.33}$$

Also arguing as in (4.6) and using (4.31) it is concluded that

$$\|z_{n+2}\|_1 \le c\frac{(m-a+\lambda)(\|\varphi\|_1) + 2.8K)}{a} \equiv k_5 c. \tag{4.34}$$

Similarly arguing as in (4.6), using (4.9) and (4.34) one establishes

$$\|y_{n+2}\|_1 \le (M-\lambda)^{-1}|g'|_\infty(\|\varphi\|_1 + k_5 + 2.8K)c. \tag{4.35}$$

Combining (4.34) and (4.35) we obtain

$$\|y_{n+2}\|_\infty \le |g'|_\infty\sqrt{\pi}(M-\lambda)^{-\frac{3}{4}}(\|\varphi\|_1 + k_5 + 1.5K)c \equiv k_6(M-\lambda)^{-3/4}c. \tag{4.36}$$

Let $u_{n+1} = v_{n+1} + x_{n+2} \in X_M$. Let a_1, α_1, be as in the statement following (4.9), by Lemma 3.5

$$\begin{aligned} \|u_{n+1}\|_\infty &\le k_4(M-1)^{\frac{3}{8}}\|u_{n+1}\|_{\frac{3}{4}} \le k_4(M-1)^{\frac{3}{8}}a_1\|u_{n+1}\|_1^{\alpha_1}\|u_{n+1}\|^{1-\alpha_1} \\ &\le k_4(M-1)^{\frac{3}{8}}(a_1(k_5+1.5K)c)^{\alpha_1}(\Gamma(\|r\| + 2|g|_\infty|\Omega|^{\frac{1}{2}}))^{1-\alpha_1} \\ &\equiv k_7(M-1)^{3/8}c^{\alpha_1}. \end{aligned} \tag{4.37}$$

Finally we estimate $\|x_{n+2}\|_1$ by estimating $\|u_{n+1}\|_1$. For the sake of simplicity in the notations we write $x = x_{n+2}$, $y = y_{n+2}$, $z = x+y$, $v = v_{n+1}$ and $u = u_{n+1} = x+v$. Since A is a selfadjoint operator in L^2, and $u_t \in X_M$ we have (4.35)

$$\int_\Omega u_t u_t = \int_\Omega g'(c\varphi + v + z)(c\varphi_t + v_t + z_t)Au_t \tag{4.38}$$

$$\le c\Gamma\|u_t\||g'|_\infty(\|\varphi_t\| + \frac{|g'|_\infty(\|\varphi\|_1 + k_5 + 2.8K)}{M-\lambda}) + \int_\Omega g'(c\varphi + u + y)u_t Au_t.$$

Let now $w : \Omega \to \mathbb{R}$ be such that $\|u_t - w_t\| \le D(\|u\|_{\frac{1}{2}} + M^{\frac{3}{2}}\|u\|)$ (See Lemma 3.2). Thus, by Lemma 3.4 and (4.36)-(4.37)

$$\begin{aligned} \int_\Omega g'(c\varphi + u + y)(u_t - w_t + w_t)Au_t &\le \Big(\int_\Omega (g'(c\varphi + u + y)w_t)^2\Big)^{\frac{1}{2}}\Gamma\|u_t\| \\ &\quad + \Gamma|g'|_\infty\|u_t\|\|u_t - w_t\| \\ &\le \Gamma(\tau\|w_t - u_t + u_t\| + |g'|_\infty\|u_t - w_t\|)\|u_t\| \\ &\le \frac{\|u_t\|^2}{4} + \|u_t - w_t\|\|u_t\|(\tau + |g'|_\infty)\Gamma \\ &\le \frac{\|u_t\|^2}{4} + D(\|u\|_{\frac{1}{2}} + M^{\frac{3}{2}}\|u\|)\Gamma(\tau + |g'|_\infty)\|u_t\| \end{aligned} \tag{4.39}$$

By (4.1) $\|u\| \leq \|v\| + \|z\| \leq \Gamma(\|r\| + 2|g|_\infty |\Omega|^{\frac{1}{2}}) \equiv k_8$ and, by (4.32) and (4.34), $\|u\|_1 \leq (k_5 + 2.8K)c$. Also by interpolation properties of Sobolev spaces we have $\|u\|_{\frac{1}{2}} \leq \|u\|^{\frac{1}{2}} \|u\|_1^{\frac{1}{2}}$ (See (3.4)). Combining these, (4.38) and (4.39) we conclude:

$$\begin{aligned} \|u_t\| \leq \frac{4}{3}\Gamma c\Big(|g'|_\infty(\|\varphi_t\| + \frac{|g'|_\infty(\|\varphi\|_1 + k_5 + 2.8K)}{M-\lambda}) \\ + (\tau + |g'|_\infty)(D(k_8^{\frac{1}{2}}|\Omega|^{1/4}(2.8K + k_5)^{1/2}c^{-\frac{1}{2}} \\ + M^{\frac{3}{2}} k_8 c^{-1})\Big). \end{aligned} \tag{4.40}$$

Therefore for c such that

$$\Gamma(\tau + |g'|_\infty)D(k_8^{1/2}(3K + k_5)^{1/2}c^{-\frac{1}{2}} + M^{\frac{3}{2}}k_8 c^{-1}) < \frac{1}{8}, \tag{4.41}$$

we obtain (See (4.7))

$$\|u_t\| \leq \frac{4}{3}c(\Gamma|g'|_\infty\|\varphi_t\| + \frac{1}{4}). \tag{4.42}$$

Now we estimate $\|u_x\|$. Multiplying v by $\Box v$, z by $\Box x$ (See (4.2)) and integrating by parts (See (2.8), (2.9) and (4.1)) yield

$$\int_\Omega u_x^2 - u_t^2 = \int_\Omega g(c\varphi + u + y)u + \lambda u^2.$$

This, (4.42), and $\|z\| \leq |g|_\infty |\Omega|^{1/2}$ yield
(4.43)

$$\|u_x\| \leq \frac{4}{3}c(\Gamma|g'|_\infty\|\varphi_t\| + \frac{1}{4}) + \{k_8(|g|_\infty|\Omega|^{1/2} + k_8\lambda)\}^{1/2} \leq \frac{4}{3}c(\Gamma|g'|_\infty\|\varphi_t\| + 1)$$

if

$$c > \{k_8(|g|_\infty|\Omega|^{1/2} + k_8\lambda)\}^{1/2}. \tag{4.44}$$

Hence (See (4.7), (4.42), (4.43) and (4.35))

$$\begin{aligned} \|v + z\|_1 &\leq \|u_x\| + \|u_t\| + \|y\|_1 \\ &\leq \frac{4}{3}c(\Gamma 2|g'|_\infty\|\varphi_t\| + \frac{5}{4} + (M-\lambda)^{-1}|g'|_\infty(\|\varphi\|_1 + k_5 + 2.8K)) \\ &\leq \frac{8}{3}c(\Gamma|g'|_\infty\|\varphi_t\| + 1) \\ &\leq (8/3)cK, \end{aligned} \tag{4.45}$$

Noting that there exists $c_1 > 0$, independent of n such that if $c > c_1$ then (4.12), (4.17), (4.20), (4.28), (4.31), $k_6(M-\lambda)^{-3/4} \leq \frac{1}{2}\min\{\epsilon, \delta_1\}$ (See (4.7)) and $k_7(M-1)^{3/8}c^{\alpha_1 - 1} \leq \frac{1}{2}\min\{\epsilon, \delta_1\}$. Thus by (4.36), (4.37) and (4.45), for $c > c_1$ (4.5) holds when n is replaced by $n+1$. This proves Lemma 4.1. ■

Proof of the Theorem 1.2. First we estimate the gap between consecutive z_n's. By hypothesis (1.8), and since $Q_m(v_n) = Q_m(v_{n-1}) = 0$, by (2.2),(2.9), and (1.8)

$$
\begin{aligned}
\|z_{n+1} - z_n\| &= \Big\|\Box^{-1} Q_m\Big(\lambda(z_{n+1} + v_n) + g(c\varphi + v_n + z_{n+1}) \\
&\qquad - \lambda(z_n + v_{n-1}) - g(c\varphi + v_{n-1} + z_n)\Big)\Big\| \\
&\leq \frac{1}{m}\|h(c\varphi + z_{n+1} + v_n) - h(c\varphi + z_n + v_{n-1})\| \\
&\leq \frac{m-a}{m}(\|v_n - v_{n-1}\| + \|z_n - z_{n+1}\|).
\end{aligned} \tag{4.46}
$$

Hence

$$
\|z_{n+1} - z_n\| \leq \frac{m-a}{a}\|v_n - v_{n-1}\| \quad \text{for each} \quad n = 1, 2, \ldots \tag{4.47}
$$

Let J be a positive integer such that

$$
J_1 \equiv (|g'|_\infty + 1)D[\Big(\frac{m + 4J^2}{4J^4}\Big)^{\frac{1}{4}} + \frac{m^{\frac{3}{2}}}{J\sqrt{2}}] \leq \frac{a}{16m\Gamma}, \tag{4.48}
$$

with D as in Lemma 3.2. Let $X \subset X_m$ be the subspace generated by the eigenfunctions $\{\alpha_{kj}, \beta_{kj};\ j \leq J,\ |k^2 - 2j^2| \leq m\}$. Let $P : L^2 \to X$ be the orthogonal projection onto X, and $Q = (I - P)P_m$. Thus by (4.2) and the mean value theorem we have

$$
\begin{aligned}
\|Q(v_n - v_{n-1})\|^2 &= \int_\Omega \Big(A(g(c\varphi + v_{n-1} + z_n) \\
&\qquad - g(c\varphi + v_{n-2} + z_{n-1})\Big)Q(v_n - v_{n-1}) \\
&= \int_\Omega A\Big(g(c\varphi + v_{n-1} + z_n) - g(c\varphi + v_{n-1} + z_{n-1}) \\
&\qquad + g(c\varphi + v_{n-1} + z_{n-1}) - g(c\varphi + v_{n-2} + z_{n-1})\Big)Q(v_n - v_{n-1}) \\
&= \int_\Omega A\Big(g'(c\varphi + v_{n-1} + \mu_1)(z_n - z_{n-1}) \\
&\qquad + g'(c\varphi + z_{n-1} + \mu_2)(v_n - v_{n-2})\Big)Q(v_n - v_{n-1}),
\end{aligned} \tag{4.49}
$$

here $\mu_1, \mu_2 : \Omega \to \mathbb{R}$ with $\|\mu_1\|_\infty \leq \max\{\|z_n\|, \|z_{n-1}\|\}$ and $\|\mu_2\|_\infty \leq \max\{\|v_{n-2}\|, \|v_{n-1}\|\}$. Let

$$
\begin{cases}
Q(v_n - v_{n-1}) = \displaystyle\sum_{\substack{|k^2 - 2j^2| \leq m \\ j > J}} a_{kj}\alpha_{kj} + b_{kj}\beta_{kj}, \\[2ex]
v = \displaystyle\sum_{\substack{|k^2 - 2j^2| \leq m \\ j > J}} (b_{kj}\alpha_{kj} - a_{kj}\beta_{kj})\frac{1}{j\sqrt{2}}.
\end{cases} \tag{4.50}
$$

Thus $v_t = Q(v_n - v_{n-1})$. Let $\tau \in (0, \min\{\frac{a}{16m\Gamma}, 1\})$. Let δ_1, δ_2 be as in Lemma 3.4. By Lemma 4.1 there exists $c_1(\delta_2)$ such that if $c > c_1(\delta_2)$ then $\|z_n\|_\infty < \frac{\delta_2 c}{2}$, $\|v_n\|_\infty < \frac{\delta_2 c}{2}$ for all $n = 1, 2, \dots$. Note that δ_2 is given by τ which in turn is determined by a, m which ultimately is defined by h. That is c_1 depends only on the function h. Let w be as in Lemma 3.2 when v is replaced by Av. Now by Lemma 3.4 and Lemma 4.1 we obtain

$$\begin{aligned}
&\int_\Omega (g'(c\varphi + v_{n-1} + \mu_1)(z_n - z_{n-1}))(Av_t - w_t + w_t) \\
&\leq \Bigg[\Big(\int_\Omega (g'(c\varphi + z_{n-1} + \mu_1)(Av_t - w_t))^2 \Big)^{1/2} \Big) \\
&+ \Big(\int_\Omega (g'(c\varphi + z_{n-1} + \mu_1) w_t)^2 \Big)^{1/2} \Bigg] \|z_n - z_{n-1}\| \\
&\leq (|g'|_\infty \|Av_t - w_t\| + \tau \|w_t\|) \|z_n - z_{n-1}\| \\
&\leq \Big((|g'|_\infty + \tau)\|Av_t - w_t\| + \tau \|Av_t\| \Big) \|z_n - z_{n-1}\| \\
&\leq \Big((|g'|_\infty + \tau) D(\|Av\|_{1/2} + m^{3/2}\|Av\|) + \tau \|Av_t\| \Big) \|z_n - z_{n-1}\|.
\end{aligned} \tag{4.51}$$

Similarly

$$\begin{aligned}
&\int_\Omega (g'(c\varphi + v_{n-1} + \mu_2)(v_{n-1} - v_{n-2}))(Av_t - w_t + w_t) \\
&\leq (((|g'|_\infty + \tau) D(\|Av\|_{1/2} + m^{3/2}\|Av\|) + \tau \|Av_t\|) \|v_{n-1} - v_{n-2}\|.
\end{aligned} \tag{4.52}$$

Hence replacing (4.47) in (4.51) (and (4.52)) and then this in (4.49) we have (4.53)

$$\|Q(v_n - v_{n-1})\|^2 \leq \frac{m}{a} \|v_{n-1} - v_{n-2}\| \Big((|g'|_\infty + \tau) D(\|Av\|_{1/2} + m^{3/2}\|Av\|) + \tau \|Av_t\| \Big).$$

By the definition of v we have

$$\|v\| \leq \frac{1}{J\sqrt{2}} \|Q(v_n - v_{n-1})\| \tag{4.54}$$

and

$$\begin{aligned}
\|v\|^2_{\frac12} &= \sum \frac{(k^2 + 2j^2)^{\frac12}}{2j^2} (b^2_{kj} + a^2_{kj}) \leq \sum (b^2_{kj} + a^2_{kj}) \frac{(m + 4j^2)^{\frac12}}{2j^2} \\
&\leq \sum (b^2_{kj} + a^2_{kj}) \Big(\frac{m + 4j^2}{4j^4} \Big)^{\frac12} \leq \|Q(v_{n-1} - v_n)\|^2 \Big(\frac{m + 4J^2}{4J^4} \Big)^{\frac12}.
\end{aligned} \tag{4.55}$$

Replacing (4.54)-(4.55) in (4.53) we conclude (See (4.48))

$$\|Q(v_n - v_{n-1})\| \leq \frac{m\Gamma}{a} \|v_{n-1} - v_{n-2}\| [J_1 + \tau] \leq \frac{1}{8} \|v_{n-1} - v_{n-2}\|. \tag{4.56}$$

Finally we estimate $\|P(v_n - v_{n-1})\|$. By (4.2) and (4.47)

(4.57)

$$\begin{aligned}
&\|P(v_n - v_{n-1})\|^2 = \\
&\int_\Omega PA(g(c\varphi + v_{n-1} + z_n) - g(c\varphi + v_{n-2} + z_{n-1}))P(v_n - v_{n-1}) \\
&= \int_\Omega PA(g'(c\varphi + z_n + \mu_1)(v_{n-1} - v_{n-2})P(v_n - v_{n-1}) \\
&+ \int_\Omega g'(c\varphi + v_{n-1} + \mu_2)AP(v_n - v_{n-1})(z_n - z_{n-1}) \\
&\leq \left(\int_\Omega (g'(c\varphi + z_n + \mu_1)AP(v_n - v_{n-1}))^2\right)^{\frac{1}{2}} \|v_{n-1} - v_{n-2}\| \\
&+ \left(\int_\Omega (g'(c\varphi + v_{n-1} + \mu_2)AP(v_n - v_{n-1}))^2\right)^{\frac{1}{2}} \frac{m-a}{a}\|v_{n-1} - v_{n-2}\| \\
&\leq \left(\int_\Omega g'(c\varphi + z_n + \mu_1)^4\right)^{\frac{1}{4}} \|AP(v_n - v_{n-1})\|_{L^4}\frac{m}{a}\|v_{n-1} - v_{n-2}\|
\end{aligned}$$

where μ_1, μ_2 are as in (4.49). Since X is a finite dimensional space, there exists a constant J_2 such that $\int_\Omega x^4 \leq J_2^4\|x\|^4$ for all $x \in X$. By Corollary 3.5 and Lemma 4.1, there exists c_2 such that if $c > c_2$ then

$$\left(\int_\Omega (g'(c\varphi + z_n + \mu_1))^4\right)^{\frac{1}{4}} J_2\frac{m}{a} < \frac{1}{8}.$$

Hence

$$\begin{aligned}
\|v_n - v_{n-1}\| &\leq \|P(v_n - v_{n-1})\| + \|Q(v_n - v_{n-1})\| \\
&\leq \frac{1}{8}\|v_{n-1} - v_{n-2}\| + \frac{1}{8}\|v_{n-1} - v_{n-2}\| \leq \frac{1}{4}\|v_{n-1} - v_{n-2}\|.
\end{aligned}$$

Thus $\{v_n\}$ converges in L^2. By Lemma 2.1 $\{z_n\}$ converges. Let $v = \lim_{n\to\infty} v_n$, and $z = \lim_{n\to\infty} z_n$ Taking limits in (4.2), we see that $z+v = A(r+g(c\varphi+z+v))$. Since $\{\|v_n\|_1\}$, $\{\|z_n\|_1\}$ are bounded we have $v, z \in H^1$. Thus $w = z + v$ is a weak solution to (2.4)-(1.5). Thus $u = w + c\varphi$ is a weak solution to (1.4)-(1.5), which proves Theorem 1.2.

References

[1] R. Adams, *Sobolev Spaces*, Academic Press, 1975.

[2] H. Brezis and L. Nirenberg, *Forced vibrations for a nonlinear wave equation*, Bull Comm. Pure Appl. Math. **31** (1978), 1-30.

[3] A. Castro and S. Unsurangsie, *A semilinear wave equation with nonmonotone nonlinearity*, Pacific Journal of Mathematics **132 No 2** (1988), 215-225.

[4] S. Chow and J. Hale, *Methods of bifurcation theory*, Springer Verlag, 1982.

[5] J. M. Coron, *Periodic solutions of a nonlinear wave equation without assumptions of monotonicity*, Math. Anal. **262** (1983), 273-285.

[6] H. Hoffer, *On the range of a wave operator with nonmonotone nonlinearity*, Math. Nachr. **106** (1982), 327-340.

[7] P. J. McKenna, *On solutions of a nonlinear wave equation question when the ratio of the period to the length of the interval is irrational*, Proceedings of the American Mathematical Society **93 No 1** (1983), 59-64.

[8] P. Rabinowitz, *Free vibrations for a semilinear wave equation*, Comm. Pure Appl. Math. **31** (1978), 165-194.

[9] M. Willem, *Density of the range of potential operators*, Proc. Amer. Math. Soc. **83** (1981).

Department of Mathematics, University of North Texas Denton, Texas 76203-5116

Departmento de Matematicas, Universidad Nacional de Colombia, Bogota Colombia

Contemporary Mathematics
Volume **208**, 1997

THE STRUCTURE OF THE SOLUTIONS TO SEMILINEAR EQUATIONS AT A CRITICAL EXPONENT

ALLAN L. EDELSON

Department of Mathematics
University of California
Davis California, 95616

Dedicated to Vic Shapiro on the occasion of his retirement

ABSTRACT. This paper is concerned with the structure of the solutions to subcritical elliptic equations related to the Matukuma equation. In certain cases the complete structure of the solution set is known, and is comparable to that of the original Matukuma equation. Here we derive sufficient conditions for a more complicated solution set consisting of; (i) crossing solutions for small initial conditions and large initial conditions; (ii) at least one open interval of slowly decaying solutions; and (iii) at least two rapidly decaying solutions. As a consequence we obtain multiplicity results for rapidly decaying, or minimal solutions.

1. Introduction. This article is concerned with the structure of the solutions to the semilinear elliptic equation

$$-\Delta u = f(|x|)u^p, x \in \mathbf{R}^n, \quad n \geq 3, \quad \frac{n+2}{n-2} > p > 1. \tag{1.1}$$

It is assumed that f is positive and has asymptotic behavior $f(|x|) \sim |x|^l, -2 < l < 0$ as $x \to \infty$. Among the important examples of related equations are the Matukuma equation

$$-\Delta u = \frac{1}{1+|x|^2}u^p, \quad x \in \mathbf{R}^3,$$

and the scalar curvature equation

$$-\Delta u = f(x)u^{\frac{n+2}{n-2}}.$$

There has been a substantial body of research into the question of the existence or non existence of positive solutions of (1.1) decaying to zero as $|x| \to \infty$. Much of this research has focussed on the range of p values between the Sobolev critical exponent $p^* = \frac{n+2}{n-2}$, and the critical exponent

$$p_* = \frac{n+2+2l}{n-2}. \tag{1.2}$$

1991 AMS Subject Classification 35J25, 35J60

The exponent p_* is critical with respect to the compact imbedding of weighted Sobolev spaces

$$H_r^1(\mathbf{R}^n) \to L^q(\mathbf{R}^n, f(|x|)dx), 0 < q < p_*,$$

(see [Eg]). For $p > p_*$ variational methods are sufficient to prove the existence of positive, decaying solutions (see [Eg], [N-S]). For $p = p_*$ those methods fail due to non compactness, and other techniques are required. This article will be primarily concerned with the structure of the set of solutions when $p = p_*$.

When f decays rapidly to zero at infinity, e.g. $l < -2$, existence and nonexistence results for (1.1) were obtained in [N]. References for the case $-2 < l < 0$, include [K-N-Y], [N-Y], [N-S], [Y-Y$_1$] and [Y-Y$_2$]. For $0 < f(0) < \infty$ and $p \neq p_*$, the existence or nonexistence of ground states is determined by the decay rate l. At the critical exponent, $p = p_*$, the existence of positive, decaying solutions depends on higher order terms in the asymptotic expansion of $f(r)$. In particular, the function

$$h(r) = r(r^{-l}f(r))_r, \tag{1.3}$$

has been useful in obtaining more precise information about the structure of the set of solutions when $0 < f(0) \leq \infty$. The following examples illustrate the role played by p and $h(r)$ in determining the existence or nonexistence of positive, entire solutions. We assume that f is radially symmetric and so the problem reduces to an ordinary differential equation. Since we are interested in nonnegative solutions we define $u^+ = max\{u, 0\}$ and let $u(r;\alpha)$ be the (unique) solution of the initial value problem

$$u'' + \frac{n-1}{r}u' + f(r)(u^+)^p = 0, \quad u(0;\alpha) = \alpha, \quad u'(0;\alpha) = 0. \tag{1.4}$$

Example I. $f(r) = r^l, -2 < l < 0$.

(1) *If* $1 < p < p_*$, *then* $u(r;\alpha)$ *has a finite zero for every* $\alpha > 0$.

(2) *If* $p_* \leq p$ *then* $u(r;\alpha)$ *is a positive, decaying solution for every* $\alpha > 0$.

In this case $h(r) = 0$, *for* $r \geq 0$.

Example II. $f(r) = O(r^\sigma)$ *as* $r \to 0$, $\sigma > -2$.

(1) *If* $p = p_*$, $f(r) \geq cr^l$ *for large* r, *and* $\int_0^R h(r)r^{n+l-1}dr \leq 0$, *for all* $R > 0$, *then for every* $\alpha > 0$, $u(r;\alpha)$ *is a positive, entire solution satisfying* $lim_{r\to\infty}u(r;\alpha) = 0$.

(2) *If* $p = p_*$, $h(r) \not\equiv 0$, *and* $\int_0^R h(r)r^{n+l-1}dr \geq 0$ *for all* $R > 0$, *then* $u(r;\alpha)$ *has a finite zero for all* $\alpha > 0$.

Example III. *Let* $\frac{n+l}{n-2} < p < p_*$, *and*

$$f(r) = \frac{(l+2)(n-2)}{(p-1)^2}\Big\{(p - \frac{n+l}{n-2}) + (\frac{l+2p}{n-2})(\frac{1}{1+r^2})\Big\}(1+r^2)^{l/2}.$$

(1) *$u(r;\alpha)$ has a finite zero for every sufficiently large $\alpha > 0$,*
(2) *$u(r;1) = (1+r^2)^{-\frac{l+2}{2(p-1)}}$ is a positive, decaying solution,*
(3) *$u(r;\alpha)$ has a finite zero for every sufficiently small $\alpha > 0$.*

Example I is contained in Proposition 4.5 of [N-Y]. The results of Example II can be found in Theorems 9.2 and 9.3 of [K-N-Y], where it is assumed that $\int_0^R h(r)r^{n+l-1}dr$ does not change sign. Example III shows the complexity of the solutions when $h(r)$ and $\int_0^R h(r)r^{n+l-1}dr$ are allowed to change sign (see also [N: Example 5.2], [N-S]). An important structure theorem for equation (1.1) is given in [K-Y-Y], where it is assumed that this integral changes sign exactly once.

It is known ([Y-Y$_1$]) that under our hypotheses all solutions of (1.4) with $\alpha > 0$, are of one of the following types:

(1) $u(r;\alpha)$ is a crossing solution, i.e.has a positive zero,
(2) $u(r;\alpha)$ is a slowly decaying solution, i.e.

$$lim_{r\to\infty} r^{n-2}u(r;\alpha) = \infty,$$

(3) $u(r;\alpha)$ is a rapidly decaying solution, i.e.

$$\overline{lim}_{r\to\infty} r^{n-2}u(r;\alpha) < \infty.$$

In sections 3 and 4 we will give criteria for existence and non existence of positive, decaying solutions when both $h(r)$ and $\int_0^R h(r)r^{n+l-1}dr$ are allowed to change sign. Then in section 5 we prove that for $p = p_*$, and under suitable hypotheses on $h(r)$, the set of solutions of (1.4) consists of:

(1) crossing solutions for sufficiently small or sufficiently large α;
(2) at least one open interval of α values for which $u(r;\alpha)$ is slowly decaying;
(3) at least two rapidly decaying solutions.

Thus under these hypotheses a structure comparable to that of Example III can be achieved for $p = p_*$. A consequence of this result is that we prove the existence of multiple minimal, i.e. rapidly decaying solutions.

2. Main Results. The following hypotheses will be assumed throughout this paper.

(f_1) $f \in C(0,\infty)$, $f(r) > 0$ for $r > 0$,
(f_2) $f(r) = O(r^l)$ as $r \to \infty, -2 < l < 0$,
(f_2') $f(r) = O(r^\sigma)$ as $r \to 0, -2 < \sigma$,

We will assume additional hypotheses chosen from the following:

(f_3) there exist positive numbers δ_1, r_2, β such that

$$h(r) < -\delta_1 r^{-\beta}, \qquad \text{for } r > r_2,$$

(f_4) $h(r) = O(r^\gamma)$, $\gamma > 0$, as $r \to 0$.
(f_5) $\int_0^\infty h(r)r^{n+l-1}dr < 0$.
(f_6) there exist positive numbers δ_1', r_2', β such that

$$|h(r)| < \delta_1' r^{-\beta}, \qquad \text{for } r > r_2',$$

(f_7) $\int_0^\infty h(r)r^{n+l-1}dr > 0$,
(f_8) $0 < r_3 = sup\{r > 0 : \int_o^r h(s)s^{n+l-1}ds \leq 0\}$.
(f_9) $\gamma(2+l) > (\sigma - l)(n+l)$.

For $p \geq p_*$ the following two theorems give sufficient conditions for the existence of positive, entire solutions when α is sufficiently small.

Theorem 1. *Assume that (f₃) and (f₄) are satisfied,* $p \geq p_*$ *and that* $0 < \beta < n+l$. *Then there exists an* $\alpha_0 > 0$ *such that for* $0 < \alpha \leq \alpha_0$, *the solution* $u(r;\alpha)$ *is a positive, entire solution which also satisfies* $lim_{r\to\infty} u(r;\alpha) = 0$.

Example 2.1. The function

$$\begin{aligned} &f(r) = (c_1 + c_2 r^2)^{\frac{\gamma}{2}}(c_3 + c_4 r^2)^{\frac{\nu}{2}}, \\ &-2 < \gamma + \nu < 0, \quad c_1c_4\gamma + c_2c_3\nu > 0, \quad c_i > 0, \end{aligned}$$

satisfies $(f_1) - (f_4)$.

Example 2.2. The function

$$f(r) = (1 + (1+r^2)^{-\frac{1}{4}})(1+r^2)^{-\frac{1}{4}}$$

satisfies (f_1)-(f_4). For $n = 3$ we have $l = -\frac{1}{2}, p_* = 4, \beta = \frac{1}{2}$, and $0 < \beta < n+l$.

This example is not covered by [K-N-Y], Theorem 9.2, since $\int_0^R h(r)r^{n+l}dr > 0$ for small R.

Example 2.3. The function

$$f(r) = \Big(\frac{9}{8} + (1+r^2)^{-1}\Big)(1+r^2)^{-\frac{3}{4}}$$

satisfies (f_1)-(f_4). For $n = 3$ we have $l = -\frac{3}{2}, p_* = 2, \beta = 2$, and $\beta > n + l$. Since $\int_0^R h(r)r^{n+l-1}dr > 0$ for all $R > 0$, it follows from Theorem 9.3 of [K-N-Y] that $u(r;\alpha)$ has a positive zero for all $\alpha > 0$.

The next result gives a variant of Theorem 1 in which the condition $\beta < n + l$ is replaced by an integral growth condition.

Theorem 2. *Assume that (f₃), (f₄) and (f₅) are satisfied, and* $p \geq p_*$. *Then there exists an* $\alpha_0 > 0$ *such that for* $0 < \alpha \leq \alpha_0$, *the solution* $u(r;\alpha)$ *is a positive, entire solution which also satisfies* $lim_{r\to\infty} u(r;\alpha) = 0$.

The next two theorems give sufficient conditions for the non existence of positive solutions when α is sufficiently large, or sufficiently small.

Theorem 3. *Assume that* (f_4), (f_6) *and* (f_9) *are satisfied, and* $p \geq p_*$. *Then there exist positive numbers* γ_*, α_1, *such that for* $\alpha > \alpha_1$ *and* $0 < \gamma < \gamma_*$ *the solution* $u(r;\alpha)$ *has a positive zero.*

Remark. For $p < p_*$ this result is proved in [N-Y, Theorem 2], without the additional hypothesis $\gamma < \gamma_*$.

Theorem 4. *Assume that* (f_4) *and* (f_7) *are satisfied,* $p \geq p_*$ *and that there exists an* r_2 *such that* $h(r) \geq 0$, *for* $r_2 < r < \infty$. *Then there exists an* $\alpha_0 > 0$ *such that for* $\alpha \in (0, \alpha_0)$ *the solution* $u(r;\alpha)$ *has a positive zero.*

Remark. A stronger version of this result is found in [Y-Y$_2$, Theorem 3].

Theorem 5. *Let* $p = p_*$ *and let* $k(r)$ *be a continuous function satisfying* (f_7), (f_9) *and also the following:*
there exist positive numbers $0 < a < b < c$, α_*, *and* $r_* > b$ *such that*

$$
\begin{aligned}
&(a) \quad k(0) = k(a) = k(b) = k(c) = 0, \quad k(r) = 0, \ \textit{for } r > c, \\
&(b) \quad k(r) > 0 \textit{ for } 0 < r < a, \textit{ and } b < r < c, \quad k(r) < 0 \textit{ for } a < r < b, \\
&(c) \quad k(r) = O(r^{\gamma}), \quad n + l - 1 < \gamma < \gamma_*, \ \textit{as } r \to 0, \\
(2.1) \qquad &(d) \quad \phi(0;\alpha_*)^{p_*+1} \int_0^a k(s)ds + \phi(b;\alpha_*)^{p_*+1} \int_a^b k(s)ds \\
&\qquad + \phi(b;\alpha_*)^{p_*+1} \int_b^{r_*} k(s)ds < -\delta^2, \\
&(e) \quad \int_0^c k(r)dr > 0,
\end{aligned}
$$

where γ_* *is given in theorem 3 and* $\phi(r;\alpha)$ *is the solution of the equation*

$$
(2.2) \qquad
\begin{aligned}
&-(r^{n-1}\phi_r)_r = r^{n-1} r^l \phi^{p_*}, \\
&\phi(0) = \alpha, \quad \phi_r(0) = 0.
\end{aligned}
$$

Then there exists a positive function $f \in C(0,\infty)$, *with* $f(r) = O(r^l)$ *as* $r \to \infty, -2 < l < 0$, *and positive numbers* $\alpha_0 < \alpha_1 < \alpha_* < \alpha_2 < \alpha_3$ *such that the solutions* $u(r;\alpha)$ *of equation (1.4) are of the following type:*

$$
(2.3) \qquad
\begin{aligned}
&(a) \quad 0 < \alpha < \alpha_0 \Rightarrow u(r;\alpha) \textit{ has a positive zero }, \\
&(b) \quad u(r;\alpha_*) \textit{ is a positive, slowly decaying solution }, \\
&(c) \quad \alpha_3 < \alpha < \infty \Rightarrow u(r;\alpha) \textit{ has a positive zero }, \\
&(d) \quad u(r;\alpha_1) \textit{ and } u(r;\alpha_2) \textit{ are rapidly decaying solutions.}
\end{aligned}
$$

It is not difficult to construct examples which satisfy the hypotheses (2.1).

3. Existence of positive, decaying solutions. The basic existence theorem for solutions to the initial value problem can be found in [N-Y]. In establishing Theorem 1 we will require several preliminary lemmas. The first is contained in Proposition 4.1 of [N-Y] and is also found in [N].

Lemma 3.1. *There exists a unique solution $u(r;\alpha) \in C((0,\infty) \cap C^2(0,\infty))$ of (1.4). The solution satisfies the integral equation*

$$u(r) = \alpha - \frac{1}{n-2}\int_0^r \Big\{1 - \Big(\frac{s}{r}\Big)^{n-2}\Big\} s f(s)(u^+(s))^p ds, \tag{3.1}$$

and also the following conditions;

$$\begin{aligned} &(a) \quad lim_{r\to\infty} r^{n-1}u_r(r) = 0 \\ &(b) \quad u_r(r) = -\int_0^r \Big(\frac{s}{r}\Big)^{n-1} f(s)(u^+(s))^p ds \le 0, \qquad for\ r > 0, \\ &(c) \quad u\ is\ non\text{-}increasing\ on\ [0,\infty). \end{aligned} \tag{3.2}$$

Lemma 3.2. *([N-Y], Prop.4.3 and (6.8)) For $R > 0$ the solution $u(r;\alpha)$ satisfies the Pohozaev type identity*

$$\begin{aligned} &\frac{n-2}{2}R^{n-1}u(R;\alpha)u'(R;\alpha) + \frac{1}{2}R^n u'(R;\alpha)^2 + \frac{1}{p+1}R^n f(R)(u^+(R;\alpha))^{p+1} \\ &= \frac{1}{p+1}\int_0^R \Big\{-\frac{n-2}{2}(p-p_*)r^{-l}f(r) + h(r)\Big\} r^{n+l-1}(u^+(r;\alpha))^{p+1} dr. \end{aligned} \tag{3.3}$$

Define r_α by.

$$r_\alpha = inf\Big\{r > 0 : u(r;\alpha) = \frac{\alpha}{2}\Big\}. \tag{3.4}$$

Lemma 3.3. *r_α satisfies the following estimates:*

$$r_\alpha = O(\alpha^{\frac{1-p}{2+l}}) \ as\ \alpha \to 0, \tag{3.5a}$$

$$r_\alpha = O(\alpha^{\frac{1-p}{2+\sigma}}) \ as\ \alpha \to \infty. \tag{3.5b}$$

Proof. By Proposition 4.1 of [N-Y] and (f_2) we have

$$\begin{aligned} \frac{n-2}{2}\alpha &= \int_0^{r_\alpha} \Big\{1 - \Big(\frac{s}{r_\alpha}\Big)^{n-2}\Big\} s f(s)(u^+(s;\alpha))^p ds \\ &\le \alpha^p \int_0^{r_\alpha} s f(s) ds. \end{aligned}$$

$$\frac{n-2}{2}\alpha^{1-p} \le \int_0^{r_\alpha} s f(s) ds.$$

By Lemma 6.1 of [N-Y] $lim_{\alpha\to 0} r_\alpha = \infty$, and therefore

$$\begin{aligned}\frac{n-2}{2}\alpha^{1-p} &\le \int_0^1 sf(s)ds + \int_1^{r_\alpha} sf(s)ds \\ &\le c + \int_1^{r_\alpha} s^{1+l}ds \le cr_\alpha^{2+l},\end{aligned}$$

which implies

$$C\alpha^{\frac{1-p}{2+l}} \le r_\alpha \text{ as } \alpha \to 0. \tag{3.6}$$

By a similar estimate we can show

$$C\alpha^{\frac{1-p}{2+l}} \ge r_\alpha, \text{ as } \alpha \to 0,$$

and this proves 3.5a.

Consider the case $\alpha \to \infty$.

$$\begin{aligned}\frac{n-2}{2}\alpha &= \int_0^{r_\alpha} \left\{1 - \left(\frac{s}{r_\alpha}\right)^{n-2}\right) sf(s)(u(s;\alpha)^+)^p ds \\ &\ge \left(\frac{\alpha}{2}\right)^p C \int_0^{r_\alpha} sf(s)ds, \\ \alpha^{1-p} &\ge \int_0^{r_\alpha} sf(s)ds,\end{aligned}$$

which implies $r_\alpha \to 0$ as $\alpha \to \infty$. (3.5b) can then be deduced.

Proof of Theorem 1. By (f_4) $h(r) > 0$ on an interval $(0, \epsilon)$. Define the quantities

$$\begin{aligned} r_0 &= inf\{r > 0 : h(r) < 0\}, \\ r_1 &= sup\{r > 0 : h(r) > 0\}, \\ \delta_2 &= \int_0^{r_1} |h(r)| r^{n+l-1} dr, \\ k &= \left(\frac{2^{-(p+1)}}{n+l-\beta}\right)\left(1 - 2^{-(n+l-\beta)}\right) > 0. \end{aligned} \tag{3.7}$$

Then r_1 exists by (f_3), and by (f_4) $0 < r_0 \le r_1 < \infty$. We apply the identity (3.3). By (3.5a) we may choose α_0 so small that

$$r_{\alpha_0}^{n+l-\beta} \ge \frac{\delta_2}{k\delta_1}. \tag{3.8}$$

For some $\alpha \in (0, \alpha_0]$ assume that $u(r;\alpha)$ satisfies the conditions

$$u(r;\alpha) > 0, \quad 0 \le r < R, \quad u(R;\alpha) = 0.$$

The left hand side of (3.3) is > 0, because $u'(R;\alpha) \neq 0$. Also, $R > r_\alpha$ by the definition of r_α, and by Lemma (3.3) we may assume $r_1 < \frac{r_\alpha}{2}$. Then

$$
\begin{aligned}
&\int_0^R h(r)r^{n+l-1}(u^+(r;\alpha))^{p+1}dr = \\
&\{\int_0^{r_1} + \int_{r_1}^{\frac{r_\alpha}{2}} + \int_{\frac{r_\alpha}{2}}^{r_\alpha} + \int_{r_\alpha}^{R}\}h(r)r^{n+l-1}(u^+(r;\alpha))^{p+1}dr \leq \\
&\delta_2\alpha^{p+1} + \int_{\frac{r_\alpha}{2}}^{r_\alpha} h(r)r^{n+l-1}(u^+(r;\alpha))^{p+1}dr \leq \\
&\delta_2\alpha^{p+1} - \delta_1\int_{\frac{r_\alpha}{2}}^{r_\alpha} r^{n+l-1-\beta}(u^+(r;\alpha))^{p+1}dr \leq \\
&\delta_2\alpha^{p+1} - \delta_1\left(\frac{\alpha}{2}\right)^{p+1}\int_{\frac{r_\alpha}{2}}^{r_\alpha} r^{n+l-1-\beta}dr \leq \\
&\alpha^{p+1}\left(\delta_2 - k\delta_1 {r_\alpha}^{n+l-\beta}\right) \leq 0.
\end{aligned}
\tag{3.9}
$$

This contradiction implies that $u(r;\alpha)$ is positive on $[0,\infty)$. The condition

$$lim_{r\to\infty}u(r;\alpha) = 0$$

is a consequence of [N; Theorem 3.10].

Example 3.4. *If*

$$f(r) = A_0r^l + o(r^l), \quad r \to \infty,$$

then the hypotheses of Theorem 1 are satisfied.

Proof. It is easy to verify that in this case $\beta = 1 < n + l$.

The next result, needed in the proof of Theorem 2, is a generalization of Lemma 3.3.

Lemma 3.5.

Define $r_{\alpha,k}$ by

$$r_{\alpha,k} = inf\{r > 0 : u(r;\alpha) = \frac{\alpha}{k}\}, \quad k > 1.$$

Then $r_{\alpha,k}$ has asymptotic behavior

$$lim_{\alpha\to 0}r_{\alpha,k} = \infty.$$

Proof. Following the proof of Lemma 3.3 we have

$$
\begin{aligned}
u(r_{\alpha,k};\alpha) = \frac{\alpha}{k} &= \alpha - \frac{1}{n-2}\int_0^{r_{\alpha,k}} \Big\{1-\Big(\frac{s}{r_{\alpha,k}}\Big)^{n-2}\Big\} sf(s)(u^+(s;\alpha))^p ds,\\
\alpha\Big(\frac{k-1}{k}\Big) &= \frac{1}{n-2}\int_0^{r_{\alpha,k}} \Big\{1-\Big(\frac{s}{r_{\alpha,k}}\Big)\Big\} sf(s)(u^+(s,\alpha))^p ds\\
&\le \frac{1}{n-2}\int_0^{r_{\alpha,k}} sf(s)(u^+(s;\alpha))^p ds\\
&\le \Big(\frac{\alpha}{k}\Big)^p \int_0^{r_{\alpha,k}} sf(s)ds,\\
\alpha^{1-p} &\le C\int_0^{r_{\alpha,k}} sf(s)ds.
\end{aligned}
$$

Therefore

$$
\begin{aligned}
\alpha^{1-p} &\le C\int_0^{r_{\alpha,k}} sf(s)ds\\
&= C\int_0^{r_{\alpha,k}} s^{l+1}ds\\
&= Cr_{\alpha,k}^{l+2}.
\end{aligned}
$$

It follows that

$$
\alpha^{\frac{1-p}{l+2}} \le Cr_{\alpha,k}
$$

and since $\frac{1-p}{l+2} < 0$, we have the stated result.

It is possible to obtain more precise estimates, analogous to Lemma 3.3, however such estimates will not be needed.

Proof of Theorem 2. Define

$$
\begin{aligned}
&(a)\quad r^0 = sup\{r>0 : \int_0^r h(s)s^{n+l-1}ds \ge 0\},\\
(3.10)\qquad &(b)\quad \Omega_0^+ = \{r\in[0,r^0] : h(r)\ge 0\},\\
&(c)\quad \Omega_0^- = \{r\in[0,r^0] : h(r)\le 0\}.
\end{aligned}
$$

By (f_4) and (f_5) $r^0 > 0$. For $\epsilon \in (0,1)$ it is a consequence of Lemma 3.5 that there exists an $\alpha_0(\epsilon)$ such that for $\alpha \in (0,\alpha_0(\epsilon))$ the conditions

$$
\begin{aligned}
&(i)(1-\epsilon)^{\frac{1}{p_*+1}}\alpha \le u(r;\alpha) \le \alpha,\ \text{for } 0\le r\le r^0,\\
(3.11)\qquad &(ii) r_\alpha > 2r^0,\\
&(iii)\int_{r^0}^{2r^0} h(r)r^{n+l-1}dr < 2^{p_*+1}\epsilon\int_{\Omega_0^-} h(r)r^{n+l-1}dr
\end{aligned}
$$

are satisfied. We will show that there exists an α_0 such that

$$\alpha \in (0, \alpha_0) \Rightarrow \int_0^R h(r) r^{n+l-1} (u^+(r;\alpha))^{p_*+1} dr < 0, \text{ for all } R > r^0.$$

In fact, for $R > r^0$,

$$\begin{aligned}
&\int_0^R h(r) r^{n+l-1} (u^+(r;\alpha))^{p_*+1} dr \\
&\leq \Big(\int_{\Omega_0^-} + \int_{\Omega_0^+} + \int_{r^0}^R\Big) h(r) r^{n+l-1} (u^+(r;\alpha))^{p_*+1} dr \\
&\leq \alpha^{p_*+1} \int_0^{r^0} h(r) r^{n+l-1} dr - \epsilon\alpha^{p^*+1} \int_{\Omega_0^-} h(r) r^{n+l-1} dr \\
&+ \int_{r^0}^R h(r) r^{n+l-1} dr + \int_{r^0}^R h(r) r^{n+l-1} (u^+(r;\alpha))^{p_*+1} dr \\
&\leq -\epsilon\alpha^{p_*+1} \int_{\Omega_0^-} h(r) r^{n+l-1} dr + \int_{r^0}^R h(r) r^{n+l-1} (u^+(r;\alpha))^{p^*+1} dr.
\end{aligned}$$

Then for $R > r^0$,

$$\begin{aligned}
&-\epsilon\alpha^{p^*+1} \int_{\Omega_0^-} h(r) r^{n+l-1} dr + \int_{r^0}^R h(r) r^{n+l-1} (u^+(r;\alpha))^{p_*+1} dr \\
&\leq -\epsilon\alpha^{p^*+1} \int_{\Omega_0^-} h(r) r^{n+l-1} dr + \int_{r^0}^{2r^0} h(r) r^{n+l-1} (u^+(r;\alpha))^{p_*+1} dr \\
&\leq -\epsilon\alpha^{p_*+1} \int_{\Omega_0^-} h(r) r^{n+l-1} dr + \Big(\frac{\alpha}{2}\Big)^{p_*+1} \int_{r^0}^{2r^0} h(r) r^{n+l-1} dr \\
&\leq \alpha^{p_*+1} \Big\{ -\epsilon \int_{\Omega_0^-} h(r) r^{n+l-1} dr + 2^{-(p_*+1)} \int_{r^0}^{2r^0} h(r) r^{n+l-1} dr < 0,
\end{aligned}$$

by (3.11iii).

As in the proof of Theorem 1, an application of the Pohozaev identity (3.3) leads to a contradiction, and once again the limit condition follows from [N; Theorem 3.10].

4. Nonexistence of positive, decaying solutions. Our proofs of the nonexistence of positive, decaying solutions are based on the Pohozaev identity. Define $w(r;\alpha) = r^{\frac{n-2}{2}} u(r;\alpha)$, and let $u(r;\alpha)$ be a positive, entire solution which decays to zero as $r \to \infty$. According to Theorem 2.2 of [K-N-Y] the following variant of the Pohozaev identity applies.

$$\begin{aligned}
&- R^2 w''(R;\alpha) w(R;\alpha) - R w(R;\alpha) w'(R;\alpha) + R^2 w'(R;\alpha)^2 \\
&- \Big(\frac{l+2}{n+l}\Big) R^n f(R) u(R;\alpha)^{p+1} \qquad (4.1)\\
&= \int_0^R h(r) r^{n+l-1} (u(r;\alpha)^+)^{p+1} dr.
\end{aligned}$$

The function w is bounded and eventually monotone, so by Lemma 4.1 of [K-N-Y] there exists a sequence $\{R_j\}$ such that as $j \to \infty$,

$$R_j \to \infty, \quad w(R_j;\alpha) \to C, \quad R_j w'(R_j;\alpha) \to 0, \quad R_j^2 w''(R_j;\alpha) \to 0. \tag{4.2}$$

Evaluating (4.1) on the sequence $\{R_j\}$ we see that the lim sup as $j \to \infty$ of the left hand side of (4.1) is ≤ 0 for $p \geq p_*$. Evaluation of the right hand side will require several preliminary results.

Lemma 4.1. *Assume that $p = p_*$, and that (f_4) and (f_9) are satisfied. Let $u(r;\alpha)$ be a solution of (1.4). Then for $\gamma > 0$ sufficiently small,*

$$lim_{\alpha\to\infty} \int_0^{r_\alpha} h(r) r^{n+l-1} (u(r;\alpha)^+)^{p+1} dr = \infty. \tag{4.3}$$

Proof. According to Lemma 3.3 $lim_{\alpha\to\infty} r_\alpha = 0$, so by (f_4) for sufficiently large α,

$$\begin{aligned}
&\int_0^{r_\alpha} h(r) r^{n+l-1} (u(r;\alpha)^+)^{p+1} dr > \\
&\quad \left(\frac{\alpha}{2}\right)^{p+1} \int_0^{r_\alpha} h(r) r^{n+l-1} dr \geq \\
&\quad c\left(\frac{\alpha}{2}\right)^{p+1} \int_0^{r_\alpha} r^{n+l-1+\gamma} dr = \\
&\quad \frac{c}{(n+l+\gamma)2^{p+1}} \alpha^{p+1} r_\alpha^{n+l+\gamma}.
\end{aligned}$$

By Lemma 3.3 this is $\geq \alpha^q$, for

$$\begin{aligned}
q &= -\{l^2 + (n+\gamma-\sigma)l + (2\gamma - \sigma n)\} = -\{\gamma(2+l) - \sigma(n+l) + l(n+l)\} \\
&= -\{\gamma(2+l) - (\sigma - l)(n+l)\},
\end{aligned}$$

and $\gamma(2+l) - (\sigma - l)(n+l) < 0$ by (f_9). In fact, it is < 0 for $\gamma < \frac{(\sigma-l)(n+l)}{(2+l)}$. Therefore the conclusion of the lemma follows. Note that a sufficient condition is $\sigma > l$ and γ sufficiently small.

We will also require the following lemma, which gives an apriori bound for positive solutions on $[r_0, \infty)$, for some $r_0 > 0$. It is based on Theorem 3.10 of [N]. Similar results can be found in [B-L], [E-R].

Lemma 4.2. *Let $u(r;\alpha)$ be a positive, entire solution of (1.1), with $p \leq p_*$, and assume that there exist positive numbers r_0, C_1 such that*

$$f(r) \geq C_1 r^l, \quad -2 < l < 0, \quad \textit{for } r > r_0.$$

Then there exists a positive constant C, independent of α, such that if $u(r;\alpha)$ is a positive, entire solution then

$$u(r;\alpha) \le C r^{\frac{2-n}{2}}, \qquad \text{for } r \ge r_0, \quad 0 < \alpha < \infty. \tag{4.4}$$

Proof. By Green's identity in any ball of radius $R > 0$,

$$\begin{aligned} \int_{B_R} f(|x|)u(|x|;\alpha)^p dx &= -\int_{B_R} \Delta u(|x|;\alpha)dx \\ &= \int_{\partial B_R} u'(r;\alpha)ds \\ &= -\omega_n R^{n-1} u'(R;\alpha). \end{aligned} \tag{4.5}$$

By (4.4) we may assume that

$$f(r) \ge C_1 r^l, \qquad \text{for } r \ge \frac{r_0}{2}.$$

Then for $R \ge \frac{r_0}{2}$, and using the monotonicity of $u(r;\alpha)$,

$$\begin{aligned} &\int_{B_R} f(|x|)(u^+(|x|;\alpha))^p dx \ge \\ &\int_{B_R - B_{\frac{r_0}{2}}} f(|x|)(u^+(|x|;\alpha))^p dx \ge \\ &\int_{B_R - B_{\frac{r_0}{2}}} C_1 |x|^l (u^+(|x|;\alpha))^p dx \ge \\ &C_1 u(R;\alpha)^p \int_{\frac{r_0}{2}}^R r^{n+l-1} dr \ge \\ &\frac{C_1}{n+l} u(R;\alpha)^p \Big\{ R^{n+l} - \Big(\frac{r_0}{2}\Big)^{n+l} \Big\}. \end{aligned}$$

Therefore for $R \ge \frac{r_0}{2}$,

$$\int_{B_R} f(|x|)(u^+(|x|;\alpha))^p dx \ge \frac{C_1}{2(n+l)} R^{n+l} u(R;\alpha)^p,$$

and by (4.5)

$$-\omega_n R^{n-1} u'(R;\alpha) \ge \frac{C_1}{2(n+l)} R^{n+l} u(R;\alpha)^p,$$

$$-\frac{u'(r;\alpha)}{u(r;\alpha)^p} \ge \frac{C_1}{\omega_n} R^{l+1} \qquad \text{for } r \ge \frac{r_0}{2}.$$

Since $l > -2$, an integration from $\frac{r_0}{2}$ to R gives

$$-\int_{\frac{r_0}{2}}^{R}\frac{u'(r;\alpha)}{u(r;\alpha)^p}dr = -\frac{u(r;\alpha)^{1-p}}{1-p}\Big|_{\frac{r_0}{2}}^{R}$$
$$= \frac{1}{p-1}\Big\{u(R;\alpha)^{1-p} - u\Big(\frac{r_0}{2};\alpha\Big)^{1-p}\Big\},$$

and therefore

$$\frac{1}{p-1}\Big\{u(r;\alpha)^{1-p} - u\Big(\frac{r_0}{2};\alpha\Big)^{1-p}\Big\} \geq \frac{C_1}{\omega_n(l+2)}r^{l+2},$$

$$u(r;\alpha)^{1-p} \geq \Big(\frac{C_1}{\omega_n(l+2)}r^{l+2} + u\Big(\frac{r_0}{2};\alpha\Big)^{1-p}\Big)(p-1) \geq \frac{C_1(p-1)}{\omega_n(l+2)}r^{l+2},$$

and

$$u(r;\alpha) \leq \Big(\frac{C_1(p-1)}{\omega_n(l+2)}\Big)^{\frac{1}{1-p}} r^{\frac{l+2}{1-p}}, \text{ for } r \geq r_0,$$

and since $p \leq p_*$, the lemma is proved.

Proof of Theorem 3. We will evaluate the right hand side of (4.1) using Lemmas 4.1 and 4.2. Note that r^0 is defined in (3.10).

$$\begin{aligned}
&\Big|\int_{r^0}^{R_j} h(r)r^{n+l-1}(u(r;\alpha)^+)^{p+1}dr\Big| \leq \\
&\int_{r^0}^{R_j} |h(r)|r^{n+l-1}(u(r;\alpha)^+)^{p+1}dr \leq \\
&\int_{r^0}^{\infty} |h(r)|r^{n+l-1}r^{\frac{2-n}{2}(p+1)}dr \leq \\
&C\delta\int_{r^0}^{\infty} r^{n+l-1+\frac{2-n}{2}(p+1)-\beta} \leq \\
&C\delta\int_{r^0}^{\infty} r^{-(1+\beta)}dr = C_0,
\end{aligned}$$

by (f_6) and Lemma 4.2. By the hypotheses $0 < \gamma < \gamma_*$ and Lemma 4.1, there exists an $\alpha_1 > 0$ such that $\alpha > \alpha_1 \Rightarrow \int_0^{r^0} h(r)r^{n+l-1}(u(r;\alpha)^+)^{p+1}dr > C_0$. For $\alpha > \alpha_1$ by (4.2) we can choose a sequence $\{R_j\}$, with $R_j \to \infty$, such that for $j > j(\alpha)$ the left hand side of (4.1) is $< C_0$, which proves the theorem.

Proof of Theorem 4. Let

$$h^+(r) = max\{0, h(r)\},$$
$$h^-(r) = min\{0, h(r)\}.$$

By (f_7) and (f_8)

$$0 < r_3 = sup\{r > 0 : \int_0^r h(s)s^{n+l-1}ds \le 0\}. \tag{4.6}$$

We choose k, ϵ, α and R_j as follows:
(a) let $0 < k < 1$, and let $\epsilon > 0$ be so small that

$$\epsilon \int_{r_0}^{r_2} h^+(r)r^{n+l-1}dr < (1-k)\int_0^{r_2} h(r)r^{n+l-1}dr, \tag{4.7i}$$

(b) let α be so small that

$$r_\alpha > r_2, \quad \text{and } (1-\epsilon)^{\frac{1}{p+1}}\alpha \le u(r;\alpha) \le \alpha, \quad \text{for } 0 \le r \le r_2, \tag{4.7ii}$$

(c) choose $R_j > r_2$ so large that

$$L(R_j,\alpha) < k\alpha^{p+1}\int_0^{r_2} h(r)r^{n+l-1}dr, \tag{4.7iii}$$

where $L(R_j,\alpha)$ is the left hand side of (4.1), and assume that $u(r;\alpha) > 0$ for $0 \le r \le R_j$. Then $\int_0^{r_3} h(r)r^{n+l-1}dr = 0$, and the right hand side of (4.1) is

(4.8)

$$\begin{aligned}
&\int_0^{R_j} h(r)r^{n+l-1}(u^+(r;\alpha))^{p+1}dr \\
&= \{\int_0^{r_0} + \int_{r_0}^{r_3} + \int_{r_3}^{r_2} + \int_{r_2}^{R_j}\}h(r)r^{n+l-1}(u^+(r;\alpha))^{p+1}dr \\
&\ge \alpha^{p+1}\Big\{(1-\epsilon)\int_0^{r_0} h(r)r^{n+l-1}dr + (1-\epsilon)\int_{r_0}^{r_3} h^+(r)r^{n+l-1}dr + \\
&\int_{r_0}^{r_3} h^-(r)r^{n+l-1}dr + (1-\epsilon)\int_{r_3}^{r_2} h^+(r)r^{n+l-1}dr + \int_{r_3}^{r_2} h^-(r)r^{n+l-1}dr\Big\} \\
&\ge \alpha^{p+1}\Big\{(1-\epsilon)\int_0^{r_0} h(r)r^{n+l-1}dr + (1-\epsilon)\int_{r_0}^{r_3} h^+(r)r^{n+l-1}dr \\
&+ (1-\epsilon)\int_{r_0}^{r_3} h^-(r)r^{n+l-1}dr + \epsilon\int_{r_0}^{r_3} h^-(r)r^{n+l-1}dr + (1-\epsilon)\int_{r_3}^{r_2} h^+(r)r^{n+l-1}dr \\
&+ (1-\epsilon)\int_{r_3}^{r_2} h^-(r)r^{n+l-1}dr + \epsilon\int_{r_3}^{r_2} h^-(r)r^{n+l-1}dr\Big\} \\
&= \alpha^{p+1}\Big\{(1-\epsilon)\int_0^{r_2} h(r)r^{n+l-1}dr + \epsilon\int_{r_0}^{r_2} h^-(r)r^{n+l-1}dr\Big\} \\
&> k\int_0^{r_2} h(r)r^{n+l-1}dr,
\end{aligned}$$

which contradicts (4.7iii).

Remark. If the condition $\int_0^r h(s)s^{n+l-1}ds \ge 0$ is satisfied for all $r > 0$, then the conclusion follows from [K-N-Y], Theorem 9.3.

5. Proof of Theorem 5. The proof is based on Theorems 3 and 4. We will construct a specific function $f(r)$ such that for $p = p_*$ the corresponding problem (1.4) has solutions with properties analogous to Example III, Section 1. This result demonstrates the complexity of the structure of the solution set which can occur when the hypothesis $r_H \le r_G$ of [Y-Y$_1$] is not satisfied.

Lemma 5.1. *Under the hypotheses (2.1) the set of slowly decaying solutions of (1.4) is open.*

Proof. By Lemma 2.6 of [K-Y-Y] a sufficient condition for the openness of slowly decaying solutions is that there exist an $r_2 > 0$ such that

$$\frac{1}{p+1} r^{l-1}(r^{-l}f(r))_r \le 0, \quad r > r_2.$$

This is equivalent to $h(r) \le 0$ for $r > r_2$, and is implied by (2.1a).

Now let $\phi(r;\alpha)$ be the solution of the equation

$$\begin{aligned} &-(r^{n-1}\phi_r)_r = r^{n-1}r^l\phi^{p_*}, \\ &\phi(0) = \alpha, \quad \phi_r(0) = 0. \end{aligned} \tag{5.1}$$

The solutions $\phi(r;\alpha)$ are all rapidly decaying, and are given explicitely by

$$\phi(r;\alpha) = \alpha\{1 + \frac{2\alpha^{p_*-1}}{(p_*+1)(n-2)^2} r^{(n-2)(p_*-1)/2}\}^{-\frac{2}{p_*-1}}. \tag{5.2}$$

Consider equation (4.1) with f(r) defined by

$$f(r) = r^l + \epsilon(p_*+1)r^l \int_0^r s^{-(n+l)}k(s)ds, \tag{5.3}$$

and define $h_\epsilon(r) = r(r^{-l}f(r))_r = \epsilon(p_*+1)r^{-(n+l-1)}k(r)$. We will compare the solutions of (5.1) with those of equation (1.4) with $f = f_\epsilon$ given by (5.3).

By the assumptions (2.1) the function $f(r)$ satisfies (f_4), (f_6), (f_7) and (f_9), and clearly $h_\epsilon(r) \ge 0$ for $r \ge c$. It follows from Theorems 3 and 4 that for each $\epsilon > 0$ there are positive numbers $\alpha_0(\epsilon)$, $\alpha_1(\epsilon)$ which satisfy conditions (2.3a) and (2.3c). We will show that for ϵ sufficiently small the family of solutions also satisfies (2.3b), i.e. are slowly decaying. In fact, for $\alpha_* > 0$ the solution $u(r;\alpha_*)$ converges uniformly on the interval $[0,c]$ to the solution $\phi(r;\alpha_*)$ as $\epsilon \to 0$. Therefore there exists an $r_* > 0$ such that for $r > r_*$,

$$P(r;u(r;\alpha_*)) = \frac{1}{p_*+1}\int_0^r s^{(n+l-1)}h_\epsilon(s)(u^+(s;\alpha_*))^{p_*+1}ds$$
$$\leq \epsilon\{\int_0^a k(s)(u^+(s;\alpha_*))^{(p_*+1)}ds + \int_a^b k(s)(u^+(s;\alpha_*))^{(p_*+1)}ds$$
$$+\int_b^r k(s)(u^+(s;\alpha_*))^{(p_*+1)}ds\}$$
$$\leq \epsilon u(0;\alpha_*)^{(p_*+1)}\int_0^a k(s)ds + \epsilon u(b;\alpha_*)^{(p_*+1)}\int_a^b k(s)ds$$
$$+\epsilon u(b;\alpha_*)^{(p_*+1)}\int_b^c k(s)ds$$
$$=\epsilon\{\phi(0;\alpha_*)^{(p_*+1)}\int_0^a k(s)ds + \phi(b;\alpha_*)^{(p_*+1)}\int_a^b k(s)ds$$
$$+\phi(b;\alpha_*)^{p_*+1}\int_b^r k(s)ds\} + o(\epsilon).$$

For ϵ sufficiently small, by (2.1d) this is $\leq -c^2$. The existence of a slowly decaying solution follows from [Y-Y$_1$, Lemma 2.5].

From the existence of a slowly decaying solution we can conclude, by the open property of the slowly decaying solutions under hypotheses 2.1, and the corresponding property for crossing solutions, that there also exist at least two rapidly decaying solutions, $u(r;\alpha_1)$ and $u(r;\alpha_2)$, with $\alpha_1 \in (\alpha_0,\alpha_*)$ and $\alpha_2 \in (\alpha_*,\alpha_3)$.

Remark. Since $p = p_*$ the solution $u(r;\alpha_*)$ is slowly decaying by the Pohozaev identity.

References

[A-R] A. Ambrosetti and P.H. Rabinowitz, *Dual variational methods in critical point theory and applications*, J. Funct. Anal. **14** (1973), 349-381.

[B-E] G. Bianchi and H. Egnell, *An ODE approach to the equation* $\Delta u + K^{\frac{n+2}{n-2}} = 0$ *in* $\mathbf{R}^n$, Math. Z. **210** (1992), 137-166.

[B-L] H. Berestycki and P.L. Lions, *Nonlinear scalar field equation I: Existence of a ground state*, Arch. Rational Mech. Anal. **82** (1983), 313-345.

[E-R] Allan L. Edelson and Adolfo J. Rumbos, *Linear and semilinear eigenvalue problems in* $\mathbf{R}^n$, Commun. in Partial Diff. Equations **18(1 & 2)** (1993), 215-240.

[Eg] Henrik Egnell, *Asymptotic results for finite energy solutions of semilinear elliptic equations*, Jour. of Differential Equations **98** (1992), 34-56.

[F-L-N] D.G. de Figueiredo, P.L. Lions, and R.D. Nussbaum, *A priori estimates and existence of positive solutions of semilinear ellptic equations*, J. Math. Pures Appl. **61** (1982), 41-63.

[G-N-N] B. Gidas, W.-M. Ni, and L. Nirenberg, *Symmetry of positive solutions of nonlinear elliptic equations in* $\mathbf{R}^n$, Adv. in Math. Supplementary Stud. **7A** (1981), 369-402.

[K-N] T. Kusano and M. Naito, *Positive entire solutions of superlinear elliptic equations*, Hiroshima Math. Jour. **16** (1986), 361–366.

[K-N-Y] N. Kawano, W.-M. Ni, and S. Yotsutani, *A generalized Pohozaev identity and its applications*, J. Math. Soc. Japan **42 No. 3** (1990), 541–563.

[K-Y-Y] Nichiro Kawano, Eiji Yanagida and Shoji Yotsutani, *Structure theorems for positive radial solutions to* $\Delta u + K(|x|)u^p = 0$ *in* $\mathbf{R}^n$, Funkcialaj Ekvacioj **36** (1993), 557-579.

[L-N] Y. Li and W.M. Ni, *On conformal scalar curvature in* $\mathbf{R}^n$, Duke Math. J. **57** (1988), 895-924.

[N] W.M. Ni, *On the elliptic equation* $\Delta u + K(x)u^{\frac{n+2}{n-2}} = 0$, *its generalizations, and applications in geometry*, Indiana Univ. Math. Jour. **31** (1982), 493-529.

[N-S] E.S. Noussair and C.A. Swanson, *Solutions of Matukuma's equation with finite total mass*, Indiana University Mathematics Journal **38, No. 3** (Fall 1989).

[N-Y] W.M. Ni and S. Yotsutani, *Semilinear elliptic equations of Matukuma-type and related topics*, Japan Journal of Applied Mathematics **5 No. 1** (Feb. 1988.), 1-32.

[Y-Y_1] E. Yanagida and S. Yotsutani, *Classification of the structure of positive radial solutions to* $\Delta u + K(|x|)u^p = 0$ *in* $\mathbf{R}^n$, Arch. Rational Mech. Anal. **124** (1993), 239-259.

[Y-Y_2] E. Yanagida and S. Yotsutani, *Existence of positive radial solutions to* $\Delta u + K(|x|)u^p = 0$, *in* $\mathbf{R}^n$, Journal of Differential Equations **115** (1995), 477-502.

Contemporary Mathematics
Volume **208**, 1997

What do the Navier–Stokes equations tell us about turbulence?

Ciprian Foias*

Introduction. The answer to the question above is: Until the early eighties, very little; since then, quite a lot. In fact, the Navier–Stokes equations now provide a corroborating testimony (of course not willingly, but under the duress of rigorous mathematical investigations) of the validity of some heuristic theories of turbulence based on physical intuition and empirical evidence. The aim of this Note is to illustrate that fact with some examples related mainly to Kolmogorov's and Kraichnan's theories of the inertial range of the energy spectrum of turbulent flows. The discussion however cannot be carried on in a purely mathematical framework because the very concept of turbulence has not yet a purely mathematical content. "Loosely speaking, turbulence is an irreversible disorder, statistically organized according to several time and space scales, of mostly fast moving fluid flows" [35]; in fact, the first mathematical difficulty seems to reside in the very large number of those scales. The universal features of turbulent behavior are due to a statistical equilibrium between the (highly nonlinear) generation of smaller structures from larger ones and the (almost linear) viscous dissipation of the very small structures. This "is the principle of the turbulence machine. Classical fluid dynamical instabilities play the role of the fuel, vortex stretching of the engine and viscous dissipation of the brakes" [8]. Hopefully the subsequent discussion will make this description of turbulence a little clearer (see especially Sec. 2.)

The examples discussed below represent modest contributions to the following program of mathematical research connected to turbulence theory: Find rigorous proofs based on the mathematical analysis of the Navier-Stokes equations of those remarkable facts in turbulence theories which have been empirically accepted and which can be given a consistent mathematical formulation. The program is both ambitious and modest. It is ambitious because there is no a priori argument that it is feasible; it may happen that turbulence is not completely captured by the

*This work was supported by grants from the Office of Naval Research NAVY N00014-91-J-1140 and the National Science Foundation DMS-9404340 and DMS-9400615.

Navier-Stokes equations and thus a good empirical insight in turbulence may be out of the reach of the mathematical analysis of the Navier-Stokes equations. The program is modest because it takes its cues from the turbulence theories already accepted by specialists.

However the work done up to now within this program provides the following striking mathematical quasi-physical fact: The rigorous mathematical reformulation of the heuristic estimates provided by some widely accepted turbulence theories always seem to represent the upper bound of the rigorous mathematical estimates. These estimates are obtained by using a lot of functional inequalities (e.g. Sobolev inequalities, etc.) and a lot of averaging. If the heuristic estimates concern real turbulent flows, then the velocity fields of those turbulent flows must force all the functional inequalities used in the rigorous mathematical inference to be near equalities. In other words, turbulent velocity fields must live for most of the time in regions of their functional phase space which are amenable to variational principles. It may turn out that this fact will lead to a genuinely original contribution to the theory of turbulence.

The content of the Note is the following. First we introduce (in §1) the rigorous mathematical framework for the Navier–Stokes equations which will be used throughout. In §2 we present Kraichnan's mechanism for producing the universal energy spectra in the inertial range, that is a heuristic presentation of both Kolmogorov's and Kraichnan's theories. Then in §3 we give some rigorous, although pedantic, mathematical definitions of some of the physical quantities playing a basic role in the Kolmogorov–Kraichnan theory. In the next three paragraphs, we give the rigorous mathematical results which agree with the physical heuristic theory, concerning especially the number of the asymptotic degrees of freedom of a turbulent flow and the exponential decay of the energy spectrum and of the velocity power-spectrum in two dimensional turbulence. The rigorous estimate (5.12), (5.13) is new. Eventually the discussion was limited only to the topics in which the author was directly involved. For other very interesting results (again meant to provide a posteriori rigorous mathematical basis for successful heuristic approaches to turbulence especially concerning the spatial feature of the flows) we refer to the work of P. Constantin and C. Doering (e.g. [8], [9], [10], [11], [12], [13]).

1. The Navier–Stokes equations

We consider the Navier–Stokes equations

$$\begin{cases} u_t + (u\cdot\nabla)u = \nu\Delta u - \nabla p + f \\ \nabla\cdot u = 0 \end{cases} \tag{1.1}$$

in $\mathbb{R}^2$ or $\mathbb{R}^3$, where the velocity flow u, the pressure p and the body force f are periodic in each variable x_i ($1 \le i \le D = 2$ or 3). To avoid technical difficulties, we assume that the period, L, is the same in each direction. Using the Galilei invariance of (1.1) we can assume without loss of generality that $\int_{[0,L]^D} f dx = 0 = \int_{[0,L]^D} u dx$. When we will consider different boundary conditions we shall make the necessary specification; otherwise we will always refer to this special periodic case. In (1.1) the coefficient ν is the kinematic viscosity > 0. The phase space H of the equation (1.1) is the closure in $L^2([0,L]^D)^D$ of the space

$$\mathcal{D} = \{u \in H^2_{\rm loc}(\mathbb{R}^D)^D : u \text{ and } \nabla u\ L\text{-periodic}, \nabla\cdot u = 0\,, \int_{[0,L]^D} udx = 0\}\,. \tag{1.2}$$

We define

$$Au = -\Delta u\,,\ B(u,v) = P[(u\cdot\nabla)v] \text{ for } u, v \in \mathcal{D}\,, \tag{1.3}$$

where P is the orthogonal projection of $L^2([0,L]^D)^D$ onto H. The equations (1.1) become

$$\frac{du}{dt} + \nu Au + B(u,u) = f \tag{1.4}$$

where for convenience we have assumed that f is divergence free, time independent and in H. We will also use the notation

$$(u,v) = \frac{1}{L^D}\int_{[0,L]^D} u(x)\cdot v(x)dx\,,\ ((u,v)) = \frac{1}{L^D}\int_{[0,L]^D} \nabla u(x) : \nabla v(x)dx \tag{1.5}$$

where for vectors $a = (a_i)_{i=1}^D$, $b = (b_i)_{i=1}^D$ (resp. tensors $c = (c_{ij})_{i,j=1}^D$, $d = (d_{ij})_{i,j=1}^D$) we set

$$a\cdot b = \sum_{i=1}^D a_i b_i \ \text{(resp. } c : d = \sum_{i,j=1}^D c_{ij}d_{ij})\,.$$

Finally

(1.6) $$|u| = (u,u)^{1/2}\ ,\ \|u\| = ((u,u))^{1/2}\ .$$

Clearly A is a positive self-adjoint operator (with eigenvalues $\frac{4\pi^2}{L^2}|k|^2$ with $k \in \mathbb{Z}^D\backslash\{0\}$) and

(1.7) $$\mathcal{D}_{A^{1/2}} = \{u \in H : \|u\| < \infty\} \text{ with } \|u\| = |A^{1/2}u| \text{ for } u \in \mathcal{D}_{A^{1/2}}\ .$$

Note also that $\frac{1}{2}|u|^2$ (resp. $\|u\|^2$) is the kinetic energy (enstrophy) per unit of mass of the flow described by u. Moreover the bilinear operator B has the following algebraic properties

(1.8) $$\begin{cases} (B(u,v),v) = 0 \quad (u,v \in \mathcal{D}), & if \mathrm{D} = 2{,}3 \\ (B(u,u),Av) + (B(u,v)+B(v,u),Au) = 0 \quad (u,v \in \mathcal{D}), & if \mathrm{D} = 2 \end{cases}$$

relevant in the energy (respectively enstrophy) balance of an actual fluid flow.

A (strong or regular) solution of (1.4) on a time interval (t_1,t_2) is a continuous function $u : (t_1,t_2) \mapsto \mathcal{D}$ (endowed with the norm $v \mapsto |Av|, v \in \mathcal{D}$) satisfying (1.4) in H for all $t \in (t_1,t_2)$. A weak solution of (1.4) on a time interval (t_1,t_2) is a continuous function $u : (t_0,t_1) \mapsto H$ endowed with its weak topology, satisfying the energy inequality

(1.9)
$$\frac{1}{2}|u(t)|^2 + \nu\int_{t_0}^{t} \|u(\tau)\|^2 d\tau \le \frac{1}{2}|u(t_0)|^2 + \int_{t_0}^{t}(f,u(\tau))d\tau \ \text{ (where } t_1 < t_0 \le t < t_2)$$

with the exception of a zero-measure set of t_0's, and the equation (1.4) in the weak sense, namely

(1.10)
$$\frac{d}{dt}(u(t),v) + \nu((u(t),v)) + (B(u,u),v) = (f,v) \text{ for all } t \in (t_1,t_0) \text{ and } v \in \mathcal{D}_{A^{1/2}}\ .$$

Given a initial value u_1 at time t_1, there exists a weak solution on (t_1,∞) such that

(1.11) $$u(t) \to u_1 \text{ in } H \text{ (i.e. } |u(t)-u_1| \to 0) \text{ for } t \searrow t_1$$

and (1.9) holds even for $t_0 = t_1$. If $D = 2$ every weak solution is regular and the initial value problem has a unique solution. Clearly in this case there exists a nonlinear semigroup $\{S(t)\}_{t\geq 0}$ of maps on H such that for any $u_1 \in H$, the function $u(t) = S(t)u_1$ is the solution of (1.4) (on $(0,\infty)$) with initial data u_1 (at $t_1 = 0$). For $D = 3$, as (basically) shown by J. Leray [50], [51] if $u_1 \in \mathcal{D}_{A^{1/2}}$ then there exists a $t_2 > t_1$ such that the weak solution is regular on (t_1, t_2),

$$\|u(t) - u_1\| \to 0 \text{ for } t \searrow t_1$$

and the initial value problem for this data u_1 has a unique solution on (t_1, t_2). (See for instance [52], [67] or [16].) Whether always $t_2 = \infty$ is the still open question raised by J. Leray more than 60 years ago. Anyway, by virtue of the facts presented above, we can consider that the (weak) solutions of (1.4) are always defined on time intervals of the form (t_1, ∞) (or $[t_1, \infty)$ if they solve an initial value problem). All these solutions always converge in the weak topology of H to a maximal weakly compact set $\mathcal{A}$ with the property that if $u_0 \in \mathcal{A}$ then there exists a (weak) solution u on $(-\infty, \infty)$ bounded in H, such that $u(0) = u_0$. By definition this set is the *global attractor* of the equation (1.4), i.e. of the Navier–Stokes equations with the corresponding periodic boundary conditions. It contains a dense open (in the weak topology of H) subset $\mathcal{A}_{\text{reg}}$ with the property that for $u_0 \in \mathcal{A}_{\text{reg}}$ there exists a time interval $(t_1, t_2) \ni 0$ such that the solution u in $\mathcal{A}$ satisfying $u(0) = u_0$ is unique on (t_1, t_2) and time analytic as $\mathcal{D}$-valued function on (t_1, t_2) [35]. In case $D = 2$, $\mathcal{A}$ is compact in $\mathcal{D}$, it attracts in $\mathcal{D}$ all solutions, $S(t)\mathcal{A} = \mathcal{A}$ for all $t \geq 0$ and $\mathcal{A} = \mathcal{A}_{\text{reg}}$. In fact it is the usual global attractor of the dissipative evolution equation (1.4) (see [68], [70] and the literature there cited).

Finally, in the sequel we will also need the concept of a stationary statistical solution of the Navier–Stokes equation [26]. A probability (Borel) measure μ on H is called a *stationary statistical solution* (of the Navier–Stokes equations in the form (1.4)) if

$$\int \|u\|^2 \mu(du) < \infty \, , \; \mu(\mathcal{A}) = 1 \tag{1.12}$$

$$\int \left[\nu((u, \Phi'(u))) + (B(u,u) - f, \Phi'(u))\right] \mu(du) = 0 \tag{1.13}$$

for all test functionals $\Phi : H \mapsto \mathbb{C}$ which are Gateaux differentiable in H at any point $u \in \mathcal{D}_{A^{1/2}}$ and $\Phi'(u)$ is bounded in $\mathcal{D}_{A^{1/2}}$ on bounded subsets of $\mathcal{D}_{A^{1/2}}$. In the case $D = 2$, the stationary statistical solutions coincide with the probability (Borel) measures which are invariant for the nonlinear semi-group $\{S(t)\}_{t\geq 0}$ (see [26], [32]).

Our existence theorem for stationary statistical solutions is a soft hybrid of both Birkhoff ergodic theorem and Krylov-Bogoliubov theory. In order to formulate it, we need to introduce a useful device from functional analysis, namely a generalized limit; this is any linear functional, denoted $\mathrm{Lim}_{t\to\infty}$, which extends the usual limit (for $t \to \infty$) to the Banach space $B(0,\infty;\mathbb{R})$ of all complex-valued functions on $(0,\infty)$, such that for $\varphi \in B(0,\infty;\mathbb{R})$

$$|\mathrm{Lim}_{T\to\infty}\varphi(T)| \leq \sup_{0<T<\infty} |\varphi(T)| \tag{1.14}$$

and

$$\mathrm{Lim}_{T\to\infty}\varphi(T) = \lim_{T\to\infty} \varphi(T) \tag{1.15}$$

whenever this last usual limit exists. Moreover for a specified real-valued $\varphi_0 \in B(0,\infty;\mathbb{R})$ we can also have

$$\mathrm{Lim}_{T\to\infty}\varphi_0(T) = \limsup_{t\to\infty} \varphi_0(t)\ .$$

(To see that such a linear functional exists, apply the Hahn–Banach theorem in $B(0,\infty;\mathbb{R})$ to the functional $\lim_{j\to\infty} \varphi(t_j)$ on $\{\varphi \in B(0,\infty;\mathbb{R}) : \lim_{j\to\infty} \varphi(t_j) \text{ exists}\}$ where $(t_j)_{j=1}^{\infty}$ is a sequence in $(0,\infty)$, such that $t_j \to \infty$ and $\lim_{j\to\infty} \varphi_0(t_j) = \limsup_{t\to\infty} \varphi(t)$. Note that if $\varphi \geq 0$ then $\mathrm{Lim}_{T\to\infty}\varphi(T) \geq 0$ too.) With this extension of the concept of limit, we have the following existence theorem:

For any (weak) solution u on $(0,\infty)$ with initial data u_0 at $t = 0$, there exists a stationary statistical solution μ such that

$$\mathrm{Lim}_{T\to\infty}\frac{1}{T}\int_0^T \Phi(u(t))dt = \int \Phi(u)\mu(du) \tag{1.16}$$

for all weakly continuous functionals on H bounded on bounded subsets of H. In case $D = 2$, (1.16) remains also valid for any continuous functional Φ on $\mathcal{D}(= \mathcal{D}_A)$ bounded on bounded subsets of $\mathcal{D}$.

The proof is a slight modification of the proofs in [26] [32] (see also [4]) and will be omitted here. For a comprehensive presentation of the rigorous mathematical approach to statistical hydrodynamics see [72].

2. Kraichnan's approach to the inertial range in turbulence theory.

For the Navier–Stokes equations with our periodic boundary conditions we have

$$u \in H \Leftrightarrow u \sim \sum_{\substack{k\in\mathbb{Z}^D \\ k\neq 0}} a_k \exp \frac{2\pi i}{L}(k\cdot x)\ ,\ a_k\cdot k = 0\ ,\ a_k = a^*_{-k} \quad \text{(for all } k\text{)}\ . \tag{2.1}$$

For k in (2.1), $\kappa = \frac{2\pi}{L}(k\cdot k)^{1/2}$ is the corresponding wave number. The component of u with wave numbers in $[\kappa_1, \kappa_2)$ is

$$u_{\kappa_1,\kappa_2} = \sum_{\kappa_1\leq\kappa<\kappa_2} a_k \exp \frac{2\pi i}{L}(k\cdot x) \tag{2.2}$$

The energy/mass (resp. enstrophy/mass) is

$$|u_{\kappa_1,\kappa_2}|^2 = \sum_{\kappa_1\leq\kappa<\kappa_2} a_k\cdot a_k^* \tag{2.3}$$

resp.

$$\|u_{\kappa_1,\kappa_2}\|^2 = \frac{4\pi^2}{L^2}\sum_{\kappa_1\leq\kappa<\kappa_2} (k\cdot k)(a_k\cdot a_k^*) \tag{2.4}$$

Note that $u_{0,\infty} = u$.

Physicists and engineers assume that for $L \gg 1$ the time averages of

$$|S(t)u_{0\kappa_1,\kappa_2}|^2\ ,\ \|S(t)u_{0\kappa_1,\kappa_2}\|^2 \ \text{and}\ |AS(t)u_{0\kappa_1,\kappa_2}|^2 \tag{2.5}$$

also exists and are constant for $u_0 \in \mathcal{A}$; moreover they assume that these values can be viewed as integrals in the wave numbers, that is

$$\int_{\kappa_1}^{\kappa_2} E(\kappa)d\kappa\ ,\ \int_{\kappa_1}^{\kappa_2} E_1(\kappa)d\kappa\ ,\ \int_{\kappa_1}^{\kappa_2} E_2(\kappa)d\kappa\ . \tag{2.5a}$$

Comparing (2.3) with (2.4) as well as with the similar formula for the last term in (2.5), we see that in this case

$$E_1(\kappa) \equiv \kappa^2 E(\kappa)\ ,\ E_2(\kappa) \equiv \kappa^4 E(\kappa)\ . \tag{2.5b}$$

The function $E(\kappa)(\geq 0)$ is called the *energy spectrum* of the turbulent flow produced by f. Also the driving force f is assumed to have no high wave number components, that is $f = f_{0,\kappa_0}$ where κ_0 is comparable in size with $2\pi/L$ (the lowest wave number).

We first consider the case $D = 2$. The equation for the enstrophy of $v = u_{\kappa_1,\kappa_2} = S(t)u_{0\kappa_1,\kappa_2}$, $\kappa_2 = 2\kappa$, $\kappa = \kappa_1 > \kappa_0$ is given by

$$\frac{1}{2}\frac{d}{dt}\|v\|^2 = \left(\frac{du}{dt}, Av\right) = -\nu|Av|^2 - (B(u,u), Av) \ .$$

Introducing $v_< = u_{0,\kappa_1}$, $v_> = u_{\kappa_2,\infty}$ we obtain

$$\begin{aligned}\frac{1}{2}\frac{d}{dt}\|v\|^2 = &-\nu|Av|^2 - (B(v_<,v_<), Av) + (B(v,v) + B(v_>,v) + B(v,v_>), Av_<) + \\ &+ (B(v,v) + B(v_<,v) + B(v,v_<), Av_>) - (B(v_>,v_>), Av)\end{aligned}$$

where we used (1.8) repeatedly. Taking the physicists' time averages (which will be denoted by $\langle\cdot\rangle$) we obtain

(2.6)
$$\begin{aligned}0 = &-\nu\langle|Av|^2\rangle - \langle(B(v_<,v_<), Av) + (B(v_> + v, v_> + v), Av_<)\rangle + \\ &+ \langle(B(v + v_<, v + v_<), Av_>)\rangle - \langle(B(v_>,v_>), A(v_< + v))\rangle \ .\end{aligned}$$

Now let us assume that at these wave numbers κ and 2κ the enstrophy is on average carried only from low wave numbers to high wave numbers (Kraichnan's cascading scenario [45]). This means that in (2.6) one should take

$$\langle(B(v + v_>, v + v_>), Av_<)\rangle \approx 0 \ , \ \langle(B(v_>,v_>), A(v_< + v))\rangle \approx 0 \ ,$$

so that

$$\nu\langle|Av|^2\rangle \approx -\langle(B(v_<,v_<), Av)\rangle + \langle(B(v_< + v, v_< + v), Av_>)\rangle \ . \tag{2.6a}$$

Thus as long as

$$\nu\langle|Av|^2\rangle \approx \nu\int_{\kappa}^{2\kappa} \chi^4 E(\chi)d\chi \approx \nu\kappa^5 E(\kappa)$$

is small compared to the two averages in the right-hand side of (2.6a), we have

$$-\langle(B(v_<,v_<), Av)\rangle \approx -\langle(B(v_< + v, v_< + v), Av_>)\rangle \ . \tag{2.7}$$

The left-hand side represents the mean enstrophy/mass passed per unit of time from the component with wave numbers $< \kappa$ to that with wave numbers in $[\kappa, 2\kappa)$. The right-hand side represents the mean enstrophy/mass passed per unit of time from the component with waves numbers $< 2\kappa$ to that with wave numbers in $[2\kappa, 4\kappa)$. Let η denote this constant dissipation of enstrophy resulting from Kraichnan's scenario. Let also κ_1 be the smallest wave number for which Kraichnan's scenario is valid. Then writing (2.6a) for $\kappa = 2^j\kappa_1$ $(j = 1, 2, \dots)$ and summing up we obtain

$$\nu\langle|Au_{\kappa_1,\infty}|^2\rangle \approx -\langle(B(u_{0,\kappa_1}, u_{0,\kappa_1})), Au_{\kappa_1,2\kappa_1})\rangle \approx \eta \, .$$

So with the assumption that $\kappa_1 \approx \kappa_0 \approx 2\pi/L, L \gg 1$,

$$\begin{aligned} \eta \approx \nu\langle|Au_{\kappa_1,\infty}|^2\rangle &\approx \nu \int_{\kappa_1}^{\infty} \chi^4 E(\chi) d\chi \approx \nu \int_{\frac{2\pi}{L}}^{\infty} \chi^4 E(\chi) d\chi \\ &= \nu\langle|Au|^2\rangle \, . \end{aligned} \tag{2.8}$$

This last quantity $\nu\langle|Au|^2\rangle$ is the dissipation (due to viscosity) of the enstrophy/mass per half unit of time of the whole fluid flow.

We can now conclude that as long as

$$\nu\kappa^5 E(\kappa) \ll \eta \tag{2.9}$$

the component with wave numbers in $[\kappa, 2\kappa)$ is just transferring enstrophy with the constant rate $\approx \eta$ from lower wave numbers to higher wave numbers.

We introduce now a remarkable mechanical device for explaining the cascading process in turbulence. This mechanism was introduced by Kraichnan in [46] and later exploited by Frisch, Nelkin and Sulem [39]. To start, observe that the wave number κ has the physical dimension of $(\text{length})^{-1}$ and the component with wave number κ is considered to represent eddies of linear size about $1/\kappa$. Thus as a function of x the component $u_{\kappa,2\kappa}$ is thought to represent the system of eddies of linear size $\in (1/2\kappa, 1/\kappa]$. The transfer of enstrophy is considered to be produced by the breaking of these eddies into eddies of linear size $\leq 1/2\kappa$. This breaking is assumed to occur after the eddy travels a distance comparable to its linear size. Since the energy/mass of the eddies with linear size in $(1/2\kappa, 1/\kappa]$ is on average about

$$\int_{\kappa}^{2\kappa} E(\chi) d\chi \approx \kappa E(\kappa) \, ,$$

so the average velocity of those eddies is about

$$V_\kappa \approx (\kappa E(\kappa))^{1/2} .$$

The time necessary for the eddies to travel their linear size is thus about

$$t_k \approx (1/\kappa)/V_\kappa = 1/(\kappa V_\kappa) \approx 1/(\kappa^{3/2} E(\kappa)^{1/2}) .$$

On the other hand, the enstrophy/mass of the eddies with linear size in $(1/2\kappa, 1/\kappa]$ is

$$\int_\kappa^{2\kappa} \chi^2 E(\chi) d\chi \approx \kappa^3 E(\kappa) .$$

According to the breaking mechanism the mean dissipation of the enstrophy/mass per unit of time should thus be

$$\eta \approx \frac{\kappa^3 E(\kappa)}{t_\kappa} \approx \kappa^3 (\kappa E(\kappa))^{3/2}$$

whence

$$E(\kappa) \approx \frac{\eta^{2/3}}{\kappa^3} . \tag{2.10}$$

According to the previous arguments, one expects (2.10) as long as

$$\kappa > \kappa_0 \text{ and (2.9) holds.} \tag{2.11}$$

Using (2.10) we find that (2.9) is equivalent to $\kappa^2 \ll \eta^{1/3}/\nu$ that is

$$\kappa \ll \kappa_\eta = \left(\frac{\eta}{\nu^3}\right)^{1/6} . \tag{2.12}$$

This κ_η is called *Kraichnan's dissipation wave number* [45]. Beyond it, one expects that the viscosity forces become dominant and break down the eddies so fast that the eddies of length $\ll \ell_\eta = 1/\kappa_\eta$ (= *Kraichnan's dissipation length*) are not expected to play any role in the dynamics of the turbulent flow.

In the three dimensional case $D = 3$, the role played by the enstrophy in the preceding argument is taken over by the energy. Consequently we now first obtain

$$\nu \int_\kappa^{2\kappa} \chi^2 E(\chi) d\chi \approx -\langle (B(v_<, v_<), v) \rangle + \langle (B(v_< + v, v_- + v), v_>) \rangle ,$$

and then

$$-\langle(B(v_<, v_<), v)\rangle \approx -\langle(B(v_< + v, v_< + v), v_>)\rangle \approx \epsilon = \nu\langle\|u\|^2\rangle$$

provided that

$$\nu\kappa^3 E(\kappa) \ll \epsilon\,.$$

Here ϵ represents the mean dissipation of the energy/mass per unit of time.

The Kraichnan mechanism now leads to the estimate

$$\epsilon \approx \frac{\kappa E(\kappa)}{t_\kappa} \approx (\kappa E(\kappa))^{3/2}\kappa\,,$$

that is

$$E(\kappa) \approx \frac{\epsilon^{2/3}}{\kappa^{5/3}} \tag{2.13}$$

for

$$\nu\kappa^{4/3}\epsilon^{2/3} \ll \epsilon\,,$$

that is

$$\kappa \ll \kappa_\epsilon = \left(\frac{\epsilon}{\nu^3}\right)^{1/4}\,. \tag{2.14}$$

This κ_ϵ is called the *Kolmogorov's dissipation wave number.* The spectra given by (2.13) and (2.10) are respectively the *Kolmogorov spectrum for 3D turbulence* [44] and the *Kraichnan spectrum for 2D turbulence* [45]. The wave number intervals where the mechanism described above holds (and hence the Kolmogorov spectrum in 3D or the Kraichnan spectrum in 2D is present) is the *inertial range* of the turbulence. (See also [58], [62]). The empirical evidence for the existence of Kolmogorov's inertial range in 3D turbulence is much stronger than that for the existence of the Kraichnan's inertial range in 2D turbulence. This is probably because, in nature, 2D turbulence is a rather rare occurrence.

3. Rigorous definitions for the dissipation waves numbers

From a mathematician's (perhaps narrow) perspective it is troubling that the mostly heuristic considerations in the preceding §2 do not provide any cue on the average operation $\langle\cdot\rangle$. The experts tacitly assume that for a function $\Phi(u)$, representing some physical quantity associated to the fluid flow u,

$$\langle\Phi(u)\rangle \approx \lim_{T\to\infty}\frac{1}{T}\int_0^T \Phi(u(t))dt \tag{3.1}$$

where $u(t)$ is almost any solution of the Navier–Stokes equation, and that the average is also given by

$$\langle\Phi(u)\rangle \approx \int \Phi(u)\mu(du) \tag{3.2}$$

with some probability measure μ on H, hopefully invariant with respect to the Navier–Stokes equations. The problem with this ergodicity assumption is that nobody has ever even come close in proving it for the Navier–Stokes equations. (For metaphors in finite dimensional dynamics and their successes, see [63] Sec. 13.5.) The theorem stated at the end of §1 allows us to bypass this difficulty. Namely, one can easily and rigorously prove that in the case $D = 2$ the set Range(η) of all

$$\mathrm{Lim}_{T\to\infty}\frac{1}{T}\int_0^T \nu|AS(t)u_0|^2 dt \tag{3.3}$$

when u_0 runs over H and $\mathrm{Lim}_{t\to\infty}$ over all possible generalized limits, coincides with the set

$$\int \nu|Au|^2\mu(du) \tag{3.4}$$

where μ run over all probability (Borel) measures invariant for $\{S(t)\}_{t\geq 0}$. It follows that

$$\mathrm{Range}(\eta) = [\underline{\eta}, \overline{\eta}] \tag{3.5}$$

and consequently the possible values of the Kraichnan dissipation wave number κ_η must satisfy

$$\left(\frac{\underline{\eta}}{\nu^3}\right)^{1/6} = \kappa_{\underline{\eta}} \leq \kappa_\eta \leq \kappa_{\overline{\eta}} = \left(\frac{\overline{\eta}}{\nu^3}\right)^{1/6}. \tag{3.6}$$

In [29] it was proven that

$$\min \kappa_{\underline{\eta}} = c_0 G^{1/6} L^{-1}\ , \ \max \kappa_{\overline{\eta}} = c_1 G^{1/3} L^{-1} \tag{3.7}$$

(here and throughout c_0, c_1 denote some absolute constants, not necessarily the same) where

$$G = \frac{L^3|f|}{\nu^2} = \frac{L^2}{\nu^2}\left(\int_{[0,L]^2} f(x)\cdot f(x)dx\right)^{1/2} \tag{3.8}$$

is the *generalized Grashoff number* (a dimensionless constant introduced in [30]) and the extrema in (3.7) are taken over all driving forces f with a large generalized Grashoff number G.

In a similar manner, in the case $D = 3$ one can prove that

$$\min \kappa_\epsilon = c_0 G_*^{1/4} L^{-1}\ , \ \max \kappa_\epsilon = c_1 G_*^{1/2} L^{-1} \tag{3.9}$$

where

$$G_* = \frac{L^2|A^{-1/2}f|}{\nu^2} = \frac{L^{1/2}}{\nu^2}\left(\int (-\Delta^{-1}f)(x)\cdot f(x)dx\right)^{1/2} \tag{3.10}$$

is a dimensionless constant [29].

4. The Landau–Lifschitz asymptotic degrees of freedom

The dissipation lengths $\ell_\eta = 1/\kappa_\eta$ and $\ell_\epsilon = 1/\kappa_\epsilon$ appear in the following beautiful heuristic Landau–Lifschitz argument [49]: Assume that one monitors the fluid in all points of a grid with mesh of length $\approx \ell_\eta$ (resp. $\approx \ell_\epsilon$) in the basic domain $[0, L]^D$ for $D = 2$ (resp. $D = 3$). The Kraichnan (resp. Kolmogorov) theory implies that the eddies of size $< \ell_\eta$ (resp. ℓ_ϵ) are rapidly annihilated by viscous dissipation and therefore are of no dynamical consequence. Hence, since all eddies of size $> \ell_\eta$ will be tracked at some grid point, one should expect that the degrees of freedom of a 2D (resp. 3D) turbulent flow be at most about

$$\left(\frac{L}{\ell_\eta}\right)^2 = (L\kappa_\eta)^2 \text{ (resp. } \left(\frac{L}{\ell_\epsilon}\right)^3 = (L\kappa_\epsilon)^3) .$$

Obviously the worst possible scenario corresponds to $\eta = \overline{\eta}$ (resp. $\epsilon = \overline{\epsilon}$), in which case the bounds are

$$(L\kappa_{\overline{\eta}})^2 \text{ (resp. } (L\kappa_{\overline{\epsilon}})^3) . \tag{4.1}$$

Taking into account that all statistical considerations in §2 ultimately concern (in view of the discussion in §3) flows u lying in the global attractor, one expects that the bound (4.1) for the degrees of freedom of a turbulent flow should be connected with the dimension of the attractor. Surprisingly enough, one can prove that this is indeed the case.

In order to present this connection, let us recall that the fractal dimension [55] of a compact set X in H is defined by

$$d_M(X) = \limsup_{\delta \searrow 0} \frac{\log N_\delta(X)}{\log \frac{1}{\delta}} \tag{4.2}$$

where for $\delta > 0$, $N_\delta(X)$ is the smallest number of balls of radii $\leq \delta$ necessary to cover X. The fractal dimension of X can be ∞ even if its Hausdorff dimension is 0 (see [1]). Moreover if $d_M(X) < \infty$ there are orthogonal projections P in H of rank $\leq 2d_M(X) + 1$ such that $(P|X)^{-1}$ is a Hölder function [31]. Therefore the fractal dimension is a better indicator than the Hausdorff dimension of the number of parameters necessary to describe a set as well as the dynamics it may carry (see also [24] Ch.X).

The basic estimate connecting the heuristic theory of Kraichnan, via the Landau-Lifschitz argument, to the rigorous mathematical theory is the following [20]:

In the case $D = 2$ the fractal dimension of the global attractor $\mathcal{A}$ satisfies the upper estimate

$$d_M(\mathcal{A}) \leq c_0 (L\kappa_{\overline{\eta}})^2 (\log(L\kappa_{\overline{\eta}}) + 1)^{1/3} \,. \tag{4.3}$$

This is a rigorous mathematical result. It is valid regardless of the spectrum of f (that is even if the force f has high wave number components). In terms of the generalized Grashoff number G it yields (by virtue of (3.7))

$$d_M(\mathcal{A}) \leq c_0 G^{2/3} (\log(c_1 G) + 1)^{1/3} \,. \tag{4.4}$$

Note also that with regard to the bound in (4.1), in (4.3) we have a supplementary logarithmic term. Remarkably, in a follow-up of [20], K. Ohkitani has shown [59] that a more careful analysis within Kraichnan's heuristic framework would have yielded, prior to the purely mathematical work [20], the logarithmic corrective term in (4.1). More recently V.X. Liu presented a proof that $d_M(\mathcal{A}) \geq c_1 G^{2/3}$ for the case when the driving force is a special eigenvector of A [53].

In the case $D = 3$ we do not yet know if the upper bound for $d_M(\mathcal{A})$ is consistent with (4.1). The first difficulty comes from the fact that it is not known if in this case too $\mathcal{A} = \mathcal{A}_{\text{reg}}$.

The best rigorous estimate for $d_M(\mathcal{A})$ in the 3D case seems to be the following:

Assume first that $\mathcal{A} \subset \mathcal{D}_{A^{1/2}}$. Then in $\mathcal{A}$ the initial value problem has a unique solution and we can define

$$\tilde{\epsilon}(u_0) = \limsup_{T\to\infty} \left[\frac{1}{T} \int_0^T \nu L^{-3} \int_{[0,L]^3} (\nabla u(t,x) : \nabla u(t,x))^{5/4} dx dt \right]^{4/5} \tag{4.5}$$

where $u(t) = u(t,x)(x \in \mathbb{R}^3)$ is the unique solution of (1.4) on $(-\infty, \infty)$ lying in $\mathcal{A}$ with initial data $u(0) = u_0$. Set

$$\tilde{\epsilon} = \sup_{u_0 \in \mathcal{A}} \tilde{\epsilon}(u_0) \ , \ \kappa_{\tilde{\epsilon}} = \left(\frac{\tilde{\epsilon}}{\nu^3} \right)^{1/4} . \tag{4.6}$$

Then

$$d_M(\mathcal{A}) \leq c_0 (L\kappa_{\tilde{\epsilon}})^3 . \tag{4.7}$$

This is a slight improvement (made possible by [25], Thm.4.1) of the result in [19], §4.5.6. Note that $\overline{\epsilon}$ (in (4.1)) $\leq \tilde{\epsilon}$ and thus the upper bound in (4.7) is larger than the one given in (4.1). For some intriguing new estimates for this hypothetical global attractor see [41].

It is appropriate to mention that, in every day life, exponential attractors are probably more relevant than the global attractor $\mathcal{A}$ itself [23]. These are outgrowths of $\mathcal{A}$, still with finite fractal dimension but attracting all solutions at an exponential rate (see [24]). However the estimates for their fractal dimensions are not yet as sharp as the ones for $d_M(\mathcal{A})$.

A mathematical concept nearer to the meaning of the degrees of freedom in the Landau–Lifschitz argument is that of determining nodes [34] [37]. The points of a fixed finite set $\mathcal{F}$ in the fluid are called determining nodes if for any two solutions u, v of the Navier–Stokes equations, the convergence

$$u(t,a) - v(t,a) \to 0 \text{ (in } \mathbb{R}^d) \text{ (as } t \to \infty) \text{ for all } a \in \mathcal{F}$$

implies

$$u(t) - v(t) \to 0 \text{ (in } H) \quad (\text{as } t \to \infty) .$$

The rigorous estimates for the minimal number N_{det} of these nodes is not yet as sharp as the estimate (4.3), but quite close, namely [43]: In the case $D = 2$, the minimal number of determining nodes satisfies

$$N_{\text{det}} \le c_0 G . \tag{4.8}$$

For similar but weaker results in the case $D = 3$, see [14].

It is worth mentioning that a theorem of Takens [66] asserts that "generically", one determining node should suffice (i.e. $N_{\text{det}} = 1$); but it is not known if the Navier–Stokes equations or any related equations are generic in the sense of Takens. There are however some interesting partial differential equations in dimension $D = 1$ for which $N_{\text{det}} \le 4$ regardless of the value of the physical constants present in the equations (see [48], [27]).

5. Kolmogorov's estimate for energy dissipation

Let us start with the following heuristic argument: If $en = \frac{1}{2}\int_0^\infty E(\kappa)d\kappa$ is the average of the energy/mass in a turbulent fluid flow with an average dissipation rate $\epsilon = \nu\int_0^\infty \kappa^2 E(\kappa)d\kappa$, then $t_\epsilon = en/\epsilon$ should represent a characteristic time for the dissipation of the energy and the characteristic mean velocity U should be $\sqrt{2en}$. The corresponding length $\ell = Ut_\epsilon$ can be viewed as the average distance travelled by the turbulent eddies until they dissipate. So

$$\epsilon \approx \frac{U^2}{t_\epsilon} = \frac{U^3}{\ell} . \tag{5.1}$$

This is the Kolmogorov estimate for the energy dissipation in a turbulent fluid flow. U and ℓ are considered to be the characteristic velocity and length respectively associated to the given particular flow so that

$$\text{Re} = \frac{U\nu}{\ell} \tag{5.2}$$

is the Reynolds number of the flow (see e.g. [49]). From (2.14), (5.1) and (5.2) it follows that

$$\ell\kappa_\epsilon \approx (\text{Re})^{3/4} . \tag{5.3}$$

Now in the original Landau-Lifschitz argument (see [49], Ch.3) the linear size of the typical box (cube in $\mathbb{R}^3$, square in $\mathbb{R}^2$) was ℓ and thus the number of the degrees of freedom comes out to be

$$(\ell\kappa_\epsilon)^3 \approx (\mathrm{Re})^{9/4}\ , \text{if } \ D=3\ , \tag{5.4}$$

$$(\ell\kappa_\epsilon)^2 \approx (\mathrm{Re})^{3/2}\ .\text{if } \ D=2\ . \tag{5.5}$$

Remarkably enough, there is a rigorous analogue of (5.5), namely: If $D = 2$ then

$$d_M(\mathcal{A}) \le c_0(\ell\kappa_\epsilon)^2 \le c_1(\mathrm{Re})^{3/2} \ \text{ with } \ \ell = L \text{ in } (5.2) \tag{5.6}$$

(where again c_0 and c_1 are some absolute constants). Moreover this Re is related to the generalized Grashoff number G (see (3.8)) by

$$\mathrm{Re} \le c_2 G^{2/3}$$

so that (5.6) yields

$$d_M(\mathcal{A}) \le c_3 G\ , \tag{5.7}$$

where c_2 and c_3 are some adimensional constants. The estimates (5.6) and (5.7) up to a logarithmic factors were first obtained in [15]; their present form was subsequently obtained in [69]. This result holds even if the periodic boundary conditions are replaced by homogeneous Dirichlet boundary conditions provided the domain filled by the fluid has a C^2-boundary and has a finite area in $\mathbb{R}^2$ and even for more general domains [42] (see also [61]). It is not yet known if the sharper estimate (4.4) holds in this last case. In the case $D = 3$, it is not known if the upper bound $(\mathrm{e})^{9/4}$ is a upper bound for $d_M(\mathcal{A})$ (see [19],§4.3.b and §4.5.c for similar but still unsatisfactory rigorous estimates).

Up to now we have only shown that the Landau–Lifschitz estimates can be given a rigorous mathematical meaning. The same turns out to be true for the Kolmorgorov estimate (5.2), due to some very beautiful mathematical work initiated by P. Constantin and C.R. Doering [10]. These authors consider a shear flow, namely the

2D Navier–Stokes equations in the rectangle $[0, L] \times [0, h]$ with periodic boundary conditions in x_1 and the Dirichlet conditions on the bottom and respectively top sides of the rectangle:

$$u\big|_{x_2=0} \equiv 0 \text{ and } u_1\big|_{x_2=h} \equiv U \,,\; u_2\big|_{x_2=h} = 0 \,. \tag{5.8}$$

Clearly, in this case the Couette flow $u = (Ux_2h^{-1}, 0)$ is a stationary solution. In [10] it is proven that

$$\epsilon = \nu \limsup_{T\to\infty} \frac{1}{T} \int_0^T \frac{1}{Lh} \int_{[0,L]\times[0,\ell]} \nabla u(t,x) : \nabla u(t,x) dx \le \frac{U^3}{h} \tag{5.9}$$

for any solution $u(t) = u(t,x)$, $x \in [0, L] \times [0, h], t \in [0, \infty)$ of the Navier–Stokes equations.

Very recently, this result was extended by X. Wang [73] to any boundary driven 2D flow. Moreover in [57] it is also established that for a boundary and force driven 2D flow in some piecewise smooth bounded domain $\Omega \subset \mathbb{R}^2$, one has the estimate

$$d_M(\mathcal{A}) \le c_0(\mathrm{Re})^{3/2} + c_1 G \tag{5.10}$$

where G is the generalized Grashoff number as defined in (3.8) while

$$\mathrm{Re} = \frac{U\ell}{\nu} \text{ with } \ell = \text{ diam}\,\Omega \,,\; U = \max\{\|u|_{\partial\Omega}\|_{L^\infty}, \ell\|\nabla_{\tan} u|_{\partial\Omega}\|_{L^\infty}\} \,. \tag{5.11}$$

For $u\big|_{\partial\Omega} \equiv 0$, the estimate (5.10) actually reduces to (5.7). However for $u\big|_{\partial\Omega} \not\equiv 0$, the estimate (5.10) is not only new, but brings a positive answer to a problem open since the late 70's; indeed the previous known rigorous estimates (e.g. [33], [2], etc.) involved Re at an unphysical high rate.

At this point, it looks rather odd that rigorous upper estimates of type (5.1) presented above have only been established for boundary problems other than the boundary free periodic case; after all, the latter case was introduced as the tamest mathematical setting for connecting the Navier–Stokes equations to conventional turbulence theory. To fill this void, below is a rigorous analogue of (5.9) for the periodic case (which unlike the cases discussed above, does not require novel mathematical insight):

Let $f \in \mathcal{D}_{A^{1/2}}$ and $L_f = (|A^{-1/2}f||A^{\frac{1}{2}}f|)^{1/2}$. Then for any stationary statistical solution of the Navier–Stokes equations (1.4) associated to a weak solution according to (1.16), one has

$$\epsilon = \nu \int \|u\|^2 \mu(du) \leq c_0 \left(\frac{L}{L_f}\right)^{\frac{D+2}{2}} \frac{U^3}{L} \tag{5.12}$$

provided

$$\mathrm{Re} = \frac{UL}{\nu} \geq c_1 \tag{5.13}$$

where

$$U = \left[\int |u|^2 \mu(du)\right]^{1/2} . \tag{5.14}$$

Proof. The (weak) solution $u = u(t)$ of (1.4) satisfies (see (1.9))

$$\mathrm{Lim}_{T\to\infty} \left[\frac{1}{T}\int_0^T \nu\|u(\tau)\|^2 d\tau - \frac{1}{T}\int_0^T (f, u(\tau))d\tau\right] \leq 0$$

and (see (1.10))

$$\mathrm{Lim}_{T\to\infty} \left[\nu \frac{1}{T}\int_0^T ((u(\tau), A^{-1}f)) + \frac{1}{T}\int_0^T (B(u(\tau), u(\tau)), A^{-1}f)d\tau\right] = (f, A^{-1}f) .$$

It follows

$$\mathrm{Lim}_{T\to\infty} \left[\frac{1}{T}\int_0^T \nu\|u(\tau)\|^2 d\tau\right] \leq \int (f, u)\mu(du) \leq |f|U$$

and

$$\begin{aligned}
|A^{-1/2}f|^2 &\leq \nu \int (u, f)\mu(du) + \mathrm{Lim}_{T\to\infty}\frac{1}{T}\int_0^T (B(u, u), A^{-1}f)d\tau \leq \\
&\leq \nu|f|U + \mathrm{Lim}_{T\to\infty}\frac{1}{T}\int_0^T (B(u, u), A^{-1}f)d\tau \leq \\
&\leq \nu|f|U + \mathrm{Lim}_{T\to\infty}\frac{1}{T}\int_0^T |(B(u(\tau), A^{-1}f), u(\tau))|d\tau \leq \\
&\leq \nu|f|U + c_0 L^{D/2}|A^{\frac{1}{2}}f|^{1/2}|A^{\frac{D-3}{2}}f|^{1/2}\mathrm{Lim}_{T\to\infty}\frac{1}{T}\int_0^T |u(\tau)|^2 d\tau .
\end{aligned}$$

Note that $|u|^2$ and $\|u\|^2$ are not functionals accepted by formula (1.16), but nevertheless it can be easily proved that

$$\epsilon = \nu \int \|u\|^2 \mu(du) \leq \mathrm{Lim}_{T\to\infty}\frac{1}{T}\int_0^T \nu\|u(\tau)\|^2 d\tau$$

and with a bit more work that

$$\mathrm{Lim}_{T\to\infty}\frac{1}{T}\int_0^T |u(\tau)|^2 d\tau = \int |u|^2 \mu(du) = \frac{U^2}{2}\,.$$

We obtain

$$\epsilon \le |f|U\ ,\ |A^{-1/2}f|^2 \le \nu|f|U + c_0 L^{D/2}|A^{\frac{1}{2}}f|^{1/2}|A^{\frac{D-3}{2}}f|^{1/2}U^2$$

So, if (5.13) holds with $c_1 = c_0^{-1}$, we have

$$\epsilon \le c_0 L^{D/2}\frac{|f||A^{\frac{1}{2}}f||A^{\frac{D-3}{2}}f|^{1/2}}{|A^{-\frac{1}{2}}f|^2}U^3 \le c_0 L^{D/2}\left(\frac{|A^{\frac{1}{2}}f|}{|A^{-\frac{1}{2}}f|}\right)^{\frac{D+2}{4}}U^3\ ,$$

whence (5.12) holds.

It is worth observing that if the driving force f is as in §2, i.e. $f = f_{0,\kappa_0}$, then $L_f\kappa_0 \ge 1$ and (5.12) becomes

$$\epsilon \le c_2(L\kappa_0)^{\frac{D+2}{2}}\frac{U^3}{L}\,. \tag{5.15}$$

Therefore if $\kappa_0 \approx 2\pi/L$, as assumed in §2 and automatically also in the heuristic discussion at the beginning of this section, then $L\kappa_0 \approx 1$ and thus (5.15) provides again a rigorous, albeit partial, confirmation of Kolmogorov's relation (5.1). If the driving force f is arbitrary, for instance if for a fixed large Grashoff number, G, we choose f such that $Af = \lambda f$, where λ is very large, then (5.1) may not hold.

In connection with the topics discussed in this section, as well as in the preceding §4, it is necessary to mention that very interesting purely mathematical work was done on the Navier–Stokes equations in thin domains (e.g. [2], [3], [40], [60], [71], [54], [75]).

6. Exponential decay of spectra

Let us return to the Kraichnan's and Kolmogorov's energy spectra; of course under the assumption that $f = f_{0,\kappa_0}$ and $G \gg \kappa_0 L$. There is as yet no mathematically rigorous confirmation based on the analysis of the Navier–Stokes equations of their form (2.10) and (2.13), respectively. Note, however, that the considerations in §§4-5 show that the wave numbers κ_η and κ_ϵ marking the upper ends of the corresponding inertial ranges can be given a rigorous meaning based on the functional geometry of the global attractor of the Navier–Stokes. Are the Kraichnan and Kolmogorov spectra the reflection of some intricate geometrical features of the global attractors? A not completely rigorous argument (see [17]) suggests that the answer to the preceeding question must also take into consideration the spectral nature of the driving force; in particular, if the driving force f is an eigenvector of A (as in the result of [53] quoted on §4) then, seemingly, the Kraichnan spectrum cannot be present.

It is amazing, on the other hand, that rigorous mathematical results can shed light on the nature of those spectra for wave numbers κ much above the wave numbers κ_η and κ_ϵ, that is in the *dissipation range* of the turbulent flow. A natural idea to come to mind is that, since $\langle\|u_{\kappa,\infty}\|^2\rangle$ for $\kappa \gg \kappa_\eta$ and $\langle|u_{\kappa,\infty}|^2\rangle$ for $\kappa \gg \kappa_\epsilon$ are very small due mainly to the viscous dissipation, these averages should behave like the Fourier components of the solutions of the linear equation

$$\frac{du}{dt} + \nu A u = f \ \text{ for } t \gg 0 \,.$$

If this were so, then it becomes mathematically clear that

$$E(\kappa) \approx \begin{cases} O\left(e^{-c_0\left(\frac{\kappa}{\kappa_\eta}\right)^2}\right) & \text{for } \kappa \gg \kappa_\eta \quad (if\text{D}=2)\,, \\ O\left(e^{-c_1\left(\frac{\kappa}{\kappa_d}\right)^2}\right) & \text{for } \kappa \gg \kappa_d \quad (if\text{D}=3)\,. \end{cases}$$

Although this does not seem to be the right idea, it is quite persistent (see [65] and the rebutal [56]). The right idea supported by theoretical arguments and excellent numerical computations (see [28], [6]) seems to be that

$$E(\kappa)/E(\kappa_\eta) \approx c_2(\kappa/\kappa_\eta)^{c_0} e^{-c_1\frac{\kappa}{\kappa_\eta}} \text{ for } \kappa \gg \kappa_\eta \ \ (\text{if } D=2) \tag{6.1}$$

$$E(\kappa)/E(\kappa_\epsilon) \approx c_2(\kappa/\kappa_\epsilon)^{c_0} e^{-c_1\frac{\kappa}{\kappa_\epsilon}} \text{ for } \kappa \gg \kappa_\epsilon \ \ (\text{if } D=3)\,, \tag{6.2}$$

where $c_1, c_2 > 0$. As yet for the values of c_0 and c_2 there is no reliable heuristic guess. The fact that for the regular solutions of the Navier–Stokes equations, the Fourier components decrease exponentially in the wave number was established in [47] Ch.9, and a bit later in [36] by a different method; subsequently explicit estimates in terms of the physical parameters (e.g. G) were given in [5] (for $D=2$) and [22] (for $D=3$). We shall present the approach considered in these last three works, because it leads to estimates which are mathematically nearer to (6.2) as well as physically less restrictive (for a discussion on this last point see [22]§5). Also in order to avoid the mathematical difficulties due to the still mysterious mathematical nature of the attractor in the case $D=3$, we shall consider only the case $D=2$.

Let us start by first giving to (6.2) (case $D=2$) a less precise but more amenable form. Recall

$$E(\kappa_\eta) \approx \frac{\eta^{2/3}}{\kappa_\eta^3} = \frac{(\eta/\nu^3)^{2/3}}{\kappa_\eta^3} = \kappa_\eta \nu^2 .$$

So, if (6.1) were valid, then for any $\alpha > 0$ and $\kappa \gg \kappa_\eta$

$$\int_\kappa^\infty \chi^{2\alpha} E(\chi) d\chi \le c_2 \kappa_\eta \nu^2 \int_\kappa^\infty e^{-c_1 \frac{\chi}{\kappa_\eta}} \chi^{2\alpha} \left(\frac{\chi}{\kappa_\eta}\right)^{c_0} d\chi \le$$

$$\le c_3 \kappa_\eta^{2(1+\alpha)} \nu^2 e^{-\frac{c_1 \kappa}{2\kappa_\eta}} \le c_5 G^{\frac{2}{3}(1+\alpha)} \frac{\nu^2}{L^{2(1+\alpha)}} e^{-c_4 \kappa L G^{-1/3}}$$

where we used (3.6), (3.7). Heuristically, for large L,

$$\int_\kappa^\infty \chi^{2\alpha} E(\chi) d\chi \approx \langle |A^{\alpha/2} u_{\kappa,\infty}|^2 \rangle .$$

Therefore to check (6.2), one should rigorously deduce from the Navier–Stokes equations (1.4) an estimate of the form

$$\int |A^{\alpha/2} u_{\kappa,\infty}|^2 \mu(du) \le c_0 G^{\frac{2}{3}(1+\alpha)} \frac{\nu^2}{L^{2(1+\alpha)}} e^{-c_1 \kappa L G^{-1/3}} \tag{6.3}$$

for any stationary statistical solution μ of the Navier–Stokes equations; recall that, since $D=2$, these μ's are just the probability Borel measures on $\mathcal{A}$ which are invariant for $\{S(t)\}_{t\ge 0}$.

The approach (called Gevrey class method) in [36], [5], [22] consists in obtaining from the equation (1.4) (or rather from its Galerkin approximations in the spaces $\{u_{0,\kappa} : u \in H\}$) an a priori estimate for a solution $u = u(t)$ of (1.4) of the norm

$$|A^{\alpha/2} e^{\beta(s\nu A)^\gamma} u(t_0+s)|^2 \tag{6.4}$$

for small $s > 0$ with $\gamma \in (0, 1/2]$. This estimate is then optimized in the dimensionless constant $\beta > 0$. Above for $\delta = \beta(s\nu)^\gamma$, $u = u(t_0+s)$, and for

$$u \sim \sum a_k e^{\frac{2\pi i}{L} k\cdot x} \text{ in } H \tag{6.5}$$

the norm is defined by

$$\|A^{\alpha/2}e^{\delta A^{\gamma}}u\|^2 = \frac{4\pi^2}{L^2}\sum_{k\in\mathbb{Z}^2\backslash\{0\}} e^{2\delta(4\pi^2 k\cdot k)^{\gamma}}(k\cdot k)^{\alpha}(a_k\cdot a_k^*) \tag{6.5a}$$

¿From (6.5a) it clearly follows, for $\theta \geq 0$,

$$|A^{\alpha/2}u_{\kappa,\infty}|^2 \leq e^{-2\theta L\kappa}|A^{\alpha/2}e^{\theta LA^{1/2}}u|^2 \tag{6.6}$$

and therefore

$$\int |A^{\alpha/2}u_{\kappa,\infty}|^2\mu(du) \leq e^{-2\theta L\kappa}\int |A^{\alpha/2}e^{\theta LA^{1/2}}u|^2\mu(du) \tag{6.7}$$

for any μ as in (6.3). We see that in order to obtain estimates of the form (6.3) it suffices to obtain good estimates for

$$\int |A^{\alpha/2}e^{\theta LA^{1/2}}u|^2\mu(du)\,.$$

In [5] it was almost explicitely established, by the Gevrey class method, that if the Grashoff number G is large enough then

$$\left\| e^{\frac{c_0 L}{G^2\log G}A^{1/2}}u\right\|^2 \leq c_1 G^2\frac{\nu^2}{L^4} \text{ for all } u\in\mathcal{A}\,.$$

In fact a rather simple improvement of the proof in [5] leads to the better estimate

$$\left\| e^{\frac{1}{G(\log c_0 G+1)}\frac{L}{2\pi}A^{1/2}}u\right\|^2 \leq c_1 G^2\frac{\nu^2}{L^4} \text{ for all } u\in\mathcal{A} \tag{6.8}$$

under the assumption that $c_1 G\geq 1$ and $G\geq L\kappa_0$. By integrating (6.8) with respect to a μ as in (6.3) and using (1.7) and (6.7) we obtain, for $\alpha = 1$,

$$\int |A^{\alpha/2}u_{\kappa,\infty}|^2\mu(du) = \int |A^{\frac{1}{2}}u_{\kappa,\infty}|^2\mu(du) \leq c_0 G^2\frac{\nu^2}{L^4}e^{-\frac{L\kappa}{\pi G(\log c_1 G+1)}} \tag{6.9}$$

which is quite off the conjectured relation (6.3), since c_0 and c_1 in (6.3) are replaced in (6.9) by $c_0 G^{2/3}$ and $(1/\pi)G^{-2/3}(\log c_1 G+1)^{-1}$, respectively.

At the present, it seems quite difficult to prove (6.3), even for $\alpha = 1$ or $\alpha = 2$. However the slightly weaker estimate (for the case $\alpha = 2, f = f_{0,\kappa_0}, G \gg 1, G \geq (L\kappa_0)^2$)

$$\int |Au_{\kappa,\infty}|^2\mu(du) \leq c_0 G^2\frac{\nu^2}{L^6}e^{-c_1\frac{\kappa L}{G^{1/2}\log G}} \tag{6.10}$$

for all stationary statistical solutions μ, may turn out to be provable.

In practice it is much easier to determine the *power-spectrum of the velocity at a given point* in the fluid than the energy-spectrum $E(\kappa)$. We recall that if $g(t)(t \in (-\infty, \infty))$ is a bounded measurable real-valued function depending of time, then its power-spectrum $P(\omega)$ is defined by

$$P(\omega) = \lim_{T\to\infty} \frac{1}{T} \left| \int_0^T e^{-i\omega t} g(t) dt \right|^2 \quad (-\infty < \omega < \infty) \tag{6.11}$$

provided that the limit exists.

In fluid dynamics, $g(t)$ is usually defined as follows: Let $u_0 \in \mathcal{D}_A$, $j = 1, 2$ and a point x_0 in the fluid be fixed. Then

$$g(t) = (S(t)u_0)_j(x_0) \text{ for } t \geq 0 \text{ and } = 0 \text{ for } t < 0\,. \tag{6.12}$$

The empirical evidence seems to suggest that in this case $P(\omega)$ decays exponentially for $|\omega| \to \infty$ (see for instance [74], [7], [64]). At least, for the two dimensional Navier-Stokes equations, this fact can be given a rigorous mathematical content [4]. First, define the power spectrum $P(\omega)$ as a Schwartz distribution by the formula (which uses the generalized limit Lim)

$$\int_{-\infty}^{\infty} P(\omega)\phi(\omega) d\omega = \operatorname*{Lim}_{T\to\infty} \int_{-\infty}^{\infty} P_T(\omega)\phi(\omega) d\omega \tag{6.13}$$

(where the test function Φ runs over the Schwartz space $\mathcal{D}(\mathbb{R})$ of all infinitely differentiable functions with compact support in $\mathbb{R}$). Then it can be proven (see [4]) that $m(d\omega) = P(\omega)d\omega$ actually defines a positive Borel measure on $\mathbb{R}$ such that

$$\int_{-\infty}^{\infty} (e^{\delta_0 \omega} + e^{-\delta_0 \omega}) m(d\omega) \leq c_0 \gamma_0 \tag{6.14}$$

It is clear that (6.14) constitutes a rigorous formulation of the fact that the power-spectrum of the fluid velocity at a given point decays exponentially for $|\omega| \to \infty$.

The parameters δ_0 and γ_0 can be expressed in terms of the physical parameters G, L, ν as follows

$$\begin{cases} \delta_0 = c_1 \frac{L^2}{\nu} G^{-2} (\log^+ c_2 G + 1)^{-1} \\ \gamma_0 = c_3 G^2 \frac{\nu^2}{L^2} (\log^+ c_2 G + 1)\,. \end{cases} \tag{6.15}$$

The above estimate for δ_0 corrects a minor misprint in the estimate (2.4.4) for δ_0 given in [4] (namely G^{-1} is replaced in (6.15) by G^{-2}) while the estimate for γ_0 is a slight improvement with respect to the one in [4]. It is worth mentioning that, even heuristically, the sharp form of the estimates (6.14), (6.15) does not seem to be known, although some inferences can be made from (6.1) or (6.2) by referring to the very old and still standing Taylor hypothesis.

Finally in connection with the topics discussed in this section we refer to the pioneering paper [39].

REFERENCES

1. A. Ben-Artzi, A. Eden, C. Foias and B. Nicolaenko, Hölder continuity for the inverse of Mañé's projection, *J. of Math. Anal. and Appl.*, **178**(1993), 22-29.

2. A.V. Babin and M.I. Vishik, Attractors of partial differential evolution equations and their dimension, *Russian Math. Survey*, **38**(1983), 151-213.

3. A.V. Babin and M.I. Vishik, *Attractors of Evolution Equations*, Nauka, Moscow (1989).

4. H. Bercovici, P. Constantin, C. Foias and O.P. Manley, Exponential decay of the power spectrum of turbulence, *J. of Stat. Phy.*, **80**(1995), 579-602.

5. D. Chae, On the ensemble average in the study of approximate inertial manifolds, II, *J. of Math. Anal. and Appl.*, **614**(1992), 337-349.

6. S. Chen, G. Doolen, J.R. Herring, R.H. Kraichnan, S.A. Orszag and Z.S. She, Far dissipation range of turbulence, *Phys. Rev. Lett.*, **70**(1993), 3051-3054.

7. E. Ching, P. Constantin, L.P. Kadanoff, A. Libchaber, I. Procaccia, and X.-Z. Wu, Transitions in convective turbulence: The role of thermal plumes, *Phys. Rev. A*, **44**(1991), 8091-8102.

8. P. Constantin, Geometric statistics in turbulence, *SIAM Review*, **36**(1994), 73-98.

9. P. Constantin, Geometric and analytic studies in turbulence, *Trends & Perspectives in Appl. Math.*, L. Sirovich (Ed.), Springer-Verlag (1994), 21-54.

10. P. Constantin and C. Doering, Variational bounds on energy dissipation in incompressible flows: Shear Flow, *Phys. Rev. E*, **5**(1994), 4087-4099.

11. P. Constantin and C. Doering, Variational bounds on energy dissipation in incompressible flows. II. Channel flow, *Phys. Rev. E*, **4**(1995), 3192-3198.

12. P. Constantin and C. Doering, Variational bounds on energy dissipation in incompressible flows: III. Convection, *Phys. Rev. E*, **53**(1996), 5957-5981.

13. P. Constantin and C. Doering, Heat transfer in convective turbulence, *Nonlinearity*, to appear.

14. P. Constantin, C. Doering and E.S. Titi, Rigorous estimates of small scales in turbulent flows, *J. of Math. Phys.*, submitted.

15. P. Constantin and C. Foias, Global Lyapunov exponents, Kaplan-Yorke formulas and the dimension of the attractors for 2D Navier–Stokes equations, *Comm. on Pure and Appl. Math.*, **38**(1985), 1-27.

16. P. Constantin and C. Foias, *Navier–Stokes Equations*, Chicago Lectures in Mathematics, University of Chicago Press, Chicago (1988).

17. P. Constantin, C. Foias and O.P. Manley, Effects of the forcing function spectrum on the energy spectrum in 2-D turbulence, *Phys. Fluids*, **6**(1994), 427-429.

18. P. Constantin, C. Foias, O.P. Manley and R. Temam, Determining modes and fractal dimension of turbulent flows, *J. of Fluid Mech.*, **150**(1985), 427-440.

19. P. Constantin, C. Foias and R. Temam, Attractors representing turbulent flows, *Memoirs AMS*, **53**, #314(1985).

20. P. Constantin, C. Foias and R. Temam, On the dimension of the attractor in two-dimentional turbulence, *Physica D*, **30**(1988), 284-296.

21. C.R. Doering and J.D. Gibbon, *Applied Analysis of the Navier-Stokes Equations*, Cambridge Univ. Press, Cambridge, (1995).

22. C.R. Doering and E.S. Titi, Exponential decay rate of the power spectrum for solutions of the Navier-Stokes equations, *Phys. of Fluids*, **7**(1995), 1384-1390.

23. A. Eden, C. Foias, B. Nicolaenko and J.S. She, Exponential attractors and their relevance to fluid dynamics systems, *Physica D*, **63**(1993), 350-360.

24. A. Eden, C. Foias, B. Nicolaenko and R. Temam, *Exponential Attractors for Dissipative Evolution Equations*, Masson, Paris, J. Wiley, Boston (1994).

25. A. Eden, C. Foias and R. Temam, Local and global lyapunov exponents, *J. of Dyn. and Diff. Equ.*, **3**(1991), 133-177.

26. C. Foias, Statistical study of Navier–Stokes Equations, II, *Rend. Sem. Mat. Univ. Padova*, **49**(1973), 9-123.

27. C. Foias and I. Kukavica, Determining nodes for the Kuramoto-Sivashinsky equation, *J. Dyn. and Diff. Equ.*, **7**(1995), 365-373.

28. C. Foias, O. Manley and L. Sirovich, Empirical and Stokes eigenfunctions and the far-dissipative turbulent spectrum, *Phys. Fluids A*, **3**(1990), 464-467.

29. C. Foias, O.P. Manley and R. Temam, Bounds for the mean dissipation of 2-D enstropy and 3-D energy in turbulent flows, *Physics Ltrs. A*, **174**(1993), 210-215.

30. C. Foias, O.P. Manley, R. Temam and Y. Treves, Asymptotic analysis of the Navier-Stokes equations, *Physica D*, **9**(1983), 157-188.

31. C. Foias and E. Olson, Finite fractal dimension and Hölder–Lipshitz parametrization, *I.U. Math. Journ.*, **45**(1996), 603-616.

32. C. Foias and R. Temam, On the stationary statistical solutions of the Navier-Stokes equations and turbulence, *Université Paris XI, U.E.R. Mathematique* (1975) Preprint #120-75-28.

33. C. Foias and R. Temam, Some analytic and geometric properties of the solutions of the evolution Navier-Stokes equations, *J. Math. Pures et Appl.*, **58**(1979), 339-368.

34. C. Foias and R. Temam, Determination of the solutions of the Navier–Stokes equations by a set of nodal values, *Math. Computation*, **48**(1984), 117-133.

35. C. Foias and R. Temam, The connection between the Navier–Stokes equations dynamical systems and turbulence theory, *Directions in Partial Differential Equations*, M.C. Randall, R.H. Rabinowitz, M.E. Turner (Ed's), Academic Press, Boston (1987), 55-73.

36. C. Foias and R. Temam, Gevrey class regularity for the solutions of the Navier-Stokes equations, *J. of Funct. Anal.*, **87**(1989), 359-369.

37. C. Foias and E.S. Titi, Determining nodes, finite difference schemes and inertial manifolds, *Nonlinearity*, **4**(1991), 135-153.

38. U. Frisch and R. Morf, Intermittency in nonlinear dynamics and singularities at complex times, *Phys. Rev. A*, **23**(1981), 2673-2705.

39. U. Frisch, M. Nelkin and P.-L. Sulem, A simple dynamical model of intermittent fully developed turbulence, *J. Fluid Mech.*, **87**(1978), 719-736.

40. J.M. Ghidaglia and R. Temam, Lower bound on the dimension of the attractor for the Navier-Stokes equations in space dimension 3, *Mechanics, Analysis and Geometry: 200 Years After Lagrange*, M. Francaviglia and D. Holms (Eds.), Elsevier, Amsterdam (1991).

41. J.D. Gibbon and E.S. Titi, Attractor dimension and small length scale estimates for the 3D Navier-Stokes equations, *Nonlinearity*, to appear.

42. A.A. Ilyin, Attractors for Navier-Stokes equations in domains with finite measure, *Nonlinear Anal., Thy., Meth. & Appl.*, **27**(1966), 605-616.

43. D.A. Jones and E.S. Titi, Upper bounds on the number of determining modes, nodes, and volume elements for the Navier-Stokes equations, *I.U. Math. Journ.*, **3**(1993), 875-887.

44. A.N. Kolmogorov, The local structure of turbulence in incompressible viscous fluid for very large Reynolds, *C.R. Acad. Sci. URSS*, **30**(1941), 301.

45. R.H. Kraichnan, Inertial ranges in two-dimensional turbulence, *Phys. Fluids* **10** (1967), 1417-1423.

46. R.H.Kraichnan, Some modern developments in the statistical theory of turbulence, *Statistical Mechanics: New Concepts, New Problems, New Applications* (ed. S.A. Rice, K.F. Freed & J.C. Light), (1972), 201-227.

47. H.-O. Kreiss and J. Lorenz, *Initial-Boundary Value Problems and the Navier-Stokes equations*, Academic Press, Boston (1989).

48. I. Kukavica, On the number of determining nodes for the Ginzburg-Landau equation, *Nonlinearity*, **5**(1992), 997-1006.

49. L.L. Landau and E.M. Lifshitz, *Fluid Mechanics*, Pergamon, New York (1959).

50. J. Leray, Éssai sur les mouvements plans d'un liquide visqueux que limitent des parois, *J. Math. Pures Appl.*, **13**(1934), 331-418.

51. J. Leray, Essai sur le mouvement d'un liquide visqueux emplissant l'espace, *Acta Math.*, **63**(1934), 193-248.

52. J.L. Lions, *Quelgues méthodes de résolutions des problèmes aux limites non linéares*, Dunod-Gauthier-Villars, Paris (1969).

53. V.X. Liu, A sharp lower bound for the Hausdorff dimension of the global attractors of the 2D Navier-Stokes equations, *Comm. Math. Phys.*, **158**(1993), 327-339.

54. V.X. Liu, Remarks on the Navier-Stokes equations on the two and three dimensional torus, *Comm. Partial Diff. Equ.*, **19**(1994), 873-900.

55. B. Mandelbrot, *The Fractal Geometry of Nature*, Freeman & Co., New York (1983).

56. O.P. Manley, The dissipation range spectrum, *Phys. Fluids A*, **4**(1992), 1320-1321.

57. A. Miranville and X. Wang, Upper bound on the dimension of the attractor for nonhomogeneous Navier-Stokes equations, *The Inst. for Sci. Comp. and Appl. Math.*, preprint #9505.

58. A.S. Monin and A.M. Yaglom, *Statistical Hydrodynamics*, Vo. I and II, MIT Press, Boston (1977).

59. K. Ohkitani, Log-corrected energy spectrum and dimension of attractor in two-dimensional turbulence, *Phys. Fluids A*, **1**(1989), 451-452.

60. G. Raugel and G. Sell, Navier-Stokes equations on thin 3D domains. I: Global attractors and global regularity of solutions, *J. Amer. Math. Soc.*, **6**(1993), 503-568.

61. R. Rosa, The global attractor for the 2D Navier-Stokes flow on some unbounded domains, *The Inst. for Sci. Comp. and Appl. Math.*, preprint #9601.

62. H.A. Rose and P.L. Sulem, Fully developed turbulence and statistical mechanics, *Journ. de Physique*, **39**(1978), 441-484.

63. D. Ruelle, Elements of Differential Dynamics and Bifurcation Theory, Academic Press, Boston (1989).

64. D.E. Sigeti, Exponential decay of power spectra at high frequency and positive Lyapunov exponents, *Physica D*, **82**(1995), 136-153.

65. L.M. Smith and W.C. Reynolds, The dissipation-range spectrum and the velocity-derivative skewness in turbulent flows, *Phys. Fluids A*, **3**(1991), 992.

66. F. Takens, Detecting strange attractors in turbulence, *Lecture Notes in Math.*, No. 898, 366-381, D.A. Rand and L.-S. Young (Eds.), Springer-Verlag, New York, (1981).

67. R. Temam, *Navier-Stokes Equations*, Theory and Num. Anal., (3rd Rev.Ed.), North Holland, Amsterdam (1984).

68. R. Temam, Navier-Stokes equations and nonlinear functional analysis, *CBMS-NSF Ref. Conf. Sem. in Appl. Math.*, SIAM, Philadelphia (1983); augmented 2nd edition (1994).

69. R. Temam, Infinite dimensional dynamical systems in fluid mechanics, *Nonlinear Functional Anal. and its Appl.*, F. Browder (Ed.), AMS Proceedings of Symposia in Pure Math., Vol. 45, (1986).

70. R. Temam, *Infinite-Dimensional Dynamical Systems in Mechanics and Physics*, Appl. Math. Sci., **68**, Springer-Verlag, New York, 1988.

71. R. Temam and M. Ziane, Navier-Stokes equations in three-dimensional thin domains with various boundary conditions, *Adv. in Diff. Equ.*, Vol. 1, **4**(1996), 499-546.

72. M.I. Vishik and A.V. Fursikov, *Mathematical Problems of Statistical Hydrodynamics*, Kluwer Academic Press, Dordrecht (1980).

73. X. Wang, Time averaged energy dissipation rate in shear driven flows, (to appear).

74. X.-Z Wu, L.P. Kadanoff, A. Libchaber and M. Sano, Frequency power spectrum of temperature fluctuations in free convection, *Phys. Rev. Lett.*, **64**(1990), 2140-2143.

75. M. Ziane, Optimal bounds on the dimension of the attractor of the Navier-Stokes equations, *I.U. Inst. for Sci. Comp. and Appl. Math.*, preprint #9617.

Ciprian Foias
Department of Mathematics
Indiana University
Bloomington, IN 47408

Contemporary Mathematics
Volume **208**, 1997

A Reminiscence and Survey of Solutions to a JPL Coding Problem

L. H. HARPER

ABSTRACT. This paper is a survey of results on the problem of minimizing the "effects" of transmission errors when linearly ordered data is communicated over a binary symmetric channel. Of particular interest is the contrast between cases with exponent r = 1 or (infinity), where the solutions are purely combinatorial, and r = 2, where the solution involves dyadic harmonic analysis.

I. Introduction

In 1962, following my first year in the graduate mathematics program at the University of Oregon, I was very fortunate to get a summer job at the Jet Propulsion Laboratory. I worked in Section 331 which was responsible for research and development of the communications systems for the unmanned space probes built by JPL. That summer I was blessed with a variety of challenging problems to work on, outstanding colleagues, a supportive environment and the pay was not bad either, about twice what I would have gotten at the previous summer job I applied for, if I had gotten it. JPL was in the midst of a crisis, having failed to get any closeup pictures of the moon in six attempts. That summer, however, the seventh try was successful, initiating the era of space exploration, and I had a ring-side seat.

Ed Posner, my boss that summer of wonders, gave me a number of good problems to work on, but one that proved particularly rewarding was to determine whether or not the coding scheme that was used in transmitting those TV pictures back from the moon could be improved. The pixels that made up the pictures were registered as shades of gray, 64 of them, from black to white. These were encoded as 6-tuples of 0's and 1's for electronic transmission. The communications systems on space probes are designed to operate near their limits of performance so occasionally a mistake would be made in transmitting a symbol, changing 0 to 1 or 1 to 0. This causes a mistake in the shade of gray for that pixel, the actual change depending on the code used. The code in use was a natural one: If shades of gray are ordered $s_0, s_1, \dots, s_{63}$ from white to black, then the 6-tuple $(v_0, \dots, v_5)$ codes for $s_{\phi(v)}$ where $\phi(v) = \sum_{i=0}^{5} v_i 2^i$, i.e. $\phi(v)$ is the integer of which v is the binary

---

1991 *Mathematics Subject Classification.* Primary: 05, 42.

Key words and phrases. Optimal codes, Dyadic harmonic analysis, discrete isoperimetric problems.

This paper is in final form and no version of it will be submitted for publication elsewhere.

represention. A nice code, but did it minimize the "effect" of those errors on the pictures?

In order to make this question precise, a model of the transmission channel and data are necessary. The model of a channel which is always addressed in these situations, because of its simplicity and applicability, is the Binary Symmetric Channel. In a BSC one of two symbols, 0 and 1, is transmitted at regular intervals. With probability $p, 0 \leq p \leq 1$, an error is made in transmission, i.e. the transmitted and received symbols are different, independent of the symbol and previous transmissions. The data, being linearly ordered, may be modeled as integers, $0, 1, \ldots, 2^d - 1$. A code is then a one-to-one function $\phi : \{0,1\}^d \rightarrow \{0, 1, \ldots, 2^d - 1\}$. If a d-tuple, v, is transmitted and w is received, then the error is $|\phi(v) - \phi(w)|$, the absolute difference of their corresponding numerical values. Natural measures of the cost or effect of such an error are $|\phi(v) - \phi(w)|^r$, the rth powers of the error, $1 \leq r \leq \infty$. We would then like to minimize the average cost of error per transmission. Kautz **[7]** posed this problem in 1954, but solved it only for $r = 1, d = 2, 3$.

II.1. Minimizing mean absolute error ($r = 1$) when p is small

With $p \ll 1/d$, the probability of making more than one error in a d-tuple is very small and may be neglected. Assuming that all d-tuples are equally likely to be transmitted, the absolute error is proportional to

$$E_1(\phi) = \sum_{H(v,w)=1} |\phi(v) - \phi(w)|$$

where $H(v, w) = \sum_{i=0}^{d-1} |v_i - w_i|$ is Hamming distance. This may be paraphrased geometrically as assigning numbers $0, 1, \ldots, 2^d - 1$ to the vertices of the d-cube so as to minimize the sum of absolute differences over all its edges.

Example $d = 2$. The 2-cube is the square. There are $4! = 24$ ways to number its vertices $0, 1, 2, 3$ but the 8 symmetries of the square partition these into $24/8 = 3$ equivalence classes. These classes are represented by the numberings of Figure 1: The sum of absolute errors over all edges for these three codes is 6, 6 & 8 respectively.

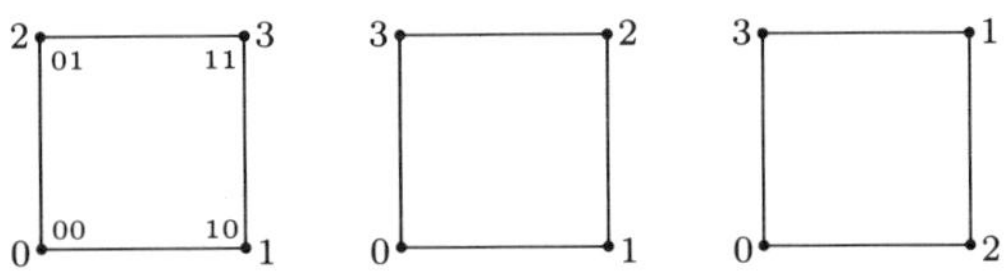

FIGURE 1. Equivalence classes

Thus 6 is the minimum absolute error for $d = 2$.

For arbitrary d, the number of codes is $(2^d)!$ which grows explosively. The number of symmetries of the d-cube is $2^d(d!)$ but there is no easy way to produce distinct representitives for the equivalence classes. For $d = 6, 2^6 = 64$ and $64! \sim 1.2 \times 10^{89}$, a large number even by today's standards. $64 \times (6!) = 46080$ is relatively inconsequential even if we could find distinct representitives for the $64!/(64 \times 6!) \simeq 2.75 \times 10^{84}$ equivalence classes. So, we see the necessity for insight if we are to solve the problem.

II.2. Reduction to an Isoperimetric Problem

In seeking to minimize $\sum\limits_{e \in E} |\phi(v) - \phi(w)|$, one notes that

i) We have a candidate for an optimum code: $\mathrm{lex}(v) = \sum\limits_{i=0}^{d-1} v_i 2^i$

ii) All we need is a lower bound equal to $E_1(\mathrm{lex})$. An obvious lower bound is

$$\begin{array}{ll} \min\limits_{\phi} & \sum\limits_{e \in E} |\phi(v) - \phi(w)| \\ \geq & \sum\limits_{e \in E} \min\limits_{\phi} |\phi(v) - \phi(w)|. \end{array}$$

But for each e in E the set of edges, the minimum of $|\phi(v) - \phi(w)|$ over all ϕ's is 1 and there is no one code for which $|\phi(v) - \phi(w)| = 1$ for all e in E.

iii) We might be able to transform our sum into another sum for which the lower bound of ii) would be stronger. One way to transform a sum is to expand the summand as a sum and interchange the order of summation: For e in $E, |\phi(v) - \phi(w)| = \sum\limits_{0 \leq \ell \leq 2^d} \chi(\phi, e, \ell)$, where

$$\chi(\phi, e, \ell) = \begin{cases} 1 & \text{if } \phi(v) < \ell \leq \phi(w) \\ 0 & \text{otherwise.} \end{cases}$$

And then,

$$\begin{aligned} \sum_{e \in E} |\phi(v) - \phi(w)| &= \sum_{e \in E} \sum_{\ell=0}^{2^d-1} \chi(\phi, e, \ell) \\ &= \sum_{\ell=0}^{2^d} \sum_{e \in E} \chi(\phi, e, \ell) = \sum_{\ell=0}^{2^d} \theta(S_\ell(\phi)), \end{aligned}$$

where $S_\ell(\phi) = \{v \text{ in } V : \phi(v) < \ell\}$, so $|S_\ell(\phi)| = \ell$.

Also, for any subset S of V, $\theta(S) = |\{e \text{ in } E : \partial(e) = \{v, w\}, v \in S \text{ and } w \notin S\}|$, ∂ being the boundary operator which gives the pair of vertices bounding each edge.

Now,

$$\begin{aligned} &\min_{\phi} \sum_{e \in E} |\phi(v) - \phi(w)| \\ &= \min_{\phi} \sum_{\ell=0}^{2^d} \theta(S_\ell(\phi)) \\ &\geq \sum_{\ell=0}^{2^d} \min_{\phi} [\theta(S_\ell(\phi))] = \sum_{\ell=0}^{2^d} \min_{|S|=\ell} \theta(S) \end{aligned}$$

since any $S \subseteq V, \quad |S| = \ell$, will be $S_\ell(\phi)$ for some ϕ.

Minimizing $\theta(S)$ over all subsets S of V, $|S| = \ell$, is a discrete analog of the classical isoperimetric problem of plane geometry: $|S|$ plays the role of area and $\theta(S)$ that of length of the boundary. Over all sets of a given "area" we wish to minimize the "boundary".

THEOREM 0.1. *For all, $\ell, 0 \leq \ell \leq 2^d$,*

$$\min_{|S|=\ell} \Theta(S) = \Theta(S_\ell(lex)).$$

Remarks on the proof: There are a number of proofs of this and related theorems in the literature. (c.f. [**45**] and [**12**]). J. D. Chavez and I have already written a monograph [**44**] expounding on our favorite methods for these problems and there is nothing more to add here.

COROLLARY 0.1. *The lexicographic code minimizes* $E_1 = \sum_{e\in E} |\phi(v) - \phi(w)|$ *over all possible codes.*

III.1. Minimizing maximum error ($r = \infty$) when p small

As in the case $r = 1$, this problem devolves to minimizing $E_\infty(\phi) = \max_{e\in E} |\phi(v) - \phi(w)|$ over all codes (one-to-one functions $\phi : V \to \{0, 1, \ldots, 2^d - 1\}$). $\max_{e\in E} |\mathrm{lex}(v) - \mathrm{lex}(w)| = 2^{(d-1)}$. A.W. Hales pointed out [**46**] that for sufficiently large d, codes which number vertices by increasing weight (number of 1's) are superior to lex. Hales' code maps v in $\{0,1\}^d$ to ℓ iff $|v| = k$, $\binom{d}{k-1} \leq \ell < \binom{d}{k}$ and v is the $\ell - \binom{d}{k-1}$th member of $\{v \in \{0,1\}^d : |v| = k\}$ with respect to the dual of lex$'$ order. (lex$'$ order is lex order with respect to the dual (reversed) order of components).

Example

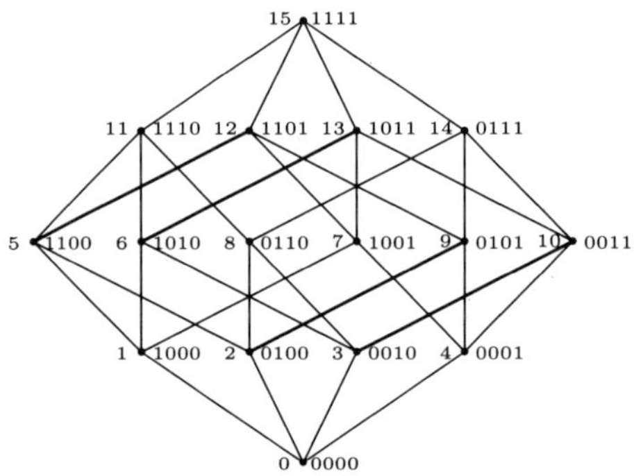

FIGURE 2. Hales code for d=4

Note that for $d = 4$ the maximum difference over all the edges is 7, less than $8 = 2^{(4-1)}$. In general, $\max_{e\in E} |\mathrm{Hales}(v) - \mathrm{Hales}(w)| = \sum_{m=0}^{d-1} \binom{m}{\lfloor m/2 \rfloor}$, which equals $2^{(d-1)}$ for $d = 1, 2, 3$ but is stricty less for $d > 3$. From Stirling's formula ($n! \sim \sqrt{2\pi n}(n/e)^n$) we have that $\binom{2n}{n} \sim 2^{(2n)}/i\sqrt{\pi n}$. Therefore

$$\sum_{m=0}^{d-1} \binom{m}{\lfloor m/2 \rfloor} \Big/ \binom{d}{\lfloor d/2 \rfloor} \sim \frac{1}{2} + \frac{1}{4} + \frac{1}{8} + \ldots = 1,$$

and

$$\binom{d}{\lfloor d/2 \rfloor} / 2^{(d-1)} \sim \sqrt{8/\pi d} \to 0 \quad \text{as} \quad d \to \infty.$$

But can we show that Hales' codes are optimal? Reasoning by analogy with $r = 1$, one may look for a boundary functional Φ, such that

$$\max_{e\in E} |\phi(v) - \phi(w)| = \max_{\ell} \Phi(S_\ell(\phi)).$$

and then $\min_{\phi} \max_{e \in E} |\phi(v) - \phi(w)| = \min_{\phi} \max_{\ell} \Phi(S_\ell(\phi)) \geqq \max_{\ell} \min_{|S|=\ell} \Phi(S)$. Such a functional does not exist, but we really do not need equality in the first relation, just an inequality, since that is all we can end up with anyway. If one relaxes the requirement for equality to

$$\max_{e \in E} |\phi(v) - \phi(w)| \geqq \max_{\ell} \Phi(S_\ell(\phi)),$$

it does exist: For $S \subseteq V$, define $\Phi(S) = |\{w \in V : \exists e \in E, \partial(e) = \{v, w\}, v \in S \text{ and } w \notin S\}|$.

THEOREM 0.2. *For any code ϕ and $0 \leq \ell \leq 2^d$,*

$$E_\infty(\phi) = \max_{e \in E} |\phi(v) - \phi(w)| \geq \Phi(S_\ell(\phi))$$

and so $\min_{\phi} \max_{e \in E} |\phi(v) - \phi(w)| \geq \max_{0 \leq \ell \leq 2^d} \min_{|S|=\ell} \Phi(S)$.

Proof: Given ϕ and $0 < \ell < 2^d$. Let $T = \{w \in \{0,1\}^d : \exists e \in E, \partial(e) = \{v, w\}, \phi(v) \leq \ell < \phi(w)\}$, the set counted by $\Phi(S_\ell(\phi))$. If $\phi(w_0) = \max\{\phi(w) : w \in T\}$ and e is an edge which puts $w \in T$, i.e. $\partial(e) = \{v_0, w_0\}$ with $\phi(v_0) \leq \ell < \phi(w_0)$, then $|\phi(w_0) - \phi(v_0)| \geq \phi(w_0) - \ell \geq |T| = \Phi(S_\ell(\phi))$ since all of the vertices of T must have values of ϕ between ℓ and $\phi(w_0)$. Again, Φ is a measure of the "boundary" of S and we have another isoperimetric problem.

THEOREM 0.3. ([46]) *For all ℓ, $0 \leq \ell \leq 2^d$,*

$$\min_{|S|=\ell} \Phi(S) = \Phi(S_\ell(Hales)).$$

As with Theorem II.3, proving Theorem III.3 is difficult. Theorem III.3 easily implies the Kruskal-Katona theorem, best known of all the discrete isoperimetric theorems.

COROLLARY 0.2. *Hales' code minimizes $E_\infty(\phi) = \max_{e \in E} |\phi(v) - \phi(w)|$ over all codes on Q_d.*

IV.1. Minimizing mean square error ($r = 2$)

Now we come to the main point of this paper. How might one prove that a given code optimizes

$$E_2(\phi) = \sum_{e \in E} [\phi(v) - \phi(w)]^2?$$

As in the cases $r = 1$ and ∞ a likely strategy is to find a way to transform the sum into an equivalent one in which each term can be optimized separately. For $d = 2$ this is easily done:

$$\sum_{e \in E} [\phi(v) - \phi(w)]^2 = 2 \sum_{v \in V} \phi^2(v) - 2 \sum_{v \text{even}} \phi(v) \sum_{w \text{odd}} \phi(w)$$

and each term is optimized by the natural binary code, lex. Similarly for d=3:

$$\sum_{e \in E} [\phi(v) - \phi(w)]^2 = 3 \sum_{v \in V} \phi^2(v) - 2 \sum_{v \text{even}} \phi(v) \sum_{w \text{odd}} \phi(w) + \sum_{v \in V} \phi(v)\phi(\bar{v})$$

$\bar{v}$ being the complement or antipode of v. Each term again being optimized by lex. But for $d > 3$ the pattern is no longer apparent. It does appear however that lex is a good candidate for the optimum code. In 1963 Mitryayev [**47**] suggested that there might be a connection with harmonic analysis on the group $G = \{0,1\}^d$. In particular

$$x_i = \frac{1 - \chi_i(x)}{2}$$

so

$$\operatorname{lex}(x) = \sum_{i=0}^{d-1} \frac{1 - \chi_i(x)}{2} 2^i$$

where $\{\chi_i\}$ is the natural basis of characters, $\chi_i(x) = \begin{cases} 1 & \text{if} \quad x_i = 0 \\ -1 & \text{if} \quad x_i = 1 \end{cases}$ Conversely, any basis $\{\chi_i\}$ of G^* gives a code by the same formula. In 1969, using dyadic harmonic analysis, Crimmins, Horwitz, Palermo & Palermo [**25**] proved that the lexicographic code minimizes mean square error in the BSC for any $p < 1/2$ and any block size, d. Actually, what they proved is much stronger but before we go into the details, let us review the basic facts of harmonic analysis.

IV.2. A quick review of harmonic analysis

Let G be a finite, additive Abelian group (in our case $G = \{0,1\}^d$) and $T = \{e^{i\theta}\}$, the circle group. A *character* of G is a homomorphism $\chi : G \to T$, i.e. $\chi(x+y) = \chi(x)\chi(y)$. G^*, the dual of G, is the set of all characters on G. With multiplication defined pointwise G^* is a group: $\chi_1\chi_2$ is a character since

$$\begin{aligned} \chi_1\chi_2(x+y) &= \chi_1(x+y)\chi_2(x+y) \\ &= [\chi_1(x)\chi_1(y)][\chi_2(x)\chi_2(y)] \\ &= [\chi_1(x)\chi_2(x)][\chi_1(y)\chi_2(y)] \\ &= \chi_1\chi_2(x)\chi_1\chi_2(y). \end{aligned}$$

For finite groups, G and G^* are actually isomorphic. $L_2(G)$ denotes the vector space of functions on G with inner product

$$\langle f_1, f_2 \rangle_G = \sum_{x \in G} f_1(x)\overline{f_2(x)}.$$

We also have $L_2(G^*)$ with its inner product

$$\langle f_1^*, f_2^* \rangle_{G^*} = \frac{1}{|G|} \sum_{\chi \epsilon G^*} f_1^*(\chi) f_2^*(\chi).$$

For $f \in L_2(G)$, $\hat{f} \in L_2(G*)$, called the *Fourier* transform of f, is defined by $\hat{f}(\chi) = \langle f, \chi \rangle_G$. We then have the Fourier series for f, $f(x) = \sum \hat{f}(\chi)\chi(x)$; and the map $\hat{}: L_2(G) \to L_2(G^*)$ is an isometry, i.e. $\langle f_1, f_2 \rangle_G = \langle \hat{f}_1, \hat{f}_2 \rangle_{G^*}$. Also, defining the operation of convolution on $L_2(G)$ by

$$f_1 * f_2(x) = \sum_{y \in G} f_1(x-y) f_2(y).$$

we have $\widehat{f_1 * f_2} = \hat{f}_1 \hat{f}_2$. The Fourier series representation of f shows that G^* generates $L_2(G)$, but it is also an orthogonal basis since

$$\langle \chi_1, \chi_2 \rangle = \begin{cases} |G| & \text{if} \quad \chi_1 = \chi_2, \\ 0 & \text{otherwise.} \end{cases}$$

IV.3 The Crimmins, Horwitz, Palermo and Palermo Calculation

Suppose that the probability of $x \in G = \{0,1\}^d$ being transmitted is $P(x)$ and, given that x was sent, the probability that y is received is $P(y|x)$. Then for the code ϕ, the "noise energy" of the channel is

$$E_2(\phi) = \sum_{x,y\in G} [\phi(x) - \phi(y)]^2 P(y|x)P(x).$$

We shall suppose that $P(x) = \frac{1}{2^d}$ i.e. that the data are uniformly distributed and that $P(y|x)$ is invariant under G. This means that for all $x, y, z \in G \quad P(y \oplus z | x \oplus z) = P(y|x)$. Letting $z = x$ we have $P(y|x) = P(x \oplus y|0) = Q(x \oplus y)$.

So

$$\begin{aligned} E_2(\phi) &= \tfrac{1}{2^d} \sum_{x,y\in G} [\phi(x) - \phi(y)]^2 Q(x \oplus y) \\ &= \tfrac{1}{2^d} \Bigg[\sum_{x,y\in G} \phi^2(x) Q(x\oplus y) - 2 \sum_{x,y\in G} \phi(x)\phi(y)Q(x\oplus y) \\ &\qquad + \sum_{x,y\in G} \phi^2(y)Q(x\oplus y) \Bigg] \\ &= \tfrac{2}{2^d} \Bigg[\sum_{x\in G} \phi^2(x) - \sum_{x,y\in G} \phi(x)\phi(y)Q(x\oplus y) \Bigg] \end{aligned}$$

Since $\sum_{x\in G} \phi^2(x)$ will be the same for every code, to minimize $E_2(\phi)$ we need only maximize

$$\sum_{x,y\in G} \phi(x)\phi(y)Q(x\oplus y).$$

Now this formulation relates nicely to $L_2(G)$ and harmonic analysis since

$$\sum_{x,y\in G} \phi(x)\phi(y)Q(x\oplus y) = \sum_{x,y\in G} \phi(x\oplus y)\phi(y)Q(x) = \langle \phi * \phi, Q\rangle_G = \langle \hat{\phi}^2, \hat{Q}\rangle_{G^*}.$$

We may number the members of $G^*, \chi_0, \chi_1, \dots, \chi_{2^d-1}$, so that

i) χ_0 is the identity.

ii) $\hat{Q}(\chi_i) \geq \hat{Q}(\chi_{i+1}) \quad i = 0, 1, \dots, 2^d - 2$.

Then we define a subsequence $0 < \alpha(1) < \alpha(2) < \dots < \alpha(d) < 2^d$ recursively by $\alpha(i) = \min\{j : \chi_j \notin [\chi_{\alpha(1)}, \dots, \chi_{\alpha(i-1)}]$, $[\chi_1, \dots, \chi_i]$ being the subgroup of G^* generated by $\chi_1, \dots, \chi_i$. $\{\chi_{\alpha(i)}, 1 \leq i \leq d\}$ is a basis for G^* so by our previous remarks

$$\phi_0 = \sum_{i=1}^{d} \left[\frac{1 - \chi_{\alpha(i)}}{2}\right] 2^{d-i}$$

is a code.

Claim. ϕ_0 is an optimal code. To prove it we must show that $\langle \hat{\phi}_0^2, \hat{Q}\rangle_{G^*} \geq \langle \hat{\phi}^2, \hat{Q}\rangle$ for any code, ϕ.

If $\{a_i\}$ and $\{b_i\}$ are sequences and we define

$$\Delta a_i = a_i - a_{i-1}$$
$$\nabla a_i = a_i - a_{i+1} = -\Delta a_{i+1}$$

then we have the following analog of integration by parts:

Lemma $\sum_{i=1}^{n} a_i \Delta b_i = a_{n+1}b_n - a_1 b_0 + \sum_{i=1}^{n} b_i \nabla a_i$.

Proof of the Lemma. Induction on n.

$\langle \hat{\phi}^2, \hat{Q} \rangle_{G^*} = \frac{1}{2^d} \sum_{i=0}^{2^d-1} \hat{\phi}^2(\chi_i)\hat{Q}(\chi_i)$, so if we let $a_i = \hat{Q}(\chi_i)$ then $\nabla a_i = \hat{Q}(\chi_i) - \hat{Q}(\chi_{i+1}) \geq 0$. Also $b_i = \sum_{j=0}^{i} \hat{\phi}^2(\chi_j)$ gives $\Delta b_i = \hat{\phi}^2(\chi_i)$, and then $b_0 = \hat{\phi}^2(\chi_0) = \langle \phi, \chi_0 \rangle^2 = \left[\sum_{x \in G} \phi(x) \right]^2$. With $n = 2^d - 1$, $b_n = \sum_{j=0}^{2^d-1} \hat{\phi}^2(\chi_j) = |G^*| \|\hat{\phi}\|^2_{G^*} = 2^d \|\phi\|^2_G = 2^d \sum_{x \in G} \phi^2(x)$, so both b_0 and b_n are independent of ϕ as are a_{n+1} and a_1. Therefore, if we can show that ϕ_0 maximizes $\sum_{j=0}^{i} \hat{\phi}^2(\chi_j)$ for $i = 1, 2, \ldots, 2^d - 1$ over all codes, ϕ, we are done.

Let $H_k = [\chi_{\alpha(1)}, \chi_{\alpha(2)}, \ldots, \chi_{\alpha(k)}] \triangleq G^*$. Since the summands of $\sum_{j=0}^{i} \hat{\phi}^2(\chi_j)$ are non negative and those for ϕ_0 are only positive when $j = \alpha(k)$ for some $k = 1, \ldots, d$, we need only show

$$\sum_{\chi \in H_k} \hat{\phi}_0^2(\chi) \geq \sum_{\chi \in H_k} \hat{\phi}^2(\chi) \quad \text{for} \quad k = 1, \ldots, d$$

Define $G_k = \{x \in G : \chi(x) = 1 \; \forall \chi \in H_k\} \triangleq G$ and $F_k = G/G_k$. If $\chi \in H_k$, then χ is constant on G_k and each of its cosets. Let $\psi_\chi(\sigma)$ denote that constant value of χ on σ for each $\sigma \in F_k$. Then ψ_χ is a character of F_k and as χ runs through H_k, ψ_χ runs through all of F_k^*. Let

$$\mathcal{E}(\sigma) = \sum_{x \in \sigma} \phi(x) \quad \text{for} \quad \sigma \in F_k.$$

Then for $\chi \in H_k$

$$\begin{aligned} \hat{\mathcal{E}}(\psi_\chi) &= \sum_{\sigma \in F_k} \mathcal{E}(\sigma)\psi_\chi(\sigma) \\ &= \sum_{\sigma \in F_k} \sum_{x \in \sigma} \phi(x)\chi(x) \\ &= \sum_{x \in G} \phi(x)\chi(x) = \hat{\phi}(\chi) \end{aligned}$$

Hence, recalling that the Fourier trasform is an isometry of $L_2(F_k)$ onto $L_2(F_k^*)$

$$\begin{aligned} \sum_{\chi \in H_k} \hat{\phi}^2(\chi) &= \sum_{\chi \in H_k} \hat{\mathcal{E}}^2(\psi_\chi) \\ &= |F_k^*| \, \|\hat{\mathcal{E}}\|^2_{F_k^*} = |F_k| \|\mathcal{E}\|^2_{F_k} \\ &= |F_k| \sum_{\sigma \in F_k} \mathcal{E}^2(\sigma) = 2^{d-k} \sum_{\sigma \in F_k} \left[\sum_{x \in \sigma} \phi(x) \right]^2 \end{aligned}$$

Now, of all the codes on G, ϕ_0 assigns the smallest values, $0, 1, \ldots, 2^k - 1$, to G_k; the next segment, $2^k, 2^k+1, \ldots, 2^{k+1}-1$, to a coset of G_k; etc. clearly this partition of $0, 1, \ldots, 2^d - 1$ maximizes

$$\sum_{\sigma \in F_k} \left[\sum_{x \in \sigma} \phi(x) \right]^2$$

and we are done.

IV.4. The Primal Example

Recall that in the Binary Symmetric Channel

$$P(y|x) = p^{H(y,x)} q^{d-H(y,x)}$$

so $Q(x) = p^{|x|} q^{d-|x|}$ where $|x| = H(x,0) = \sum_{i=0}^{d-1} x_i$. Let $\{\delta^{(i)}\}$ be the standard basis of $G = \{0,1\}^d$, i.e. $\delta_j^{(i)} = \begin{cases} 1 & \text{if } i = j \\ 0 & \text{otherwise.} \end{cases}$ Then for any $x \in G, x = \sum_{i=0}^{d-1} x_i \delta^{(i)}$ and for any $\chi \in G^*, \chi(x) = \Pi_{i=0}^{d-1} \chi(\delta^{(i)})^{x_i}$ and $\hat{Q}(\chi) = \langle Q, \chi \rangle_G = \sum_{x \in G} Q(x)\chi(x)$

$$= \sum_{x \in G} p^{\sum x_i} q^{d - \sum x_i} \prod_{i=1}^d \chi(\delta^{(i)})^{x_i}$$
$$= \prod_{i=1}^d (q + p\chi(\delta^{(i)})) = (q-p)^{\beta(\chi)}$$

where $\beta(\chi) = |\{i : \chi(\delta^{(i)}) = -1\}$

Now $p < \frac{1}{2}$, so $q - p > 0$. The maximum of $\hat{Q}(\chi)$, $\chi \in G^* - \{\chi_0\}$ is then achieved when $\beta(\chi) = 1$. There are d such characters

$$\chi_j(\delta^{(i)}) = \begin{cases} -1 & \text{if } j = i \\ 1 & \text{otherwise,} \quad 1 \le j \le d \end{cases}$$

and $\{\chi_j\}$ is the standard basis of G^*. Thus lex $= \sum_{i=1}^{d} \frac{1-\chi_j}{2} 2^{d-i}$ is an optimal code.

IV.5. A lesson learned the hard way

In the Fall of 1961, shortly after I arrived for graduate work at the University of Oregon math department, Andrew Moursund, longtime department head, took me aside and asked about my professional goals. "I want to be an applied mathematician", I said, continuing that I liked physics but my way of thinking seemed more in tune with math. "Well then", he pronounced to me, "first you learn mathematics, then you apply it!" Looking back now, 35 years later, I can see that Prof. Moursund's advice was absolutely perfect, the product of a deep understanding of the natures of pure and applied mathematics. But I was skeptical at the time, thinking that he had a vested interest in having me forget about applied problems while in his department. Four years later, having studied the beautiful abstractions of transfinite numbers, nowhere differentiable curves, unmeasureable sets, the axiom of choice, groups, rings, fields and categories, and having completed a thesis on sets of uniqueness for Walsh-Fourier series (under the guidance of Professor Shapiro), it was even less clear to me what academic mathematics had to do with physical reality.

On the other hand, I was not entirely happy with combinatorics either. The problems which I had been able to solve during my Summers at JPL and Bell Labs had required only simple tricks, usually a single idea would suffice. Did I want to spend the rest of my life in such a shallow subject? At least academic mathematics had depth to it. Any new idea in analysis, algebra or topology might resonate with a classic. I considered going into probability theory as a compromise.

Gian-Carlo Rota, my postdoctoral mentor at the Rockefeller University (1965-67) convinced me to stay with combinatorics. His thesis was that the task of combinatorial theorists is to identify the fundamental problems of the subject and create or find the appropriate tools for their solution. He, however, was a profound scholar of mathematics, able to digest the results of centuries of work in all areas. I was not so gifted in this respect and had been attracted to combinatorics just because it was mainly problem-solving. How I might contribute to Rota's vision of Combinatorial Theory was not clear to me until 1969 when the paper by Crimmins, Horwitz, Palermo & Palermo appeared.

In reading the CHP&P paper I felt profoundly embarrassed at having missed their result. I was very familiar with the methods, having taken a Rudin-style course in abstract harmonic analysis and studied the (infinite) dyadic group, $\{0,1\}^{\omega}$ as background for my thesis. In the academic year 1964-65 I was working on my thesis with Professor Shapiro at UCR and also consulting one day per week at JPL. There were many times during that year when I spent one day thinking about harmonic analysis on $\{0,1\}^{\omega}$ and the next thinking about the JPL coding problem with exponent 2, and yet it never occured to me to apply the one to the other. I knew of the optimality of lex for $d = 2$ and 3, as shown at the beginning of Section IV.1, and was vaguely aware that such an expansion is based on symmetry, but did not explicitly notice that all the sets being summed over; evens and odds, antipodal pairs, etc., are cosets of subgroups of the finite dyadic group, $\{0,1\}^{d}$. I was so embarrassed by my oversight that I neglected to tell Professor Shapiro about it. Looking back though, I see that this was a turning point in my understanding of mathematics. Even though I did not continue on in harmonic analysis, my work under Professor Shapiro's guidance had a profound effect on my subsequent research.

Having absorbed the facts, (in the previous paragraph) I then faced the central question "What caused me to miss these clues, which seem so obvious in retrospect?" Skepticism is healthy but I had carried it too far. I had failed to appreciate the power and flexibility of harmonic analysis but even more damaging was the way that I had compartmentalized my thinking. I resolved never to make those mistakes again. Here then was one classical theory, harmonic analysis, which I understood very well and which applied to solve a familiar combinatorial problem. The gulf between classical mathematics and combinatorics seemed immeasureably less than it had. All one needed, apparently, was the notion that such-and-such theory might apply to so-and-so problem in order to start the process of discovery.

Sometime around 1976 I was telling Professor Rota about the methods which I and others had developed to solve discrete isoperimetric problems when he remarked that it reminded him of Steiner symmetrization. With the above lessons in mind I was eager to investigate the connection so he recommended looking at Polya & Szego's book [**43**], *Isoperimetric Inequalities in Mathematical Physics.* Polya & Szego recount how symmetrization was invented by Jakob Steiner in 1838 to prove the classical isoperimetric theorem in the plane. In 1945 they (Polya & Szego) used it to show that the sphere has the lowest electrostatic capacity of all conducting bodies with a fixed volume. Thus reproving a theorem of Szego's which had proved a 1903 conjecture of Poincarre which had extended an 1890 conjecture of Lord Rayley. This stimulated a systematic study of symmetrization and its applications leading to the solution of a number of problems in mathematical physics. Symmetrization, the analytic technique, with its variety of applications, has served and continues to serve as a role model for the development of combinatorial "Steiner operations".

The theory and some applications of Steiner operations have been made into a monograph by myself and J.D.Chavez [**44**]. The literature on discrete isoperimetric problems has a large quotient of difficult and even incorrect proofs (of which the present author has contributed his share). The theory of Steiner operations, however, is made up of transparant constructions and demonstrations rendering the difficult easy and the formerly impossible merely difficult.

V.0. A Survey of Extensions, Variations and Applications

The papers which followed our three basic papers and refer to at least one of them in the bibliography fall into three different subject areas; combinatorial optimization, electrical engineering (including computer science) and functional analysis (including probability theory). We briefly summarize their contributions by area.

V.1. Combinatorial Optimization

Before listing the individual papers, we would like to point out that an important survey of discrete isoperimetric and related combinatorial optimization problems was published by K. Engel and H.-D.O.H Gronau in 1985 [**1**]. Their title *Sperner Theory in Partially Ordered Sets* honors the other historical root of the subject, namely Hilbert's solution of Gordon's Problem (the Hilbert Basis Theorem). In the effort to extend Hilbert's result and simplify its proof, E. Sperner made several outstanding contributions, and identified its combinatorial essence. A second version of the Engel-Gronau monograph, authored soley by Engel, is currently in press [**2**].

1964 In [**20**] J.H.Lindsey extends the results of our Section II to any product of complete graphs (the d-dimensional cube being the d-fold product of K_2). Posner suggested this extension to me after I had completed my original paper, but I was wary of the additional technical difficulties. Lindsey, an undergrad at CalTech, worked in JPL Section 331 the following Summer ('63), so Posner passed the project on to him. Lindsey pushed my program through, overcoming the difficulties, and was awarded a prize for undergraduate research at CalTech.

1965 In [**14**] K. Steiglitz & A.J.Bernstein note that the lexicographic code, lex:$Q_d \to \{0, 1, \ldots, 2^d-1\}$, is optimal for any set of real data, $\{a_0 \leq a_1 < \cdots \leq a_{2^d-1}\}$. Later [**13**], Bernstein points out that my proof of Theorem II.3 had overlooked a case and he fills in the gap.

1969 J.B.Kruskal [**3**] points out the interesting similarity between his (1963) isoperimetric theorem on the (complex of) the d-simplex and mine on the (graph of) the d-cube. He speculates that mine should extend to the complex of all faces of the d-cube.

B. Lindstrom ([**4**], 1971) verifies this conjecture but his proof is also not complete.

In [**19**] R.L.Graham generalizes the Steiglitz-Bernstein Theorem to any "complete partition" of $\{a_0 \leq a_1 \leq \cdots \leq a_{2^d-1}\}$.

1971 In [**15**] G.F.Clements finds sets of lattice points which contain a maximal number of edges. This extends a variant of Theorem II.3 but not of its Corollary, i.e. it does not produce an optimal code.

1975 J. Chvatalova [**9**] found an encoding for a product of two paths which minimizes the bandwidth.

1978 In [**16**] M.R.Garey et al show that the bandwidth problem (i.e. minimizing $\max|\phi(v) - \phi(w)|$ over all numberings (codes) on $G = (V, E, \partial)$) is NP-complete, even if G is restricted to trees of vertex-degree ≤ 3.

1981 In [**12**] P. Frankl and Z. Furedi prove a beautiful extension of the isoperimetric theorem for Φ : Let $\mathcal{A}$ and $\mathcal{B}$ be set systems on $X(|X| = n)$ and $H(\mathcal{A}, \mathcal{B}) = \min\{H(A, B) : A \in \mathcal{A} \text{ and } B \in \mathcal{B}\}$. Then $|\mathcal{A}| = k$ and $|\mathcal{B}| = l$ implies

$$H(\mathcal{A}, \mathcal{B}) \geq H(S_k(Hales), S_\ell(Hales^*)).$$

1983 For a set system, $\mathcal{A}, |\mathcal{A}| = m$, D.E.Daykin and P. Frankl [**10**] study $\lambda(n, m)$, the maximum number of containments, $A \subseteq B$ with $A, B \in \mathcal{A}$. Using the Kruskal-Katona Theorem (a special case of our Theorem III.3) they show that $\lambda(n, m) = (1 + o(1))$ if and only if $m \rightarrow 1$.

My student, H.S.Moghadam, extends Chvatalova's Theorem to a product of any number of paths [**6**]. This was one of the first real achievements of the theory of Steiner operations.

1984 J. Korner & V. Wei [**11**] extend our solution of the isoperimetric problem for Φ to the graph of even vertices of the d-cube with edges between vertices of Hamming distance 2.

1985 H.J.Tiersma [**17**] gives a simple proof of the Korner-Wei Theorem.

1986 In [**18**] G. Mitchison & R. Durbin find the optimal code (minimizing $\sum |\phi(v) - \phi(w)|$) for an $m \times m$ grid. Unlike the d-cube, the grid does not have a nested family of sets, one of each cardinality, which minimize . θ. This makes the Mitchison-Durbin arguement even more delicate and, again unlike Chvatalova's Theorem, it has never been extended to higher dimensions.

1989 In [**21**] P. Frankl gives a lower bound on the size of a simplicial complex (on p points) generated by an antichain. The bound is related to our isoperimetric theorem for Φ, for which he provides a relatively short proof.

1990 B. Bollobas & I. Leader [**22**] solve a problem of D.L.Wang & P.Wang, proving an isoperimetric inequality on the discrete torus.

1991 Bollobas & Leader [**23**] reprove the Chvatalova-Moghadam Theorem using a novel extension of compression.

G.Gierz & F.Hergert [**36**] observe that our solution of the bandwidth problem for the d-cube is also the solution of the bandwidth (scheduling) problem for the Boolean algebra on d generators. The Chvatalova-Moghadam Theorem solves the bandwidth (scheduling) problem for the product of chains by the same reasoning. This raises the intriguing possibility of extending the theory of Steiner operations to complexes (ranked posets).

1993 Bollobas & Leader [**24**] solve a problem closely related to our isoperimetric problems: They characterize, for each D, the maximum size of a set of diameter D in the grid.

V.2. Electrical Engineering and Computer Science

The difference between mathematics and applications is sometimes hazy but we believe that there is a difference and that it is a distinction worth making.

1970 G.G.Apple & P.M.Wintz [**26**] show how to efficiently calculate the Fourier transforms required by CHP&P. The mathematics of this was known to Gauss in the early 19th century and when rediscovered in the 1950's it was called the Fast Fourier Transform. However it is always interesting to have a new application pointed out.

Assuming the result of CHP&P, T.R.Crimmins & H.M.Horwitz [**27**] show how to select the coset leaders of a group code so as to minimize mean- square error.

1974 In [**28**] G.A.Wolf & G.Redinbo extend the results of [**25**] & [**27**] to allow the decoding function to take arbitrary real values. They show that the optimal decoding function is given by conditional expection.

1976 T.R.Crimmins restricts the Wolf-Redinbo problem so that the range of the decoding function, is $S = \{0, 1, \ldots, 2^d - 1\}$, same as the domain of the encoding function, . Their conclusion is that for ϕ and η which minimize expected mean-square error, $\eta = \phi^{-1}$.

1978 In [**31**] A.L.Rosenberg develops a model for encoding a "data structure" into a "storage structure" so as to optimize "computational efficiency". Both the data and storage "structures" are weighted graphs and an encoding is an embedding of the one into the other. In [**32**], [**33**] & [**34**] Rosenberg and his coauthors treat special cases of this general problem and derive heuristic principles of computer design from their results.

1979 G.S.Bloom & S.W.Golomb [**29**] present a survey of optimal graph-numbering techniques and their applications in electrical engineering.

1982 In [**35**] T.Lengauer points out that for a graph $G = (V, E,)$ the parameter $\max \theta(S_\ell(\phi))$ is relevant to VLSI design. He gives a polynomial time algorithm for calculating it within a factor of 2 on trees.

1991 If $\phi : G \to H$ is an embedding of a graph G into a graph H, the dilation cost (bandwidth) of ϕ is $\max d(\phi(v), \phi(w))$, $d(v', w')$ being the distance (minimum length of any path in H) from v' to w'. This problem arises in VLSI design. The dilation is proportional to maximum delay in any wire. T.-H. Lai & A.P. Sprague [**38**] show how to embed the d-cube into a grid so that the dilation is within 26% of the minimum.

1993 A paper by E. Ma & L. Tao [**37**] constructs optimal and near-optimal (with respect to dilation cost) embeddings when G & H are meshes (grids) & tori.

V.3. Functional Analysis and Probability Theory

Relatively recently some combinatorialists and analysts have realized that discrete isoperimetric theorems could be applied to continuous problems and that analytic methods could be applied to discrete isoperimetric problems.

1985 N. Alon & V.D. Milman [**40**] develop a general method for obtaining asymptotic discrete isoperimetric inequalities. Their method uses the second smallest eigenvalue of the incidence matrix of the graph and is the discrete version of a method used previously for Riemannian manifolds.

1988 M. Talagrand [**41**] extends the notion of compression to (continuous) probability spaces and obtains isoperimetric inequalities from it.

1991 Let $p_c(d)$ be the critical probability for percolation in $\mathbb{Z}^d$. In [**38**] D. Gordon shows that $\lim 2dp_c(d) \to 1$, affirming an outstanding conjecture. Gordon's proof uses Moghadam's isoperimetric theorem.

In [**42**] I.Leader surveys the various methods for proving isoperimetric inequalities: martingale techniques, eigenvalue analysis and combinatorics (which often give the sharpest results) and applies them to the "concentration of measure" phenomenon.

VI. Closing Comments

In looking back over the previous sections, it is striking that there has been very little overlap or cross-fertilization between the papers which followed from the solutions of the JPL Coding Problem for $r = 1$ & ∞ and those from the solution for $r = 2$. Because the methods seem so disparate, finding a common generalization (one which would work for all exponents, $r, 1 \leq r \leq \infty$) would require effort but will surely be rewarded with interesting insights. Lindsey extended the result of Section I to any product of complete graphs (whose symmetry groups are products of symmetric groups).Is it possible to extend the result of Section III to products of symmetric groups? Chvatalova & Moghadam extended the result of Section II to products of Z_n's (whose symmetry group is the dihedral graoup, D_n). Is it possible to extend the result of Section III to any product of dihedral groups? Another intriguing challenge is to extend the results of Sections I, II & III by having Q_d be embedded into Z_{2^d} (or the 2^dth roots of unity). For $r = 1$ & 2 there is then an obvious candidate for the optimal code, namely the Grey code. Just as the lexicographic code reiterates, in the subcube $v_d = 1$, the order of the corresponding vertices in the subcube $v_d = 0$, the Grey code reverses that order. For $r = \infty$ there is no obvious way to improve upon Hales' code when embedding Q_d into Z_{2^d} but neither is it obvious how to prove that there is no such improvement.

Lastly, I would like to thank those who were my mentors in the development of this material: Ed Posner (posthumously), Gian-Carlo Rota and Vic Shapiro. Their objective contributions have already been noted. But also the patient encouragement, sharing of knowledge and skillful listening which they bestowed so generously has lighted my way.

References

[1] K. Engel & H.-D. O.F. Gronau, *Sperner Theory in Partially Ordered Sets*, Teubner, Leipzig (1985), 232 pgs.

[2] K. Engel; *Sperner Theory*, to be published by Cambridge University Press.

[3] J.B. Kruskal; The number of s-dimensional facts in a complex; An analogy between the simplex and the cube, J. Comb. Th. 6(1969) 86-89.

[4] D. L. Wang and P. Wang, Discrete isopermetric problems, SIAM J. Appl. Math. 32 (1977), 860-870.

[5] J. Chvatalova, A.K. Dewdoney, N.E. Gibbs and K.R. Korfhage; The bandwidth problem for graphs, a collection of recent results, Res. Rep. 2-1, Dept. of Computer Sci., Univ. of Western Ontario, London, Ontario, Canada (1995).

[6] H.S. Moghadam; Compression operators and a solution to the bandwidth problem of the product of n paths, Doctoral Dissertation Univ. of California, Riverside (1982).

[7] W.H. Kautz; Optimal encoding for digital computers, in Conv. Rec. of Inst. of Radio Engineers 2(1954) 47-57.

[8] B. Lindström; The optimal number of faces in a cubical complex, Arkivfur Mathematik 8 (1971), 245-257.

[9] J. Chvatalova; Optimal labelling of a product of two paths, Disc. Math 11(1975), 249-253.

[10] D.E. Daykin and P. Frankl; On Kruskals cascades and counting containments in a set of subsets, Mathematica 30 (1983) 134–141.

[11] J. Körner and V. Wei; Odd and even Hamming spheres also have minimum boundary, Disc. Math. 51 (1984), 147-165.

[12] P. Frankl and Z. Furedi, A short proof for a theorem of Harper about Hamming-spheres, Disc. Math 34 (1981) 311-313.

[13] A.J. Berstein; Maximally connected arrays on the n-cube SIAM J. Appl. Math 15(1967), 1485-1489.

[14] K. Steiglitz and A.J. Bernstein; Optimal binary ordering of ordered numbers, SIAM J. Appl. Math. 13 (1965) 441-443.

[15] G.F. Clements; Sets of lattice points which contain a maximal number of ideals, Proc. AMS 17 (1971), 13-15.

[16] M.R. Garey, R. L. Graham, D.S. Johnson and D.E. Knuth; Complexity results for bandwidth minimization, SIAM J. Appl. Math. 34, (1978), 477-495.

[17] H. J. Tiersma; A note on Hamming spheres, Dis. Math. 54 (1985), 225-228.

[18] G. Mitchison and R. Durbin; Optimal numberings of an $N \times N$ array, SIAM J. Alg. and Disc. Math. (1986), 571-582.

[19] R. L. Graham; Ann. NY Acad. Sci 175(1970) 170-186.

[20] J. H. Lindsey; Assignments of numbers to vertices, Amer. Math. Monthly 71 (1964), 508-516.

[21] P. Frankl; A lower bound on the size of a complex generated by an antichain, Disc Math. 76(1989), 51-56.

[22] B. Bollobas and I. Leader; An isoperimetric inequality on the discrete torus, SIAM J. Disc. Math. 3 (1990) 32-37.

[23] B. Bollobas and I. Leader; Compressions and isoperemetric inequalities, J. Comb. Th. - A 56(1991), 47-62.

[24] B. Bollobas and I. Leader; Maximal sets of a given diameter in the grid and the torus, Disc. Math. 122(1993), 17-35.

[25] T.R. Crimmins, H.M. Horwitz, C.J. Palermo and R.V. Palermo; Minimization of mean-square error for data transmitted via group codes, IEEE Trans. Inf. Th. IT-15 (1969), 72-78.

[26] G.G. Apple and P.A. Wintz; Calculation of Fourier Transforms on finite Abelian groups, IEEE Trans. Inf. Th. (correspondence, March 1970), 233-234.

[27] T.R. Crimmins & H.M. Horwitz; Mean-square error coset leaders for group codes, IEEE Trans. Inf. Th. IT-16 (1970), 429-432.

[28] G.A. Wolf & G. Redinbo; The optimum mean-square estimate for decoding binary block codes, IEEE Trans Inf. Th. IT-20 (1974), 344-351.

[29] G.S. Bloom & S.W. Golomb; Applications of numbered undirected graphs, Proc. IEEE 65(1977), 562-520.

[30] C.R. Redinbo; Optimum symbolily symbol mean-squared error channel coding, IEEE Trans. Inf. Th. IT-25 (1979), 387-415.

[31] A.L. Rosenberg; Data encodings and their costs, Acta Informatica 9(1978), 273-292.

[32] A.L. Rosenberg and L. Synder; Bounds on the costs of data encoding, Math. Systems Theory 12 (1978), 9-39.

[33] A.L. Rosenberg; Encoding data structures in trees, J. ACM 26(1979), 668-689.

[34] A.L. Rosenberg, R. Wood and Zv Gail; Storage representations for tree-like data structures, Math. Systems Theory 13 (1979), 105-130.

[35] T. Langauer; Upper and lower bounds on the complexity of min-cut linear arrangements problem on trees, SIAM J. Alg. Disc. Math. 3 (1982), 99-113.

[36] G. Gierz & F. Hergert; The bandwidth problem for distributive lattices of breadth 3, Disc. Math. 88 (1991), 157-177.

[37] E. Ma and L. Tau; Embeddings among meshes and tori, J. Parallel and Distributed Computing 18(1993), 44-55.

[38] T.-H. Lai and A.P. Sprague; Placement of processors of a hypercube, IEEE Trans. Computers 40(1991) 714-722.

[39] D.M. Gordon; Percolation in high dimension, J. London Math. Soc. (2) 94(1991), 373-384.

[40] N. Alon and V.D. Milman; λ_1, isoperimetric inequalities for graphs and superconcentrators, J. Comb. Th. B 38(1985), 73-88.

[41] M. Talagrand; An isoperimetric inequality on the cube and the Khintchine-Kahane inequalities, Proc. AMS 104(1988), 905-909.

[42] I. Leader; Discrete isoperimetric inequalities, Proc. Symposia in Appl. Math., vol. 44 (B. Ballobos, Ed.) AMS, Providence RI (1991), 57-80.

[43] G. Polya and G. Szegö; *Isoperimetric Inequalities in Mathematical Physics*, Ann. Math. Study No. 27, Princeton Univ. Press, Princeton, N.J. (1951)

[44] L.H. Harper and J.D. Chavez; Global Theories of combinatorial Optimization I: Discrete Isoperimetric Problems and Steiner Operations, to be published by Cambridge University Press.

[45] L.H. Harper; Optimal Assignments of Numbers to Vertices J. Soc. Indust. Appl. Math., vol. 12 1(1964) 131-135.

[46] L.H. Harper; Optimal Numbering and Isoperimetric Problems on Graphs, J. Comb. Theory, vol. 1, 3(1966) 385-393.

[47] Y.V. Mitryayev; Transmission of telemetry data by means of group codes, Radio Engrg. Electron (USSR), vol. 8 (1963).

MATHEMATICS DEPARTMENT, UNIVERSITY OF CALIFORNIA, RIVERSIDE, CA 92521
E-mail address: harper@math.ucr.edu

Contemporary Mathematics
Volume **208**, 1997

Weak Limit sets of Differential Equations

Morris W. Hirsch

Dedicated to Professor Victor L. Shapiro

ABSTRACT. Let E be a Banach space. We investigate the limit set $\mathcal{L}[u]$ and the weak limit set $\mathcal{L}_{\mathsf{w}}[u]$ of solutions $u : \mathbf{R}_+ \to E$ to differential equations $Tu = f$ where T is the composition of n first order differential operators of the form $a_k D + b_k I$. Here D is differentiation, I is the identity; the coefficients $a_k, b_k : \mathbf{R}_+ \to \mathbf{R}$ are continuous and nonvanishing with $\int |a_k/b_k| = \infty$; and $f : \mathbf{R}_+ \to E$ is continuous.

Theorem *Assume $n = 1$ and $a_1/b_1 > 0$. If K is a nonempty closed convex set such that* $\operatorname{dist}(f(t)/b_1(t), K) \to 0$, *then $\mathcal{L}[u] \subset K$.*

For the following results, assume the b_k are bounded and bounded away from zero.

Theorem $\mathcal{L}[u] \subset \mathcal{L}_{\mathsf{w}}[u] \subset M\overline{\mathrm{co}}(\mathcal{L}_{\mathsf{w}}[f])$, *where $M \subset \mathbf{R}$ is a certain compact interval associated to T.*

Let B denote the set of maps $g : \mathbf{R}_+ \to E$ such that for every continuous linear functional L on E, the composition $L \circ g : \mathbf{R}_+ \to \mathbf{R}$ has compact (possibly empty) limit set.

Theorem *If $f \in \mathsf{B}$, then $u \in \mathsf{B}$.*

Applications are made to machine learning, nonlinear autonomous equations, and real eigenfunctions of the operator T.

1. Introduction

A fundamental problem in dynamical systems is to obtain information on the limit sets of solutions to differential equations. Generally we have to deal with nonlinear equations $dx/dt = F(t,x)$. But sometimes information can be got about a particular solution $u(t)$ from the observation that it satisfies the first order linear nonhomogeneous equation $du/dt = f(t)$ where $f(t) = F(t, u(t))$, provided we know something about F and have some *a priori* knowledge of u. In fact a situation of this kind, described next, motivated the present paper. Too simple to be called an "application" of our results, this example is presented first as motivation for the general results.

The basic question attacked in this paper is to obtain information on the limiting behavior of a solution to an nth order nonautonomous, linear, nonhomogeneous differential equation in a Banach space E of the form:

$$Tu = f, \tag{1}$$
$$T = (a_n D + b_n I) \circ \cdots \circ (a_1 D + b_1 I) \tag{2}$$

[1]This research was partially supported by National Science Foundation grant DMS-9424382.

1991 *Mathematics Subject Classification.* Primary 34A26 34A30 34C10 34G10 34L15; Secondary 68T05 47E05 34L15.

Key words and phrases. Differential Equations, Differential Operators, Disconjugate Equations, Dynamical Systems, Eigenfunctions, Limit Sets, Machine Learning, Weak Limits.

where D is the differentiation operator and I the identity. The unknown function u and the driving function f are maps from $[t_0, \infty)$ to E. We always assume the coefficients are continuous nonvanishing maps $a_k, b_k : [t_0, \infty) \to \mathbf{R}$ such that $\int |a_k/b_k| = \infty$. In particular, we would like to obtain information on the limit set $\mathcal{L}[u]$, and the weak limit set $\mathcal{L}_{\mathsf{w}}[u]$, in terms of f and the coefficients.

Set $\sigma_i = \operatorname{sgn}(a_i/b_i)$ and $\sigma = \sigma_1 \cdots \sigma_n$. Define $J_i \subset \mathbf{R}$ to be the closed interval $\mathcal{L}[b_i]$, and define the interval

$$J = J_1 \cdots J_n = \{s_1 \cdots s_n : s_k \in J_k, \ \ k = 1, = \dots, n.\}$$

When $0 \notin J$ we define J^{-1} to be the interval consisting of the reciprocals of numbers in J, and set $J_1/J = J_1 J^{-1}$.

Define B_E to be the set of continuous functions $g : [t_0, \infty) \to E$ such that for each $L \in E^*$ (the dual space of E), $\mathcal{L}[L \circ g]$ is bounded (perhaps empty).

The closed convex hull of a set $A \subset E$ is denoted by $\overline{\text{co}}\,(A)$.

Here is a special case of Theorem 4.7:

THEOREM 1.1. *Assume each J_i is a finite interval (or a point) not containing zero. Let $v : [t_0, \infty) \to E$ be a solution to the differential equation $Tv = f$. Then:*

(a): $\mathcal{L}_{\mathsf{w}}[v] \subset \sigma J^{-1} \overline{\text{co}}\,(\mathcal{L}_{\mathsf{w}}[f])$.
(b): *If $f \in \mathsf{B}_E$ then $v \in \mathsf{B}_E$.*

After discussing the original motivating problem we present generalities on limit sets in Banach spaces in Section 2. Limit sets to solutions of differential equations in $\mathbf{R}$ are treated in Section 3. Equations in arbitrary Banach spaces are covered in Section 4, together with an application to nonlinear dynamical systems.

J. Muldowney has kindly pointed out that the higher order equations considered in Section 3 include the classical disconjugate equations (Pólya [**6**], Coppel [**2**]); and that some of the results of that section are similar to Proposition 3.2 of Muldowney [**5**].

1.1. A Design Problem in Machine Learning. The following problem came up in dynamical systems models of machine learning (see Baird and Hirsch [**4**], Hirsch [**3**]). Suppose we are designing a system, e.g., a neural network for pattern recognition, involving a differential equation in Euclidean n-space $\mathbf{R}^n$ which we write in the not uncommon form

$$\frac{dx}{dt} = -x + F(x). \tag{3}$$

There is a certain attractor $A \subset \mathbf{R}^n$ for the flow generated by F that we would like to reach in the following way: start in some given initial state s, add a *teaching signal* I to Equation (3), and run the initial value problem:

$$\frac{dy}{dt} = -y + F(y) + I, \tag{4}$$

in the hope that $y(t)$ will end up in the basin of attraction of A at some time t_1. If we then set I to 0, the trajectory of Equation (3) starting at $y(t_1)$ will proceed towards A.

How should I, which could depend on t, be chosen? We want I to be simple and robust— not sensitive to mathematical details— so that it can be implemented by a machine, or for biological modeling, by some more primitive part of the nervous system.

In general this problem is quite difficult to solve in terms of a formula for F; basins of attractions are notoriously difficult to locate. (See e. g. Chiang *et al.* [**1**] for a contribution and references to this problem.)

But we do not need to find a general solution. F does not arise from nature; we have considerable freedom in designing it, subject to the constraints of our network. And a successful solution of the mathematical problem for a class of differential equations might suggest better ways to design networks.

If this idea is to succeed, some part of the basin of A needs to be easy to locate. We assume the basin contains the closed δ-neighborhood $N_\delta(K)$, $\delta > 0$ of some convex set K which is known to us. Moreover we assume δ is a uniform upper bound for $||F(x)||$. While these are drastic assumptions, it is easy to find plausible network equations admitting attractors that satisfy them (e. g. Baird and Hirsch [**4**], Hirsch [**3**]). We then have the following result, where the distance from z to K is denoted by dist (z, K).

THEOREM 1.2. *Assume the function $I : [0, \infty) \to \mathbf{R}^n$ is such that*

$$\lim_{t\to\infty} \text{dist}\,(I(t), K) = 0.$$

If $x(t), 0 \leq t < \infty$ is a solution to Equation (4), then

$$\lim_{t\to\infty} \text{dist}\,(x(t), N_\delta(K)) = 0.$$

PROOF. Fix the solution x and set $f(t) = F(x(t)) + I(t)$. Then x is a solution to

$$\frac{dx}{dt} = -x + f(t). \tag{5}$$

Now

$$\text{dist}\,(f(t), K) \leq ||F(x(t))|| + \text{dist}\,(I(t), K) \leq \delta + \text{dist}\,(I(t), K).$$

Therefore

$$\lim_{t\to\infty} \text{dist}\,(f(t), N_\delta(K)) = 0.$$

Define

$$V : \mathbf{R}^n \to \mathbf{R},\ z \mapsto \text{dist}\,(z, N_\delta(K)).$$

Then V is a convex Lipschitz map, and $\lim_{t\to\infty} V(f(t)) = 0$.

LEMMA 1.3.

$$\limsup_{h\to 0} \frac{V(x(t+h) - V(x(t)))}{h} \leq -V(x(t)) + V(f(t)).$$

To prove the lemma, we use Equation (5) to obtain a Taylor estimate, abbreviating $x = x(t), f = f(t)$:

$$x(t+h) = x + hx'(t) + o(h) = (1-h)x + hf(t) + o(h).$$

Using the Lipschitz property of V and convexity, we get:

$$\frac{V(x(t+h)) - V(x)}{h} = \frac{V((1-h)x + hf) + o(h) - V(x)}{h} \tag{6}$$

$$\le \frac{(1-h)V(x) + hV(f) - V(x) + o(h)}{h} \tag{7}$$

$$= -V(x) + V(f) + \frac{o(h)}{h}. \tag{8}$$

Letting $h \to 0$ proves the lemma.

Now fix $\epsilon > 0$ and choose $t_0 \ge 0$ so large that $V(f(t)) < \epsilon$ if $t \ge t_1$. Then the lemma and the Comparison Principle (Proposition 3.1 below) imply $V(t_1+s) \le v(s)$ for all $s \ge 0$, where $v(s)$ solves the initial value problem

$$\frac{dv}{dt} = -v + \epsilon, \quad v(0) = V(t_1).$$

Thus

$$V(t_1 + s) \le (1 - e^{-s})\epsilon + e^{-s}V(t_1), \tag{9}$$

proving that $\limsup_{t\to\infty} V(t) \le \epsilon$. Since this holds for all $\epsilon > 0$, the proof of Theorem 1.2 is complete. □

2. Limit sets of curves in Banach spaces

Let E denote a real Banach space, with dual space E^* . The topological vector space consisting of E with its weak topology is denoted by E_{w}. Recall that a subset $X \subset E$ is weakly compact (i.e., compact in E_{w}) if and only if X is bounded and weakly closed. In particular, every bounded closed convex subset of E is weakly compact.

By a *curve* in E we mean a continuous map $g : [t_0, \infty) \to E$; unless otherwise indicated, the domain of any curve will be $[t_0, \infty)$.

The *limit set* $\mathcal{L}[g] \subset E$ of a curve g consists of all limits of sequences $g(t_k)$, $t_k \to \infty$. This set may, of course, be empty.

The *weak limit set* $\mathcal{L}_{\mathsf{w}}[g]$ is the set of weak limit points of sequences $g(t_k)$, $t_k \to \infty$. Equivalently, $\mathcal{L}_{\mathsf{w}}[g]$ is the set of limits of sequences $g(t_k)$, $t_k \to \infty$, in E_{w}. Evidently $\mathcal{L}_{\mathsf{w}}[g]$ is weakly closed. It contains $\mathcal{L}[g]$, but it may be much larger.

LEMMA 2.1.

(a): *Suppose the image of g lies in a compact subset $C \subset E$. Then $\mathcal{L}[g] = \mathcal{L}_{\mathsf{w}}[g]$; and this set is a nonempty, compact connected subset of C.*

(b): *If $\mathcal{L}_{\mathsf{w}}[g]$ is bounded, it is weakly compact.*

(c): *Suppose the image of g lies in a closed bounded convex set $K \subset E$. Then $\mathcal{L}_{\mathsf{w}}[g]$ is a nonempty, weakly compact, weakly connected subset of K.*

PROOF. The first statement in (a) holds because E and E_{w} induce the same topology on compact subsets. To prove the second statement, we note that

$$\mathcal{L}[g] = \bigcap_{t \ge t_0} \operatorname{clos} g([t, \infty)),$$

where clos denotes closure. This exhibits $\mathcal{L}[g]$ as the intersection of a nested family of compact connected subsets of C; a standard result in topology implies the second statement in (a). In (b) $\mathcal{L}_{\mathsf{w}}[g]$ lies in some closed ball B. Since B is weakly compact and $\mathcal{L}_w[g]$ is weakly closed, it follows that $\mathcal{L}_w[g]$ is weakly compact. For (c) we note that K is weakly compact, so reasoning as in (a) completes the proof. □

EXAMPLE 2.2. Assume E is separable and infinite dimensional, and let $\{e_k\}_{k\in\mathbf{N}}$ be a dense sequence of linearly independent unit vectors. Let $h : [0, \infty) \to E$ map each natural number $k \in \mathbf{N}$ to e_k, with h affine on $[k-1, k]$. Define $g(t) = h(t)/||h(t)||$, $0 \le t < \infty$. Then the image of g lies in the (boundary) unit sphere $S = \{x \in e : ||x|| = 1\}$. It is clear that $\mathcal{L}[g]$ is empty. On the other hand, $\mathcal{L}_w[g] = \{0\}$ because $g(t) \to 0$ in E_w.

EXAMPLE 2.3. Example 2.2 can be changed slightly to yield a curve h in S which converges weakly to 0, but has a unique limit point p, necessarily in S. Then $\mathcal{L}_w[h]$ is a compact connected subset of the closed unit ball B containing 0 and p. Therefore $\mathcal{L}_w[h]$ must contain a point of norm λ, for each $\lambda \in [0, 1]$.

EXAMPLE 2.4. Example 2.2 can be changed to a curve r in S which converges weakly to 0, but has precisely two limit points p, q. In this case $\mathcal{L}_w[r]$ is nonempty, weakly compact and weakly connected, while $\mathcal{L}_w[r] = \{p, q\}$, which is nonempty, compact and disconnected.

We say a closed subset $P \subset E$ *absorbs* g if P is nonempty and

$$\lim_{t\to\infty} \mathrm{dist}\,(g(t), P) = 0.$$

If P absorbs g then P contains $\mathcal{L}[g]$, but it need not contain $\mathcal{L}_w[g]$.

In Example 2.2, the unit sphere S absorbs g, but it does not contain $\mathcal{L}_w[g]$. The closed unit ball B, which is convex, does contain $\mathcal{L}_w[g]$.

We say P *weakly absorbs* g if:

$$\sup L(P) \ge \overline{\lim}\,(L \circ g)\ \forall L \in E^*.$$

Clearly this holds when P absorbs g. Two equivalent formulations are:

$$\inf L(P) \le \underline{\lim}\,(L \circ g)\ \forall L \in E^*,$$

and

$$L(P) \supset \mathcal{L}[L \circ g]\ \forall L \in E^*.$$

In Example 2.2, the origin weakly absorbs g, but does not absorb g.

LEMMA 2.5. *The following statements (i), (ii), (iii) are equivalent:*

(i): *The image of g lies in a compact subset of E.*

(ii): *$\mathcal{L}[g]$ is a nonempty compact set that absorbs g.*

(iii): *Some nonempty compact set absorbs g.*

When the image of g lies in a locally compact set (e.g., $E = \mathbf{R}^n$), then these are equivalent to:

(iv): *$\mathcal{L}[g]$ is a nonempty compact set.*

PROOF. The implications (i) $\Rightarrow$ (ii) and (ii) $\Rightarrow$ (iii) are obvious. If (iii) holds then it is easy to see that the closure of the image of g is compact, proving (i). Now assume the image of g lies in a locally compact set X and $\mathcal{L}[g]$ is a nonempty compact set. Then there exists a compact set $N \subset X$ which is a neighborhood of $\mathcal{L}[g]$ in X, so that $\mathcal{L}[g]$ lies in the relative interior N^0 of N in X. Since the compact set $\partial N = N \setminus N^0$ is disjoint from $\mathcal{L}[g]$, it follows that there exists t_1 such that $g(t) \notin \partial N$ for all $t > t_1$. Because $\mathcal{L}[g]$ is a nonempty set in N, this implies $g(t) \in N$ for all $t \ge t_1$, and this yields (iv). □

LEMMA 2.6. *If P weakly absorbs g, then $\mathcal{L}_{\mathsf{w}}[g] \subset \overline{\mathrm{co}}\,(P)$.*

PROOF. Set $K = \overline{\mathrm{co}}\,(P)$. Evidently $N_\epsilon(K)$ weakly absorbs g for every $\epsilon > 0$. It follows that if $H \subset E$ is any supporting hyperplane for the convex set $N_\epsilon(K)$, then $\mathcal{L}_{\mathsf{w}}[g]$ and K lie on the same side of H. The hyperplane separation theorem therefore implies $\mathcal{L}_{\mathsf{w}}[g] \subset N_\epsilon(K)$. Since this holds for every $\epsilon > 0$, the lemma is proved. □

The *weakly absorbing hull* $\mathcal{A}[g]$ of g is the smallest set that weakly absorbs g, namely

$$\mathcal{A}[g] = \{x \in E : Lx \in \mathcal{L}[L \circ g] \,\forall\, L \in E^*\}.$$

This set is convex and weakly closed, and is contained in every convex weakly closed set that weakly absorbs g.

PROPOSITION 2.7. *Let g be a curve in E. Then:*

(a): $\mathcal{L}_{\mathsf{w}}[g] \subset \mathcal{A}[g]$.

(b): *If the image of g lies in a compact set, then $\mathcal{A}[g] \subset \overline{\mathrm{co}}\,(\mathcal{L}[g])$.*

PROOF. Lemma 2.6 implies (a). For (a) we note that compactness implies

$$\mathcal{L}[L \circ g] \subset L(\mathcal{L}[g]) \subset L(\overline{\mathrm{co}}\,(\mathcal{L}[g])).$$

Therefore $x \in \mathcal{A}[g]$ implies $Lx \in L(\overline{\mathrm{co}}\,(\mathcal{L}[g]))$ for all $L \in E^*$, and the hyperplane separation theorem shows $x \in \overline{\mathrm{co}}\,(\mathcal{L}[g])$. □

3. Differential Equations in R

For notational convenience we adjoin endpoints $\pm\infty$ to the real line $\mathbf{R}$, obtaining the *extended line* $\overline{\mathbf{R}} = [-\infty, \infty]$. For any function $g : [t_0, \infty) \to \mathbf{R}$ we allow only finite values for $Dg(t) = g'(t)$, but the following inferior and superior limits are taken in $\overline{\mathbf{R}}$:

$$\begin{aligned}
\overline{\lim}\, g &= \limsup_{t\to\infty} g(t), \\
\underline{\lim}\, g &= \liminf_{t\to\infty} g(t), \\
\overline{D}g(t) &= \limsup_{h\to 0} \frac{g(t+h) - g(t)}{h}, \\
\underline{D}g(t) &= \liminf_{h\to 0} \frac{g(t+h) - g(t)}{h},
\end{aligned}$$

$$Dg(t) = g'(t).$$

We define $\mathcal{L}[g]$ to be the closed interval $[\underline{\lim}\, g, \overline{\lim}\, g] \subset \overline{\mathbf{R}}$.

We have already used a special case of the following well-known result:

PROPOSITION 3.1 (COMPARISON PRINCIPLE). *If continuous functions $x, y : [t_0, \infty) \to \mathbf{R}$ satisfy $x(t_0) \le y(t_0)$ and $\overline{D}x \le \underline{D}y$, then $x \le y$.*

Now let $u, f : [t_0, \infty) \to \mathbf{R}$ be continuous functions.

LEMMA 3.2.

(a): *If $\overline{D}u \le -u + f$, then $\overline{\lim}\, u \le \overline{\lim}\, f$.*

(b): *If $\underline{D}u \ge -u + f$, then $\underline{\lim}\, u \ge \underline{\lim}\, f$.*

(c): *Suppose $Du = -u + f$. Then $\mathcal{L}[u] \subset \mathcal{L}[f]$.*

PROOF. The proof of (a) is similar to that of Theorem 1.1. Part (b) follows from (a) by considering $-u$. Part (c)(i) follows from (a) and (b). □

We shall also need the following more subtle fact:

LEMMA 3.3. *Assume $Du = u + f$. Then:*

(a): *Either $\overline{\lim}\, u \leq \overline{\lim}\,(-f)$, or else $u(t) \to \infty$.*
(b): *Either $\underline{\lim}\, u \geq \underline{\lim}\,(-f)$, or else $u(t) \to -\infty$.*
(c): *If $|u(t)| \not\to \infty$, then $\mathcal{L}[u] \subset= \mathcal{L}[-f] = -\mathcal{L}[f]$.*
(d): *If $f(t) \to \pm\infty$ then $u(t) \to \pm\infty$, respectively.*

PROOF. To prove (a), assume $\overline{\lim}\, u > \overline{\lim}\,(-f)$. There exist t_1, β and $\epsilon > 0$ such that for all $t \geq t_1$:

$$u(t) > \beta + \epsilon,\ f(t) > -\beta,$$

so

$$u'(t) = u(t) + f(t) > \epsilon,$$

whence $u(t) \to \infty$. The proofs of (b) and (d) are similar, and (c) follows from (a) and (d). □

3.1. Nonconstant coefficients. We fix some interval $[t_0, \infty)$ and consider differential operators acting on functions $u : [t_0, \infty) \to \mathbf{R}$. A first order operator is *allowable* if it has the form $aD + bI$, where D is differentiation, I is the identity operator, and $a, b : [t_0, \infty) \to \mathbf{R}$, satisfy the following two conditions:

HYPOTHESIS 3.4.

(i): *$a(t) \neq 0, b(t) \neq 0$ for all $t \in [t_0, \infty)$.*
(ii): $\displaystyle\int_{t_0}^{\infty} \frac{|b(t)|}{|a(t)|} dt = \infty.$

An nth order differential operator T is allowable if it is the composition of n allowable first order operators. An *allowable differential equation* is one of the form $Tu = f$ where T is an allowable operator and $f : [t_0, \infty) \to \mathbf{R}$ is continuous.

Consider an allowable first order equation:

$$a\frac{dv}{dt} + bv = f \tag{10}$$

PROPOSITION 3.5. *Suppose $a/b > 0$, and let $v : [t_0, \infty) \to \mathbf{R}$ be a solution to Equation (10). Then:*

(a): *$\mathcal{L}[v] \subset \mathcal{L}[f/b]$.*
(b): *If $f(t)/b(t) \to c \in \overline{\mathbf{R}}$ then $v(t) \to c$.*

PROOF. Equation (10) is equivalent to

$$\frac{a}{b}\frac{dv}{dt} = -v + \frac{f}{b}. \tag{11}$$

We simplify further by replacing the time variable t with

$$s = \int_{t_0}^{t} \frac{|b(s)|}{|a(s)|} dt, \tag{12}$$

noting that s runs from 0 to ∞ by Hypothesis 3.4 (ii). We express t as a function $t(s)$, then set $w(s) = v(t(s))$ and $h(s) = f(t(s))/b(t(s))$. Now Equation (11) is equivalent to:

$$\frac{dw}{ds} = -w + h.$$

The proof of (a) is completed by applying Lemma 3.2(c), and (b) follows. □

PROPOSITION 3.6. *Assume $a/b < 0$ and let $v : [t_0, \infty) \to \mathbf{R}$ be a solution to Equation (10). Then:*

(a): *Either $\overline{\lim}\, v \leq \overline{\lim}\,(-f/b)$, or else $v(t) \to \infty$.*
(b): *Either $\underline{\lim}\, v \geq \underline{\lim}\,(-f/b)$, or else $v(t) \to -\infty$.*
(c): *If $f(t)/b(t) \to \pm\infty$, then $v(t) \to \pm\infty$, respectively.*

PROOF. Making the change of variable (12), we obtain the equivalent equation

$$\frac{dw}{ds} = w + h,$$

where $h(s) = f(t(s))/b(t(s))$, $0 \leq s < \infty$. The proof is completed by applying Lemma 3.3. □

In some cases there is a convenient estimate for the limit set of v:

LEMMA 3.7. *Let $v : [t_0, \infty) \to \mathbf{R}$ be a solution to Equation (10). Assume $\mathcal{L}[b]$ is a compact interval J not containing* 0. *Set $\sigma =$ $\mathrm{sgn}(a/b)$. Then:*

$$\mathcal{L}[v] \subset \sigma \mathcal{L}[f]/\mathcal{L}[b].$$

PROOF. From Propositions 3.5 and 3.6 we see that $\mathcal{L}[v] \subset \sigma\mathcal{L}[f/b]$, and the assumptions on $\mathcal{L}[b]$ imply $\mathcal{L}[f/b] \subset \mathcal{L}[f]/\mathcal{L}[b]$. □

Define $\mathsf{B} = \mathsf{B}([t_0, \infty), \mathbf{R})$ to be the set of continuous functions $g : [t_0, \infty) \to \mathbf{R}$ whose limit sets are bounded (possibly empty). Thus $g \in \mathsf{B}$ provided $\mathcal{L}[g]$ is a compact interval (possibly a point), or $|g(t)| \to \infty$. Roughly speaking, $g(t)$ cannot traverse arbitrarily large intervals infinitely often.

The following result shows that B is closed under solutions to certain kinds of linear differential equations:

PROPOSITION 3.8. *Assume $f/b \in \mathsf{B}$ and let $v : [t_0, \infty) \to \mathbf{R}$ be a solution to Equation (10). Then: $v \in \mathsf{B}$.*

PROOF. This is a consequence of Propositions 3.5 and 3.6. □

3.2. Higher Order Equations. Consider now an allowable nth order linear differential operator T of the form:

$$T = T_n \circ \cdots \circ T_1, \tag{13}$$

$$T_i = a_i D + b_i I,\ i = 1, \ldots, n \tag{14}$$

Suppose $n \geq 2$. Given a function $f : [t_0, \infty) \to \mathbf{R}$, we recursively define the *domain* of T to be the set of differentiable functions $v : [t_0, \infty) \to \mathbf{R}$ such that $T_1 v$ is in the domain of $T_n \circ \cdots \circ T_2$. In particular, every C^n function is in the domain of T. In what follows, the notation Tv implies the tacit assumption that v is in the domain of T.

THEOREM 3.9. *Let v be a solution to the allowable differential equation $Tv = f$. Then:*

(a): *Assume each interval $\mathcal{L}[b_i]$ is finite and does not contain zero. Set $\sigma_i = \operatorname{sgn}(a_i/b_i)$ and $\sigma = \sigma_1 \cdots \sigma_n$. Then:*

$$\mathcal{L}[v] \subset \sigma\mathcal{L}[f]/\mathcal{L}[b_1]\cdots\mathcal{L}[b_n].$$

(b): *Assume*

$$f/b_n \ldots b_i \in \mathsf{B},\ i = 1, \ldots, n.$$

Then $v \in \mathsf{B}$. In particular, $v \in \mathsf{C}$ provided the $\mathcal{L}[b_i]$ are as in (a) and $f \in \mathsf{B}$.

PROOF. The proof is by induction on n, the case $n = 1$ being covered by Propositions 3.5, 3.6 and 3.8. Assume $n \geq 2$ and define the allowable differential operator

$$S = T_n \circ \cdots \circ T_2.$$

The equation

$$Sw = f \tag{15}$$

is allowable and has the solution w where

$$(a_1 D + b_1)v = w. \tag{16}$$

By the induction hypothesis applied to Equation (15), in case (a),

$$\mathcal{L}[w] \subset \tau\mathcal{L}[f]/\mathcal{L}[b_n]\cdots\mathcal{L}[b_2] \tag{17}$$

where $\tau = \sigma_1 \cdots \sigma_{n-1}$. And $w \in \mathsf{B}$ in case (b).

Now v is a solution to Equation (16), or equivalently, to

$$(\frac{a_1}{b_1}D + I)v = \frac{w}{b_1}. \tag{18}$$

so $v \in \mathsf{B}$ by Proposition 3.8. In case (a), Lemma 3.7 shows

$$\mathcal{L}[v] \subset \sigma_1\mathcal{L}[w/b_1] \subset \sigma_1\mathcal{L}[w]/\mathcal{L}[b_1].$$

Therefore Equation (17) implies

$$\mathcal{L}[v] \subset \frac{\sigma_1\tau\mathcal{L}[f]/\mathcal{L}[b_2]\cdots\mathcal{L}[b_n]}{\mathcal{L}[b_1]} \tag{19}$$

$$= \sigma\mathcal{L}[f]/\mathcal{L}[b_1]\cdots\mathcal{L}[b_n]. \tag{20}$$

In case (b), Proposition 3.8 shows $v \in \mathsf{B}$. The induction is complete. □

4. Differential Equations in Banach Spaces

Now we extend the preceding results to Banach spaces. The notion of an allowable first order differential operator $aD + bI$ on functions $u : [t_0, \infty) \to E$ is formally the same as before, namely $a, b : [t_0, \infty) \to \mathbf{R}$ satisfy Hypothesis 3.4. A composition $T = T_n \circ \cdots \circ T_1$ of allowable first order operators is also called allowable. A differential equation of the form $Tu = f$ is allowable if T is an allowable operator and $f : [t_0, \infty) \to \mathbf{R}$ is continuous.

We first treat allowable first order equations

$$a\frac{dw}{dt} + bw = f. \tag{21}$$

THEOREM 4.1. *Let w be a solution to Equation (21), and assume $a/b > 0$.*

(a): *If $V : E \to \mathbf{R}$ is convex and locally Lipschitz, then*

$$\overline{\lim}\,(V \circ w) \leq \overline{\lim}\,(V \circ f/b).$$

(b): *For all $L \in E^*$,*

$$\mathcal{L}[L \circ w] \subset \mathcal{L}[L \circ (f/b)].$$

PROOF. We can make the change of variable (12), and replace f by f/b. Thus it suffices to consider the special case:

$$\frac{dw}{dt} = -w + f.$$

To prove (a), imitate the proof of Theorem 1.1 to obtain

$$\overline{D}\,(V \circ w) \leq \overline{D}\,(V \circ f).$$

Now apply Theorem 3.2(a).

Part (b) is proved by applying (a) twice: first to $V = L$ to get $\overline{\lim}\,(L \circ w) \leq \overline{\lim}\,(L \circ f)$, then to $V = -L$ to get $\underline{\lim}\,(L \circ w) \leq \underline{\lim}\,(L \circ f)$. □

COROLLARY 4.2. *Let w satisfy Equation (21) with $a/b > 0$. Suppose $K \subset E$ is a closed nonempty convex set that absorbs (or weakly absorbs) f/b. Then K absorbs (respectively, weakly absorbs) w. In particular:*

$$\mathcal{L}[w] \subset \overline{\mathrm{co}}\,(\mathcal{L}[f/b]). \tag{22}$$

PROOF. We refer to Section 2 for the definition of "absorbs". Define $V : [t_0, \infty) \to \mathbf{R}$, $V(z) = \mathrm{dist}\,(z, K)$. Then V is convex and continuous, and therefore locally Lipschitz (Van Tiel [**7**], theorem 5.21). Notice that K absorbs a curve $x : [t_0, \infty) \to E$ if and only if $\overline{\lim}\,(V \circ x) = 0$. Therefore Theorem 4.1(a) implies that if K absorbs f/b, then K absorbs w; and this implies (22). If K weakly absorbs f/b, then Theorem 4.1(b) implies K weakly absorbs w. □

THEOREM 4.3. *Let w satisfy Equation (21) with $a/b > 0$. If $\mathcal{L}_{\mathsf{w}}[f/b]$ is bounded, then*

$$\mathcal{L}[w] \subset \mathcal{L}_{\mathsf{w}}[w] \subset \overline{\mathrm{co}}\,(\mathcal{L}_{\mathsf{w}}[f/b]).$$

PROOF. Since the first inclusion is universal, it suffices to consider weak limits. Note that $\mathcal{L}_{\mathsf{w}}[f/b]$ is compact by Lemma 2.1(b). Given $L \in E^*$, consider the curve

$$v = L \circ w : [t_0, \infty) \to \mathbf{R},$$

which satisfies the equation

$$a\frac{dv}{dt} + bv = L \circ f.$$

If $p \in \mathcal{L}_{\mathsf{w}}[w]$, then $Lp \in \mathcal{L}[L \circ (w)]$. Therefore Proposition 3.5 (a)(b) implies

$$L(\mathcal{L}_{\mathsf{w}}[w]) \subset \mathcal{L}[(L \circ f)/b)], \tag{23}$$

$$\subset L\mathcal{L}_{\mathsf{w}}[(f/b)] \tag{24}$$

the last inclusion following from weak compactness of $\mathcal{L}_{\mathsf{w}}[f/b]$. Thus we have shown that $\mathcal{L}_{\mathsf{w}}[f/b]$ weakly absorbs $\mathcal{L}_{\mathsf{w}}[w]$. The proof is completed by Lemma 2.6. □

Next we consider the case $a/b < 0$.

PROPOSITION 4.4. *Let $w : [t_0, \infty) \to E$ be a solution to Equation (21), and assume $a/b < 0$.*

(a): *Let $L \in E^*$ be such that $|Lw(t)| \not\to \infty$. Then:*

$$\mathcal{L}[L \circ w] \subset -\mathcal{L}[L \circ (f/b)].$$

(b): *If $\mathcal{L}_{\mathsf{w}}[f/b]$ is bounded, then $\mathcal{L}_{\mathsf{w}}[w] \subset -\overline{\mathrm{co}}\,(\mathcal{L}_{\mathsf{w}}[f/b])$.*

PROOF. Part (a) is proved by applying Proposition 3.6(a)(b) to the curve $v = L \circ w : [t_0, \infty) \to \mathbf{R}$, which satisfies the equation

$$a\frac{dv}{dt} + bv = L \circ f.$$

The proof of (b) is similar to that of Theorem 4.3. □

If $K \subset E$ is convex and $J \subset \mathbf{R}$ is an interval not containing zero, then we define the set

$$K/J = \{\lambda^{-1}x : \lambda \in J, x \in K\},$$

and note that it is convex.

COROLLARY 4.5. *Let $w : [t_0, \infty) \to E$ be a solution to Equation (21). Assume:*

(a): *$\mathcal{L}_{\mathsf{w}}[f/b]$ is bounded,*
(b): *$\mathcal{L}[b]$ is a finite interval not containing* 0.

Set $\sigma =$ $\mathrm{sgn}(a/b) \in \{\pm 1\}$. Then:

$$\mathcal{L}_{\mathsf{w}}[w] \subset \sigma\overline{\mathrm{co}}\,(\mathcal{L}_{\mathsf{w}}[f])/\mathcal{L}[b].$$

PROOF. Follows from Propositions 4.3 and 4.4, because (b) implies

$$\overline{\mathrm{co}}\,(\mathcal{L}_{\mathsf{w}}[f/b]) \subset \overline{\mathrm{co}}\,(\mathcal{L}_{\mathsf{w}}[f])/\mathcal{L}[b].$$

□

4.1. Application to Autonomous Nonlinear Dynamical Systems. Let $F : E \to E$ be continuous and let $\Lambda \subset E$ be a compact invariant set for the equation

$$dx/dt = F(x). \tag{25}$$

For $\gamma \in \mathbf{R}$ define the map

$$G_\gamma : E \to E, \;\; x \mapsto x + \gamma F(x).$$

THEOREM 4.6. *Suppose there exists a solution $u : [t_0, \infty) \to \Lambda$ to Equation (25) that is dense in Λ (e.g., Λ is a periodic orbit). Then*

$$\Lambda \subset \overline{\mathrm{co}}\,(G_\gamma(\Lambda))$$

for all $\gamma \in \mathbf{R}$.

PROOF. Fix $\gamma \neq 0$. Let u be a solution curve that is dense in Λ, and set

$$f(t) = \gamma^{-1}G_\gamma(u(t)).$$

Then u satisfies the equation

$$\frac{du}{dt} + \gamma^{-1}u = f(t).$$

By density of u, Corollary 4.2 and continuity of G_γ:

$$\begin{aligned}\Lambda &= \mathcal{L}[u]\\ &\subset \overline{\text{co}}\,(\mathcal{L}[\gamma f])\\ &= \overline{\text{co}}\,(\mathcal{L}[G_\gamma \circ u])\\ &= \overline{\text{co}}\,(G_\gamma(\mathcal{L}[u]))\\ &= \overline{\text{co}}\,(G_\gamma(\Lambda)).\end{aligned}$$

□

4.2. Higher order operators in Banach spaces. Define $\mathsf{C}_E = \mathsf{C}([t_0,\infty),E)$ to be the set of continuous functions $g:[t_0,\infty)\to E$ such that $L\circ g\in \mathsf{C}([0,\infty),\mathbf{R})$ for each $L\in E^*$. Thus $g\in \mathsf{C}_E$ if and only if for each $L\in E^*$, either $\mathcal{L}[L\circ g]$ is a nonempty compact interval, or $|Lg(t)|\to\infty$ and $\mathcal{L}[L\circ g]$ is empty.

For $i=1\dots,n$ let $a_i,b_i:[t_0,\infty)\to\mathbf{R}$ be given, set $T_i=a_iD+b_iI$, and consider the n'th order linear differential operator

$$T = T_n\circ\cdots\circ T_1$$

where D denotes differentiation and I the identity.

Define $J_i\subset\mathbf{R}$ to be the closed interval $\mathcal{L}[b_i]$, and $J=J_1\cdots J_n$. Set $\sigma_i=\operatorname{sgn}(a_i/b_i)$ and $\sigma=\sigma_1\cdots\sigma_n$.

THEOREM 4.7. *Let $f:[t_0,\infty)\to\mathbf{R}$ be continuous, and assume each pair of functions a_i,b_i satisfy Hypothesis 3.4. Let $v:[t_0,\infty)\to E$ be a solution to the differential equation $Tv=f$.*

(a): *Assume each J_i is a finite interval not containing zero. Then:*

$$\mathcal{L}_{\mathsf{w}}[v]\subset \sigma J^{-1}\overline{\text{co}}\,(\mathcal{L}_{\mathsf{w}}[f]).$$

(b): *Assume*

$$f/b_n\dots b_i\in \mathsf{C}_E,\ i=1,\dots,n.$$

Then $v\in\mathsf{C}_E$. In particular, $v\in\mathsf{B}_E$ provided $f\in\mathsf{C}_E$ and the J_i are as in (a).

PROOF. The proof, similar to that of Theorem 3.9, is by induction on n, using Corollary 4.5. □

We call a real number λ an *eigenvalue* of T with *eigenfunction* $u:[t_0,\infty)\to E$ if u is not identically zero, and

$$Tu=\lambda u. \tag{26}$$

Assume the hypothesis of Theorem 4.7. Let Equation (26) hold, and consider the finite interval

$$M=[\alpha,\beta]=\sigma\lambda J^{-1}=\{\sigma\lambda s\in\mathbf{R}: s^{-1}\in J\}.$$

If $0\in M$ then 0 is an endpoint of M. Then:

THEOREM 4.8.

(a): *If $\lambda=0$ then $\mathcal{L}_{\mathsf{w}}[u]=\{0\}$.*

(b): *If $\alpha<0$ then either $\mathcal{L}_{\mathsf{w}}[u]=\{0\}$ or else 0 is an internal point of $\overline{\text{co}}\,(\mathcal{L}_{\mathsf{w}}[u])$.*

(c): *If $|\alpha|>1$ then $0\in\overline{\text{co}}\,(\mathcal{L}_{\mathsf{w}}[u])$.*

(d): *If $|\beta|<1$ then $\mathcal{L}_{\mathsf{w}}[u]$ is unbounded.*

PROOF. Let K denote the nonempty closed convex set $\overline{\mathrm{co}}\,(\mathcal{L}_{\mathsf{w}}[u])$. Then

$$K \subset [\alpha, \beta]K \tag{27}$$

by Theorem 4.7 applied to the equation $Tu = f$, taking $f(t) = \lambda u(t)$. This proves (a), because $\alpha = \beta = 0$ if $\lambda = 0$. The other parts are also consequences of (27). □

COROLLARY 4.9. *Let the operator T be as in Theorem 4.7. If the eigenfunction u in Equation (26) is bounded and $\overline{\mathrm{co}}\,(\mathcal{L}_{\mathsf{w}}[u])$ does not contain the origin, then $\lambda \neq 0$ and $0 < \alpha \leq \beta \leq 1$.*

References

1. H.-D. Chiang, M.W. Hirsch, and F. Wu, *Stability regions of nonlinear autonomous dynamical systems*, IEEE Transactions on Automatic Control **33** (1988), 16–27.
2. W. A. Coppel, *Disconjugacy*, Lecture Notes in Mathematics No. 220, Springer-Verlag, New York, 1971.
3. M. W. Hirsch, *On-line training of memory-driven attractor networks*, (S. W. Ellacott, ed.), Proceedings of Conference on Mathematics of Neural Networks and Applications, Oxford University July 3-7, 1996, Special issue of Annals of Mathematics and Artificial Intelligence, in press.
4. M. W. Hirsch and B. Baird, *Computing with dynamic attractors in neural networks*, Biosystems **34** (1995), no. 1-3, 173–195, Special issue: Biological and Evolutionary Information Processing.
5. J. S. Muldowney, *On an inequality of Čaplygin and Pólya, proceedings of royal irish academy*, Proc, Royal Irish Academy **76** (1976), 85–99.
6. G. Pólya, *On the mean-value theorem corresponding to a given linear homogeneous differential equation*, Trans. Amer. Math. Soc **24** (1924), 312–324.
7. J. van Tiel, *Convex analysis*, John Wiley and Sons, New York, 1984.

DEPARTMENT OF MATHEMATICS, UNIVERSITY OF CALIFORNIA, BERKELEY, CALIFORNIA 94720-3840

E-mail address: hirsch@math.berkeley.edu

Contemporary Mathematics
Volume **208**, 1997

TOWARDS A NONCOMMUTATIVE FRACTAL GEOMETRY? LAPLACIANS AND VOLUME MEASURES ON FRACTALS

MICHEL L. LAPIDUS[1-2]

ABSTRACT. We review some recent work of the author (and his collaborators) regarding harmonic analysis and partial differential equations on fractals; more specifically, the vibrations of "fractal drums" and the spectral distribution of Laplacians on (suitable) self-similar fractals. We also discuss how this work was combined by the author with techniques from Connes' noncommutative geometry to construct "volume measures" on such fractals, including an analogue in this context of the Riemannian volume measure. Further, we announce new results regarding the nature of these volume measures and, in special cases, their relationship with a suitable notion of Hausdorff measure. In addition, we consider the notion of "complex dimension" of (self-similar) fractals and its relationship with oscillatory phenomena in our setting. We also propose and discuss a number of conjectures and open problems in this area, aimed at further developing geometric analysis on fractals and–in the long-term–at laying out some of the foundations for what may be coined "noncommutative fractal geometry".

Although we will stress mostly the mathematical aspects in this paper, we mention that this work—along with its proposed extensions—has natural physical motivations and applications, including transport in porous media, percolation networks, the vibrations of fractal resonators (such as cetain microwave cavities), as well as diffusion and wave propagation in fractally structured media.

Dedicated to Professor Victor L. Shapiro,
colleague and long-time friend,
for his contagious enthusiasm
and his dedication to Mathematics

---

[1]This research was partially supported by the National Science Foundation under Grants DMS-9207098 and DMS-9623002. It was presented in part at a plenary lecture at the Schauder Memorial Symposium on "Nonlinear Partial Differential Equations and Applications" held at the Banach Center in Warsaw, Poland, in July 1994, the Workshops on "C^*-Algebras and Dynamical Systems" and on "Cyclic Cohomology and Noncommutative Geometry" held at the Fields Research Institute for Mathematical Sciences, Canada, in October 1994 and June 1995, as well as at the Institut des Hautes Etudes Scientifiques (IHES) during the academic year 1995-96.

[2]The author was a Visiting Professor at the Institut des Hautes Etudes Scientifiques during the Summer of 1995, in December 1995 and from March through September 1996 while much of this paper was written up.

1991 *Mathematics Subject Classification.* Primary: 28A80, 35P20, 58B30, 58G25; Secondary: 28A12, 58G15, 58G32, 78A40, 81Q20.

Key words and phrases. Vibrations of fractal drums, analysis and PDEs on fractals, self-similar sets, Laplacians and volume measures on fractals, spectral dimensions, Hausdorff and Minkowski dimensions, Dixmier trace, quantized calculus, classical vs. quantum space, noncommutative geometry, complex dimensions of fractals, oscillatory phenomena.

This paper is in final form and no version of it will be submitted for publication elsewhere.

1. Introduction

As is well-known, historically, the study of vibrations and of the classical partial differential equations (PDEs) of mathematical physics—such as the Laplace, the Helmholtz, the heat and the wave equations—has been an extremely rich source of developments in harmonic analysis. More recently, we have witnessed a number of fruitful interactions between harmonic analysis and the study of nonlinear PDEs, as is illustrated by many articles in this volume.

In the present paper, we would like to focus our attention on the case where the equation is linear *but the medium is intrinsically nonlinear*; that is, not necessarily embedded in Euclidean space or a more general Riemannian manifold, and in any case, neither locally diffeomorphic nor even locally homeomorphic to some Euclidean (linear) space.

We review, in particular, work of the author (and his collaborators) regarding harmonic analysis and PDEs *on* fractals. (*See* especially Sects. 3-5.) We also announce (in Sect. 6.2.) new (joint) results ([KiLa2]) on this subject, as well as discuss (in Sect. 8) several conditional results (building upon our recent joint work in [La-vF]). Further, we propose a number of open problems and conjectures aimed at suggesting directions of future research in this area and–in the long-term–at laying out some of the foundations for what may be coined (as in our previous paper [La4]) "*noncommutative fractal geometry*". (*See* especially Sects. 6.1 and 7 below.)

More specifically, we consider the vibrations of *"fractal drums"* (interpreted here as *"drums with fractal membrane"* rather than as *"drums with fractal boundary"* as in our related work [La1-3 and 5, LaFl, LaPo1-3, LaMa1-2, HeLa1-2, La-vF, LaPa, LaNeReGr, GrLa]) and study the spectral asymptotics of (Dirichlet and Neumann) "Laplacians" on a suitable (*i.e.*, "finitely ramified" or p.c.f.) self-similar fractal F. We describe (in Sect. 3) the partial analogue of Hermann Weyl's classical formula [We] obtained in this context (jointly with Jun Kigami) in [KiLa1]. (*See* Theorem 3.1 below.) Moreover, we explain (in Sect. 5) how–using in particular the results of [KiLa1] and certain methods from Connes' noncommutative geometry [Co1, 3-4] (namely, the notion of Dixmier trace)—the author constructed in [La4] an analogue of Riemannian volume" (or, more generally, of "volume measures") on the fractal F, and thereby obtained a much closer counterpart of Weyl's formula in this context. (*See* Theorem 5.1.) The aforementioned new results [KiLa2] announced in Sect. 6.2 specify the nature of the above "Riemannian volume measure" and its relationship with (a suitably defined) Hausdorff measure for a special subclass of (p.c.f., self-similar) fractals. (*See* Theorem 6.9, which will be expanded upon and established in [KiLa2].)

At this point, it may be helpful for some readers to recall Weyl's classical formula [We] for Laplacians on Riemannian manifolds. (*See, e.g.*, [Ho] for a modern exposition.) Let Δ denote the (Dirichlet or Neumann) Laplace-Beltrami operator on X, a compact, connected Riemannian manifold X with dimension m and (finite) volume denoted Vol (X). [For notational simplicity, we adopt here and throughout this paper the geometer's convention according to which Δ (rather than $-\Delta$) is a nonnegative operator.] It is well-known that the spectrum of Δ is discrete and consists of an infinite sequence of eigenvalues $\{\lambda_j\}_{j=1}^{\infty}$ written (according to multiplicity) as follows: $0 \leq \lambda_1 \leq \lambda_2 \leq \cdots \leq \lambda_j \leq \cdots \uparrow \infty$, as $j \to \infty$. Let $N(\lambda)$, the *"eigenvalue counting function"*, denote the number of positive eigenvalues λ_j not exceeding $\lambda > 0$. Then Weyl's formula states that

$$N(\lambda) = e_m \mathrm{Vol}(X)\big(\lambda^{m/2} + o(1)\big), \quad \text{as} \quad \lambda \to +\infty, \tag{1.1}$$

where e_m is a constant depending only on $m = \dim(X)$ and $o(1)$ stands for a function which vanishes as $\lambda \to +\infty$. (Physically, $\sqrt{\lambda}$ corresponds to the frequency of the associated "drum"; further, quantum-mechanically, the large eigenvalues correspond to high energies and hence to small scales, as was used, in particular, in the case of "drums with fractal boundary" in [La1].)

Now, let us return to the case of (suitably defined) Laplacians on fractals. When F is a (p.c.f., self-similar) fractal in *"general position"* (in a geometric, dynamical and analytical sense), we obtain in [La4] a formula which looks formally much like (1.1) (except for the analogue of the constant e_m). [*See* Equation (5.3) below (from [La4]), which specifies Equation (3.2) (from [KiLa1]) and corresponds to the *"nonlattice case"* of Theorems 3.1 and 5.1 (case (i)).] On the other hand, when F is *"sufficiently symmetric"* (in the above sense), (periodic) *oscillations* may occur in the leading asymptotics of $N(\lambda)$ and the counterpart of formula (1.1) has to be modified appropriately. [*See* Equations (3.5) (from [KiLa1]) and (5.4) (from [La4]) below, which correspond to the *"lattice case"* of Theorems 3.1 and 5.1 (case (ii)).]

It is natural to wonder what is the analogue of m, the dimension of the smooth manifold X, in the present context of a fractal F. It turns out that it is a *"spectral exponent"* d_S (as shown in [KiLa1]) defined by formulas (3.4) and (3.3) below and depending on the underlying geometric, measure-theoretic and analytic structures of F. [More specifically, in (3.3)-(3.4), *d_S is expressed in terms of the* (characteristic constants of the) *given self-similar measure* (or "mass distribution") *and self-similar Dirichlet form* (or "energy functional") *on F*.] The largest possible value of d_S is denoted d_S^* and called (in [KiLa1]) the *"spectral dimension"* of F. By [KiLa1] and [Ki2], it can be expressed in terms of the (Hausdorff or Minkowski) "fractal dimension" of F, with respect to a suitable *"intrinsic"* metric (associated to the energy functional on F). (*See* Sect. 4.)

Special classes of (self-similar) fractals are used throughout (the rigorous part of) this paper, but we also intend in future work to tackle broader classes of "fractals", discussed, *e.g.*, in [La3, Part II] and including "approximately self-similar fractals" as well as random (or "statistically self-similar") fractals, which play an important role in physical applications.

Although we will mostly discuss the mathematical aspects in this paper, we point out that this work—along with its proposed extensions—has many natural physical motivations and potential applications (*see, e.g.,* [Be1-2, Dh, AO, R, RT, HHW, ...]), including transport in porous media and percolation networks, vibrations of fractal resonators and standing waves in (rough) microwave cavities, phase transitions in condensed matter physics, as well as diffusions and wave propagation in fractally structured media. Further, it is directly relevant to several aspects of current technology—such as oil recovery, computer microchips, radar imaging, quasi-crystals, polymer synthesis, and the possible manufacturing of new extremely performant heat or sound insulators, called "fractal aerogels". (*See, e.g.,* the aforementioned physical references and the books or survey articles [DaWi, EvPeVo, FedAh, HavBu, Liu, Ma, NYO, Sap, Sc].)

We have stressed so far in this introduction the more concrete aspects of this paper because they should seem more familiar to most readers. However, as was alluded to above, parts of this paper (especially Sects. 6.1 and 7) are of a more speculative nature and deal with a partial analogy between geometric analysis on smooth Riemannian manifolds (as well as their noncommutative versions) and the present setting. It would aim, in particular, at developing new tools to probe the fine geometric structure of fractals—such as, for example, a *"quantized (differential and integral) calculus"* (à la Connes and Sullivan [Co4, Chapt. IV; CoSu]) and a suitable "pseudodifferential calculus" on (certain classes of) fractals. In the long-term, we would like to find in this context analogues of "tangent space", "phase space", "geodesic flow", and (possibly noncommutative) "curvature". In short, one of our ultimate goals would be to obtain a suitable definition and understanding–as well as make an effective use–of the rather elusive notion of *"fractal manifold"*, possibly viewed as a *"quantum" (or noncommutative* [Co3,4]) *space.* (*See* Sect. 7.3.)

We close this introduction by putting this paper in a broader context and pointing out a useful analogy between two types of *"fractal drums"* which is used in several places in this work, particularly in Sects. 3, 5 and 8. Roughly speaking, the work of the author (and his collaborators) in the area of *"spectral and fractal geometry"* [La2,3]–motivated in part by appealing conjectures of the physicist Michael V. Berry in [Be1,2]–has been concerned with two classes of problems; namely, the study of the vibrations of *"drums with fractal boundary"* (*i.e.*, Laplacians on open sets $\Omega \subset \mathbb{R}^n$ with fractal boundary $\Gamma = \partial\Omega$, as in [La1-3 and 5, LaFl, LaPo1-3, LaMa1-2, HeLa1-2, La-vF1-4]) and of *"drums with fractal membrane"* (Laplacians *on* fractals themselves, as in [La3-5, KiLa1-2]). [Mathematical references on the latter can be found throughout the rest of this paper and additional ones are given in [KiLa1-2, La3-5], while many references on the former (including, *e.g.*, [BrCa, Ce, EHa, FlVa, Ge]) can be found in [La1,3 and 5, LaPo2-3, HeLa2, LaNeRoGr, La-vF1]). (In addition, a third, but related problem–physically motivated in part by [Sap] and the relevant references therein–aims, in the long-term, at understanding why and how "fractal shapes" (such as coastlines, blood vessels, lungs and trees) occur in nature; *see, e.g.*, the articles [LaPa, LaNeRoGr, GrLa] which investigate this question by means of 'traditional' mathematical methods as well as of computer graphics and numerical modelizations.)]

In [La3, Part II], *an analogy was developed between the study of "drums with (self-similar) fractal boundary" and that of "drums with (self-similar) fractal membrane"*; further, it was used to formulate several problems and conjectures, the solution of some of which is presented here. (*See, e.g.*, Remark 3.2(a) below.)

On the other hand, the study of *"fractal strings"* (*i.e.*, drums with one-dimensional fractal boundary", with $n = 1$ above)–initiated in [La1-2, LaPo1-2]–has led the author and his collaborators to discover unexpected connections between aspects of "spectral and fractal geometry" and number theory; specifically, the theory of the Riemann zeta-function $\zeta = \zeta(s)$. (*See* [La2-3, LaPo1-3, LaMa1-2, HeLa1-2, La-vF].) In particular, the question–à la Mark Kac [Kc], but interpreted quite differently as a suitable *inverse spectral problem* in this context–"*Can on hear the shape of a fractal sting?"* has led us to obtain a new spectral characterization of the Riemann hypothesis in [LaMa1,2]. Further, the work in [LaPo] and [LaMa] (joint with Carl Pomerance and Helmut Maier, respectively,) has suggested that a suitable notion of *"complex dimension"* should exist and be helpful in analyzing *oscillatory phenom-*

ena in the context of fractal and spectral geometry. At least for "fractal strings" and their higher-dimensional analogue ("fractal sprays" [LaPo3, La2], with $n \geq 2$), this program has been recently carried out in [La-vF] (joint with Machiel van Frankenhuysen). It has enabled us, in turn, to obtain a deeper understanding (and extend parts) of the work in [La2-3, LaPo, LaMa, HeLa] as well as, for example, to obtain new results concerning the distribution of the critical zeros of the Riemann zeta-function $\zeta = \zeta(s)$ [and of its higher-dimensional analogue, the n-dimensional Epstein zeta-function (*see, e.g.,* [La2, Sects. 4 and 5]), with $n \geq 2$]. More specifically, as an application of the theory developed in [La-vF], we show in [La-vF1] (to be announced in [La-vF2-4]) that ζ (and more generally ζ_n, with $n \geq 1$, as well as many other zeta-functions, such as all Epstein and Dedekind zeta-functions) can *not* have a sequence of critical zeros forming an arithmetic progression.

Moreover, in [La-vF1,3], the spectral characterization of the Riemann hypothesis obtained in [LaMa] is reinterpreted in terms of the notion of complex dimension and is extended to a large class of zeta-functions, including all those for which the (generalized) Riemann hypothesis is expected to hold. In addition to all Epstein zeta-functions [Te], this class includes, for example, the Dirichlet L-series and Dedekind zeta-functions, as well as more generally, all Hecke L-series associated with algebraic number fields [PS].

Most importantly from our present perspective, the idea of 'complex dimension' made rigorous and developed in [La-vF1-3] is turned into a useful tool to study oscillations in the geometry and the spectrum of fractal strings. Furthermore, several key results in [La-vF1-3] suggest that it can be given a concrete geometric meaning (by analogy with aspects of Riemannian geometry) and can be used to propose a new definition of 'fractality' in this context. (See especially [La-vF1,3].)

Motivated by these recent developments in [La-vF], we will propose in Sect. 8 below to use once again the aforementioned analogy between the two kinds of "fractal drums" in order to begin the study of the *"complex dimensions" of fractals* themselves (viewed intrinsically) and of the associated *oscillatory phenomena in the spectrum* (and the geometry) *of "drums with fractal membrane"*. This should enable us in later work to shed new light on–and extend–several of the results and/or conjectures from [La3, KiLa1, La4, KiLa2] discussed here. Moreover, it should help us establish further ties between the various themes developed throughout this paper.

2. Self-Similar Fractals and Dirichlet Forms

From the point of view of geometric analysis, we proposed in [La4, §2] to define a *"self-similar (s.s.) fractal"* $F = (F, \mathcal{S}, \mu, \mathcal{E})$ as a compact, connected, metrizable space F equipped with: (i) a *"s.s. structure"* $\mathcal{S}$ (as in, *e.g.*, [Ki1, KiLa1, La4]), (ii) a *"s.s. measure"* μ, and (iii) a *"s.s. Dirichlet form"* $\mathcal{E}$ on $L^2(\mu) := L^2(F, \mu)$. Intuitively, $\mathcal{S}$ specifies the underlying (symbolic) *dynamical* and *geometric* structures of F, μ yields the *"mass distribution"* on F, and $\mathcal{E}$ is a suitable *"energy functional"* on $L^2(\mu)$ which specifies the *"conformal geometry"* of F. More precisely:

(i) (*Self-Similar Structure*.) Let $\Sigma = A^{\mathbb{N}}$ be the set of 'words' $\omega = \omega_1\omega_2\cdots (\omega_j \in A)$ on the 'alphabet' $A := \{1, \ldots, N\}$, with $N \geq 2$. For $k = 1, \ldots, N$, let the map $W_k : F \to F$ be continuous and $1-1$. Then $\mathcal{S} = (A, \{W_k\}_{k=1}^N)$ is a *s.s. structure* on F if there exists a continuous, onto map $\Pi : \Sigma \to F$ such that for each $k = 1, \ldots, N, W_k \circ \Pi = \Pi \circ \sigma_k$, where the kth shift $\sigma_k : \Sigma \to \Sigma$ is defined by

$\sigma_k(\omega_1\omega_2\cdots) = k\omega_1\omega_2\cdots$.

(ii) (*Self-Similar Measure.*) A (Borel) probability measure μ on F is called *s.s.* (relative to $\mathcal{S}$) with *"weights"* $\{b_k\}_{k=1}^N (b_k \geq 0, \sum_{k=1}^N b_k = 1)$ if

$$\mu = \sum_{k=1}^{N} b_k \mu \circ W_k^{-1};$$

i.e. (with $C(F)$ denoting the space of real-valued continuous functions on F),

$$\int_F f d\mu = \sum_{k=1}^{N} b_k \int_F f \circ W_k d\mu, \text{ for all } f \in C(F). \tag{2.1}$$

(iii) (*Self-Similar Dirichlet Form* [La4].) A Dirichlet form $\mathcal{E}$ (with domain $\mathcal{F} \subset L^2(\mu)$) is *s.s.* (relative to $\mathcal{S}$) with *"harmonic constants"* $\{c_k\}_{k=1}^N$ ($0 < c_k < 1$, for $k = 1, \ldots, N$) if

$$\mathcal{E} = \sum_{k=1}^{N} c_k^{-1} \mathcal{E} \circ W_k^{-1};$$

i.e.,

$$\mathcal{E}(u) = \sum_{k=1}^{N} c_k^{-1} \mathcal{E}(u \circ W_k), \quad \text{for all } u \in \mathcal{F}. \tag{2.2}$$

(In short, a *"Dirichlet form"* is a closed, nonnegative unbounded quadratic form on $L^2(\mu)$ which satisfies the "Markov property"; *see, e.g,* [KiLa1, Definition 4.1, p. 115] and, for a detailed treatment of the theory, Fukushima's book [Fu1].)

It follows from (i) that F is a union of "scaled copies" of itself; *i.e.*, $F = \cup_{k=1}^N W_k(F)$. However, we stress that even if F is contained in some Euclidean space $\mathbb{R}^n$, a s.s. fractal is viewed here purely *intrinsically,* without any *a priori* choice of metric (compare [Hu; Fc, §9]). Further, roughly speaking, F is *p.c.f.* [Ki1] (*i.e.*, postcritically finite or "finitely ramified") if these copies $W_k(F)$ intersect (at most) at finitely many points. (*See, e.g.,* [KiLa1, Definition 1.2, p. 99] for a more precise statement.) For instance, the Sierpinski gasket is p.c.f. whereas the Sierpinski carpet is not. [For a picture of the standard 2-dimensional Sierpinski gasket (resp., carpet), *see, e.g.,* [Fc, Fig. 0.3, p. xvi] or [KiLa1, Fig. 1, p. 95] (resp., [Ma, Plate 145, p. 144]).] We refer to [Ki1; KiLa1, §3] for many other examples of p.c.f. fractals, including self-similar trees [Ha] and a class of highly symmetric self-similar fractals called "nested fractals" [Li]. For simplicity, *we assume here that F is p.c.f.* (because this is also the situation about which we have the most detailed analytical information at present) but we intend in later work to deal with a broader class of fractals.

In [KiLa1, §6], we have shown (extending results in [Fu2]) that many natural energy functionals on F (called *"regular"* and defined as limits of suitably renormalized *discrete* Dirichlet forms on approximating graphs) are s.s., in the sense of (iii) above. For example, for Lindstrom's "nested fractals" [Li] (roughly, p.c.f. fractals $F \subset \mathbb{R}^n$ with high symmetry), we have $b_1 = \ldots = b_N = 1/N$ and $c_1 = \ldots = c_N = 1/\alpha$, where the *" renormalization constant"* α is a suitable positive eigenvalue of a *nonlinear "renormalization operator"*.

Intuitively, the identity (2.2) means that the "energy flow" is distributed throughout the fractal F in a self-similar manner. For example, if all the c_k's are equal (as is the case for the standard Sierpinski gasket or, more generally, for the "nested fractals"), then the "energy" is (suitably) *equidistributed* at all scales.

Given a (p.c.f.) self-similar structure $\mathcal{S}$ on F, it is not difficult to show (by using the classical Banach fixed point theorem, as in [Hu]) that there exists a unique s.s. measure with given weights $\{b_k\}_{k=1}^N$; *see* [La4, Theorem 2.18, p. 147].

Remark 2.1. Although this is clearly a more delicate matter, one would expect to be able to use some kind of fixed point theorem from nonlinear analysis to deal with s.s. Dirichlet forms of given "harmonic constants" $\{c_k\}_{k=1}^N$, as was suggested in [La4, Remark 2.15(a), p. 146]. A partial answer to this problem was recently provided by Christophe Sabot [Sa]; perhaps surprisingly, Equation (2.2) does not always admit a solution (in some sense, it does, however, have a "degenerate" solution in a suitable compactification of a space of Dirichlet forms). For the inhomogeneous Sierpinski gasket, for example, one can determine exactly when a solution to (2.2) does not exist (or is degenerate) and when it exists and is unique. Further, for "nested fractals", (2.2) always has a unique solution. However, several interesting related questions remain to be further investigated in this context.

3. Laplacians on Fractals and Weyl's Formula

Let $F = (F, \mathcal{S}, \mu, \mathcal{E})$ be a (regular, p.c.f.) s.s. fractal, as above. *Let $\Delta = \Delta_\mu$ be the corresponding (positive) Dirichlet (or Neumann) Laplacian on F. [In short, these Laplacians are defined as suitably renormalized limits of finite-difference operators acting on finite graphs approximating F; see* [Ki1; KiLa1, §1]. Further, in [KiLa1,§5], the associated boundary value problems are defined in terms of the (regular) Dirichlet form $\mathcal{E}$.] It is shown in [KiLa1] that the spectrum of Δ_μ is discrete and that its eigenvalues $\{\lambda_j\}_{j=1}^\infty$ satisfy $0 < \lambda_1 \leq \lambda_2 \leq \cdots$, where $\lambda_j \to +\infty$ as $j \to \infty$. For $\lambda > 0$, let

$$N(\lambda) = \#\{j \geq 1 : \lambda_j \leq \lambda\} \tag{3.1}$$

denote the associated *"eigenvalue counting function"; i.e.,* the number of eigenvalues $\leq \lambda$ (counted according to multiplicity). [For the Neuman problem, we really work with $\Delta_\mu + 1$ (say) instead of Δ_μ.] We can now state the main result of [KiLa1] (rephrased as in [La4]) which provides a partial analogue of Weyl's asymptotic formula for Laplacians on (p.c.f.) fractals; *see* [KiLa1, Theorem 2.1 and Corollary 2.5, pp. 104-105]. (*Theorem 3.1 will be revisited and completed in Theorem 5.1. below.*)

Recall that an additive subgroup of $\mathbb{R}$ is either dense in $\mathbb{R}$ or discrete (and hence, necessarily of the form $T\mathbb{Z}$, with $T > 0$). We will apply this basic fact to the subgroup G defined below, and call the corresponding dichotomy the nonlattice and lattice case, respectively.

Theorem 3.1. (Kigami and Lapidus, [KiLa1]). [A Partial Analogue of Weyl's Formula on Fractals.] *We have the following dichotomy:*

(i) (*"Nonlattice case": the additive group $G := \sum_{k=1}^N (\ln \gamma_k)\mathbb{Z}$ is dense in $\mathbb{R}$). Then as $\lambda \to +\infty$,*

$$N(\lambda) = \mathcal{C}\lambda^{d_S/2}(1 + o(1)), \tag{3.2}$$

for some positive constant $\mathcal{C}$. *(That is,* $N(\lambda) \sim \mathcal{C}\lambda^{d_S/2}$ *as* $\lambda \to +\infty$.*) Here,*

$$\gamma_k := \sqrt{b_k c_k} \quad \text{for} \quad k = 1, \ldots, N \tag{3.3}$$

and the "spectral exponent" $d_S = d_S(\mu)$ *is the unique positive number such that*

$$\sum_{k=1}^{N} \gamma_k^{d_S} = 1. \tag{3.4}$$

(ii) *("Lattice case": G is a discrete subgroup of* $\mathbb{R}$, *say* $G = T\mathbb{Z}$ *with* $T > 0$). *Then as* $\lambda \to +\infty$,

$$N(\lambda) = \lambda^{d_S/2}(g(\ln \sqrt{\lambda}) + o(1)), \tag{3.5}$$

where g is a positive, periodic function of period T (the positive generator of G). *Further,* $g \in L^1_{\text{loc}}(\mathbb{R})$ *is bounded away from* 0 *and* $+\infty$; *so that*

$$N(\lambda) \asymp \lambda^{d_S/2} \quad \text{as} \quad \lambda \to +\infty; \tag{3.6}$$

i.e., there exist positive constants c_1 *and* c_2 *such that* $c_1\lambda^{d_S/2} \leq N(\lambda) \leq c_2\lambda^{d_S/2}$, *for all* λ *sufficiently large.*

We note that the periodic function g obtained in [KiLa1] is nonconstant for the Sierpinski gasket [FuSh] and the "modified Koch curve" [M], as well as more generally, for all "nested fractals" [BKi] (except for the trivial ones, such as $[0,1]$); *see* also [Sh]. (All these are, of course, examples of lattice case fractals. Further, still in the lattice case, g can be a highly discontinuous function.)

Remark 3.2. (a) Theorem 3.1 specifies, corrects and proves a conjecture of the author stated for *"drums with fractal membrane"* in [La3, Conjecture 5, Remark 5.11 (b), (c), pp. 189-190], itself motivated by our earlier work and conjectures for *"drums with fractal boundary"* in [La1-3, LaPo1-2] (*see* especially [La3, Conjecture, pp. 163-164]), as well as by work of Lalley [L] in a related context.

(b) The proof of Theorem 3.1—given in detail in [KiLa1, Sects. 2, 4 and 5]—relies in particular on (a refinement of) the classical renewal theorem from probability theory ([Fel], [L]), as well as on methods from symbolic dynamics, spectral theory and the calculus of variations. We note that in order to define precisely the Dirichlet and Neumann boundary value problems (associated with the Laplacian on F), we use in [KiLa1, §5] a suitable version of Gauss' theorem (or Stokes' formula) in the present situation of Laplacians on s.s. fractals. (*See* [Ki1] and [KiLa1, Theorem 1.13, p. 103].)

(c) A number of examples illustrating Theorem 3.1 are given in [KiLa1, §3]. As noted above, the Sierpinski gasket (in any dimension and with its standard mass distribution) and more generally, the "nested fractals", correspond to the "lattice case". (*See*, respectively, [KiLa1], Example 2, pp. 110-112 and Example 5, pp. 113-114.) Moreover, the "nonlattice case" is the *"generic"* case; more precisely, if we let μ vary, it follows from Theorem 3.1 that for a fixed s.s. structure $\mathcal{S}$ and Dirichlet form $\mathcal{E}$, we are in the nonlattice case for almost every choice of s.s. measure μ (that is, for Lebesgue-a.e. $(b_1, \ldots, b_N)$ in the $(N-1)$-simplex $\sum_{k=1}^{N} b_k = 1$). (*See* [La4, Corollary 3.17, p. 154].) Natural examples of nonlattice s.s. fractals include the "modified Sierpinski gasket" ([KiLa1], Example 3 and Figure 2, pp. 111-112) and a large class of self-similar trees ([KiLa1], Example 4 and Figure 3, pp. 112-113).

4. Spectral and Fractal Dimensions

Fix ($\mathcal{S}$ and) $\mathcal{E}$, and let the s.s. measure μ *vary*. Then, by [KiLa1, Theorem A.2, p. 121], $d_S^* := \max_\mu d_S(\mu)$ is achieved for a unique s.s. measure μ^*, with weights defined by [*caution*: S (and δ^*)–thus also Eq. (4.3)–differ from [KiLa1] (and [Ki2])]

$$b_k = \sqrt{c_k}^S \qquad \text{for} \quad k = 1, \dots, N, \tag{4.1}$$

where $S > 0$ is the unique positive solution of

$$\sum_{k=1}^{N} \sqrt{c_k}^S = 1. \tag{4.2}$$

The positive number $d_S^* = d_S(\mu^*)$ is called (in [KiLa1]) the "spectral dimension" of $F = (F, \mathcal{E})$ and μ^* is called (in [La4]) the "natural s.s. measure" of F. It turns out that (by [KiLa1, Theorem A.2] and then [Ki2])

$$d_S^* = \frac{2S}{S+2} \tag{4.3}$$

and $S = d_H(\delta^*) = d_M(\delta^*)$ is the Hausdorff (= Minkowski $d_M(\delta^*)$, in this case) dimension of F with respect to a suitable ("*intrinsic*") metric δ^* on F (associated with the energy functional $\mathcal{E}$); namely, for $x, y \in F$,

$$\delta^*(x, y) := \sup\{|u(x) - u(y)| : \sqrt{\mathcal{E}(u)} \leq 1, u \in \mathcal{F}\}. \tag{4.4}$$

Further, the maps $W_k : F \to F$ are *contractions* (with Lipschitz constants $\sqrt{c_k}$) on the (bounded, complete) metric space (F, δ^*); however, we stress that even for the Sierpinski gasket $F \subset \mathbb{R}^2$, they are *nonlinear* maps (*see* [Ki2]).

We mention that for (regular) p.c.f. fractals, we always have $d_S^* < 2$ (and hence $d_S = d_S(\mu) < 2$ for any s.s. measure μ); *see* [KiLa1]. Therefore, in some sense, such a fractal is "between one and two-dimensional" (Recall that we disregard here the degenerate case when F is totally disconnected, such as the Cantor set.)

Problem 4.1. (Sobolev Embedding Theorem on Fractals.) *Assume that* $\mu = \mu^*$, *for simplicity. Establish a suitable analogue of the Sobolev embedding theorem in this context. Further, connect the "Sobolev exponents" to the above exponent* S (*called the "similarity dimension" of* $(F, \mathcal{E})$ *in* [KiLa1]) *or, equivalently, to the "spectral dimension"* d_S^* *of* $(F, \mathcal{E})$, *given by* (4.2) *and* (4.3), *respectively.*

It should follow from this "Sobolev embedding theorem on fractals" that, since (by [KiLa1]) $1 \leq d_S^* < 2$ for (regular) p.c.f. self-similar sets, we must have $\mathcal{F} \subset C(F)$ (as is indeed the case in [Ki1, KiLa1]), where $\mathcal{F}$ is the domain of the Dirichlet form $\mathcal{E}$; *see* [La4]. (Since $\mathcal{F}$ is the domain of the nonnegative, unbounded self-adjoint operator $\sqrt{\Delta}$, we can think of $\mathcal{F}$ as being the counterpart for the s.s. fractal F of the standard Sobolev space H^1 for Euclidean space.)

We note that the counterpart of Problem 4.1 for "infinitely ramified" fractals—such as the Sierpinski carpet (in 2 and especially 3 or higher dimensions), studied probabilistically in [BB1,2]—would also be of great interest. In that case (*e.g.*, for the 3-dimensional Sierpinski carpet), we can have $d_S^* > 2$ and therefore (by analogy with the classical Sobolev embedding theorem), *we no longer expect* that the "Sobolev space" $\mathcal{F}$ be composed of continuous functions.

Remark 4.2. (a) When $\mu = \mu^*$, the natural s.s. measure on $(F, \mathcal{E})$, the dichotomy lattice/nonlattice case can equivalently be expressed in terms of $\{c_k\}_{k=1}^N$ instead of $\{\gamma_k\}_{k=1}^N$. (Hence, for example, the nonlattice case means that $\sum_{k=1}^N (\ln c_k)\mathbb{Z}$ is dense in $\mathbb{R}$. This fact—noted in [La3, Remark 5.16(a), (b), p. 197]—follows easily from Equations (3.3), (3.4) and (4.1)-(4.3).) Since, as mentioned above, W_k has Lipschitz constant $\sqrt{c_k}$ with respect to δ^*, this shows that in this important situation ($\mu = \mu^*$), the original conjecture made in [La3, Remark 5.11 (b), (c), pp. 189-190] is correct provided one measures distances in F relative to the "intrinsic" metric δ^* defined by (4.4).

(b) Various examples of computation of the "similarity dimension" S and of the "spectral dimension" d_S^* of $(F, \mathcal{E})$ are provided in [KiLa1], Section 3 and Appendix, esp. p. 122. For the Sierpinski gasket, for instance, we have

$$S = \ln 3/(\ln 5 - \ln 3) \quad \text{and} \quad d_S^* = \ln 9/\ln 5. \tag{4.5}$$

(*See* [KiLa1, Example 2, p. 122].) Naturally, in this special case, the latter value of d_S^* coincides with that obtained previously in the physical (*e.g.*, [AO, HHW, R, RT]) and mathematical (*e.g.*, [BP, FuSh, Li]) literature. More generally, for the N-Sierpinski gasket $F \subset \mathbb{R}^{N-1}$ (*see* [KiLa1, Example 2, pp. 109-111]), with $N \geq 2$, we have [KiLa1, Eq. (A.6), p. 122]

$$S = \ln N/(\ln(N+2) - \ln N) \quad \text{and} \quad d_S^* = 2\ln N/\ln(N+2). \tag{4.6}$$

[The case where $N = 3$ (resp., $N = 2$) corresponds to the standard 2-dimensional Sierpinski gasket (resp., to the unit interval in $\mathbb{R}^1$); clearly, $S = d_S^* = 1$ when $N = 2$.] It follows that $1 < d_S^* < 2$ for all $N \geq 3$ and that $d_S^* \uparrow 2$ as $N \uparrow \infty$ (without ever reaching 2). This lends credence to the aforementioned phrase: "*F is between one and two-dimensional*".

5. Construction of Volume Measures on Fractals via the Dixmier Trace

We briefly recall the notion of "*Dixmier trace*" [Di], which is a key tool in Connes' noncommutative geometry [Co1-6]. Given a compact (nonnegative and self-adjoint, say) operator R on a Hilbert space $\mathcal{H}$, with eigenvalues $\{\rho_j(R)\}_{j=1}^\infty \downarrow 0$, we write that $R \in \mathcal{L}^{1+}$ (the "Matsaev ideal" [Co4]) if the sequence $(\ln J)^{-1}\Sigma_{j=1}^J \rho_j(R)$ is bounded. Let $R \in \mathcal{L}^{1+}, R \geq 0$. *Then, roughly speaking, the "Dixmier trace" of R is defined by*

$$Tr_w(R) = \operatorname*{Lim}_w \, (\ln J)^{-1} \sum_{j=1}^J \rho_j(R), \tag{5.1}$$

where "Lim_w" is a suitable notion of limit (of bounded sequences) *with nice scale-invariance (i.e., renormalization) properties.* (*See, e.g.*, [Co4, Chapter IV] or [La4, §4.1] for more details.) Further $Tr_w(R)$ extends to a finite, positive (*non normal*) unitary trace on $\mathcal{L}^{1+}$, which vanishes on the trace class operators. [*Intuitively, $Tr_w(R)$ captures the semiclassical (spectral) information contained in R.*] We can now state the main result of [La4], which completes and specifies Theorem 3.1; it follows from [La4, Theorem 4.41 and Corollary 4.45, pp. 169-171].

[Recall from our brief discussion in the introduction that the classical Weyl formula [We, Ho] for Laplacians on a *smooth* compact Riemannian manifold M involves the "*volume*" of M; namely,

$$\mathrm{vol(M)} = \lim_{\lambda \to +\infty} \lambda^{-m/2} N(\lambda),$$

where $m := \dim M$. (*See* Equation (1.1) above.) Further recall that a Radon measure [Ch] on the compact (metrizable) space F is a continuous linear functional on $C(F)$; in addition, it is well-known that the positivity of such a functional implies its continuity.]

Theorem 5.1. (Lapidus, [La4]). ["Volume Measures" on Fractals and Weyl's Formula, Revisited.] *Let $\nu = \Phi(\mu)$ be defined by*

$$\nu(f) = Tr_w(f\Delta_\mu^{-ds/2}) \; (:= Tr_w(M_f R_\mu)), \quad \textit{for all} \quad f \in C(F), \tag{5.2}$$

where M_f denotes the (bounded) multiplication operator by f on $\mathcal{H} := L^2(\mu)$, and the operator $R_\mu := \Delta_\mu^{-ds/2}$ is defined via the functional calculus. Then $\nu = \Phi(\mu)$ is a well-defined, (nonzero) positive Radon measure on F. [We still denote by ν the associated (finite, positive) Borel measure.] *Moreover, its "total mass" $\nu(F)(= \nu(1) = Tr_w(R_\mu))$ is given by (compare with Theorem 3.1 above):*

(i) *In the "nonlattice case",*

$$\nu(F) = \lim_{\lambda \to +\infty} \lambda^{-ds/2} N(\lambda) = \mathcal{C}, \tag{5.3}$$

where the (positive and finite) *constant $\mathcal{C}$ is as in Theorem* 3.1(i).

(ii) *In the "lattice case",*

$$\nu(F) = \lim_{\Lambda \to +\infty} \frac{1}{\ln \Lambda} \int_1^\Lambda \lambda^{-ds/2} N(\lambda) \frac{d\lambda}{\lambda} = \frac{1}{T} \int_0^T g(t)dt, \tag{5.4}$$

the mean-value of the T-periodic function g in Theorem 3.1(ii).

[*Further, ν is independent of the choice of (Dirichlet or Neumann) boundary conditions.*]

Remark 5.2. (a) Here and henceforth, following common practice, we identify the function f (in $C(F)$) and the bounded, self-adjoint operator of multiplication by f (in $\mathcal{L}(L^2(\mu))$), denoted M_f and defined for every $\varphi \in L^2(\mu) = L^2(F, \mu)$ by

$$(M_f \varphi)(x) = f(x)\varphi(x), \tag{5.5}$$

for μ-a.e. $x \in F$.

(b) Using results from [KiLa1] (including (3.2) and (3.6) above), we can show in [La4] that the self-adjoint operator $\Delta_\mu^{-ds/2}$ is compact (and in $\mathcal{L}^{1+}$) and hence (by (a)) that $M_f \Delta_\mu^{-ds/2}$ is also compact, self-adjoint and in $\mathcal{L}^{1+}$, so that its Dixmier trace in (5.2) is well-defined and (for $f \geq 0$) is given by (5.1).

(c) The proof of Theorem 5.1–given in [La4,§4]–relies on Theorem 3.1 above (from [KiLa1]) as well as on several key properties of the Dixmier trace, including its *positivity* and its unitarity. Moreover, in order to calculate (as in (5.3) and

(5.4)) the *"logarithmic divergencies"* (*and especially average out as in* (5.4) *the oscillations occurring in* (3.5) *in the lattice case*), we obtain in [La4] a refinement of a known lemma connecting the notion of Dixmier trace and that of "noncommutative residue" [Wo] (*e.g.*, [Co4, Proposition 4, p. 306]), which should be useful in related situations. (*See* [La4, Proposition 4.17, pp. 163-164] and Remark 5.3(a) below.)

Intuitively, the operator $R_\mu = \Delta_\mu^{-d_S/2}$ can be thought of as an elliptic *pseudodifferential operator* (p.d.o.) on the "fractal" F, of order $-d_S$ (compare [Co1]). Further, heuristically, the (square-root of the) "*Green's operator*" Δ_μ^{-1} corresponds to the "*infinitesimal metric element*" $|d\sigma|$ on F (especially when $\mu = \mu^*$, as in §4 above), and so the definition (5.2) of ν can be rewritten as follows:

$$\nu(f) = \fint f|d\sigma|^{d_S}, \quad \text{for all} \quad f \in C(F), \tag{5.6}$$

where the symbol $\fint$ denotes the "*Connes integral*" (of the type appearing on the cover of [Co4]). (*Caution*: Although ν induces naturally a positive Borel measure in the usual manner, it would be wrong in general to replace the continuous function f in (5.2) or (5.6) by the characteristic function of an arbitrary Borel measurable subset of F. Indeed, for such a function f, the number $\nu(f)$ may no longer be given by the expression involving the Dixmier trace on the right-hand side of (5.2) or (5.6); *see, e.g.*, [La4, Remark 4.48(d), pp. 171-172].)

A construction similar to that used above—but applied to the Dirac operator on a *smooth* (compact, spin) Riemannian manifold (instead of the square-root of the Laplacian on a s.s. fractal)—had earlier enabled A. Connes [Co1; Co4, §VI.1] to recover the usual "Riemannian volume (measure)" from purely operator-theoretic (spectral) data, within the context of noncommutative geometry. (*See* also [CoLo].) In view of Theorem 5.1 above and the author's earlier work (*see* especially [La3, Part II]), this motivates in part the following questions, raised, in particular, in [La4] (with the exception of Question $\mathbf{Q}_4$) and supplementing those in [La3, Sects. 5 and 6].

$\mathbf{Q}_1$. *What is the analogue of "Riemannian volume" on a "fractal" F, and what are its geometric properties? Further, do the above "volume measures" $\nu = \Phi(\mu)$ have a natural geometric interpretation in this context?*

$\mathbf{Q}_2$. *Is there a suitable analogue of the Dirac operator* (or other elliptic p.d.o.) *on F?*

$\mathbf{Q}_3$. *What is a suitable substitute for the notion of "geodesic flow"* (*or of the "horocycle flow"*) *on F?*

$\mathbf{Q}_4$. *What is an appropriate way to define and describe the "symmetries" of F, as well as their relationships with the oscillations observed in Theorems* 3.1 *and* 5.1?

We would like to consider these questions not only in the context of p.c.f. fractals, but for other classes of "fractals" to which we conjecture that our results (in [KiLa1] and [La4]) can be suitably extended. (*See, e.g.*, [La4], Conjecture 3.37, p. 159 and §5.3, pp. 188-189.) In addressing Questions $\mathbf{Q}_3$ and $\mathbf{Q}_4$, we would expect to establish contact with [BdFUr] (and the relevant references therein), as well as with the (heuristic) notion of "complex dimensions" appearing in our earlier (joint) work on "fractal strings and the Riemann hypothesis" [LaMa1,2] (*see* also [LaPo1,2]) and presently being rigorously developed and applied in the related context of "fractal strings" [La1-3, La5, LaPo, LaMa, HeLa] in our joint work with Machiel van Frankenhuysen [La-vF].

Question $\mathbf{Q}_1$ will be addressed in the next section while Questions $\mathbf{Q}_2$ - $\mathbf{Q}_4$ will be pursued and completed in §7 below. (*See* also §8 for aspects of Question $\mathbf{Q}_4$.)

Remark 5.3. (The "Length" of Fractal Boundaries.) (a) In [La4, §4.2], we also show that $\nu(F)$, the "*volume*" of F (in Theorem 5.1), can be expressed as a "*generalized* (*noncommutative*) *residue*" [Wo] (of the analytic continuation) of the "*spectral zeta-function*" of R_μ, $\zeta_R(s) := \sum_{j=1}^{\infty} \lambda_j^{-s/2}$. In particular, it does *not* depend on the averaging procedure used in defining the Dixmier trace of R_μ in (5.1). (*See* [La4], Proposition 4.17, pp. 163-164 and Remark 4.39(a), p. 168.) In view of [La2-3, LaPo1-3], this sheds new light on the corresponding (but even more delicate) problem for "*drums with fractal boundary*" (*cf.* Question $\mathbf{Q}_1$): *What is a suitable analogue of the "length" of a "fractal" boundary* $F = \partial\Omega$ *of a* (*bounded*) *open set* $\Omega \subset \mathbb{R}^n$ $(n \geq 2)$*?*

(b) ($n = 1$.) In the special case of "*fractal strings*" [La1-3 and 5, LaPo, LaMa, HeLa] (*i.e.*, one-dimensional drums with fractal boundary), the proper notion of "length" (of the boundary $F = \partial\Omega$) was shown in [LaPo1-2] to be directly related to the Minkowski content of $F = \partial\Omega$ (*see* esp. [LaPo2, Theorems 2.1, 2.2 and Corollary 2.3, p. 46]), and was later reinterpreted (at least in the case of a Cantor-type set F) in [Co4, §IV.3.ε, esp. pp. 327-328] in terms of the notion of Dixmier trace and a version in this context of the "quantized calculus" (*see* §7.2 below). (Compare, *e.g.*, [Co4, Example 23, p. 328] and [LaPo2, Example 45, p. 65].) This led to a computation of the analogue in the context of ([CoSu], [Co4, §IV.3]) of the constant $\mathcal{C}$ for limits sets of Fuchsian groups of the second kind; *see* [Co4, §IV.3.ε], *as well as* Remark 6.4 below. To our knowledge, in the general case of quasi-Fuchsian groups studied in ([CoSu], [Co4, §IV.3]), the problem of computing (the analogue of) $\mathcal{C}$ remains open.

6. An Analogue of Riemannian Volume on Fractals

In this section, we discuss several conjectures from [La4, §5.1] concerning the (natural) volume measures obtained in Theorem 5.1 above. (*See* §6.1.) We also present new (joint) results from [KiLa2] establishing some of those conjectures for a certain class of fractals (including the Sierpinski gasket in any dimension).(*See* §6.2.)

6.1. Conjectures.

We now return to the setting of §4. Let μ^* be the "natural s.s. measure" on $F = (F, \mathcal{S}, \mathcal{E})$, with weights $\{c_k^S\}_{k=1}^N$, where the c_k's are the "harmonic constants" of $\mathcal{E}$. Thus μ^* satisfies (2.1) with $b_k := c_k^{S/2}$ for $k = 1, \ldots, N$. We would like to mimic this situation, except that we will allow for the possibility of *variable*—rather than constant—weight functions.

Let $\nu^* = \Phi(\mu^*)$ be the "*natural volume measure*", constructed as in [La4] (*see* Theorem 5.1 above). Let $\nu_1 := \nu^*/\nu(F)$ be the associated probability measure, with $\nu(F)$ $(\in (0, +\infty))$ given as in Theorem 5.1. *We proposed in* [La4] *to view* ν^* (*or equivalently,* ν_1) *as an analogue of Riemannian volume on the p.c.f. fractal* F (*cf. Question* $\mathbf{Q}_1$ *above*).

Conjecture 6.1. *The (normalized) "natural volume measure" ν_1 is "approximately self-similar"* (in the sense of [La4]), *with "weight functions"* $\{\Psi_k\}_{k=1}^N$ ($\Psi_k \geq 0$, $\sum_{k=1}^N \int_F \Psi_k d\nu_1 = 1$) *depending only on* $\mathcal{E}$*; namely* (*cf.* formula (2.1)),

$$\int_F f d\nu_1 = \sum_{k=1}^N \int_F \Psi_k (f \circ W_k) d\nu_1, \quad \textit{for all} \quad f \in C(F). \tag{6.1}$$

Next, recall from §4 that the maps $W_k : F \to F$ are (globally) Lipschitz with respect to the *"natural (or intrinsic) metric"* $\delta^* = \delta^*_{\mathcal{E}}$, given by (4.4). Further, the "*similarity* dimension" S of $(F, \mathcal{E})$ is also given by $S = d_H = d_M$, where d_H (resp., d_M) denotes the Hausdorff (resp., Minkowski) dimension of (F, δ^*). We now propose to relate the values $\Psi_k(x)$ of the weight function Ψ_k ($k = 1, \ldots, N$) to the *"local Lipschitz constant"*,

$$|W_k'(x)| := \lim_{y \to x} \delta^*(W_k(x), W_k(y))/\delta^*(x, y) \tag{6.2}$$

(assumed to exist for ν_1-a.e. x in F), of the map W_k with respect to the metric δ^*.

Conjecture 6.2. *For $k = 1, \ldots, N$, we have $\Psi_k = |W_k'|^S (= |W_k'|^{d_H} = |W_k'|^{d_M})$ in formula* (6.1).

Remark 6.3. (a) An observation of Christina Q. He [He] implies that if Conjecture 6.2 is (strictly) true, then ν_1 is equal to μ^* (and hence ν is proportional to the natural s.s. measure μ^*). Indeed, in view of (6.2) and since the map W_k has Lipschitz constant $\sqrt{c_k}$, we must have $|W_k'(x)| \leq c_k^{S/2}$ ν_1-a.e. Then, much as in [La4, Remark 5.18(d), p. 177], we let $f \equiv 1$ in (6.1) and use formulas (4.2) and (4.1) to deduce that

$$1 = \sum_{k=1}^N \int |W_k'(x)|^S d\nu_1(x) \leq \sum_{k=1}^N c_k^{S/2} = 1.$$

Thus

$$\sum_{k=1}^N \int (c_k^{S/2} - |W_k'(x)|^S) d\nu_1(x) = 0$$

and because the integrand is nonnegative, it follows that

$$\Psi_k(x) = |W_k'(x)|^S \equiv c_k^{S/2} \quad \nu_1\text{-a.e.}$$

Substituting back into (6.1), we see that ν_1 is self-similar with weights $c_k^{S/2}$ and hence (by the aforementioned uniqueness theorem [La7, Theorem 2.18, p. 147]), coincides with μ^*.

(b) Note that the above reasoning presented in (a) relies on the fact that the limit in (6.2) exits pointwise in the standard sense. This may not always be the case, however, and other intepretations of the symbol "$|W_k'|$" are also possible, as is discussed in various places in [La4], especially at the end of §5.1. (See also Remark 6.6 below.)

We will see in §6.2 that at least in some important cases, we have recently established in [KiLa2] a suitable version of Conjectures 6.1 and 6.2. For now, we would like to comment on some aspects of these conjectures.

Remark 6.4. (Conformal Measure and Quantized Calculus.) (a) According to formula (6.1), with Ψ_k as in Conjecture 6.2 for $k = 1, \ldots, N, \nu_1$ is an analogue—in the present setting of (p.c.f.) s.s. fractals—of the Patterson-Sullivan measure τ (also called *conformal measure*) [Pa, Su] on limit sets L of Fuchsian or Kleinian groups. [In short, a Fuchsian (resp., Kleinian) group is a suitable discrete group of isometries of the hyperbolic plane (resp., space); *see, e.g.,* [Ba,Bd] and the references therein. Typically, its *"limit set"* L will be "approximately self-similar" and have nonintegral "fractal" (Hausdorff, Minkowski, or packing) dimension.]

(b) In ([CoSu], [Co4, §IV.3]), Connes and Sullivan reconstruct τ by means of the new notion of *"quantized (differential) calculus"* (arising from noncommutative geometry [Co4]), in the case of limit sets L of quasi-Fuchsian groups (such as certain Julia sets). Very briefly, their argument goes as follows: (i) They first define a positive (Radon) measure κ on L by means of "integrals" involving certain operators. [These "integrals" are of the same nature as in [Co1, 3-4] and Theorem 5.1 above, but their "integrand" is expressed somewhat differently due to the special nature of "*quantized differentials*" and the existence of (an analogue of) the "*Dirac operator*" on L; *see* §7.2 below.] (ii) Using the properties of "quantized differentials" (especially, the "*chain rule*" and the "*change of variable formula*"), as well as the unitarity of the Dixmier trace (§5), they then show that the measure κ is "*q-conformal*", with q equal to the Hausdorff dimension of L; namely, for all transformations β in the group, $\kappa \circ \beta^{-1} = |\beta'|^q \kappa$ or equivalently,

$$\int_L (f \circ \beta) d\kappa = \int_L |\beta'|^q \, f d\kappa, \text{ for all } \quad f \in C(L). \tag{6.3}$$

They thus deduce from Sullivan's *uniqueness* theorem in [Su] that (up to normalization) $\kappa = \tau =$ the (q-dimensional) Hausdorff measure on $L \subset \mathbb{C}$.

A possible strategy (inspired in part by the recent work in [MauUr] and the relevant references therein) for proving the above conjectures can be succintly stated as follows. First show that there exists a *"q-semiconformal measure"* ι on F [*i.e.*, a probabiliy measure satisfying an (iterated) analogue of (6.1), with weight functions Ψ_k as in Conjecture 6.2]. If, in addition, Ψ_k is assumed to be continuous for $k = 1, \ldots, N$, then show that such a measure is unique. Actually, ι is an eigenmeasure (with eigenvalue 1) of the dual of the "Perron-Frobenius-Ruelle operator" (*see, e.g.*, [Bd, Bo, Rue]) acting on $C(F)$. Moreover, under appropriate hypotheses (which, once suitably abstracted from [MauUr], should also hold here), ι is necessarily *"q-conformal"* (as in Remark 6.4(b) above, but with β replaced by $W_k, k = 1, \ldots, N$). In our context, an extension of this method—based on nonlinear analysis (especially, the Schauder fixed point theorem), ergodic theory and dynamical systems—should also enable us to prove the following conjecture.

Conjecture 6.5. (*Up to measure equivalence*) ν_1 *is "uniquely ergodic"* (with continuous weights) *with respect to the semigroup* $\mathcal{G} := \{W_\omega : \omega \in \Sigma_m, m \geq 0\}$ *spanned by* $\{W_k\}_{k=1}^N$. *Further,* ν_1 *is a "Gibbs equilibrium measure"* (in a sense adapted from Bowen [Bo] and Ruelle [Rue]).

[Here, $\Sigma_m = A^m$ denotes the set of words of length m, with letters in the alphabet $A = \{1, \ldots, N\}$. Further, for $\omega = \omega_1 \cdots \omega_m \in \Sigma_m$, we let $W_\omega := W_{\omega_1} \circ \cdots \circ W_{\omega_m}$.]

Remark 6.6. As was alluded to earlier, a more refined interpretation of Conjecture 6.5, as well as of Conjectures 6.1 and 6.2, is provided in [La4, Conjecture 5.35,

p. 183]. It is based on the twin notions of *"Radon-Nikodym cocyles"* and *"G-measures"* (roughly, "abstract Riesz products" or weak limits of infinite product measures [Ke, BD]); *cf.* [La4, §5.1, esp. pp. 181-183] and the relevant references therein.

The next conjecture would establish further ties with ([Su], [CoSu]). We state it here in sufficient generality so as to allow more flexibility for extensions to other settings, such as the Sierpinski carpet (especially in dimension higher than 2). (*See, e.g.*, [Fc, Chaps. 2-3] and [La1, §3] for the various notions of fractal dimensions and measures used below.)

Conjecture 6.7. ("Natural Volume Measure" and Hausdorff Measure.) *There exists a "Connes metric" δ (see Definition 7.4 below) on F with respect to which the following properties hold:*

(i) *The Hausdorff and Minkowski (and hence also packing) dimensions coincide with S.*

(ii) *The* (S-dimensional) *Hausdorff (resp., packing) measure of F is positive and finite.*

(iii) *The* (S-dimensional) *normalized Hausdorff and packing measures coincide with ν_1.* (Hence, by Theorem 5.1, the "natural volume measure" $\nu^* = \Phi(\mu^*) = \nu(F)\nu_1$ is proportional to Hausdorff measure on (F, δ), with proportionality constant depending only on $\mathcal{E}$.)

A natural candidate—in the present case of p.c.f. self-similar fractals—for the above metric δ is the *"intrinsic metric"* $\delta^* = \delta^*_{\mathcal{E}}$ (defined by formula (4.4) above). (Note the striking analogy between the definition of δ^* in (4.4) and that of δ in (7.1) below.) Then, in view of [KiLa1, Ki2, La4], it (mainly) suffices to prove (iii).

Remark 6.8. (*Volume Measures and Multifractal Measures.*) Let $\nu = \Phi(\mu)$ be the "volume measure" associated with an *arbitrary* s.s. measure μ (*see* Theorem 5.1). We still conjecture that ν (once normalized) is "approximately s.s." (as in Conjecture 6.1, but with weights depending also on μ). Further (following a suggestion of Dan Mauldin [Mau]), we expect it to be a "*multifractal measure*" on F which is deducible from $\nu^* = \Phi(\mu^*)$ by means of the "orthogonal transition kernels" studied in [MauPW]. [The refinement (given in [La4, Conjecture 5.35]) of Conjectures 6.5, 6.1 and 6.2 could also be useful in this context; *see* Remark 6.6 above.]

6.2. New Results. We close this section by stating new results recently obtained in collaboration with Jun Kigami [KiLa2]. (Their precise statement and complete proof will be provided in [KiLa2].) They establish, in particular, a suitable version of Conjectures 6.1 and 6.2, and go some way towards establishing Conjecture 6.7, for a subset of a restricted but important class of p.c.f. self-similar fractals, called here the *"decimable fractals"*. These fractals have played a significant role as physical [R, RT] and mathematical [FuSh, M, Sh] models. Indeed, their usefulness stems from the fact that the eigenvalues and the eigenfunctions of the associated Dirichlet (or Neumann) Laplacian can be computed rather explicitly (by means of a suitable algorithm). [Our results in [KiLa2] are somewhat more general than stated in Theorem 6.9, as will be briefly commented upon in Remark 6.10(d) below.]

Theorem 6.9. ([KiLa2]). *Let $\nu^* = \Phi(\mu^*)$ be the "natural volume measure" constructed in* [La4] *(see Theorem 5.1 above). Further, let $\nu_1 = \nu^*/\nu^*(F)$ be the associated probability measure, as before.*

(a) *Assume, for simplicity, that F is the Sierpinski gasket (in dimension 2 or higher).*

Then

$$\nu_1 = \mu^*, \tag{6.4a}$$

the natural self-similar measure on F, and hence

$$\nu^* = \nu^*(F)\mu^*, \tag{6.4b}$$

where $\nu^*(F)$ is the (finite and nonzero) *constant calculated in Theorems 5.1 and 3.1.* [Recall that (by [La4]) $\nu^*(F)$ is independent of the averaging procedure and in fact, depends only on $\mathcal{E}$.]

In particular, it follows that ν_1 coincides with the natural (normalized) *Hausdorff measure on F* (with respect to the Euclidean metric).

(b) *More generally, the same results* (suitably interpreted) *hold for certain other "decimable fractals" F, such as the modified Sierpinski gasket* [KiLa1, pp. 111-112], *the modified Koch curve and Vicsek's fractal* (each equipped with their standard harmonic structure). *In particular, $\nu_1 = \mu^*$ and so* (by [Ki2]) *ν_1 is equivalent to the Hausdorff measure with respect to the "intrinsic" metric δ^* .*

Remark 6.10. (a) Here, we say that (the p.c.f., self-similar fractal) F is "*decimable*" whenever the "*decimation method*" (which goes back at least to Paul Lévy in probability theory and was later rediscovered in this context and used rather cleverly by the physicists Rammal and Toulouse [R, RT]) is applicable. As was alluded to above, this method enables one to compute recursively the spectrum of the Laplacian by means of a renormalization argument and of an appropriate dynamical system interpretation.

(b) At the beginning of this subject, the Sierpinski gasket was essentially the only deterministic self-similar fractal used in physical models or amenable to mathematical (mostly probabilistic) treatment; *see, e.g.*, [Dh, AO, Go, Ku1, BP]. Other examples of decimable fractals include the modified Koch curve (studied, *e.g.*, in [M]) and more generally, the self-similar fractals considered in [Sh].

(c) The method used in [KiLa2] to establish Theorem 6.9 above relies, in particular, on specific information about the *localized eigenfunctions* and eigenvalues ([AO], [BKi] and [Ki4]) of the Laplacian. It also enables us to express the proportionality constant $\nu^*(F)$ in (6.4b) (as well as in Theorems 5.1 and 3.1) in terms of the localized eigenvalues of Δ.

(d) In general, our joint work in [KiLa2] suggests that the volume measure $\nu = \Phi(\mu)$ (constructed in Theorem 5.1 [La4]) is closely related (and possibly proportional) to the self-similar measure η with weights

$$\eta_k := \gamma_k^{d_S} = (b_k c_k)^{d_S/2} \quad \text{for} \quad k = 1, \ldots, N. \tag{6.5}$$

In particular, if as above, we let $\mu = \mu^*$, the natural s.s. measure on F, then by (4.1) [KiLa1], we have $b_k = c_k^{S/2}$ with S given by (4.2), and so η_k is a power of c_k alone, for $k = 1, \ldots, N$. Specializing still further, if we assume that $(F, \mathcal{E})$ is sufficiently symmetric (that is, the energy flow is "equidistributed" at all scales, as is the case for the standard Sierpinski gasket), then $c_1 = \cdots = c_N$ ($= 1/\alpha$, where α is the *"renormalization constant"*) and so $\eta_1 = \ldots = \eta_N = 1/N$; that is, $\eta = \mu^*$.

Note, however, that even if $\eta = \nu^*$ in this case (a fact which we have *not* proven at this stage in [KiLa2], except for the decimable fractals of Theorem 6.9 above) and F is embedded in some Euclidean space $\mathbb{R}^n$, then ν^* *does not in general* coincide (up to a multiplicative constant) with the Hausdorff measure with respect to the Euclidean metric on F. [This contradicts an early conjecture made in [Be1,2]. (*See*, however, (e) below.) We note that at the time when the papers ([Be1], [Be2]) were written, the notion of "spectral dimension" had not yet been recognized, even in a physical sense, and so the spectral exponents obtained, for example, in [KiLa1] (see Theorem 3.1 and §4 above) were not guessed correctly.]

We mention in closing this comment that for certain inhomogeneous Sierpinski gaskets (and hence for μ not necessarily equal to μ^*), we can show in [KiLa2] that $\nu := \Phi(\mu) = \nu(F)\eta$, where η is the s.s. measure defined by (6.5).

(e) A discussion with Pierre Mathieu [Mat] indicates that if F is a *"nested fractal"* (a highly symmetric p.c.f. fractal embedded in some Euclidean space $\mathbb{R}^n$, as in [Li], with $c_1 = \cdots = c_N$), then the measure $\nu^* = \Phi(\mu^*)$ *is absolutely continuous with respect to* μ^*. [Recall that in this case, μ^* is the self-similar measure defined by $b_1 = \cdots = b_N = 1/N$ and coincides with the natural Hausdorff measure on F (with respect to the Euclidean metric); *see* [KiLa1, Example 5, pp. 113-114 and p. 122].) This fact can be established by adapting and localizing the construction of ν^* in [La4] and by using, in particular, the "Nash-Moser inequalities" on F, a version of which has recently been obtained for "nested fractals" in [FiHK].

Next, using the Radon-Nikodym theorem, we deduce that $\nu^* = \theta\mu^*$, for some (suitable) density function θ on F, and we can then show that the measure ν^* is "*approximately self-similar*", in the sense of (6.1). *We thus conclude, in particular, that Conjecture 5.1 holds for "nested fractals".*

7. Towards a Noncommutative Fractal Geometry?

We now propose to investigate several conjectures and open problems connecting analysis on fractals and spectral geometry on the one hand, and aspects of noncommutative geometry and the associated "quantized calculus" [Co1-6, CoSu] on the other hand. Eventually, we would expect parts of these subjects to merge and give rise to a new subarea, coined here (as in [La4]) *"noncommutative fractal geometry"*.

It should be clear to the reader that much of this program (and section) is of a speculative nature and that many of the proposed problems are of a long-term nature.

In the following, a "fractal" F will be viewed at first in Sections 7.1 and 7.2 as a *"classical"* (or *commutative*) space [that is, identified with the commutative algebra $C(F)$], and later on, in Section 7.3, as a *"quantum"* (or *noncommutative*) space. It seems to us that these points of view are somewhat complementary and can both be helpful to make further progress in this subject.

7.1. Spectral Triples, Dirac Operators and Connes Metrics.

In order to make our presentation more accessible to readers who are not experts in the theory of operator algebras, we first recall several definitions and examples. (We refer to [Co1-4] for a much more detailed discussion.)

Definition 7.1. A "*spectral triple*" (or "unbounded Fredholm module") is a triple $(\mathcal{A}, \mathcal{H}, D)$, where $\mathcal{A}$ is a C^*-algebra, $\rho : \mathcal{A} \to \mathcal{L}(\mathcal{H})$ is a $*$-representation in the Hilbert space $\mathcal{H}$, and D is an unbounded self-adjoint operator on $\mathcal{H}$, with compact resolvent, such that $\forall a \in \mathcal{A}$, the *commutator* $[D, \rho(a)]$ is bounded (*i.e.*, lies in $\mathcal{L}(\mathcal{H})$). Moreover, this spectral triple is said to be *p^+-summable* if $Q := |D|^{-1} \in \mathcal{L}^{p+}$ (*i.e.*, if $p > 1$, Q is compact with $\rho_j(Q) = O(j^{-1/p})$ *as* $j \to \infty$). (In the following, we will ignore ρ and simply write, *e.g.*, $[D, a]$ instead of $[D, \rho(a)]$.)

Remark 7.2. Henceforth, we will use the term "spectral triple" adopted in recent papers (such as [Co5,6]) rather than the earlier terms "K-cycle" or "unbounded Fredholm module" (used, *e.g.*, in [Co3-4, CoLo]).

The following provides the basic example of "spectral triple" used in classical differential geometry. (*Caution*: For a mathematician, the "Dirac operator" is elliptic. For a physicist, it is not and, in fact, is defined somewhat differently.)

Example 7.3. (Dirac Operators on Riemannian Manifolds.) Let D be the classical Dirac operator acting on M, a smooth compact (spin) Riemannian manifold. (Recall that D is an elliptic self-adjoint operator with spectrum unbounded in both directions.) Let $\mathcal{A} = C(M)$, acting by multiplication operators on $\mathcal{H} = L^2(M; Sp)$ (the square-integrable spinors on M). Then $(\mathcal{A}, \mathcal{H}, D)$ is a p^+-summable spectral triple with $p = m$, the dimension of M.

Intuitively, *the* (classical) *Dirac operator can be thought of as the square-root of the Laplacian, with a sign ambiguity.* We will pursue this analogy below (*see* especially Conjecture 7.5). More generally, let X be a compact metric space and let $(\mathcal{A}, \mathcal{H}, D)$ be a spectral triple, where $\mathcal{A} = C(X)$. Then, D (resp., $D^2 = |D|^2$) can be thought of as an abstract analogue of the Dirac (resp., Laplace) operator on X. Further, *the involution* $\mathcal{I} = \mathrm{sgn}(D)$, *the sign of* D, *may be viewed as an abstract analogue of the* (classical) *Hilbert transform* from harmonic analysis (*cf.* [CoSu; Co4, §IV.3]). Phrased differently: *In some sense,* $|D|$ *represents the metric aspects of the geometry of* X *whereas the "phase"* $\mathcal{I} = D|D|^{-1}$ *represents its conformal geometry* (the measurement of "angles" rather than of "distances" [Co2-6]).

Next, we introduce the notion of *"Connes metric"* in the case of most immediate interest to us when the algebra $\mathcal{A}$ is commutative.

Definition 7.4. Let X be a compact metrizable space, and let $(\mathcal{A}, \mathcal{H}, D)$ be a spectral triple with $\mathcal{A} := C(X)$. Then the associated *"Connes metric"* δ on X is defined by

$$\delta(x, y) := \sup\{|a(x) - a(y)| : \|[D, a]\| \leq 1, a \in \mathcal{A}\}, \tag{7.1}$$

for all $x, y \in X$.

[Here, $\|[D, a]\|$ denotes the norm of the bounded linear operator $[D, a]$, the "commutator" of D and a.]

In the case of a Riemannian manifold $X = M$ *studied in Example 7.3 above,* δ *coincides with the Riemannian metric on* M (*see* [Co2; Co4, §VI.1; CoLo]). *Hence we can also recover the underlying Riemannian structure from purely operator-theoretic data.*

We now return to the setting of §5 and specify, in particular, Conjecture 6.7 above.

Conjecture 7.5. (Dirac Operators and Laplacians on Fractals.) *There exists a spectral triple* $(\mathcal{A}, \mathcal{H}, D)$, *with* $\mathcal{A} := C(F)$ *and* D *a suitable analogue of the "Dirac operator" on* F, *satisfying the following properties:*

(i) *The absolute value,* $|D|$*, of* D *is equal to* $\sqrt{\Delta}$*, the (positive) square-root of the Laplacian* $\Delta = \Delta_{\mu^*}$.

(ii) *The spectral triple* $(\mathcal{A}, \mathcal{H}, D)$ *is* p^+*-summable, with* $p := d_S^* = d_S(\mu^*)$*, the "spectral dimension" of* $(F, \mathcal{E})$ *defined in §4.* (We can derive (ii) from (i) and [La4].)

(iii) *Conjecture 6.7 holds for* δ*, the Connes metric associated with* $(\mathcal{A}, \mathcal{H}, D)$ (as in Definition 7.4). *Further, the measure* $\nu^* = \nu^*(F)\nu_1$ *(constructed in Theorem 5.1, with* $\mu := \mu^*$*) can also be recovered by substituting* D *for* $\sqrt{\Delta}$ *in* (5.2).

7.2. Quantized Calculus. We begin by asking a general question that should play a central role in this subject. (In our present context, we will specify it below in Problem 7.7.)

$\mathbf{Q}_5$. (Quantized Differential and Integral Calculus.) *Can we construct, develop and use in a fruitful way a* (canonical) *"quantized calculus" on fractals, that would play the role of ordinary differential (and integral) calculus on smooth manifolds?*

This calculus would provide a replacement for nonstandard analysis and (aspects of) distribution theory that would enable us, in particular, to differentiate and integrate highly nondifferentiable functions (in a way compatible with products). (*See* [Co3] and [Co4, esp. Chapt. IV] for various examples.) [Its] *"greater power and flexibility has opened up exciting new 'semi-classical' applications in both physics and mathematics, when the space of interest exhibits a fine structure beyond the reach of ordinary differential geometry"* [Co5]. This should certainly also be the case in our present situation.

Remark 7.6. Roughly speaking, the "*quantized differential*" dh of a ("bad") function h will in general be an *operator* rather than a function or even a distribution. (Further, $dh := [\mathcal{I}, h]$, the commutator of $\mathcal{I}$ and the operator M_h representing h.) This permits us to use the spectral theorem to form various functions of dh. Moreover, the (Connes) "*integral*" of T, denoted $⨍T$, is defined as the Dixmier trace of T, much as in §5 above. (Here, T is assumed to be an "infinitesimal of order 1"; that is, a compact operator in $\mathcal{L}^{1+}$.) This notion is invariant by perturbations of T by "infinitesimals of order > 1", and in particular by trace class operators.

We next specify Question $\mathbf{Q}_5$ in our setting by stating a problem in which we establish closer ties with the recent work of Connes and Sullivan ([CoSu], [Co4, §IV.3]) on the "*quantized calculus*" on limit sets of quasi-Fuchsian groups (such as certain Julia sets, namely, those which are also quasicircles). [Recall from Remark 6.4 that, in our context, ν^* is viewed as an analogue of the "Patterson-Sullivan measure" on such limit sets.]

Problem 7.7. (Quantized Calculus on Fractals.) *Assume that Conjecture 7.5 is true. Let* $\mathcal{I} = \mathrm{sgn}(D) = D|D|^{-1}$ *denote the sign of* D*, defined via the functional calculus. Let* Z *be a "model map" for the "fractal"* F*. Extend the setting of Conjecture 7.5 as well as the "quantized calculus" of Connes and Sullivan so as to make sense, in particular, of the "quantized differential" of* Z,

$$dZ := [\mathcal{I}, Z] \tag{7.2}$$

(*as well as of the formula below*). [*This extended calculus should be equivariant with respect to the action of the semigroup* $\mathcal{G} := \{W_\omega : \omega \in \Sigma_m, m \geq 0\}$ *spanned by the* W_k*'s* $(k = 1, \ldots, N)$ *and should also satisfy an analogue of the "chain rule" in* *this context.] Then show that the "natural volume measure" $\nu^* = \Phi(\mu^*)$ on $(F, \mathcal{E})$ constructed in Theorem 5.1 can also be defined as follows (cf.* (5.2)):

$$\nu^*(f) = \int_F f d\nu^* := Tr_w(f(Z)|dZ|^S), \quad \textit{for all} \quad f \in C(F), \tag{7.3}$$

where (by Conjecture 7.5) $S = d_H$, the Hausdorff dimension of (F, δ).

[Here, we assume implicitly that the <u>operator</u> dZ $(:= \mathcal{I}Z - Z\mathcal{I})$ is compact and that $|dZ|^S \in \mathcal{L}^{1+}$. Further, the bounded *operators* $|dZ|^S$ and $f(Z)$ (= the operator of multiplication by $f \circ Z$, acting in $L^2(F, \mu)$) are defined via the functional calculus. Of course, $|dZ|$ denotes the absolute value of dZ. (*Compare with the theory of distributions where expressions like $|dZ|^S$ are in general meaningless; see* [Co4].).]

Remark 7.8. (a) A natural candidate for the "model map" Z is the continuous (but far from "smooth") surjection $\Pi : \Sigma \to F$, where $\Sigma = A^{\mathbb{N}}$ and $A = \{1, \dots, N\}$ (*see* §2 above). [Then Π would be the analogue of the boundary value $Z : S^1 \to L$ of a conformal equivalence (provided by a Riemann map), where S^1 is the unit circle and $L \subset \mathbb{C}$ is the limit set studied in [CoSu]. (Given this analogy, it would be natural and easier to begin by considering the special case when $F \subset \mathbb{R}^2 = \mathbb{C}$.)]

(b) As was noted in [La4, Remark 5.51(c), p. 187], we would expect the signum operator $\mathcal{I} = \text{sgn}(D)$ to be the pull-forward by Π of a "dyadic martingale" (or its analogue) on the Cantor group $\Sigma = \{1, \dots, N\}^{\mathbb{N}}$ when $N = 2$ (≥ 2). Indeed, such "martingales" are natural analogues of the Hilbert transform or equivalently, of the "conjugate operator" in this context. [*See, e.g.*, [SWaS, Chapt. 3] and the relevant references therein, as well as, for related material, the interesting survey article [Wa] on "*Dyadic Harmonic Analysis*" by Bill Wade (one of Vic Shapiro's first Ph.D. students) to be published in this volume.]

[*Warning*: It is relatively easy to construct a conjugate-type operator $\mathcal{I}$ satisfying *some* of the required properties. A major difficulty, however, is to construct such an operator respecting faithfully the geometric and dynamical structure of the (p.c.f. self-similar) fractal $F = (F, \mathcal{S}, \mu^*, \mathcal{E})$.]

Remark 7.9. Perhaps a "noncommutative" version of Problem 7.7 (and Conjecture 7.5) would be better suited in this context, as we will discuss in §7.3. (*See* Problem 7.11 below.) In that case, the abelian algebra $C(F)$ is replaced by a noncommutative algebra $\mathcal{A}$, and formula (7.1) defining the metric δ is replaced by the following "dual" formula [Co2]:

$$\delta(\varphi, \psi) = \sup\{|\varphi(a) - \psi(a)| : \|[D, a]\| \leq 1, a \in \mathcal{A}\}. \tag{7.4}$$

for any "states" φ, ψ on $\mathcal{A}$.

7.3. Fractal Manifolds: Classical versus Quantum Spaces.

We now would like to discuss more directly the *noncommutative* aspects of geometric analysis on fractals, and close this section by making several suggestions for the direction of future research. Many of the questions or problems below will be formulated sufficiently generally so as to apply to other classes of fractals, such as the "infinitely ramified" s.s. fractals (including, in particular, the Sierpinski carpet in 2 or more dimensions).

So far, the (p.c.f., self-similar) fractal F has been viewed as a *classical* (or commutative) space, with associated commutative algebra $\mathcal{A} = \mathcal{A}ce$ (= $C(F)$, for example).

In the following, we will also consider it as a *quantum* (or noncommutative) space [Co3,4], for which the abelian algebra $\mathcal{A}ce$ is replaced by a suitable *noncommutative* (von Neumann or C^*-) algebra $\mathcal{A}$ representing F. *Heuristically, this noncommutative algebra $\mathcal{A}$ plays the role of the "algebra of coordinates" of the underlying "quantum space"* [Co4].

We think that both points of view (classical vs. quantum space) complement each other nicely and will be useful to reach some of our long-term goals. [Our discussion will be much more extensive (if not detailed) than in [La4, §5.2.2].]

Example 7.10. (Well-Known Quantum Spaces.) A typical example of a *noncommutative* (or "*quantum*") space X is the *set of Penrose tilings* (equivalence classes of aperiodic tilings of the plane); *cf.* [Co4, §II.3]. It can be modelized as the quotient space of a Cantor set by a suitable equivalence relation. In this case, from the classical point of view, the topology of X (and the associated commutative algebra $C(X)$ of continuous functions on X) is trivial since X is *not* Hausdorff. *However, from the noncommutative point of view, the "topology" of X is far from trivial*: It gives rise to a *noncommutative* C^*-algebra $\mathcal{A}$ (the groupoid convolution algebra) which is an increasing limit of finite-dimensional algebras (an approximately finite or AF-algebra). The corresponding K-groups, $K_0(\mathcal{A})$ and $K_0^+(\mathcal{A}) = \{(u,v) \in \mathbb{Z}^2 : u + \rho v \geq 0\}$, involve the golden ratio $\rho = (1+\sqrt{5})/2$ which can then be viewed as a *topological invariant* of X. Further, their counterpart in more general situations agrees (via the Murray-von Neumann theory of "continuous dimensions" and von Jones' theory of subfactors of von Neumann algebras) with conjectured values for the density of "patterns" of (generic choices of) more general quasiperiodic tilings (*cf.* [Co3, p. 49]).

[Other well-known examples of quantum spaces relevant to our program include the *set of leaves* (or "leaf space") *of a foliation* on a differentiable manifold (such as the Kronecker foliation on the 2-torus), for which, in particular, an analogue of the Atiyah-Singer index theorem [Gi] is available (*cf.* [Co4, Sects. II.8 and II.9]).]

Intuitively, a (self-similar) *fractal can be thought of as a classical space* (say, a suitable topological space F) *with redundancies, with points identified via an appropriate equivalence relation on* F. Hence, for example, in our present setting, we proposed in [La4, §5.2.2] to think of the "quantum space" associated with (the p.c.f., self-similar set) F as the quotient space of the compact space F with a suitable equivalence relation $\mathcal{R}$ defined in terms of the maps Π and $\{W_k\}_{k=1}^N$ introduced in §2 above. There are two natural noncommutative algebras that take into account the topological *as well as* the dynamical structure of a (s.s.) fractal F. They are associated with the following two equivalence relations $\mathcal{R}_1$ and $\mathcal{R}_2$ on F: $x\mathcal{R}_1 y$ (resp., $x\mathcal{R}_2 y$) if and only if there exist $m \geq 1$ and $\omega_q \in \Sigma_m, q = 1, 2$ (resp., $m_q \geq 1$ and $\omega_q \in \Sigma_{m_q}, q = 1, 2$) such that $W_{\omega_1} x = W_{\omega_2} y$. (*See* [La4, p. 188], with a misprint in part (i) corrected as above.) One typically associates to $\mathcal{R}_1$ an AF (*i.e.*, "approximately finite") algebra, defined as an inductive limit of finite-dimensional algebras, whereas one associates to $\mathcal{R}_2$ a (richer) Cuntz-Krieger algebra; *see* [CuKr, Co4]. (In a more concrete context–closer to that of [St1-2] on self-similar measures than to the present one–related algebras are considered in [JoPe].) For our purpose, a third possibility may be better suited; it will be discussed below, following the statement of Problem 7.11.

We are now ready to consider the counterpart for F, viewed as a quantum space, of Conjecture 7.5 and Problem 7.7 (on Dirac-type operators and the quantized calculus). (*See* also Remark 7.9 above.)

Problem 7.11. *Extend Conjectures 6.7 and 7.5 as well as Problem 7.7 to the noncommutative setting. In particular, find a suitable spectral triple* $\mathcal{T} = (\mathcal{A}, \mathcal{H}, D)$ *that is equivariant with respect to the action of* $\mathcal{G}$ *(the semigroup spanned by the maps* $W_k, k = 1, \ldots, N$*, defining* F*).*

A natural choice [Co7] for $\mathcal{A}$ would be the "*crossed product algebra*" $\mathcal{A} = \mathcal{A}ce \rtimes \mathcal{G}$ of the commutative algebra $\mathcal{A}ce (= C(F)$, for instance) by the action of $\mathcal{G}$ on F. A definite advantage of this construction is that it provides us with topological, analytic and geometric objects which are automatically *equivariant* with respect to $\mathcal{G}$; *see, e.g.,* [Co4, §II.6-II.8 and §II.C].

We will adopt this point of view below (and for technical reasons, will also use a counterpart of $\mathcal{A}$ for von Neumann algebras).

Remark 7.12. At first, in order to avoid additional mathematical difficulties, one could assume that F is embedded in some Euclidean space $\mathbb{R}^n$ and that all the maps $W_k (k = 1, \ldots, N)$ are invertible (which is the case of the Sierpinski gasket and most concrete s.s. fractals); so that $\mathcal{G}$ is a group rather than a semigroup.

In noncommutative geometry, a spectral triple $\mathcal{T}$ is a fundamental object that enables one to "define" and explore a quantum space X from many different points of view. It is used, in particular, to study the noncommutative topology and differential geometry of X, as well as (*in certain situations*) the notions of *curvature* (and connection) on X. [In some sense, this is reminiscent of Einstein's philosophy according to which the notion of geometric space and that of (physical) fields acting on this space are intimately connected.] We will next ask a series of questions that will aim at obtaining analogous results in our present situation. (*We will always assume in the following that an appropriate version of Problem 7.11 has already been solved and that, in particular, we have at our disposal a suitable spectral triple* $\mathcal{T} = (\mathcal{A}, \mathcal{H}, D)$.)

The following question extends, in particular, the scope of Question $\mathbf{Q}_4$ (from §5) to the noncommutative setting.

$\mathbf{Q}_6$. (Quantum Symmetries, Spectral Asymptotics and Oscillatory Phenomena.) *What is a suitable definition of the group of (quantum) symmetries of a fractal (both from a geometric and analytic point of view)? Further, can we understand the delicate oscillatory phenomena occurring (both for the geometry and, as in Theorems 3.1 and 5.1, the spectral asymptotics) by means of this symmetry group* $\mathcal{Q}$ *as well as of Connes' noncommutative analogue of the Radon-Nikodym theorem and classification of (type III) von Neumann algebras* [Co4, §V.5]?

Here, *a quantum space such as a (s.s.) fractal is viewed* (as above) *as a classical space with redundancies and carries a natural dynamics*—that provided (via the Tomita-Takesaki theory) by the one-parameter *group of modular automomorphisms* associated with the (von Neumann) algebra $\mathcal{A}$ (and a suitable KMS state on $\mathcal{A}$); *cf.* [Co3; Co4, §V.5]. (For technical reasons, we consider here the von Neumann (rather than the C^*-) crossed product algebra $\mathcal{A} = L^\infty(\mu) \rtimes \mathcal{G}$, where μ is a s.s. measure on F; *cf.* [Co4, §V.A].) We then expect the type of $\mathcal{A}$ to be given by the following conjecture:

Conjecture 7.13. *For generic data (the "nonlattice case" of* [KiLa1, La3-4], *as defined in Theorems 3.1 and 5.1 above), the von Neumann algebra $\mathcal{A}$ is of type* III_1, *whereas for highly symmetric situations (the "lattice case"* of the above references), *it is of type* III_λ (with $0 < \lambda < 1$).

A potential candidate for the *"symmetry group"* $\mathcal{Q}$ is the automorphism group of $\mathcal{A}$; it is of a very different nature in each of the situations described in the above conjecture. More generally, it could be a suitable *quantum group* naturally associated with $\mathcal{T} = (\mathcal{A}, \mathcal{H}, \mathcal{D})$. Further, we expect the periodic function g occurring in the "lattice case" in the statement of Weyl's spectral law (*see* Theorem 3.1(ii), Equation (3.5)) to be related to Connes' noncommutative analogue of the Radon-Nikodym derivative. (Understanding precisely when g is nonconstant and hence *oscillations do occur* in the spectral asymptotics is a part of the above problem; *see* [AO, BKi, Ki4, KiLa2] regarding the *"localized eigenfunctions"* of Laplacians on F.)

In addition, as was mentioned following Question $\mathbf{Q}_4$, in addressing Question $\mathbf{Q}_6$, the notion of *"complex dimension"* ([La2-3, La5, LaPo1-2, LaMa1-2] and esp. [La-vF]) should play a key role in analyzing the nature of the oscillations; *see* also §8 below.

Our next question is closely related to Question $\mathbf{Q}_2$ above (from §5) and also supplements it in this setting.

(*See* also [La3, §5], especially Question $5.\hat{5}$, p. 179, where related questions were asked, but with a somewhat different perspective.)

$\mathbf{Q}_7$. (Pseudodifferential Calculus and Index Theorems.) *Can we develop a suitable "pseudodifferential calculus" on fractals and* (in the long-term) *use it to obtain a counterpart of the Atiyah-Singer index theorem in this context? In particular, what is the concrete form of the (noncommutative) "local index formula" obtained for general spectral triples by Connes and Moscovici in* [CoMo]?

Answering the following question, in conjunction with Question $\mathbf{Q}_3$ (from §5), would go a long way towards understanding (and precisely defining) the rather elusive notion of "*fractal manifold*".

$\mathbf{Q}_8$. (Tangent Space, Phase Space, and (Noncommutative) Curvature.) *Is there an appropriate notion of "tangent space" (as well as of "cotangent space" or "phase space") in this context? Can we define the (possibly noncommutative) "curvature" of a "fractal manifold" F? If so, can it be used to express the spectral asymptotics of Laplacians (and other p.d.o) on F in terms of this curvature [and its (quantized) differentials]?*

In addressing Question $\mathbf{Q}_8$, we would expect to establish contact with the notion of *"tangent groupoid"* to a (smooth) manifold [Co4, §II.5].

Our final (formal) question is certainly of great difficulty—although it is very natural in our setting.

$\mathbf{Q}_9$. (Quantum Spaces of Noninteger Dimension.) *Can we extend the axioms of noncommutative geometry proposed in* [Co6] *so as to include spaces of* (possibly) *noninteger dimension, such as the fractals discussed in the present paper? Further, can some of the significant aspects of noncommutative geometry, such as cyclic cohomology* [Co4, Chapt. III], *be extended to such spaces as well?*

Many other questions could be raised–such as using the theory of *"asymptotic morphisms"* and *deformation of algebras* ([Co4, §II.B], "E-theory") to further develop geometric analysis on fractals, or defining Laplacians (and other elliptic operators) on a (s.s.) fractal F, by means of renormalized limits of discretized operators on *simplicial complexes* (rather than on finite graphs as for p.c.f. fractals) approximating F. (This would be especially important for "infinitely ramified fractals"—such as the 3-dimensional Sierpinski carpet studied probabilistically in [BB2] but for which a good analytical definition of the Laplacian has been hard to find. Further, it will entail obtaining a suitable *analogue of the Sobolev embedding theorem* in this context; *cf.* [BB2, La4] and Problem 4.1 above. We would expect to establish contact, in particular, with [BB1-2, Ki3, KiLa1, La3-4, Ku2, KuZh, W1]. The fact that *noncommutative geometry enables us to blend discrete and continuous structures* [Co4] should be very useful in this endeavor.

In the long-term, it is expected that the study of these and related questions should help us to better probe analytically (and spectrally) the fine geometric structure of fractals.

8. Complex Dimensions of Fractals and Oscillatory Phenomena

We close this paper by proposing to extend (and apply) our recent joint work with Machiel van Frankenhuysen [La-vF] ([La-vF1], to be announced in [La-vF2-4]) on "complex dimensions" of "fractal strings" (one-dimensional "drums with fractal boundary") to the present setting of "drums with fractal membrane". This will pursue the analogy drawn (and used) throughout [La3, Part II] between "drums with (self-similar) fractal boundary" and "drums with (self-similar) fractal membrane". (The theory developed in [La-vF] is motivated in part by the results and the methods of proof as well as by the questions raised in [LaPo1-3, LaMa1-2, La2-3]. In order to avoid repetitions, we refer the interested reader to the last part of the introduction (§1 above) and to [La-vF1] for further background on this subject.)

Let $F = (F, \mathcal{S}, \mu, \mathcal{E})$ be a p.c.f., self-similar fractal F, as in Sects. 2-5 above. (Recall that the s.s. set F is viewed intrinsically and need not be embedded in some Euclidean space.) In particular, μ is a s.s. measure on F with (constant) "weights" $\{b_k\}_{k=1}^N$ and $\mathcal{E}$ is a s.s. Dirichlet form with "harmonic constants" $\{c_k\}_{k=1}^N$. (*See* §2.) In the following, the self-similar structure $\mathcal{S}$ and energy functional $\mathcal{E}$ will be held fixed; therefore, we will often write (F, μ) or F when no ambiguity may arise. (Further, when $\mu = \mu^*$, the natural s.s. measure studied in Sects. 4 and 6, we will denote indifferently $(F, \mathcal{S}, \mu^*, \mathcal{E})$ by (F, μ^*), $(F, \mathcal{E})$, or simply F.)

As in Sects. 3-5, let $\{\gamma_k\}_{k=1}^N$ be the positive numbers defined by $\gamma_k = \sqrt{b_k c_k}$ for $k = 1, \ldots, N$ (*see* Eq. (3.3)); also, let d_S, the *"spectral exponent"* of $F = (F, \mathcal{S}, \mu, \mathcal{E})$, be defined (as in Eq. (3.4)) as the unique *real* (and thus necessarily positive) number such that

$$\sum_{k=1}^{N} \gamma_k^{d_S} = 1. \tag{8.1}$$

[Recall that by construction, we have $0 < \gamma_k < 1$ for all $k = 1, \ldots, N$; so that the above existence and uniqueness statement follows from elementary calculus.]

Finally, let $\Delta = \Delta_\mu$ be the (Dirichlet or Neumann) Laplacian of F, as in Sects. 3-5.

Now, consider the *"spectral zeta-function"* ([Gi], [La2, §5], [La3, LaPo1-3, LaMa1-2]) of Δ (or of F, in short); that is,

$$\zeta_F(s)(=\zeta_{F,sp}(s)) := \sum_{j=1}^{\infty} \lambda_j^{-s/2}, \tag{8.2}$$

for $s \in \mathbb{C}$ with $Re\ s$ sufficiently large.

It follows easily from Theorem 3.1 above (*see* Equations (3.2) and (3.6)) that ζ_F is well-defined and has an analytic continuation to the open half-plane $Re\ s > d_S$. (In particular, ζ_F is pole-free for $Re\ s > d_S$.) Following common practice, we will continue to denote by $\zeta_F(s)$ this analytic continuation (or any further meromorphic continuation to a larger domain). (Note that by Theorem 3.1, d_S must be a singularity of ζ_F.)

Remark 8.1. (a) In formula (8.2), for notational convenience, we use the normalization of ζ_F used, *e.g.*, in ([La2, §5] of [La3]) rather than the standard one in, *e.g.*, [Gi]. According to [Gi], we should substitute $\sum_{j=1}^{\infty} \lambda_j^{-s}$ for $\sum_{j=1}^{\infty} \lambda_j^{-s/2} = \sum_{j=1}^{\infty} (\sqrt{\lambda_j})^{-s}$ on the right-hand side of (8.2). From our point of view, however, we prefer to express the various functions involved in term of the *"frequencies"* $\sqrt{\lambda_j}$ rather than the eigenvalues λ_j of the "fractal drum" associated with F.

(b) Of course, ζ_F (and hence also its poles) depends on $\mathcal{S}, \mu, \mathcal{E}$, as well as on the (Dirichlet or Neumann) boundary conditions defining the Laplacian Δ (as in [KiLa1]). This will be understood implicitly troughout the rest of this section.

(c) Strictly speaking, what we will define below (in Definition 8.3) are the *spectral* "complex dimensions" of F. If we used instead in Definition 8.3 the poles of (a meromorphic extension of) a suitable "geometric" (rather than "spectral") zeta-function, we would obtain the set of *geometric* "complex dimensions" of F. (*See* [La-vF1] where analogous concepts are used in the context of "fractal strings".) For clarity, except towards the end of this section (*cf.* Remark 8.12(b), Conjecture 8.13, Problems 8.14, 8.19, and especially Conjecture 8.16), we will not make this distinction between "geometric" and "spectral" complex dimensions, and hence deal with $\mathcal{D}_F (= \mathcal{D}_{F,sp})$ as given just after Definition 8.3 below.

(d) We stress, however, that the closest analogue of the results of [La-vF1] (announced in [La-vF2]) on the structure of the complex dimensions of "self-similar fractal strings" would be obtained by considering the *geometric* (rather than the spectral) complex dimensions of $F = (F, \mu)$. In the important case where $\mu = \mu^*$, we will do precisely this at the end of this section (*see* Conjecture 8.16) and the conjectured results will be analogous to those stated below in the spectral case. (*See*, however, Remark 8.15 below, which—by analogy with the study of the geometry and the spectrum of self-similar fractal strings conducted in [La-vF1-3]—we expect to apply especially to the spectral rather than the geometric situation.) For the sake of exposition, we will first deal with the spectral situation and leave aside some of the possible technical difficulties and necessary adjustments briefly discussed in Remark 8.15 below.

(e) For simplicity, we will limit ourselves in this section to the case of p.c.f., self-similar fractals. However, provided certain analytical difficulties are overcome (*see, e.g.*, [La4]), the theory sketched in this section should also apply to the broader classes of (self-similar) fractals mentioned in this paper.

In the following discussion, we will exclude the degenerate situation where (the p.c.f., self-similar set) F *is* (equivalent to) *the unit interval* $[0,1]$.

Our first conjecture is suggested by the results of [La-vF1, §3] (announced in [La-vF2]) giving the structure of the complex dimensions of self-similar "fractal strings", as well as by formula (8.1) above (from Theorem 3.1 and [KiLa1]).

Conjecture 8.2. (Meromorphic Continuation of ζ_F.) *The spectral zeta-function ζ_F has a meromorphic extension to all of* $\mathbb{C}$, *given by*

$$\zeta_F(s) = \frac{\kappa(s)}{1 - \sum_{k=1}^{N} \gamma_k^s}, \quad \textit{for } s \in \mathbb{C}, \tag{8.3}$$

where κ is an entire function.

We stress that the following results (with the exception of Proposition 8.4 below) *are all* **conditional** *upon the truth of Conjecture 8.2.*

Corollary (and Definition) 8.3. ("Complex Dimensions" of a Self-Similar Fractal.) *The complex dimensions of $F = (F, \mathcal{S}, \mu, \mathcal{E})$ —defined, by analogy with* [La-vF], as the *poles of* (the meromorphic continuation of) *the spectral zeta-function ζ_F—are the* (*complex*) *solutions of the equation*

$$\sum_{k=1}^{N} \gamma_k^s = 1, \quad s \in \mathbb{C}. \tag{8.4}$$

[Compare with formula (8.1) above.]

Henceforth, *we will denote by $\mathcal{D} = \mathcal{D}_F (= \mathcal{D}_{F,sp})$ the set of complex dimensions of F* (*i.e.*, according to Definition 8.3, *the set of poles of* ζ_F). (When needed, we will also denote by ω the elements of $\mathcal{D}$.) Note that $\mathcal{D}$ *is really a set with* (integer) *multiplicity*; that is, the *multiplicity* of ω in $\mathcal{D}$ is, by definition, equal to that of ω, viewed as a pole of ζ_F.

According to the discussion preceding formula (8.1), the only real dimension of (F, μ) is (positive and) given by the "spectral exponent" d_S; namely, $\mathcal{D} \cap \mathbb{R} = \{d_S\}$. Further, since–by Theorem 3.1 ([KiLa1])–ζ_F is analytic for $Re\ s > d_S$, we have

$$d_S = \max\{Re\ \omega : \omega \in \mathcal{D}\}. \tag{8.5}$$

(These two facts do *not* depend on Conjecture 8.2.) Recall that in particular, when $\mu = \mu^*$, the natural s.s. measure, as in Sects. 4 and 6, we have $d_S = d_S^*$, the *"spectral dimension"* of $F = (F, \mathcal{E})$, and d_S^* is given by

$$d_S^* = \frac{2d_M}{d_M + 2}, \tag{8.6}$$

where d_M is the fractal (Minkowski or Hausdorff) dimension of (F, δ^*), with δ^* denoting the "intrinsic" metric of $(F, \mathcal{E})$, as in (4.4). (*See* §4 above.)

We summarize part of the above discussion by the following (unconditional) result.

Proposition 8.4. *The unique complex dimension of F on the real axis Re $s = 0$ is necessarily equal to $d_{\mathcal{S}}$, the "spectral exponent" of $F = (F, \mathcal{S}, \mu, \mathcal{E})$. Further, all the complex dimensions of F* (i.e., the poles of ζ_F) *are located to the left of or on the vertical line Re $s = d_{\mathcal{S}}$.*

The following (conditional) "structure theorem" is a consequence of Conjecture 8.2 and of the aforementioned results of [La-vF1, §3]. [Recall that the dichotomy lattice case vs. nonlattice case for (a p.c.f., self-similar set) F was defined in Theorem 3.1 (and also used, *e.g.*, in Theorem 5.1) above.]

For our purpose, it is convenient to note (before stating Theorem 8.5) that in the *"lattice case"*, there exists a number r, with $0 < r < 1$, called (by analogy with [La-vF1]) the *"multiplicative generator"* of $F = (F, \mathcal{S}, \mu, \mathcal{E})$. It is the generator, r, of the multiplicative group $\Pi_{k=1}^{N} \gamma_k^{\mathbb{Z}}$ such that $0 < r < 1$. Without loss of generality, we can assume that $0 < \gamma_N \leq \cdots \leq \gamma_1 < 1$. Then, there exist N relatively prime positive integers $t_1, \ldots, t_N$ (with $0 < t_1 \leq \ldots \leq t_N$) such that

$$\gamma_k = r^{t_k} \quad \text{for} \quad k = 1, \ldots, N. \tag{8.7}$$

Theorem 8.5. (Structure of the Complex Dimensions.)

(i) *In the "nonlattice case", the only complex dimension of F on the vertical line Re $s = d_{\mathcal{S}}$ is real and thus necessarily equal to $d_{\mathcal{S}}$. In addition, the pole $d_{\mathcal{S}}$ is simple. Moreover, there is an infinite sequence of simple complex dimensions coming arbitrarily close to the line Re $s = d_{\mathcal{S}}$.*

(ii) *In the "lattice case", the complex dimensions of F lie on finitely many lines Re $s = \sigma_k$ (with $\sigma_k \in \mathbb{R}$, $q \geq 1$ and $k = 1, \ldots, q$), where*

$$\sigma_q \leq \ldots \leq \sigma_2 \leq \sigma_1 := d_{\mathcal{S}}. \tag{8.8}$$

More precisely, there exist complex numbers z_k (with $z_1 := \sigma_1 = d_{\mathcal{S}}$ and $\sigma_k =$ Re z_k for $k = 2, \ldots, q$) such that the complex dimensions on the line Re $s = \sigma_k$ are exactly of the form

$$s = z_k + i\left(\frac{2\pi}{\ln r^{-1}}\right)\ell, \quad \textit{for} \quad \ell \in \mathbb{Z}, \tag{8.9}$$

where r is the "multiplicative generator" of F (as defined just before the statement of this theorem). (*By analogy with* [La-vF], *we will call the positive number* $\mathbf{p} := 2\pi/\ln r^{-1}$ *the "oscillatory period" of F.*)

In addition, all the poles on the line Re $s = d_{\mathcal{S}}$, $s = d_{\mathcal{S}} + i\mathbf{p}\ell$ ($\ell \in \mathbb{Z}$), are simple.

[Note that, except for $k = 1$, we must have $Im\ z_k \neq 0$ by Proposition 8.4. Further, observe that in the notation of Theorems 3.1 and 5.1, we have $T = \ln r^{-1}$ and hence $\mathbf{p} = 2\pi/T$. (Recall that T is the positive generator of the additive group $\sum_{k=1}^{N} (\ln \gamma_k)\mathbb{Z}$.)]

Corollary 8.6. (Rationality of ζ_F in the Lattice Case.) *In the "lattice case", $\zeta_F = \zeta_F(s)$ is a rational function of the variable $Z := r^s$. Moreover, its poles* (i.e., the complex dimensions of F) *are obtained by solving the algebraic equation* (of degree t_N):

$$\sum_{k=1}^{N} Z^{t_k} = 1, \quad \textit{where} \quad Z = r^s \quad (Z \in \mathbb{C}). \tag{8.10}$$

Further, given a (complex) *solution Z of* (8.10), *the multiplicity of the corresponding complex dimensions s* (necessarily on a same vertical line) *is equal to that of Z.*

Our next result also follows from Conjecture 8.2 and [La-vF1, §3]. In view of Theorem 8.5(ii) and Corollary 8.6, it is especially of interest in the "nonlattice case".

Theorem 8.7. *Under the above hypothese, $\mathcal{D}_F$, the set of complex dimensions of F, consists of an infinite sequence (without limit point in $\mathbb{C}$) of complex numbers, located in complex conjugate pairs in a bounded vertical strip $a \leq Re\ s \leq d_S$, for some $a \in \mathbb{R}$.*

Remark 8.8. (a) The results of [La-vF1,2] would also enable us to specify the asymptotic density of $\mathcal{D}_F$, but we will omit the corresponding statement here.

(b) In view of the definition of ζ_F in (8.2) and since the eigenvalues λ_j are real, it is clear that if $\omega \in \mathcal{D}_F$, then so does its complex conjugate $\bar{\omega}$. Moreover, $\mathcal{D}_F$—being the set of poles of a meromorphic function—is a discrete subset of $\mathbb{C}$.

Example 8.9. For the standard (2-dimensional) Sierpinski gasket F, we are in the "lattice case" and

$$\mathcal{D} = \mathcal{D}_F = \left\{ d_S^* + i\left(\frac{2\pi}{\ln 5}\right)\ell : \ell \in \mathbb{Z} \right\}, \tag{8.11}$$

with the "spectral dimension" d_S^* given by (4.5); namely, $d_S^* = \ln 9/\ln 5$.

More generally, by Remark 4.2(b), an entirely analogous statement holds for the N-Sierpinski gasket (with $N \geq 3$), except with $\ln 5$ replaced by $\ln(N+2)$ and d_S^* now given by (4.6) in the counterpart of (8.11). Note that the *"oscillatory period"* $\mathbf{p} := 2\pi/\ln(N+2)$ tends to zero as $N \to \infty$.

[When $N = 3$, a related result–partly motivated by [La3, Part II] and [KiLa1]–seems to have been obtained independently (and in a different language) by Simon Waddington [W2].]

Naturally, using the examples discussed in Section 3 and the Appendix of [KiLa1], we could provide many further illustrations of Theorems 8.5 and 8.7, as well as of Corollary 8.6, most explicitly in the "lattice case".

We now propose a new definition of "fractality" in this context:

Definition 8.10. (a) We will say that $F = (F, \mathcal{S}, \mu, \mathcal{E})$ is "*(intrinsically) fractal*" if it has at least one *nonreal* complex dimension ω such that $Re\ \omega > 0$ (in which case, by Remark 8.8(b) above, $\mathcal{D} = \mathcal{D}_F$ contains at least two such distinct complex conjugate dimensions, ω and $\bar{\omega}$).

(b) Further, we will say that F is "*strongly (intrinsically) fractal*" if it has a complex dimension ω, with $\omega \neq d_S$, on the vertical line $Re\ s = d_S$, where d_S is the "spectral exponent" of F (given by (8.1)). [Of course, this is a stronger notion than that introduced in (a).] Finally, if F is (intrinsically) fractal but is *not* "strongly fractal" (so that $\mathcal{D} \cap \{s \in \mathbb{C} : Re\ s = d_S\} = \{d_S\}$), then it will be said to be a "*smooth fractal*".

We deduce at once from Theorem 8.5 the following corollary.

(In the nonlattice case, the fact that F admits a nonreal complex dimension with *positive* real part follows from the result stated in Theorem 8.5(i) according to which there are complex dimensions of F arbitrarily close to the vertical line $Re\ s = d_S$.)

Corollary 8.11. *Let F be a p.c.f., self-similar set* (that is not isomorphic to the unit interval $[0,1]$, as assumed throughout). *Then F is "intrinsically fractal"* (in the sense of Definition 8.10(a)). *More precisely, in the "lattice case", F is "strongly fractal", whereas in the "nonlattice case", it is a "smooth fractal"* (in the sense of Definition 8.10(b)).

Remark 8.12. (a) When F is the unit interval $[0,1]$ (with its standard s.s. structure, measure and energy functional), as in [KiLa1, Example 1, pp. 109–110], we have $d_S^* = 1$ and, in fact, $\mathcal{D} = \{1\}$ consists of a single point, even though we are in the "lattice case". (This follows easily from the definitions.) Hence F is *not* (intrinsically) "fractal" in this degenerate situation.

(b) In the case of "drums with fractal boundary" (rather than in the present case of "drums with fractal membrane"), studied, *e.g.*, in [La-vF] and in [La1-3, LaPo, LaMa, HeLa], an analogous definition of "fractality" can be given. [Then, the (spectral) "complex dimensions" are defined as the poles of (the meromorphic continuation of) the spectral zeta-function of the (Dirichlet or Neumann) Laplacian on the associated open set $\Omega \subset \mathbb{R}^n$ with boundary $F := \partial\Omega$, as is done (without using the present terminology) in [La2-3, LaPo, LaMa] and most explicitly in [La-vF]. Alternately, one can also define the (geometric) "complex dimensions" as the poles of a corresponding (geometric) zeta-function; *cf.* [La-vF1].] This new definition of *"fractality"* presents the advantage of eliminating some of the obviously spurious examples such as [La1, Examples 5.1 and 5.1$'$, pp. 512-515]. On the other hand, it enables us to include, as desired, the examples of 'borderline' (self-affine) fractals, such as the graph F of the Cantor-Lebesgue function (also called the *"Devil staircase"* in [Ma]). The latter example was eliminated by "the" traditional definition of "fractality" [according to which the (Hausdorff or Minkowski) fractal dimension must be strictly greater than the topological dimension]. Indeed, for the "Devil staircase", all three notions of dimensions coincide and are equal to one, whereas the reader may find it interesting to guess (and possibly verify) what are its "complex dimensions". [Compare the discussion in [Ma, Plate 83, p. 82]; *see* also [La1]. (If–by analogy with what is done in [La-vF] for "fractal strings"–they are defined as the poles of the corresponding *"geometric"* (rather than *"spectral"*) zeta-function, we can check that the (geometric) "complex dimensions" of the "Devil staircase" F are all simple and given by

$$\mathcal{D} = \left\{1 + i\left(\frac{2\pi}{\ln 3}\right)\ell : \ell \in \mathbb{Z}\right\}. \tag{8.12}$$

Consequently, F is definitely "fractal", in our new sense; *cf.* [La-vF1,3].)]

(c) Still in the context of "drums with fractal boundary" (as in (b) above), the notion (and the term) of "*smooth fractal*" was used by the author in earlier (unpublished) work presented at the Special Session on "Differential Equations" of the Regional Meeting of the American Mathematical Society held in October 1990 at the University of North Texas in Denton. More precisely, $F = \partial\Omega$ was then said to be a (spectrally) "smooth fractal" if the asymptotic second term of the eigenvalue counting function $N(\lambda)$ did *not* present oscillations (*i.e.*, was monotonic). [Similarly, the boundary $F = \partial\Omega$ was said to be a (geometrically) "smooth fractal" if the (leading) asymptotics of a suitable geometric counting function–such as the volume of the (one-sided) ε-neighborhoods of $\partial\Omega$ in $\mathbb{R}^n$–did *not* present oscillations (*i.e.*, for the aforementioned counting function, if $F = \partial\Omega$ was "Minkowski measurable" in the sense of, *e.g.*, [BrCa, La1-3, LaPo1-2, LaMa].] In addition, a whole hierarchy of fractals (in a geometric or in a spectral sense) was defined according to the nature of the oscillations in associated (geometric or spectral) counting functions; however, we will not need this generalization here and hence we will omit a more detailed discussion of it.

We can now return to several of the main themes of this paper. (*See*, in particular, Question $\mathbf{Q}_4$ from Sect. 5, as well as Question $\mathbf{Q}_6$ and Conjecture 7.13 from Sect. 7.3.)

[At this point, some readers may wish to briefly review Definition 8.3 in conjunction with Remark 8.1(c) where the distinction between *"geometric"* and *"spectral"* complex dimensions was introduced. We will further discuss this subject following Remark 8.15 below.]

Conjecture 8.13. (Complex Dimensions of Fractals and Oscillatory Phenomena.) *We conjecture that our joint results* (possibly suitably adapted) *in* [La-vF] *apply in our present situation to yield "explicit formulas"* (in a sense analogous to that of classical number theory) *that connect the oscillations in the spectrum* (*or in the geometry*) *of* (p.c.f., self-similar) *fractals to the structure of their spectral* (*or geometric*) *complex dimensions.*

Further, these "explicit formulas" can be used (and possibly extended) *to describe the connections between the* (potential) *oscillations in the geometry and in the spectrum of such fractals* [*as well as of more general classes of* (*deterministic or random*) *self-similar fractals*].

Problem 8.14. *Specify the above conjecture and show that the distributional and pointwise "explicit formulas" obtained in* [La-vF1] *can be transposed to the present context of "drums with fractal membrane". In particular, in the "lattice case", deduce a more precise version of Weyl's law from Theorem 3.1*(ii) ([KiLa1]) *that involves the periodic "oscillations" of the eigenvalue counting function* $N(\lambda)$ (*or of related counting functions, as in* [La-vF]). *Similarly, in the "nonlattice case", sharpen Theorem 3.1*(i) *that yields a monotic leading term in the statement of Weyl's asymptotic law.*

Moreover, connect (*in the spirit, for example, of* [La3, Part II], [LaPo], [LaMa], *and* [La-vF1]) *the* (possible) *oscillations in the spectrum of a "drum with self-similar fractal membrane" to the* (potential) *oscillations in the "geometry" of the associated self-similar set.*

[In order to define the latter geometric oscillations, we must choose an appropriate "geometric" counting function and consider–as is suggested in several places in Sects. 4–7 above–a suitable metric on F. For instance, one possible choice of "geometric" counting function $\mathcal{N}(x)$ is the smallest number of balls of radius x^{-1} (relative to that metric) needed to cover F.]

Remark 8.15. (a) It is possible that some of the complex dimensions are "screened" (or "masked") in the sense of [La-vF1-3]; so that only $\mathcal{D}_W = \mathcal{D} \cap W$ is relevant to the above study, where $W \subset \mathbb{C}$ is a suitable domain (called the *"window"* in the above papers) containing a neighborhood of $\{s \in \mathbb{C} : Re\ s \geq d_S\}$. Intuitively, $\mathcal{D}_W$ can be thought of as the set of *"visible" complex dimensions* of F (see [La-vF1-3]).

(i) Firstly, instead of being entire, the function $\kappa = \kappa(s)$ occurring in (8.3) may only have an analytic continuation to some open half-plane $Re\ s > \sigma_0$, say, with σ_0

satisfying $-\infty \leq \sigma_0 < d_S$. Hence, $\mathcal{D}$ must be replaced by $\mathcal{D} \cap \{s \in \mathbb{C} : Re\ s > \sigma_0\}$ in this case. [We know in several examples (and we conjecture in general) that even for the present *spectral* situation, we must have at least $\sigma_0 \leq 0$. *A priori*, $\zeta_F(s)$ is analytic for $Re\ s > d_S$; so that formula (8.3) must first be verified for $Re\ s > d_S$ and is then seen to hold by meromorphic continuation for $Re\ s > \sigma_0$. (It is also possible that $\kappa(s)$ admits a meromorphic continuation to a neighborhood of $Re\ s \geq \sigma_0$, in which case the corresponding poles must be added to the complex solutions of (8.4) in the expression of $\mathcal{D}$.)]

(ii) Secondly, the "growth conditions" for the (spectral) zeta-function $\zeta_F(s)$ assumed in the hypotheses of the "explicit formulas" obtained in [La-vF1-3] may only be satisfied in some domain $W \subset \mathbb{C}$ (say, *e.g.*, $W = \{s \in \mathbb{C} : Re\ s > \sigma_0^+\}$); so that, as is the case in [La-vF] when working with *spectral* (rather than geometric) counting functions associated with "self-similar fractal strings", these "explicit formulas" are expressed in terms of $\mathcal{D}_W := \mathcal{D} \cap W$, the set of "visible" complex dimensions (instead of all of $\mathcal{D}$). [We refer to [La-vF1-3] for a precise formulation and for various examples of this *"screening effect"*.]

(b) For technical reasons, it may also be necessary to consider not only the usual spectral (or geometric) counting functions, but also various smoothed out versions, such as their "integrated forms" considered in [La-vF1-3]. In addition, it would be interesting to apply the *distributional* "explicit formula" of [La-vF1,2] to deduce a distributional expression for the *"density of (spectral) states"* (in physicists' language, but now viewed as a positive measure or distribution) in terms of the (spectral) complex dimensions of F; see Problem 8.19 below. [An analogous distributional "explicit formula" is obtained in [La-vF1,2] for the "density of states" of self-similar (fractal) strings, both in the geometric and spectral situations.]

We now close this discussion by considering in more detail the *geometric* (rather than the spectral) situation. For simplicity, we will limit ourselves to the important case where $\mu = \mu^*$, the natural self-similar measure on F associated with $\mathcal{E}$. (*See* Sects. 4 and 6 above.) Recall from Sect. 4 that μ^* has weights $b_k = \sqrt{c_k}^{\,S}$, for $k = 1, \ldots, N$, where S is the unique real (and hence necessarily positive) solution of Equation (4.2). In addition, $S = d_M(\delta^*)$, the Minkowski dimension of F equipped with the "intrinsic" (or "effective resistance") metric δ^* given by (4.4). Further recall from Remark 4.2(a) above (and [La3]) that the dichotomy "lattice vs. nonlattice case" can be expressed equivalently either in terms of the "harmonic constants $\{c_k\}_{k=1}^N$ or in terms of $\{\gamma_k\}_{k=1}^N$.

Let $\mathcal{N}(x)$ be a suitable geometric counting function, and let $\zeta_{F,g} = \zeta_{F,g}(s)$ be the corresponding *geometric* zeta-function of $F = (F, \mathcal{E}) = (F, \mathcal{S}, \mu^*, \mathcal{E})$. By analogy with [La-vF], $\zeta_{F,g}(s)$ can be defined as the Mellin transform of $d\mathcal{N}(x)$, the Lebesgue-Stieltjes measure associated with $\mathcal{N}$; namely, $\zeta_{F,g}(s) = \int_0^\infty x^{-s} d\mathcal{N}(x)$. Further, in our present situation where $\mu = \mu^*$, $\mathcal{N}(x)$ can be defined, for example, as the smallest number of balls of radius x^{-1} (relative to the "intrinsic metric" δ^*) needed to cover F. (Other choices for $\mathcal{N}(x)$ are possible and may turn out to be more convenient in this context.) Finally, we let $\mathcal{D}_{F,g}$ denote the set of *"geometric complex dimensions"*; that is, the set of poles (of a meromorphic continuation) of $\zeta_{F,g}$.

We are now ready to state the counterpart of Conjecture 8.2 (and of its consequences) for the present *geometric* (rather than spectral) situation. Actually, in view of the methods and definitions of [La-vF1-3]—and as was alluded to in Remark 8.1(d)—it may be more natural (and easier to establish) than the conditional results obtained above in the spectral context.

Conjecture (and Conditional Theorems) 8.16. *Let $F = (F, \mathcal{E})$ be a p.c.f. self-similar fractal, equipped with its natural self-similar measure μ^*. Then, with the above notation, the analogue of Conjecture 8.2 holds for the geometric zeta-function $\zeta_{F,g}$, except for $\{\gamma_k\}_{k=1}^N$ replaced by $\{\sqrt{c_k}\}_{k=1}^N$. More precisely, $\zeta_{F,g}$ has a meromorphic continuation to all of $\mathbb{C}$, given by*

$$\zeta_{F,g}(s) = \frac{\tau(s)}{1 - \sum_{k=1}^{N} \sqrt{c_k}^{\,s}}, \quad \text{for} \quad s \in \mathbb{C}, \tag{8.13}$$

where τ is an entire function.

Hence $\mathcal{D}_{F,g} = \{\omega \in \mathbb{C} : \sum_{k=1}^N \sqrt{c_k}^{\,\omega} = 1\}$ and $S = d_M(\delta^*) = \max\{Re\ \omega : \omega \in \mathcal{D}_{F,g}\}$. It then follows from the results of [La-vF1,2] *that the counterpart of the above (conditional) results–namely, Proposition 8.4, Theorems 8.5 and 8.7, Corollaries 8.6 and 8.11–holds in the present context, except for $\{\gamma_k\}_{k=1}^N$ replaced by $\{\sqrt{c_k}\}_{k=1}^N$, ζ_F $(= \zeta_{F,sp})$ by $\zeta_{F,g}$, and $\mathcal{D}_F$ $(= \mathcal{D}_{F,sp})$ by $\mathcal{D}_{F,g}$.*

Further, F has infinitely many (geometric) complex dimensions and they all lie in a bounded strip $\sigma^*_{left} \leq Re\ s \leq S$.

(i) *In the "nonlattice case" (i.e., $\sum_{k=1}^N (\ln c_k)\mathbb{Z}$, or equivalently $\sum_{k=1}^N (\ln \gamma_k)\mathbb{Z}$, is dense in $\mathbb{R}$), S is simple and is the unique pole of $\zeta_{F,g}$ on the line $Re\ s = S$. Further, there is an infinite sequence of simple (geometric) complex dimensions of F coming arbitrarily close to the line $Re\ s = S$.*

(ii) *In the "lattice case" (i.e., $\sum_{k=1}^N (\ln c_k)\mathbb{Z}$, or equivalently $\sum_{k=1}^N (\ln \gamma_k)\mathbb{Z}$, is discrete), there are infinitely many poles of $\zeta_{F,g}$ (i.e., geometric complex dimensions of F) on the line $Re\ s = S$. More generally, there exist finitely many poles $\omega_1 (= S), \omega_3, \dots, \omega_q$ and a positive real number* **p**–*called the oscillatory period of $(F, \mathcal{E})$–such that*

$$\mathcal{D}_{F,g} = \{\omega_j + i\mathbf{p}\ell : \ell \in \mathbb{Z}, j = 1, \dots, q\}. \tag{8.14}$$

Remark 8.17. (a) Since $\zeta_{F,g}(s)$ is real-valued for s real, nonreal geometric complex dimensions—like their spectral counterpart—come in complex conjugate pairs.

(b) By analogy with the results obtained in [La-vF1,2] for self-similar strings, we may expect that the function $\tau(s)$ is proportional to $(\nu^*(F))^s$, where $\nu^* = \Phi(\mu^*)$ is the *"volume measure"* constructed in [La4] (see Theorem 5.1 above) and $\nu^*(F)$, the total mass of ν^*, can be thought of as an analogue of the "Riemannian volume" of F (see Sect. 6). (*See* Equation (5.3) or (5.4), with $\nu = \nu^*$, for an expression of $\nu^*(F)$ in the nonlattice or lattice case, respectively.) Hence, after a suitable volume normalization, we could choose $\tau(s) \equiv 1$.

(c) It is easy to check that $\omega = \omega_g$ is a solution of $\sum_{k=1}^N \sqrt{c_k}^{\,s} = 1$ if and only if $\omega_{sp} = (2/(S+2))\omega_g$ is a solution of (8.4). This induces a similar correspondence between the (visible) geometric and spectral complex dimensions of $(F, \mathcal{E})$. In particular, to $S \in \mathcal{D}_{F,g}$, the "fractal dimension" of (F, δ^*), is associated $d_S^* = 2S/(S+2)$, the "spectral dimension" of $(F, \mathcal{E})$.

Definition (and Corollary) 8.18. A (p.c.f.) self-similar set F is said to be *"geometrically* (intrinsically) *fractal"* (or briefly, *"geometrically fractal"*) provided it has at least one *nonreal* geometric complex dimension ω with $Re\ \omega > 0$. Further, F is "*geometrically strongly fractal*" if it has a *nonreal* geometric complex dimension of real part S, and is said to be a "*geometrically smooth fractal*", otherwise.

Then, it follows from (the conditional) Theorem 8.16 that *a p.c.f. self-similar set is always "geometrically fractal" in the above sense* (except in the trivial case when it is isomorphic to the unit interval $[0,1]$). *Moreover, it is a "geometrically smooth fractal" in the nonlattice case, and is "geometrically strongly fractal" in the lattice case.*

Our last problem relies on Conjecture 8.13, Problem 8.14 and Remarks 8.15(b) and 8.17(c), as well as on the methods and results of [La-vF1-3]. Its satisfactory resolution would provide, in particular, new interpretations of several of the conjectures of [La3, Part II] in the present context of "drums with (self-similar) fractal membrane".

Problem 8.19. *Define precisely and study—much as is done for fractal strings in* [La-vF1,2]*—the "spectral operator"* ([La-vF1,2])*, that expresses the spectrum of a p.c.f. self-similar set* $(F,\mathcal{E}) = (F,\mathcal{S},\mu^*,\mathcal{E})$ *in terms of its geometry, and sends the "density of geometric states"* (*the positive measure* $d\mathcal{N}(x)$*, viewed as the distributional derivative of the geometric counting function* $\mathcal{N}(x)$) *onto the "density of spectral states"* (*the positive measure* $\sum_{j=1}^{\infty}\delta_{\sqrt{\lambda_j}}$*, viewed as the distributional derivative of the "frequency counting function"* $N_1(x) := N(\sqrt{\lambda})$). *More precisely, it should send each term associated with an admissible geometric complex dimension* ω_g *in the distributional explicit formula* (in the sense of [La-vF1,2]) *for the former onto the corresponding term associated with the spectral complex dimension* $\omega_{sp} = (2/(S+2))\omega_g$ *in the distributional formula for the latter.* (See Remark 8.17(c) above.) *Further, it should send the "error term" of the former explicit formula onto that of the latter.* [Compare ([La-vF1], [La-vF2,§3].)

Finally, define accurately the "inverse spectral operator" ([La-vF1,2])*, which expresses the geometry of* $(F,\mathcal{E})$ *in terms of its spectrum by sending back the "density of spectral states" onto the "density of geometric states", and determine to which extent it exists.* [Some notable differences with ([La-vF1], [La-vF2, §3]) are expected in this last case.]

In the long-term, according to our previous discussion in Sect. 7, the conjectures and conditional results of this section should yield natural examples of noncommutative (or "quantum) spaces with *complex* (nonreal) dimensions. [This would be of great interest in view, in particular, of the recent work in [CoMo] (one aspect of which uses in a noncommutative setting a notion suggested by the results and heuristics of [LaPo1-2, LaMa1-2, La2-3], now developed in [La-vF] in the context of "fractal strings"). We point out, in particular, Question $\mathbf{Q}_7$ from Sect. 7.3, as well as [Co6] and Question $\mathbf{Q}_9$ above extended in a natural way in the present context.] In turn, this would broaden considerably the scope of the questions raised in Sects. 5–7 (and especially in Sect. 7.3) above. *It would also extend and deepen significantly the relationships between fractal and noncommutative geometry.* It is hoped that both of these subjects will be enriched by the further investigation of some of these problems and will continue in the foreseeable future to cross-fertilize each other.

Acknowledgements. I am grateful to Professors Alain Connes, David Ellwood, Jerry Kaminker, Dan Kucerovsky, Dennis Sullivan and Machiel van Frankenhuysen for helpful conversations, as well as to Professor Gerald W. Johnson for valuable comments on a preliminary version of this paper. I would also like to thank the Institut des Hautes Etudes Scientifiques (IHES) in Bures-sur-Yvette, France, for providing me with a perfect research environment during my stays as a visiting professor in the Summer of 1995, December 1995 and from March through September 1996, during which much of this work was written up. Finally, I wish to thank Professors Jean-Pierre Bourguignon, Alain Connes and Dennis Sullivan for their kind hospitality at the IHES.

REFERENCES

[AO] S. ALEXANDER AND R. ORBACH, *Density of states on fractals: "fractons"*, J. Physique Lettres **43** (1982), L625–L631.

[BB1] M. T. BARLOW AND R. F. BASS, *Construction of Brownian motion on the Sierpinski carpet*, Ann. Inst. Henri Poincaré **25** (1989), 225-257.

[BB2] ___, *Coupling and Harnack inequalities for Sierpinski carpets*, Preprint, 1993.

[BKi] M. T. BARLOW AND J. KIGAMI, *Localized eigenfunctions of the Laplacian on p.c.f. self-similar sets*, J. London Math. Soc. (to appear).

[BP] M. T. BARLOW AND E. A. PERKINS, *Brownian motion on the Sierpinski gasket*, Prob. Th. Rel. Fields **79** (1988), 543-623.

[Ba] A. F. BEARDON, *The Geometry of Discrete Groups,* Springer-Verlag, Berlin, 1983.

[Bd] T. BEDFORD, M. KEANE AND C. SERIES (eds.), *Ergodic Theory, Symbolic Dynamics and Hyperbolic Spaces,* Oxford Univ. Press, Oxford, 1991.

[Be1] M. V. BERRY, *Distribution of modes in fractal resonators*, in: Structural Stability in Physics (W. Güttinger and H. Eikemeier, eds.), Springer-Verlag, Berlin, 1979, pp. 51-53.

[Be2] ___, *Some geometric aspects of wave motion: wavefront dislocations, diffraction catastrophes, diffractals*, in: Geometry of the Laplace Operator, Proc. Symp. Pure Math., vol. 36, Amer. Math. Soc., Providence, 1980, pp. 13-38.

[Bo] R. BOWEN, *Equilibrium States and the Ergodic Theory of Anosov Diffeomorphisms*, Lecture Notes in Math., vol. 470, Springer-Verlag, Berlin, 1975.

[BrCa] J. BROSSARD AND R. CARMONA, *Can one hear the dimension of a fractal?*, Commun. Math. Phys. **104** (1986), 103-122.

[BD] G. BROWN AND A. H. DOOLEY, *Odometer actions on G-measures, Ergodic Theory and Dynamical Systems* **11** (1991), 279–307.

[Ce] A. M. CAETANO, Some domains where the eigenvalues of the Dirichlet Laplacian have non-power second term asymptotic estimates, J. London Math. Soc. (2) **43** (1991), 431-450.

[Ch] G. CHOQUET, *Lectures on Analysis*, vol. I, *Integration and Topological Vector Spaces*, W. A. Benjamin, Readings, 1969.

[Co1] A. CONNES, *The action functional in non-commutative geometry*, Commun. Math. Phys. **117** (1988), 673–683.

[Co2] ____, *Compact metric spaces, Fredholm modules, and hyperfiniteness*, Ergodic Theory and Dynamical Systems **9** (1989), 207–220.

[Co3] ____, *Géométrie Non Commutative*, InterEditions, Paris, 1990.

[Co4] ____, *Noncommutative Geometry*, Academic Press, New York, 1994.

[Co5] ____, *Noncommutative geometry and reality*, J. Math. Phys. **36**, No. 11 (1995), 6194-6231.

[Co6] ____, *Gravity coupled with matter and the foundation of noncommutative geometry*, Commun. Math. Phys. (to appear).

[Co7] ____, Private communication, June 1995.

[CoLo] A. CONNES AND J. LOTT, *Particle models and noncommutative geometry*, Nuclear Phys. B **18B** (1990), 29–47.

[CoMo] A. CONNES AND H. MOSCOVICI, *The local index formula in noncommutative geometry*, Geometric and Functional Analysis **5** (1995), 174-243. (Special issue dedicated to Misha Gromov.)

[CoSu] A. CONNES AND D. SULLIVAN, *Quantized calculus on S^1 and quasi Fuchsian groups* (in preparation); and [Co4, §IV.3].

[CuKr] J. CUNTZ AND W. KRIEGER, *A class of C^*-algebras and topological Markov chains*, Invent. Math. **63** (1981), 25-40.

[DaWi] M. DAOUD AND C. WILLIAMS (eds.), *La Juste Argile*, Les Editions de Physique, Les Ulis, France and Cambridge, Mass., 1995.

[Dh] D. DHAR, *Lattices of effectively nonintegral dimensionality*, J. Math. Phys. **18** (1977), 577-585.

[Di] J. DIXMIER, *Existence de traces non normales*, C. R. Acad. Sci. Paris **262** (1966), 1107–1108.

[EHa] W. D. EVANS AND D. J. HARRIS, *Fractals, trees and the Neumann Laplacian*, Math. Ann. **296** (1993), 493-527.

[EvPeVo] C. J. G. EVERTSZ, H.-O. PEITGEN AND R. F. VOSS (eds.), *Fractal Geometry and Analysis: The Mandelbrot Festricht* (Proc. of an Intern. Symposium in honor of Benoît B. Mandelbrot's 70th birthday, Curaçao, Dutch Antilles, Feb. 1995), World Scientific, Singapore, 1996.

[Fc] K. FALCONER, *Fractal Geometry: Mathematical Foundations and Applica-*

tions, J. Wiley, Chichester, 1990.

[FedAh] J. FEDER AND A. AHARONY (eds.), *Fractals in Physics: Essays in honour of Benoit B. Mandelbrot* ('s 65th birthday), North-Holland, Amsterdam, 1990.

[Fel] W. FELLER, *An Introduction to Probability Theory and its Applications,* vol. II, Chapt. XI, J. Wiley, New York, 1966.

[FiHK] P. J. FITZSIMMONS, B. M. HAMBLY AND T. KUMAGAI, *Transition density estimates for Brownian motion on nested fractals*, Commun. Math. Phys. **165** (1994), 595-620.

[FlVa] J. FLECKINGER-PELLÉ AND D. G. VASSILIEV, *An example of a two-term asymptotics for the "counting function" of a fractal drum*, Trans. Amer. Math. Soc. **337** (1993), 99-116.

[Fu1] M. FUKUSHIMA, *Dirichlet Forms and Markov Processes,* Kodansha and North Holland, 1980.

[Fu2] ____, *Dirichlet forms, diffusion processes and spectral dimensions for nested fractals*, in: Ideas and Methods in Mathematical Analysis, Stochastics and Applications, Proc. Conf. in Memory of Hoegh-Krohn, vol. 1 (S. Albeverio *et al.*, eds.), Cambridge Univ. Press, Cambridge, 1993, pp. 151-161.

[FuSh] M. FUKUSHIMA AND T. SHIMA, *On a spectral analysis for the Sierpinski gasket*, Potential Analysis **1** (1992), 1–35.

[Ge] J. GERLING, *Untersuchungen zur Theorie von Weyl-Berry-Lapidus*, Graduate Thesis (Diplomarbeit), Dept. of Physics, Universität Osnäbruck, Germany, 1992.

[Gi] P. B. GILKEY, *Invariance Theory, the Heat Equation, and the Atiyah-Singer Index Theorem*, 2nd. ed., Publish or Perish Inc., Wilmington, 1984.

[Go] S. GOLDSTEIN, *Random walks and diffusions on fractals*, in: IMA Math. Appl., vol. 8 (H. Kesten, ed.), Springer-Verlag, New York, 1987, pp. 121–129.

[GrLa] C. A. GRIFFITH AND M. L. LAPIDUS, *Computer graphics and the eigenfunctions for the Koch snowflake drum*, Preprint IHES/M/96/59, Institut des Hautes Etudes Scientifiques, Bures-sur-Yvette, France, 1996; to appear in "Progress in Inverse Spectral Geometry" (S. I. Andersson and M. L. Lapidus, eds.), Birkhäuser-Verlag, Basel.

[Ha] M. HATA, *On the structure of self-similar sets*, Japan J. Appl. Math. **3** (1985), 381–414.

[HHW] K. HATTORI, T. HATTORI AND H. WATANABE, *Gaussian field theories on general networks and the spectral dimensions*, Progr. Theoret. Phys. Suppl. **92** (1987), 108–143.

[HavBu] S. HAVLIN AND A. BUNDE, *Percolation* II, Chapt. 3 in: Fractals and Disordered Systems, Springer-Verlag, Berlin, 1991, pp. 97-149.

[He] C. Q. HE, Private communication, 1995.

[HeLa1] C. Q. HE AND M. L. LAPIDUS, *Generalized Minkowski content and the vibrations of fractal drums and strings*, Mathematical Research Letters **3** (1996), 31-40.

[HeLa2] ____, *Generalized Minkowski content, spectrum of fractal drums, fractal strings and the Riemann zeta-function*, Preprint IHES/M/95/47, Institut des Hautes Etudes Scientifiques, Bures-sur-Yvette, France, 1995, 96 pages; to appear in *"Memoirs of the American Mathematical Society"*, vol. **127**, No. 608, May 1997.

[Ho] L. HÖRMANDER, *The Analysis of Linear Partial Differential Operators,* vols. III & IV, Springer-Verlag, Berlin, 1985.

[Hu] J. E. HUTCHINSON, *Fractals and self-similarity*, Indiana Univ. Math. J. **30** (1981), 713–747.

[JoPe] P. E. T. JORGENSEN AND S. PEDERSEN, *Harmonic analysis and fractal limit-measures induced by representations of a certain C^*-algebra*, J. Functional Anal. **125** (1994), 90-110.

[Kc] M. KAC, *Can one hear the shape of a drum?*, Amer. Math. Monthly (Slaught Mem. Papers, No. 11) **73** (1966), 1-23.

[Ke] M. KEANE, *Strongly mixing g-measures*, Invent. Math. **16** (1972), 309-324.

[Ki1] J. KIGAMI, *Harmonic calculus on p.c.f. self-similar sets*, Trans. Amer. Math. Soc. **335** (1993), 721–755.

[Ki2] ____, *Effective resistances for harmonic structures on p.c.f. self-similar sets*, Math. Proc. Cambridge Philos. Soc. **115** (1994), 291-303.

[Ki3] ____, *Harmonic calculus on limits of networks and its applications to dendrites*, J. Functional Anal. **128** (1995), 48-86.

[Ki4] ____, Work in preparation.

[KiLa1] J. KIGAMI AND M. L. LAPIDUS, *Weyl's problem for the spectral distribution of Laplacians on p.c.f. self-similar fractals*, Commun. Math. Phys. **158**, No. 1 (1993), 93–125.

[KiLa2] ____, *Volume, Hausdorff and spectral measures on self-similar fractals* (in preparation).

[Ku1] S. KUSUOKA, *A diffusion process on a fractal*, in: Probabilistic Methods in Mathematical Physics, Proc. of Taniguchi Intern. Symp. (K. Ito and N. Ikeda, eds.) Kinokuniya, Tokyo, 1987, pp. 251-274.

[Ku2] ____, *Dirichlet forms on fractals and products of random matrices*, Publ. RIMS **25** (1989), 659-680.

[KuZh] S. KUSUOKA AND X. Y. ZHOU, *Dirichlet forms on fractals: Poincaré constant and resistance*, Prob. Th. Rel. Fields **93** (1992), 169–196.

[L] S. P. LALLEY, *Packing and covering functions of some self-similar fractals*, Indiana Univ. Math. J. **37** (1988), 699-709.

[La1] M. L. LAPIDUS, *Fractal drum, inverse spectral problems for elliptic operators and a partial resolution of the Weyl-Berry conjecture*, Trans. Amer. Math. Soc. **325** (1991), 465-529.

[La2] ___, *Spectral and fractal geometry: From the Weyl-Berry conjecture for the vibrations of fractal drums to the Riemann zeta-function*, in: Differential Equations and Mathematical Physics, (C. Bennewitz, ed.), Proc. Fourth UAB Intern. Conf. (Birmingham, March 1990), Academic Press, New York, 1992, pp. 151-182.

[La3] ___, *Vibrations of fractal drums, the Riemann hypothesis, waves in fractal media, and the Weyl-Berry conjecture*, in: Ordinary and Partial Differential Equations, vol. IV (B. D. Sleeman and R. J. Jarvis, eds.), Proc. Twelth Intern. Conf. (Dundee, Scotland, UK, June 1992), Pitman Research Notes in Mathematics Series **289**, Longman Scientific and Technical, London, 1993, pp. 126–209.

[La4] ___, *Analysis on fractals, Laplacians on self-similar sets, noncommutative geometry and spectral dimensions*, Topological Methods in Nonlinear Analysis **4**, No. 1 (1994), 137-195.

[La5] ___, *Fractals and vibrations: Can you hear the shape of a fractal drum?* Fractals **3**, No. 4 (1995), 725-736. (Special issue in honor of Benoît B. Mandelbrot's 70th birthday. Reprinted in [EvPeVo], pp. 321-332.)

[LaFl] M. L. LAPIDUS AND J. FLECKINGER-PELLÉ, *Tambour fractal: vers une résolution de la conjecture de Weyl-Berry pour les valeurs propres du Laplacien*, C. R. Acad. Sci. Paris Sér. I Math. **306** (1988), 171-175.

[LaMa1] M. L. LAPIDUS AND H. MAIER, *Hypothèse de Riemann, cordes fractales vibrantes et conjecture de Weyl-Berry modifiée*, C. R. Acad. Sci. Paris Sér. I Math. **313** (1991), 19-24.

[LaMa2] ___, *The Riemann hypothesis and inverse spectral problems for fractal strings*, J. London Math. Soc. (2) **52**, No. 1 (1995), 15-34.

[LaNeReGr] M. L. LAPIDUS, J. W. NEUBERGER, R. RENKA AND C. A. GRIFFITH, *Snowflake harmonics and computer graphics: Numerical computation of spectra on fractal domains*, Intern. J. Bifurcation & Chaos **6**, No. 7 (1996), 1185-1210.

[LaPa] M. L. LAPIDUS AND M. M. H. PANG, *Eigenfunctions of the Koch snowflake domain*, Commun. Math. Phys. **172**, No. 2 (1995), 359-376.

[LaPo1] M. L. LAPIDUS AND C. POMERANCE, *Fonction zêta de Riemann et conjecture de Weyl-Berry pour les tambours fractals*, C. R. Acad. Sci. Paris Sér. I Math. **310** (1990), 343-348.

[LaPo2] ___, *The Riemann zeta-function and the one-dimensional Weyl-Berry conjecture for fractal drums*, Proc. London Math. Soc. (3) **66**, No. 1 (1993),

41–69.

[LaPo3] ____, *Counterexamples to the modified Weyl-Berry conjecture on fractal drums*, Math. Proc. Cambridge Philos. Soc. **119**, No. 1 (1996), 167-178.

[La-vF1] M. L. LAPIDUS AND M. VAN FRANKENHUYSEN, *Complex dimensions of fractal strings, oscillatory phenomena, and explicit formulas for geometric and spectral zeta-functions*, IHES Preprint (in preparation). (To be announced in [La-vF2-4] below.)

[La-vF2] ____, *Complex dimensions of fractal strings and explicit formulas for geometric and spectral zeta-functions*,Preprint IHES/M/97/34, Institut des Hautes Etudes Scientifiques, Bures-sur-Yvette, France, 1997.

[La-vF3] ____, *Complex dimensions and oscillatory phenomena, with applications to the geometry of fractal strings and to the critical zeros of zeta-functions*, Preprint IHES/M/97/ , Institut des Hautes Etudes Scientifiques, Bures-sur-Yvette, France, 1997.

[La-vF4] ____, (i) *Complex dimensions of fractal strings and oscillatory phenomena*; (ii) *Zeta-functions and explicit formulas for the geometry and spectrum of fractal strings*, Abstracts #918-35-537 and 918-35-539, *Abstracts Amer. Math. Soc.* **18**, No. 1 (1997), pp. 82-83. [Presented (by M. L. L. and M. v. F., resp.) at the Annual Meeting of the American Mathematical Society, San Diego, Calif., Jan. 11, 1997, Special Session on "Analysis, Diffusions and PDEs on Fractals"; AMS meeting 918, event code AMS SS M1.]

[Li] T. LINDSTROM, *Brownian motion on nested fractals*, Mem. Amer. Math. Soc. **83**, No. 420 (1990), 1-128.

[Liu] S. H. LIU, *Fractals and their applications in condensed matter physics*, Solid State Phys. **39** (1986), 207-273.

[M] L. MALOZEMOV, *Spectral theory of the differential Laplacian on the modified Koch curve*, in: Geometry of the Spectrum (R. Brook *et al.*, eds.), Contemporary Math., vol. 173, Amer. Math. Soc., Providence, 1994, pp. 193-224.

[Ma] B. B. MANDELBROT, *The Fractal Geometry of Nature,* rev. and enl. ed., W. H. Freeman, New York, 1983.

[Mat] P. MATHIEU, Private communication, September 1995.

[Mau] R. D. MAULDIN, Private communication, June 1994.

[MauPW] R. D. MAULDIN, D. PREISS AND H. V. WEIZSÄCKER, *Orthogonal transition kernels*, Ann. Prob. **11** (1983), 970-988.

[MauUr] R. D. MAULDIN AND M. URBAŃSKI, *Dimensions and measures in infinite iterated functions systems*, Proc. London Math. Soc. (3) **73** (1996), 105-154.

[NYO] T. NAKAYAMA, K. YAKUBO AND R. L. ORBACH, *Dynamical properties of fractal networks: Scaling, numerical simulations, and physical realizations*,

Rev. Modern Phys. **66**, No. 2 (1994), 381-443.

[PS] A. N. PARSHIN AND I. R. SHAFAREVICH (eds.), *Number Theory* I & II, Encycl. Math. Sciences, vols. 49 & 62, Springer-Verlag, Berlin, 1995 & 1992.

[Pa] S. J. PATTERSON, *The limit set of a Fuchsian group*, Acta Math. **136** (1976), 241–273.

[R] R. RAMMAL, *Spectrum of harmonic excitations on fractals*, J. Physique **45** (1984), 191-206.

[RT] R. RAMMAL AND G. TOULOUSE, *Random walks on fractal structures and percolation clusters*, J. Physique Lettres **44** (1983), L13-L22.

[Rue] D. RUELLE, *Thermodynamic Formalism*, Addison-Wesley, Reading, 1978.

[Sa] C. SABOT, *Diffusions sur les Espaces Fractals*, Thèse de Doctorat, Université Paris VI, France, 1995.

[Sap] B. SAPOVAL, *Fractals*, Aditech, Paris, 1990.

[SWaS] F. SCHIPP, W. R. WADE AND P. SIMON, *Walsh Series: An Introduction to Dyadic Harmonic Analysis*, Adam Hilger, Bristol, 1990.

[Sc] M. SCHROEDER, *Fractals, Chaos, Power Laws*, W. H. Freeman, New York, 1991.

[Sh] T. SHIMA, *On eigenvalue problems for Laplacians on p.c.f. self-similar sets,* Japan J. Appl. Math. (to appear).

[St1] R. S. STRICHARTZ, *Fourier asymptotics of fractal measures*, J. Functional Anal. **89** (1990), 154–187.

[St2] ___, *Self-similar measures and their Fourier transforms*, I–III. I, Indiana Univ. Math. J. **39** (1990), 797–817; II, Trans. Amer. Math. Soc. **336** (1993), 335-361; III, Indiana Univ. Math. J. **42** (1993), 367-411.

[Su] D. SULLIVAN, *Entropy, Hausdorff measures old and new, and limit sets of geometrically finite Kleinian groups*, Acta Math. **153** (1984), 259–277.

[Te] A. TERRAS, *Harmonic Analysis on Symmetric Spaces*, vol. I, Springer-Verlag, New York, 1985.

[W1] S. WADDINGTON, *The Dirichlet problem for Laplacians on post critically finite self-similar fractals,* Preprint, 1996.

[W2] ___, Private communication, March 1996.

[Wa] W. R. WADE, *Dyadic harmonic analysis*, in: Harmonic Analysis and Nonlinear Differential Equations: A volume in honor of Victor L. Shapiro (M. L. Lapidus, L. H. Harper and A. J. Rumbos, eds.), Contemporary Mathematics, vol. 208, Amer. Math. Soc., Providence, 1997 (in press).

[We] H. WEYL, *Das asymptotische Verteilungsgesetz der Eigenwerte linearer partieller Differentialgleichungen*, Math. Ann. **71** (1912), 441-479.

[Wo] M. WODZICKI, *Local invariants of spectral asymmetry*, Invent. Math. **75** (1984), 143–178.

MICHEL L. LAPIDUS

UNIVERSITY OF CALIFORNIA, DEPARTMENT OF MATHEMATICS, SPROUL HALL, RIVERSIDE, CA 92521-0135, USA (permanent address).

E-mail address: lapidus@math.ucr.edu

and

INSTITUT DES HAUTES ETUDES SCIENTIFIQUES, 35, ROUTE DE CHARTRES, 91440 BURES-SUR-YVETTE, FRANCE.

E-mail address: lapidus@ihes.fr

Contemporary Mathematics
Volume **208**, 1997

Some Remarks on Global Nonexistence for Nonautonomous Abstract Evolution Equations

Howard A. Levine   Patrizia Pucci   James Serrin

ABSTRACT. In this paper we consider the problem of non-continuation of solutions of the initial value problem for abstract evolution equations of the form

$$Pu_{tt} + Q(t)u_t + A(t,u) = F(t,u), \quad t \in J = [0,\infty)$$

where P and Q are linear self-adjoint operators, and $A(t,u)$ and $F(t,u)$ are respectively a linear operator in u (typically of differential type) and a nonlinear driving force. We also consider the case in which the linear dissipation, $Q(t)u_t$ is replaced by a nonlinear dissipation, $Q(t,u,u_t)$. The results extend earlier works of Levine and Levine and Serrin.

§1. Introduction.

In a series of papers [1–3] by the first author the problem of non–continuation was studied for abstract evolution equations of the type

$$Pu_{tt} + Qu_t + A(t,u) = F(u), \quad t \in J = [0,\infty), \tag{1.1}$$

where P and Q are linear self-adjoint operators, and $A(t,u)$ and $F(u)$ are respectively a linear operator in u (typically of differential type) and a nonlinear driving force. In [4] the operator A was allowed some degree of nonlinearity, but at the expense of dropping dependence on t.

In a recent paper [5] this work was extended to evolution equations of the form

$$(P(u_t))_t + Q(t,u,u_t) + A(u) = F(u), \tag{1.2}$$

where all the operators are allowed to be nonlinear, though with A independent of time.

The purpose of the present paper is to generalize the above results, and to some extent to unify them, by admitting *the same degree of nonlinearity* for the operators A and F in both (1.1) and (1.2), by letting both A *and* F depend on time, and finally by allowing time–dependent operators Q in (1.1).

Our results are established in two theorems, the first (Section 2) corresponding to the ideas and methods of [1–3], and the second (Section 3) to those of [5]. We end with the presentation of various examples, of interest in themselves, illustrating the range of our conclusions when interpreted in concrete cases. In this respect it almost goes without saying that the results of both Sections 2 and 3 apply to the wave operator $u_{tt} - \Delta u$, or its degenerate counterpart u_{tt}–div$(|Du|^{s-2}Du)$, and to the corresponding parabolic operators $u_t - \Delta u$ and u_t–div$(|Du|^{s-2}Du)$.

While the results of Section 2 are, in some ways, more special than those of Section 3, we note in Section 4 several situations of interest covered by Theorem 1 but not by Theorem 2.

§2. The case with linear dissipation.

Let X be a Banach space, and X' its dual space. If $x \in X$ and $x' \in X'$, we shall write $\langle x', x\rangle_X$ to denote the natural pairing of x and x', that is $\langle x', x\rangle_X = x'(x)$.

Let V be a Hilbert space. An operator $P : V \to V'$ will be called *symmetric* if

$$\langle Pv, w\rangle_V = \langle Pw, v\rangle_V \quad \text{for all } v, w \in V,$$

and *non–negative definite* if

$$\langle Pv, v\rangle_V \geq 0 \quad \text{for all } v \in V.$$

It is easy to check that a symmetric operator must be linear and, moreover, continuous by the uniform boundedness theorem.

In this section we study the evolution equation

$$Pu_{tt} + Q(t)u_t + A(t, u) = F(t, u), \qquad t \in J, \tag{2.1}$$

where P is symmetric and non–negative definite from V into V'. We suppose that the *dissipation operator* $Q(t)$ is, for each $t \in J$, symmetric and non–negative definite from an appropriate Hilbert space Y into its dual Y'. In addition, assume $Q \in C(J \to B(Y, Y'))$, that is $\langle Q(\cdot)v, w\rangle_Y : J \to \mathbb{R}$ is continuous for each $v, w \in Y$. Note that $P \equiv 0$ and $Q \equiv 0$ are specifically allowed.

Finally, the operators A and F are given by

$$A : J \times W \to W', \qquad F : J \times X \to X',$$

with W, X Banach spaces and W', X' their duals. In order to define the energy $\mathcal{E}u$ of a solution of (2.1), see below, it is necessary that there exist C^1 potentials

$$\mathcal{A} : J \times W \to \mathbb{R}, \qquad \mathcal{F} : J \times X \to \mathbb{R},$$

such that for each fixed t the operators A and F are the Fréchet derivatives with respect to u of $\mathcal{A}$ and $\mathcal{F}$, respectively; by normalization we can take $\mathcal{A}(t, 0) \equiv 0$, $\mathcal{F}(t, 0) \equiv 0$.

Now suppose that there is given a nontrivial subspace G of V, W, X and Y – not necessarily closed. Let

$$K = \{\varphi : J \to G \mid \varphi \in C(J \to W) \cap C(J \to X) \cap C^1(J \to V) \cap AC(J \to Y)\}.$$

We say that u is a (*strong*) *solution* of (2.1) if

(a) $u \in K$;

(b) Distribution Identity:

$$\langle Pu_t(\tau), \varphi(\tau)\rangle_V \Big|_0^t = \int_0^t \{\langle Pu_t(\tau), \varphi_t(\tau)\rangle_V - \langle Q(\tau)u_t(\tau), \varphi(\tau)\rangle_Y - \langle A(\tau, u(\tau)), \varphi(\tau)\rangle_W + \langle F(\tau, u(\tau)), \varphi(\tau)\rangle_X\} d\tau$$

for all $t \in J$ and $\varphi \in K$;

(c) Energy Conservation:

$$\mathcal{E}u(t) - \mathcal{E}u(0) \leq -\int_0^t \{\langle Q(\tau)u_t(\tau), u_t(\tau)\rangle_Y - \mathcal{A}_t(\tau, u(\tau)) + \mathcal{F}_t(\tau, u(\tau))\}d\tau,$$

where

$$\mathcal{E}u(t) = \tfrac{1}{2}\langle Pu_t(t), u_t(t)\rangle_V + \mathcal{A}(t, u(t)) - \mathcal{F}(t, u(t)), \quad t \in J, \tag{2.2}$$

is the total energy of u.

For our first result we assume even more that $Q \in C^1(J \to B(Y, Y'))$ with $Q_t(t) : Y \to Y'$ being non–positive definite and (necessarily) symmetric for all $t \in J$.

Theorem 1. *Suppose there is a constant $q > 2$ such that, for all $(t, u) \in J \times G$ for which $\mathcal{A}(t, u) < \mathcal{F}(t, u)$, we have*

$$\langle A(t, u), u\rangle_W - \langle F(t, u), u\rangle_X \leq q\{\mathcal{A}(t, u) - \mathcal{F}(t, u)\} \tag{2.3}$$

and

$$\mathcal{A}_t(t, u) - \mathcal{F}_t(t, u) \leq 0. \tag{2.4}$$

Then there is no solution u of (2.1) *on J such that $\mathcal{E}u(0) < 0$.*

Proof. We first show that, along any solution u of (2.1) with $\mathcal{E}u(0) < 0$, we have

$$\mathcal{A}(t, u(t)) - \mathcal{F}(t, u(t)) < 0 \tag{2.5}$$

and

$$\mathcal{E}u(t) - \mathcal{E}u(0) \leq -\int_0^t \langle Q(\tau)u_t(\tau), u_t(\tau)\rangle_Y d\tau \tag{2.6}$$

for all $t \in J$. Indeed, since $\mathcal{E}u(0) < 0$, by (2.2) and the fact that P is a non–negative definite operator, we obtain

$$\mathcal{A}(0, u(0)) - \mathcal{F}(0, u(0)) < 0.$$

If (2.5) fails to hold for all $t \in J$, then there exists a first value $T > 0$ such that

$$\mathcal{A}(T, u(T)) - \mathcal{F}(T, u(T)) = 0.$$

Consequently by (2.4)

$$\mathcal{A}_t(t, u(t)) - \mathcal{F}_t(t, u(t)) \leq 0 \quad \text{for all } t \in [0, T).$$

Then by the energy conservation law (c) we get, since $Q(t)$ is a non–negative definite operator,

$$\mathcal{E}u(t) \leq \mathcal{E}u(0) < 0 \quad \text{in } [0, T].$$

Hence by (2.2)

$$\mathcal{A}(T, u(T)) - \mathcal{F}(T, u(T)) < 0,$$

which is the required contradiction. Consequently (2.5) holds, and in turn by (c) and (2.4) also (2.6) is verified.

The inequalities (2.5) and (2.6) being shown, we now define, corresponding to any solution u of (2.1) on J,

(2.7)
$$\mathcal{I}(t) = \langle Pu(t), u(t)\rangle_V + \int_0^t \{\langle Q(\tau)u(\tau), u(\tau)\rangle_Y + (\tau - t)\langle Q_t(\tau)u(\tau), u(\tau)\rangle_Y\}d\tau \\ + (T_0 - t)\langle Q(0)u(0), u(0)\rangle_Y + \beta(t+t_0)^2,$$

where t_0, T_0, β are positive constants which will be fixed later (see [3, p.145]). Then one finds, from the assumption that P, $Q(t)$, $Q_t(t)$ are linear, continuous and symmetric,

$$\begin{aligned}\mathcal{I}'(t) &= 2\langle Pu(t), u_t(t)\rangle_V + \langle Q(t)u(t), u(t)\rangle_Y - \langle Q(0)u(0), u(0)\rangle_Y \\ &\quad - \int_0^t \langle Q_t(\tau)u(\tau), u(\tau)\rangle_Y + 2\beta(t+t_0) \\ &= 2\langle Pu(t), u_t(t)\rangle_V + 2\int_0^t \langle Q(\tau)u(\tau), u_t(\tau)\rangle_Y d\tau + 2\beta(t+t_0).\end{aligned}$$

With the help of the distribution identity (b), with $\varphi = u \in K$, it follows next that

$$\begin{aligned}\tfrac{1}{2}\mathcal{I}''(t) &= \{\langle Pu_t(t), u_t(t)\rangle_V - \langle Q(t)u_t(t), u(t)\rangle_Y \\ &\qquad - \langle A(t,u(t)), u(t)\rangle_W + \langle F(t,u(t)), u(t)\rangle_X\} \\ &\qquad + \langle Q(t)u(t), u_t(t)\rangle_Y + \beta \\ &= \langle Pu_t(t), u_t(t)\rangle_V - \langle A(t,u(t)), u(t)\rangle_W + \langle F(t,u(t)), u(t)\rangle_X + \beta.\end{aligned}$$

This may be simplified by using (2.3) and (2.2), namely

$$\begin{aligned}\tfrac{1}{2}\mathcal{I}''(t) &\geq \langle Pu_t(t), u_t(t)\rangle_V - q\{\mathcal{A}(t,u(t)) - \mathcal{F}(t,u(t))\} + \beta \\ &= \left(1 + \frac{q}{2}\right)\langle Pu_t(t), u_t(t)\rangle_V - q\mathcal{E}u(t) + \beta.\end{aligned}$$

Now assume for contradiction that u has the property $\mathcal{E}u(0) < 0$. Using (2.6) to eliminate $\mathcal{E}u(t)$ from the previous inequality and choosing $\beta = 2|\mathcal{E}u(0)|$, we obtain the crucial estimate

$$\mathcal{I}''(t) \geq (q+2)\{\langle Pu_t(t), u_t(t)\rangle_V + \beta\} + 2q\int_0^t \langle Q(\tau)u_t(\tau), u_t(\tau)\rangle_Y d\tau. \tag{2.8}$$

Take t_0 so large that $\mathcal{I}'(0) = 2\langle Pu(0), u_t(0)\rangle_V + 2\beta t_0 > 0$. Then, using the fact that P, $Q(t)$ are non–negative definite yields

$$\mathcal{I}'',\ \mathcal{I}',\ \mathcal{I} > 0 \quad \text{on } J.$$

We assert that

$$\mathcal{I}\mathcal{I}'' - \alpha\mathcal{I}'^2 \geq 0 \quad \text{on } [0, T_0], \tag{2.9}$$

where $\alpha = (q+2)/4$. Indeed, put

$$\mathbb{A} = \langle Pu(t), u(t)\rangle_V + \int_0^t \langle Q(\tau)u(\tau), u(\tau)\rangle_Y d\tau + \beta(t+t_0)^2,$$

$$\mathbb{C} = \langle Pu_t(t), u_t(t)\rangle_V + \int_0^t \langle Q(\tau)u_t(\tau), u_t(\tau)\rangle_Y d\tau + \beta,$$

and $\mathbb{B} = \frac{1}{2}\mathcal{I}'$. Since $Q(t)$ is non–negative definite and $Q_t(t)$ non–positive definite for each $t \in J$, we see that

$$\mathbb{A} \le \mathcal{I} \qquad \text{on } [0, T_0]. \tag{2.10}$$

Moreover, by (2.8) and the fact that $2q > q+2$,

$$\mathbb{C} \le \mathcal{I}''/(q+2) \qquad \text{on } J. \tag{2.11}$$

Now observe that, for all $(\xi, \eta) \in \mathbb{R}^2$ and $t \in J$,

$$\begin{aligned} \mathbb{A}\xi^2 + 2\mathbb{B}\xi\eta + \mathbb{C}\eta^2 = {} & \langle \xi Pu(t) + \eta Pu_t(t), \xi u(t) + \eta u_t(t)\rangle_V \\ & + \int_0^t \langle \xi Q(\tau)u(\tau) + \eta Q(\tau)u_t(\tau), \xi u(\tau) + \eta u_t(\tau)\rangle_Y d\tau \\ & + \beta\{(t+t_0)\xi + \eta\}^2 \ge 0, \end{aligned}$$

because P, $Q(t)$ are linear, symmetric and non–negative definite. Thus $\mathbb{AC}-\mathbb{B}^2 \ge 0$. In turn (2.9) is valid by virtue of (2.10), (2.11) and the fact that $\mathbb{A}$, $\mathbb{C} > 0$.

Of course $\alpha > 1$ since $q > 2$ by assumption. The inequality (2.9) can be written as $(\mathcal{I}^{-\alpha}\mathcal{I}')' \ge 0$, so

$$\frac{\mathcal{I}'(t)}{\mathcal{I}^\alpha(t)} \ge \frac{\mathcal{I}'(0)}{\mathcal{I}^\alpha(0)} > 0 \quad \text{for } t \in [0, T_0].$$

This is a Riccati inequality with blow–up time

$$T < \frac{1}{\alpha - 1}\frac{\mathcal{I}(0)}{\mathcal{I}'(0)}.$$

Consequently, if T_0 is chosen as the right hand side of the above inequality, we have a contradiction. In fact, since $\mathcal{I}(0)$ depends on T_0, this gives an easily solved linear equation for T_0, the solution being *positive* for all t_0 large enough, e.g., whenever

$$\beta t_0 > \frac{2}{q-2}\langle Q(0)u(0), u(0)\rangle_Y - \langle Pu(0), u_t(0)\rangle_V.$$

(An optimal choice for t_0, to minimize T_0, is easily determined, but is unnecessary for our purposes.) This completes the proof of the theorem.

Corollary. *Assume that* $\mathcal{A}(t,u) \geq 0$ *on the entire space* $J \times G$. *Then the assertion of Theorem 1 remains true if* (2.3) *is replaced by the property*

(2.12) $(t,u) \in J \times G$ *and* $\mathcal{F}(t,u) > 0$ *implies* $\mathcal{F}_t(t,u) \geq 0$ *and* $(\mathcal{A}/\mathcal{F})_t(t,u) \leq 0$.

Proof. We show that (2.12) implies (2.4). Thus let $(t,u) \in J \times G$ be such that $\mathcal{A}(t,u) < \mathcal{F}(t,u)$. Then $\mathcal{F}(t,u) > 0$ and so (2.12) can be applied, that is

$$\mathcal{A}_t(t,u) - \mathcal{F}_t(t,u) \leq \frac{\mathcal{F}_t(t,u)}{\mathcal{F}(t,u)}\{\mathcal{A}(t,u) - \mathcal{F}(t,u)\} \leq 0.$$

Special Case. $\mathcal{A}(t,u) = a(t)\hat{\mathcal{A}}(u)$ and $\mathcal{F}(t,u) = b(t)\hat{\mathcal{F}}(u)$. If $a \geq 0$ and $b > 0$ on J, and $\hat{\mathcal{A}} \geq 0$ on G, then (2.12) holds provided a/b is non–increasing and b is non–decreasing on J.

Remark. The special case of (2.1) when $F = F(u)$, $Q(t) \equiv 0$, $A(t,u) = A(t)u$ is a linear non–negative definite symmetric operator in W, with W a Hilbert space, was treated in [2]. In this case, condition (2.4) is implied by condition (A-iii) of Theorem 1 of [2]. Indeed, we clearly have

$$\mathcal{A}(t,u) = \int_0^1 \langle \tau A(t)u, u\rangle_W d\tau \geq 0$$

and moreover

$$\mathcal{A}_t(t,u) = \int_0^1 \langle \tau A_t(t)u, u\rangle_W d\tau \leq 0$$

by (A-iii). Hence $\mathcal{A}_t(t,u) - \mathcal{F}_t(t,u) = \mathcal{A}_t(t,u) \leq 0$ for all $(t,u) \in J \times W$, as required.

A similar remark applies to the parabolic subcase of (2.1) treated in Theorem 1 of [1], for which $P \equiv 0$ and the corresponding space V is no longer needed.

§3. The case with general dissipation.

Here we study the more general evolution equation

$$Pu_{tt} + Q(t,u,u_t) + A(t,u) = F(t,u), \qquad t \in J, \tag{3.1}$$

where P, A, F are as in Section 2, while Q is a nonlinear damping operator. We introduce the set

$$S = J \times X \times Y$$

and suppose $Q \in C(S \to X')$. Here Y no longer need be a Hilbert space.

By a (*strong*) *solution* of (3.1) we mean a function u such that

(a) $u \in K$;
(b) Distribution Identity:

$$\langle Pu_t(\tau), \varphi(\tau)\rangle_V \Big|_0^t = \int_0^t \{\langle Pu_t(\tau), \varphi_t(\tau)\rangle_V - \langle Q(\tau, u(\tau), u_t(\tau)), \varphi(\tau)\rangle_X$$
$$- \langle A(\tau, u(\tau)), \varphi(\tau)\rangle_W + \langle F(\tau, u(\tau)), \varphi(\tau)\rangle_X\}d\tau$$

for all $t \in J$ and $\varphi \in K$;
(c) Energy Conservation:

$$\mathcal{E}u(t) - \mathcal{E}u(0) \leq -\int_0^t \{\mathcal{D}(\tau, u(\tau), u_t(\tau)) - \mathcal{A}_t(\tau, u(\tau)) + \mathcal{F}_t(\tau, u(\tau))\} d\tau,$$

where $\mathcal{E}u$ is defined as in (2.2).

The function $\mathcal{D} = \mathcal{D}(t, u, v) \in C(S \to \mathbb{R}_0^+)$ is the *dissipation rate.* Clearly it has the property that

$$\mathcal{D}(\cdot, u(\cdot), u_t(\cdot)) \in L^1_{\text{loc}}(J) \tag{3.2}$$

along any solution u of (3.1).

We assume that *there is an exponent $m > 1$, a real constant κ and a positive function $\delta = \delta(t)$ of class of $C^1(J)$ such that*

$$\|Q(t, u, v)\|_{X'} \leq [\delta(t) \cdot \|u\|_X^\kappa]^{1/m} [\mathcal{D}(t, u, v)]^{1/m'} \quad \textit{on } S,$$

where m' is the Hőlder conjugate of m.

For a complete discussion of this abstract formulation, see [5] and also [6, 7].

We now present a generalization of Theorem 1 of [5], covering the case when A and F depend on t as well as u. In this result, condition (2.3) is replaced by the following three growth conditions:

There exists $q > 0$ such that

$$\langle A(t, u), u \rangle_W \leq q\mathcal{A}(t, u) \quad \textit{for all } (t, u) \in J \times G. \tag{3.3}$$

For all $\varepsilon > 0$ there are there are positive constants $c_1 = c_1(\varepsilon)$, $c_2 = c_2(\varepsilon)$ and an exponent $p > 1$ such that

$$c_1\mathcal{F}(t, u) \leq c_2\|u\|_X^p \leq \langle F(t, u), u \rangle_X - q\mathcal{F}(t, u) \tag{3.4}$$

whenever $(t, u) \in J \times G$ and $\mathcal{F}(t, u) \geq \varepsilon$.

There exists a positive constant c_3 such that

$$\|Pv\|_{V'}^2 \leq c_3 \langle Pv, v \rangle_V \qquad \text{for all } v \in V. \tag{3.5}$$

Finally, we suppose that the space X is continuously embedded in V.

Theorem 2. *Suppose $\delta'(t) = o(\delta(t))$ as $t \to \infty$. Assume that $p > \max\{2, m + \kappa\}$ and*

$$\min\{1, \delta^{(1+\theta)/(1-m)}\} \notin L^1(J),$$

for some constant θ, with $0 < \theta < \theta_0$, where

$$\theta_0 = \begin{cases} \dfrac{p - m - \kappa}{pm - (p - m - \kappa)} & \textit{if } \dfrac{\kappa}{p} + \dfrac{m}{2} \geq 1 \\[2ex] \dfrac{p - 2}{2p - (p - 2)} & \textit{if } \dfrac{\kappa}{p} + \dfrac{m}{2} < 1. \end{cases}$$

Then no solution u of (3.1) *can exist on J whenever* (2.4) *holds and* $\mathcal{E}u(0) < 0$.

Proof. We first observe, in view of (2.4), that along any solution u of (3.1) the conditions (2.5) and

$$\mathcal{E}u(t) - \mathcal{E}u(0) \leq - \int_0^t \mathcal{D}(\tau, u(\tau), u_t(\tau))d\tau \tag{3.6}$$

hold on J, see the proof of Theorem 1 above. The proof of the theorem therefore reduces exactly to that of Theorem 1 of [5], or more precisely to the case where $k(t) = 1$, $\rho = \delta^{1/(m-1)}$ on J, and P is linear – these latter specializations are not essential, but have been made only for the sake of simplicity. Note also that the Remark at the end of Section 3 of [5] has been used to treat the case when κ is not zero.

§4. Examples.

A number of concrete examples relative to linear operators A were given in Section III of [2], to which we refer the reader. Example VI of [2, p.16] in particular deserves special mention since for precision the space Y as well as W must be chosen as $H_0^1(\Omega)$.

Other concrete operators $A(u)$, not appearing in [2], are noted in [4] and in Section 6 of [6], notably the degenerate Laplacian $-\mathrm{div}(|Du|^{p-2}Du)$ and the polyharmonic operator $(-\Delta)^L$, where $L \geq 1$ is an integer, and still further examples are given in Section 4 of [5].

We conclude with a discussion of several model cases, the first being

$$u_{tt} - \Delta u = f(t, x, u), \quad x \in \Omega, \ t \in J, \tag{4.1}$$

where

$$f(t, x, u) = g(t, x)u + c|u|^{p-2}u, \quad c > 0, \ p > 2.$$

For the conclusions below we ask that

$$\frac{\partial g}{\partial t} \geq 0 \quad \text{on } J \times \Omega, \qquad \|g(t, \cdot)\|_{L^{p/(p-2)}} \leq c_0 \quad \text{for all } t \in J. \tag{4.2}$$

The natural spaces associated to (4.1) are $V = L^2(\Omega)$, $W = H_0^1(\Omega)$, $X = L^p(\Omega)$ and $G = L^p(\Omega) \cap H_0^1(\Omega)$. (The remaining spaces S and Y are unneeded, since we have omitted the damping terms from the equation. Note moreover that in applications of Theorem 2 the domain Ω must be bounded, in order that $X = L^p(\Omega)$ be embedded in $V = L^2(\Omega)$.)

The operator P corresponding to (4.1) is given by $\langle Pv, w \rangle_V = (v, w)_{L^2}$; clearly P is symmetric and positive definite, and satisfies (3.5). Moreover $A(u) = -\Delta u$, so

$$\langle A(u), u \rangle_{H_0^1} = \|u\|_{H_0^1}^2, \qquad \mathcal{A}(u) = \frac{1}{2}\|u\|_{H_0^1}^2$$

and $q\mathcal{A}(u) \geq \langle A(u), u \rangle_{H_0^1}$ whenever $q \geq 2$. On the other hand,

$$\mathcal{F}(t, u) = \frac{1}{2} \int_\Omega g(t, x)|u|^2 dx + \frac{c}{p}\|u\|_{L^p}^p. \tag{4.3}$$

Hence, (2.4) is satisfied on $J \times G$ by $(4.2)_1$ – we assume suitable regularity of $g = g(t,x)$ so that $\mathcal{F}_t(t,u)$ can be calculated on $J \times G$ by differentiation under the integral sign in (4.3).

Moreover,

$$\langle F(t,u), u\rangle_{L^p} = \int_\Omega g(t,x)|u|^2 dx + c\|u\|^p_{L^p}.$$

Therefore (2.3) is verified on $J \times G$ and Theorem 1 can be applied, *provided that*

$$\left(\frac{q}{2}-1\right)\int_\Omega g(t,x)|u|^2 dx \le c\left(1-\frac{q}{p}\right)\|u\|^p_{L^p}, \tag{4.4}$$

with $2 < q \le p$. This obviously occurs whenever $g(t,x) \le 0$ a.e. in $J \times \Omega$, but fails when $g(t,x) > 0$ on $J \times \Omega$ and $|u(x)|$ is sufficiently small.

Nevertheless non–continuation still occurs. To see this, observe from (2.2) and (3.6) that necessarily, for all solutions u of (2.1) satisfying $\mathcal{E}u(0) < 0$, we have

$$\mathcal{F}(t,u(t)) = \tfrac{1}{2}\|u_t(t)\|^2_{L^2} + \tfrac{1}{2}\|u(t)\|^2_{H^1_0} - \mathcal{E}u(t) \ge -\mathcal{E}u(t) \ge -\mathcal{E}u(0) = |\mathcal{E}u(0)|, \quad t \in J.$$

Thus we only need to verify (4.4) whenever $(t,u) \in J \times G$ is such that

$$\mathcal{F}(t,u) \ge |\mathcal{E}u(0)| = \varepsilon, \qquad \text{that is} \qquad \frac{1}{2}\int_\Omega g(t,x)|u|^2 dx + \frac{c}{p}\|u\|^p_{L^p} \ge \varepsilon.$$

In particular, this implies by Hőlder's inequality and $(4.2)_2$

$$\frac{c_0}{2}\|u\|^2_{L^p} + \frac{c}{p}\|u\|^p_{L^p} \ge \varepsilon.$$

Consequently we need to check (4.4) only when $\|u\|_{L^p} \ge \lambda(\varepsilon) = \lambda > 0$, where

$$\frac{c_0}{2}\lambda^2 + \frac{c}{p}\lambda^p = \varepsilon.$$

In fact, rather than (4.4), consider the stronger inequality

$$\left(\frac{q}{2}-1\right)c_0\|u\|^2_{L^p} \le c\left(1-\frac{q}{p}\right)\|u\|^p_{L^p};$$

this however holds for $\|u(t)\|_{L^p} \ge \lambda$ provided that

$$c\left(1-\frac{q}{p}\right)\lambda^{p-2} - \left(\frac{q}{2}-1\right)c_0 \ge 0.$$

Thus it is enough if q obeys

$$2 < q \le \frac{c\lambda^{p-2}+c_0}{2c\lambda^{p-2}+pc_0}\,2p = q_1;$$

note that $q_1 > 2$ since $p > 2$ by assumption. Of course $q_1 = p$ when $c_0 = 0$, while $q_1 < p$ when $c_0 > 0$ and $p > 2$.

This example can also be treated by means of Theorem 2. With the above calculations in mind, we take $q = 2$. Then (2.4) is verified as above, while (3.4) becomes

$$c_1 \left(\frac{1}{2} \int_\Omega g(t,x)|u|^2 dx + \frac{c}{p} \|u\|_{L^p}^p \right) \le c_2 \|u\|_{L^p}^p \le c \left(1 - \frac{2}{p}\right) \|u\|_{L^p}^p.$$

This holds if $c_2 = c(1 - 2/p)$ and if also the following inequality is satisfied,

$$c \left(1 - \frac{2}{p} - \frac{c_1}{p}\right) \|u\|_{L^p}^p - \frac{c_0 c_1}{2} \|u\|_{L^p}^2 \ge 0.$$

As before, this is required only when $\mathcal{F}(t,u) \ge |\mathcal{E}u(0)| = \varepsilon$ on $J \times G$, i.e. when $\|u\|_{L^p} \ge \lambda(\varepsilon) = \lambda > 0$. This gives the condition

$$c_1 = \frac{2(p-2)c\lambda^{p-2}}{2c\lambda^{p-2} + pc_0}.$$

Obviously $c_1 > 0$ since $p > 2$ by assumption. Here, as in the applications of Theorem 1, it remains crucial that $c > 0$.

Next consider the problem

$$u_{tt} - \operatorname{div}(|Du|^{s-2} Du) = f(t,x,u), \quad x \in \Omega, \ t \in J,$$

where $s > 1$, $s \neq 2$, and

$$f(t,x,u) = g(t,x)|u|^{s-2}u + c|u|^{p-2}u, \quad c \ge 0, \ p > s.$$

We assume the analogue of (4.2), namely

$$\frac{\partial g}{\partial t} \ge 0 \quad \text{on } J \times \Omega, \qquad \|g(t,\cdot)\|_{L^{p/(p-s)}} \le c_0 \quad \text{for all } t \in J.$$

Here the appropriate spaces are $V = L^2(\Omega)$, $W = W_0^{1,s}(\Omega)$, $X = L^p(\Omega)$ and $G = L^2(\Omega) \cap L^p(\Omega) \cap W_0^{1,s}(\Omega)$. Of course, as before, in applications of Theorem 2 the domain Ω must be bounded.

Using Theorem 1 when $s > 2$, and choosing $q = s$, we get non–continuation whenever $c \ge 0$. Using Theorem 2 with $q = s$, we get non–continuation for $c > 0$ and $p > \max\{2, s\}$.

Hence, when $s > 2$ and $c = 0$ the result given by Theorem 1 is unobtainable from Theorem 2. Conversely, when $1 < s < 2$ and $c > 0$ the result given by Theorem 2 is unobtainable from Theorem 1.

Another function f for which Theorem 1 applies, but not Theorem 2, is given by

$$f(u) = |u|^{p-1}, \quad p > 2.$$

Here

$$\langle F(u), u \rangle_{L^p} - q\mathcal{F}(u) = \left(1 - \frac{q}{p}\right) \int_\Omega |u|^{p-1} u dx,$$

so we can take $q = p$ for Theorem 1. On the other hand, for Theorem 2 the condition

$$c_2\|u\|_{L^p}^p \leq \left(1 - \frac{q}{p}\right) \int_\Omega |u|^{p-1} u dx$$

cannot in general be satisfied unless one assumes *a priori* that $u \geq 0$.

Theorems 1 and 2 also differ in their requirements concerning the dissipation operator Q. We leave such comparisons to the reader.

References

[1] H.A. Levine, *Some nonexistence and instability theorems for solutions of formally parabolic equations of the form* $Pu_t = Au + \mathcal{F}(u)$, Archive Rational Mech. Anal. **51** (1973), 371–386.

[2] H.A. Levine, *Instability and nonexistence of global solutions of nonlinear wave equations of the form* $Pu_{tt} = Au + \mathcal{F}(u)$, Trans. Amer. Math. Soc. **192** (1974), 1–21.

[3] H.A. Levine, *Some additional remarks on the nonexistence of global solutions to nonlinear wave equations*, SIAM J. Math. Anal. **5** (1974), 138–146.

[4] H.A. Levine, *Nonexistence of global solutions of nonlinear wave equations*, in *Improperly Posed Boundary Value Problems*, Res. Notes Math. **1**, 94–104; Pitman, London, 1975.

[5] H.A. Levine & J. Serrin, *Global nonexistence theorems for quasilinear evolution equations with dissipation*, Archive Rational Mech. Anal., to appear.

[6] P. Pucci & J. Serrin, *Asymptotic stability for non–autonomous dissipative wave systems*, Comm. Pure Appl. Math. **XLIX** (1996), 177–216.

[7] P. Pucci & J. Serrin, *Stability for abstract evolution equations*, in *Partial Differential Equations and Applications – Collected papers in Honor of Carlo Pucci*, P. Marcellini, G. Talenti & E. Vesentini eds., Lecture Notes in Pure and Appl. Math. 1996, M. Dekker, Inc. New York, 279–288.

Contemporary Mathematics
Volume **208**, 1997

Cartwright and Littlewood on Van der Pol's Equation

Shawnee L. McMurran and James J. Tattersall

ABSTRACT. Van der Pol's experiments with nonlinear oscillators during the 1920s and 1930s stimulated mathematical interest in nonlinear differential equations arising in radio research. The problems caught the attention of British mathematicians M.L. Cartwright and J.E. Littlewood, initiating a collaboration that lasted more than ten years. Cartwright and Littlewood's analysis of the Van der Pol equation and its generalizations led them to explore some interesting topological methods, including the development of a fixed point theorem for continua invariant under a homeomorphism of the plane. They were among the earliest mathematicians to apply Poincaré's transformation theory to the analysis of dissipative systems. Their research is among the earliest in large parameter theory and played a role in the development of the modern theory of dynamical systems and chaos theory.

Newton's work in celestial mechanics during the seventeenth century may have sparked the development of dynamic systems analysis, but progress was slow during the following two centuries. Analytical techniques for solving dynamics problems such as the three-body problem eluded mathematicians. A breakthrough came at the end of the nineteenth century when Poincaré realized that a geometric approach might be more productive than an analytical one. His technique of reducing a dynamical problem into a transformation problem provided an efficient method of analyzing differential equations. During the 1920s George Birkhoff generalized some of Poincaré's results and investigated applications to dynamics. At the time, Balthazar van der Pol's research on the behavior of nonlinear oscillators had caught the attention of mathematicians. The problems were challenging since many techniques developed for the analysis of conservative systems in celestial mechanics could not be applied to oscillators described by systems that were sometimes dissipative and often neither dissipative nor conservative. Transformation theory provided a key to analyzing such systems. Developments in nonlinear oscillation theory have played a significant role in the modern theory of dynamical systems and chaos theory, as well as radio, radar and laser technology. In addition, investigations into the van der Pol equation have played a vital role in both the theory of relaxation oscillations and

---

1991 *Mathematics Subject Classification.* 34-03, 01A60, 34A34; Secondary: 34C10, 34C15, 70K40, 54H25.

Key words and phrases. Cartwright, Littlewood, van der Pol, oscillator theory, nonlinear dynamics, dissipative systems, large parameter theory, relaxation oscillations, fixed point theorems.

This paper is in final form and no version of it will be submitted for publication elsewhere.

bifurcation theory. Two pioneers in nonlinear oscillation theory were Cambridge mathematicians John E. Littlewood and Mary L. Cartwright.

Cartwright and Littlewood began work on the van der Pol equation just before World War II. Their collaboration lasted approximately ten years and produced some of the earliest rigorous work in large parameter theory. Four papers [20,21,23,24] were published jointly and several other papers based on joint work were published individually. They were among the first mathematicians to apply the ideas of Poincaré and Birkhoff to nonlinear oscillation theory, combining topological and analytical methods in order to obtain results. In addition, their work on van der Pol's equation helped inspire the construction of Smale's horseshoe diffeomorphism [34]. The topological methods and ideas on fixed points, recurrent sets and rotation numbers developed by mathematicians such as Cartwright, Littlewood, Liénard, Levinson, Smith, Birkhoff, Denjoy, and Charpentier were invaluable to the development of modern dynamical theory.

Cartwright first met Littlewood in June of 1930 at her D.Phil. examination at Oxford. The following October she began a three-year research fellowship at Girton College, Cambridge where she attended several of Littlewood's courses and seminars. She occasionally wrote or spoke with Littlewood on topics pertaining to his lectures and she considered him a very great mathematician. She recalled impressing Littlewood with her solution to a problem he had presented in his theory of functions class concerning multivalent functions. The polished verion of her proof was published [2] and Littlewood refers to her paper in his *Lectures on the Theory of Functions.* Cartwright speculates that her work on this problem prompted Littlewood and Hardy to write a joint letter to Girton resulting in a Faculty Assistant Lectureship and the renewal of her research fellowship for a fourth year.

In January of 1938, the Radio Research Board of the Department of Scientific and Industrial Research issued a memorandum [33] to the London Mathematical Society appealing to pure mathematicians for help with the solutions of nonlinear differential equations connected to radio oscillators and the thermionic valves (vacuum tubes) used to control the flow of electricity in the circuitry of transmitters and receivers. Radio engineers wanted to find stable periodic oscillations for corresponding physical systems. However, according to Cartwright, the engineers really only needed a description of the general behavior of solutions [15]. The objective was to determine how existence, frequency, amplitude and stability of oscillations varied with the parameters of the physical systems. In particular, discerning precise frequencies and discovering any possible instabilities due to changes in parameters were important goals. A precise estimate of the period was also needed for autonomous systems [10].

According to the memorandum, the need had "arisen for a more complete understanding of the actual behavior of certain assemblages of electrical apparatus, including thermionic valves." Until this time, analysis of the physical system designed for generation, transmission, and reception of electric waves was based on the theory of linear differential equations with constant coefficients. Behavior of systems was determined from the explicit solutions to these equations. In the case of thermionic valves, the characteristic of the vacuum tube had been approximated by a plane, therefore the existing linear theory of differential equations had been sufficient for mathematical analysis. A closer approximation of the actual physical system required that bends of the characteristic be taken into account. A nonlinear theory of electric oscillations was needed [35]. The alternative methods of numerical

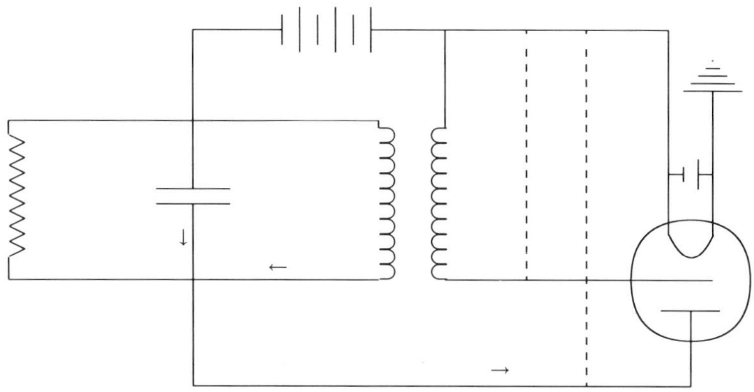

FIGURE 1

and graphical analysis were not practical techniques for analyzing the system. Before World War II, engineers had no efficient method of generating the large number of numerical solutions necessary to determine how solutions vary with the parameters of a physical system. Graphical solutions were impractical since individual parameters lost their identity.

Radio engineers hoped mathematicians could provide a practical theory for nonlinear differential equations that had the same relative simplicity as that for linear differential equations. The engineers desired "an explicit formulation of the dependent variable in terms of the independent variable and the parameters of the system, preferably a formulation in terms of known or tabulated functions in order that the solution and its variation with the different parameters may be evaluated in actual numbers without a prohibitive labour of computation" [33]. Physicists and engineers encountering the systems in their experimental work had attempted some research on simple nonlinear equations such as

$$\frac{d^2y}{dx^2} + f(y)\frac{dy}{dx} + ky = 0$$

where $f(y)$ is a polynomial. However, mathematical papers on the systematic study of nonlinear differential equations were sparse. The memorandum suggested that the lack of published work in the field might have been due to a "deficiency of liaison between mathematicians on the one hand and engineers and physicists on the other" and anticipated that the nonlinear theory might be complicated. They wanted to know if the latter were the case in order to prevent a "waste of time and energy spent in pursuit of a will-o'-the-wisp" [33]. The response from mathematicians soon convinced the engineers that the nonlinear phenomena were quite complicated and no easily applicable general theory could be expected [9].

Radio engineers found few practical applications for the theories developed by mathematicians and concluded that experimental methods would be more effective. Mathematicians continued their research since they found the nonlinear systems interesting and suggestive. The techniques they developed provided part of the foundation of modern dynamical theory.

One basic electronic building block of early twentieth century circuitry was the triode oscillator. (See Figure 1.) To get the differential equation for the triode oscillator in its simplest form the load resistance is represented by a parallel resistance. Under certain assumptions on the currents [35], the differential equation for the free oscillations of a triode oscillator is

$$\frac{d^2v}{dt^2} + \frac{1}{C}\frac{d}{dt}\left\{\frac{v}{R} + \psi(kv)\right\} + \frac{1}{LC}v = 0 \tag{1}$$

where $i = \psi(kv)$. Van der Pol approximates $i = \psi(kv)$ over the domain of the characteristic in which the oscillation is to take place by the third degree polynomial

$$i = \psi(kv) = -\alpha v + \beta v^2 + \gamma v^3.$$

If one does not take into account the curvature of the characteristic then the solution is an exponentially increasing oscillation. Nonlinear terms are necessary to obtain an oscillation of constant amplitude. If the triode is working on an inflection point of its characteristic then the coefficient β of the second order term is zero. This term is not essential for an analysis of most of the oscillation phenomena discussed by van der Pol. By appropriately scaling the variables t and ν and making suitable substitutions, equation (1) is simplified to the familiar van der Pol equation

$$\ddot{v} - k(1 - v^2)\dot{v} + v = 0. \tag{2}$$

According to van der Pol, equation (2) may be regarded as the basic equation for the triode oscillator [35]. The forced van der Pol equation,

$$\ddot{v} - k(1 - v^2)\dot{v} + v = b\lambda k \sin \lambda t \tag{3}$$

is considered the basic equation for an oscillator with an impressed periodic electromotive force. Although the models are not completely realistic, they retain the essential features of the situation.

The problems presented by the Radio Research Board emphasized the steady state solution. The objective was to determine which parameter values would lead to periodic or almost periodic solutions and how the frequency or frequencies varied with the parameters of the circuit. Cartwright thought the problems appeared interesting and began her investigation using the references provided. One of the most important references in the memorandum was van der Pol's survey [35]. It included eighty-seven references and provided a connected account of the main lines of development in nonlinear oscillation theory since 1920. The paper investigated the behavior of solutions to several differential equations including those modeling triode oscillators with free oscillations, coupled circuits, and relaxation oscillators with and without impressed electromotive force. He explored general approximation methods for second order equations with small nonlinearities and aimed to treat the subject from a physical point of view without stressing mathematical rigor. He also described Soviet efforts to apply the ideas of Poincaré to the singular points of two dimensional autonomous systems.

Analytical investigations of relaxation oscillations had only tentatively begun at the time of van der Pol's survey. When the nonlinear term of equation (2) or (3) is large, the equation represents a relaxation oscillator. Its periodic solutions are called relaxation oscillations. Resonance is practically absent for relaxation oscillations, so an external electromotive force cannot appreciably influence the amplitude of the relaxation oscillator, but it can easily influence the time period. In addition to their applications to radio work, relaxation oscillations can be used to describe diverse physical phenomena for which sinusoidal oscillations fail to provide satisfactory explanations. Examples include heartbeats, aerodynamic phenomena associated with eddies, epidemics, periodic occurrences of economical crises, some pulsating motion of stars, periodic bursts of stream in geysers, and sudden displacement of tectonic plates. The questions raised by investigations of relaxation oscillators particularly interested Cartwright.

Cartwright brought several interesting problems that she found in the engineering literature to Littlewood's attention. She thought he might help her understand the dynamics of the situation [18]. He seemed to understand the dynamical aspects of radio work and he immediately solved or suggested methods of solutions for most of the problems. However, he found certain problems more difficult, and therefore more interesting.

The first problem attacked by Cartwright and Littlewood concerned the amplitude of the stable periodic solution of equation (2), the unforced van der Pol equation. Using the isocline method, van der Pol had obtained graphical solutions for the cases $k = 0.1$, 1, and 10. ([35], p.1077 or [32], p.839) Each graph appeared to show convergence to a periodic solution with amplitude two. Littlewood showed that when k is small and positive, equation (2) has one periodic solution of amplitude $2 + O(k)$. Cartwright and Littlewood showed in [8] that when k is large all solutions of equation (2), except the trivial solution $x = 0$, converge to a periodic solution whose amplitude tends to 2 as $k \to \infty$.

Cartwright and Littlewood next attacked van der Pol's equation with a forced oscillation in the form

$$\ddot{x} - k(1 - x^2)\dot{x} + x = bk\lambda \cos \lambda t. \tag{4}$$

Cartwright and Littlewood were particularly interested in the possibility that equation (4) could have two stable periodic solutions with different periods when k was large. Van der Pol and van der Mark had constructed a system based on equation (4) with k large in which they produced subharmonics of the forced oscillation [36]. When the relative frequency of the driving force λ was increased and decreased they observed that two subharmonics of different orders, one even and one odd, occurred for the same values of the parameters. Although the existence of an even subharmonic indicated that the experimental system was unsymmetrical, van der Pol thought it likely that a similar phenomenon should occur for the symmetrical equation (4). The suggestion provided a starting point for Cartwright and Littlewood [12].

Cartwright and Littlewood agree that their most analytically interesting results came from their investigation of equations with two or more stable periodic solutions [16]. They concentrated on van der Pol's equation since it seemed to be the simplest type of equation having two stable periodic oscillations with periods prime to each other. Their results led to some rather interesting mappings. A paper by Levinson [27] stimulated their interest in the mappings and considerably influenced their work. In his paper, Levinson described a general topological approach to periodic second order differential equations. Cartwright and Littlewood expanded their research to the generalized form of van der Pol's equation and their topological interpretation of the problem led to [23], a paper on fixed point theorems. Cartwright and Littlewood also explored second order equations with positive damping. Littlewood originally thought the problems would be straightforward but discovered that some were more difficult than he had anticipated. He told Cartwright, "All details have a nasty way of ramifying into difficulties."[13] Their results were published in

[21], a paper Cartwright claims was much corrected by Littlewood. Her own technique was to "polish, polish, polish" before publishing and it was due in part to incessant changes from both parties that they separated. At that time they agreed that Littlewood would finish the hardest problem and Cartwright would complete the rest of their work. Her results appear in [3,4,5,6,7,8].

Cartwright published a significant portion of her joint work with Littlewood under her name alone. She explained, "With one exception he would not agree to letting his name appear as co-author unless he had vetted the final version" [17]. The exception was [23], their paper on fixed-point theorems. Littlewood gave way on condition that the paper was checked by someone accepted by him. H.D. Ursell was approved and, in the process, he added a result of his own to the paper.

Cartwright and Littlewood published [20], their first paper, at the end of the war. The paper was a preliminary survey of their results regarding van der Pol's equation with forcing term and k large. Littlewood made a special effort to have it published promptly to support Cartwright's candidacy for the Royal Society. The full proofs of the results announced in [20] were not published until twelve years later [28, 29]. Cartwright insisted that Littlewood publish [28] and [29] under his name alone, although she completed the final details of the proof corrections when Littlewood visited America. In each manuscript he affirms that the work was based on joint efforts.

[28] is an intermediate paper designed as an exposition of the main lines of the arguments introduced in [20]. It investigates the behavior of solutions to van der Pol's equation for k large and $0 < b < 2/3$. It remains unemcumbered by some of the more complicated details contained in [29]. [29] investigates the generalized van der Pol equation

$$\ddot{y} + kf(y)\dot{y} + g(y) = bkp(\phi), \qquad \phi = t + \alpha.$$

It is one of Littlewood's most intricate papers. He once said of the work, "It is very heavy going and I should never have read it had I not written it myself." [1]

In [20], Cartwright and Littlewood explained that $b_0 = 2/3$ is a critical value for the parameter b in equation (4). Littlewood carried the study of the case $b > 2/3$ and k large far enough to convince himself that equation (4) has a periodic solution of period $2\pi/\lambda$ toward which all other solutions tend. He never completed the proof for this case. It was completed independently in [31] by Noel Lloyd, whom Cartwright had examined for his D.Phil. In the case $0 < b < 2/3$ and k large[1] subharmonic motions of large orders (compared with k) appeared with the orders decreasing as b increased from 0 to 2/3. In addition, complications arose in the stationary solutions and discontinuous recurrent motions developed. Both stable and unstable periodic solutions of at least period $2\pi n/\lambda$ appeared along with a single unstable periodic solution of period $2\pi/\lambda$. Cartwright and Littlewood showed that the interval $(0, 2/3)$ could be partitioned into three sets of intervals, i.e.,

$$(0, 2/3) = \mathbf{S}_1 + \mathbf{S}_2 + \mathbf{S}_3$$

where $\mathbf{S}_1$ consisted of a set of intervals $\mathbf{B}_1$, $\mathbf{S}_2$ consisted of a set of intervals $\mathbf{B}_2$ that alternate with $\mathbf{B}_1$, and $\mathbf{S}_3$ was a set of short transitional intervals between $\mathbf{B}_1$ and $\mathbf{B}_2$. The length of the intervals of $\mathbf{S}_3$ could be made arbitrarily small by increasing

[1] By k large we mean k is large enough to exclude a linear approximation for equation (4). For example, $k = 5$ is considered large. Experiments showed subharmonics of orders between 100 and 200 corresponding to $k = 5$.

k and appeared to be exponentially short in terms of k. Adjacent $\mathbf{B}_1$ and $\mathbf{B}_2$ are approximately equal in length where the length is on the order of $1/k$. For $b \in \mathbf{S}_1$ there existed a single set of subharmonics of order $2n+1$, where n is on the order of $(2/3-b)\cdot k$, such that almost every solution of (4) converged to some member of the set. In addition, there was a set of nonstable subharmonics of order $2n+1$. Note that given one periodic solution Γ of period $2m\pi$, a set m of trajectories is obtained by translating Γ by $t = 0, 2\pi, \ldots, (m-1)2\pi$. For $b \in \mathbf{S}_2$ there existed two stable sets of subharmonics of order $2n+1$ and $2n-1$ and any solution outside a certain set of measure zero tends to one of these stable solutions. In this case Cartwright and Littlewood also showed existence of an infinite set of nonstable periodic motions having a great variety of "structures" and an infinite set of non-periodic limiting trajectories described as discontinuous recurrent. As b increased from 0 to 2/3, alternating between $\mathbf{B}_1$ and $\mathbf{B}_2$ and jumping the transition zones, the behavior of equation (4) switched from one mode to the other and the value of n decreased by 1 with each pair of switches. The results confirmed experimental results such as those of Donald Herr [26]. His experiments showed that the greater the relative frequency of the driving force, the higher the order of odd subharmonics it was possible to produce. Herr's experiments also showed that in going from drifting oscillations to the synchronized state that synchronization was suddenly effected on the nearest odd subharmonic.[2]

At the time of their collaboration, Cartwright and Littlewood could determine very little about the narrow transition zones of $\mathbf{S}_3$. They were particularly interested in the strange invariant sets created within these zones. Littlewood's paper [29] appears to contain the first detailed account of a strange invariant set and [30] was the first to give any specific information on behavior in the transition zones. It eventually became evident that the van der Pol oscillator provided an example of a stable chaotic system. In 1959 Norman Levinson described this example to Stephen Smale, leading Smale to discover a new way to envision the "full complexity of dynamical systems" [25]. Dissipation in the van der Pol oscillator meant that the system's shape in phase space would contract. A complex combination of transformations was necessary to represent the system in phase space fully, hence Smale's construction of the horseshoe diffeomorphism. In *Chaos*, Gleick recounts that van der Pol had a glimpse of chaotic behavior in his oscillator, but had no basis for understanding what he had found. Gleick explains that radio engineers had no oscilloscopes in the 1920s. They monitored circuits by listening to changing tones in a telephone handset. As van der Pol changed the current that fed his circuit, he noticed an irregular noise in the telephone receiver before the frequency tone jumped to lock onto the next lower value. Van der Pol was unaware that he was hearing the conflicting pulls of a higher and lower frequency. He merely thought that what he heard was a "subsidiary phenomenon" [36].

Cartwright and Littlewood observed that the existence of two stable sets of subharmonics of different orders led to complex topological structures. At the boundary between the domains of attraction of the two sets there exists an intricate fine structure of highly nonstable trajectories. The inner and outer edges of the boundary have different rotation numbers under the diffeomorphism corresponding to advancing the solutions of (4) through one period. Therefore, the boundary

[2] A synchronized oscillation is one in which the period is some fixed integer multiple of the impressed electromotive force. A drifting oscillation is one which never achieves the periodic synchronized steady state condition.

cannot be a Jordan curve. Cartwright and Littlewood concluded that the boundary was most likely an indecomposable continuum, which, according to Littlewood, is "about the dirtiest thing one can say about a set of points" [30]. Except for its instability, the structure resembles that of a strange attractor. Although the strange invariant set is not an attractor, the machinery that produces it is the same as that normally used for producing strange attractors.

The homeomorphism of the plane defined by following the trajectories of van der Pol's equation through one period maps the set onto itself. This phenomenon suggested to Cartwright and Littlewood that subharmonics and nonstable periodic solutions of the equation corresponded to fixed points of the set under the homeomorphism or an iterate of it. Existence of these fixed points could be deduced from suitable fixed point theorems. This idea led Cartwright and Littlewood to investigate fixed point theorems for continua invariant under a diffeomorphism of the plane. Such theorems enabled them to deal with van der Pol's equation and the general equation

$$\ddot{y} + f(y)\dot{y} + g(y) = p(t) \tag{5}$$

where p is periodic and f, p, and g are smooth and well-behaved enough to ensure existence and uniqueness of solutions everywhere.

Cartwright and Littlewood's analysis of equation (5) required fixed point theorems that were, in some respects, more general than those available from existing algebraic fixed point theory. Singular invariant curves could not be ruled out, even by reasonable regularity arguments. Fixed point theorems for plane continua that were acyclic but otherwise unrestricted were required. The search for these fixed points led Cartwright and Littlewood to the theorems of [23]. Their analysis used Birkhoff's results for existence of an analytic transformation of the plane into itself with a complicated invariant curve K. The solutions of van der Pol's equation set up such a transformation. Cartwright and Littlewood used these transformations to show existence of stable periodic subharmonic motions, unstable motions, and recurrent motions each associated with some subset of K. They appear to be the first mathematicians to apply Birkhoff's work to a simple differential equation. Their technique is briefly described below.

Consider equation (5). Assume p is periodic with period λ; f and p are continuous; g satisfies a Lipschitz condition; $g(\xi)/\xi \geq 1$ for large ξ; and $\liminf_{y\to\pm\infty} f > 0$ so that (5) has positive damping for large $|y|$. Note that equation (4) satisfies these conditions. A *representative point* of a trajectory is defined as the point $P = (\xi, \eta)$ where $\xi = y(t_0)$ and $\eta = y'(t_0)$. If $P' = (\xi', \eta')$, where $\xi' = y(t_0 + \lambda)$ and $\eta' = y'(t_0 + \lambda)$, then the transformation $T \colon P \to P'$ is one-to-one, continuous and bounded as $t \to \infty$. T is also orientation preserving in the open ξ, η-plane. The fixed points of T correspond to periodic solutions of equation (5) with period λ. A subharmonic solution of order m corresponds to any point y satisfying the condition that $y = T^m(y)$ and $y \neq T^n(y)$ for $n = 1, 2, \ldots, m-1$. Levinson showed that there exists a closed Jordan curve $J = J(y)$ through any point y sufficiently near the point at infinity such that $T(J)$ lies in $D^{(i)}(J)$, the interior domain of J. This implied the existence of an acyclic bounded continuum $I_L = \prod_{n=1}^{\infty} T^n(D^{(i)}(J)) = \prod_{n=1}^{\infty} D_n^{(i)}(J)$ invariant under T. If the plane is made compact by the addition of the point at infinity, then the complement $C(I_L)$ is a single simply connected domain. It follows from Brouwer's fixed point theorem that $D_n^{(i)}$ contains at least one fixed point for each n. Hence I_L contains a fixed point corresponding to a solution of equation (5) with period λ. However, this solutions is sometimes unstable. This is the case for van der Pol's equation (4).

A fixed point P_0 and its corresponding periodic motion are stable if $T^m P \to P_0$ as $m \to \infty$ for all P in a neighborhood of P_0. Therefore, the points of greatest interest belong to the class K of limiting trajectories, i.e., the class whose representative points are the set of limit points of $T^m(\xi, \eta)$ as $m \to \infty$. This class is invariant under T and the trajectories are bounded in $-\infty < t < \infty$. The simplest case is if K consists of a single point. Then there exists a stable periodic motion of order one toward which every solution converges. The next simplest case occurs if there exist a finite number of periodic motions to some one of which every other solution converges. In general, however, ruling out potentially troublesome topological possibilities can be very difficult.

Since Cartwright and Littlewood were interested in finding stable periodic solutions, they proceeded to investigate fixed point theorems for points corresponding to subharmonic solutions of (5). In [23] they explain that since $C(I_L)$ is a simply connected domain it can be mapped $(1,1)$ and conformally on $|z| > 1$ if the point at infinity is made to correspond to itself. Denote this mapping by μ. Then $\tau = \mu T \mu^{-1}$ sets up a $(1,1)$ continuous mapping of $|z| > 1$ onto itself and τ remains $(1,1)$ and continuous on $|z| = 1$. Now if τ leaves one point fixed on $|z| = 1$, then every point on $|z| = 1$ is either fixed or tends to a fixed point. A point on $|z| = 1$ may correspond to a single point on the frontier $F\big(C(I_L)\big) = F(I_L)$ of $C(I_L)$ or to a prime end defined according to Carathéodory's theory by a chain of cross cuts converging to a point on $F(I_L)$. The cross cuts determine either a single point or a continuum P on $F(I_L)$. According to Cartwright and Littlewood, this raises the question: If P corresponds to a fixed point on $|z| = 1$, do P and/or $F(I_L)$ necessarily contain fixed points? To address this question they needed a fixed point theorem that would apply to a set I that was invariant under more general conditions than I_L, in particular, a bounded invariant continuum whose complement is a single simply connected domain that may not have the Jordan curve tending to it. This was Theorem 5 of [23] and it states:

THEOREM 0.1. *If T is a $(1,1)$ continuous and orientation preserving transformation of the whole plane into itself which leaves a bounded continuum I invariant, and if $C(I)$ is a single simply connected domain, then I contains a fixed point.*

Cartwright says this is the deepest theorem of [23]. It is more general than Brouwer's fixed point theorem since I need not be locally connected [11]. However, the conditions on T are more restrictive. In addition to Theorem 1, [23] includes theorems for fixed points on the frontier of an invariant set I.

As with Cartwright and Littlewood, American and Soviet interest in the mathematical analysis of electrical circuits was inspired by applications connected with the war. Unfortunately, a great deal of concurrent foreign mathematical research was not available to Cartwright and Littlewood until the conclusion of World War II. American mathematicians including Lefschetz, Levinson, and Minorsky, were working toward a general mathematical theory. The Americans put greater emphasis on the amplitude and period of oscillations, rather than frequencies and parameters. They were particularly interested in the boundedness of solutions to the nonlinear differential equations associated with electrical circuits. Soviet mathematicians had made progress on nonlinear systems with several degrees of freedom, on the stability of regulating systems, and on other related topics. Kryloff, Bogoliuboff, Mitropolsky and other Russian mathematicians were analyzing frequency demultiplication in sinusoidal systems with the help of perturbation theory and employing averaging methods to analyze equations such as van der Pol's.

In America, Lefschetz expressed interest in the work of Cartwright and Littlewood. He invited Cartwright to lecture on nonlinear differential equations in the United States from January to June of 1949. She spent three weeks at Stanford, one week at UCLA, and the rest at Princeton. Cartwright published her series of lectures at Princeton [6]. The paper provides a detailed overview of her joint work with Littlewood and includes pertinent work of others in the field. It is written on a less formal level and includes more details than most of their other papers. In [6] Cartwright discusses both solved and unsolved problems associated with equation (4) and its generalizations. She discusses relevant topological background and analytical devices used to show that the topological transformations associated with the differential equation satisfy various required hypotheses. She also discusses nearly linear oscillations using known methods and the work of herself and Littlewood with difference equations.

Cartwright found it remarkable that van der Pol's experimental results guided her and Littlewood to so many unsolved problems in topology [9]. However, she does not find it surprising when one considers the qualitative behavior of solutions as a topological problem. She stressed that the topological interpretation of a problem can give one valuable insight, even when one is unfamiliar with topological methods ([14],[19]). The original ideas of Cartwright and Littlewood led them to produce several papers, in particular [8], [20], [21], [22], [28], and [29], that are among the earliest fully rigorous publications in large parameter theory. Work such as theirs helped build the firm foundation on which the modern theory of dynamical systems is being built.

References

1. J. C. Burkill, *John Edensor Littlewood*, Bull. London Math. Soc. **11** (1979), 59–103.
2. M. L. Cartwright, *Some inequalities in the theory of functions*, Math. Annalen **111** (1935), 98–118.
3. M. L. Cartwright, *Forced oscillations in nearly sinusoidal systems*, J. Inst. Elec. Engrs. III **95** (1948), 88–96.
4. M. L. Cartwright, *Topological aspects of forced oscillations*, Research **1** (1948), 601–666.
5. M. L. Cartwright, *On non-linear differential equations of the second order III. The equation $\ddot{x} - k(1-x^2)\dot{x} + x = p \cdot k\lambda\cos(\lambda t + \alpha)$, k small and λ near* 1, Proc. Camb. Phil. Soc. **45** (1949), 495–501.
6. M. L. Cartwright, *Forced oscillations in nonlinear systems*, Contributions to the theory of non-linear oscillations, ed. S. Lefschetz, Annals of Math. Studies **20** (Princeton) (1950), 149–241.
7. M. L. Cartwright, *Forced oscillations in nonlinear systems*, J. Research Nat. Bur. Standards **45** (1950), 514–518.

8. M. L. Cartwright, *Van der Pol's equation for relaxation oscillations*, Contributions to the theory of non-linear oscillations II, ed. S. Lefschetz, Annals of Math. Studies **29** (Princeton) (1952), 3–18.
9. M. L. Cartwright, *Non-linear vibrations: a chapter in mathematical history*, Presidential address to the Mathematical Association, January 3rd, 1952, Math. Gazette **36** (1952), 81.
10. M. L. Cartwright, *Some aspects of the theory of non-linear vibrations*, Proc. International Congress of Mathematicians, Amsterdam, 1954, Vol. 3, 71–76, North Holland, Amsterdam, 1956.
11. M. L. Cartwright, *Some decomposition theorems for certain invariant continua and their minimal sets*, Fund. Math **48** (1959/60), 229–250.
12. M. L. Cartwright, *Balthazar van der Pol*, J. Lond. Math. Soc. **35** (1960), 367–376.
13. M. L. Cartwright, *From non-linear oscillations to topological dynamics*, Presidential address to the London Mathematical Society, 1963, J. Lond. Math. Soc. **39** (1964), 193–210.
14. M. L. Cartwright, *Topological problems of nonlinear mechanics*, Konferenz über Nichtlineare Schwingungen, 1964, Berlin, Abh. Deutsch. Akad. Wiss. Berlin Kl. Math. Phys. Tech. (1965) No. 1, 135–142.
15. M. L. Cartwright, *Mathematics and Thinking Mathematically*, Samuel Newton Lecture at Goucher College, January 30th, 1969, Amer. Math. Monthly **77** (1970), 20–28.
16. M. L. Cartwright, *Some points in the history of nonlinear oscillations*, Bull. Inst. Math. Appl. **10** (1974), 329–333.
17. M. L. Cartwright, *Some exciting mathematical episodes involving J.E.L.*, Papers presented at the Symposium on Excitement in Mathematics (Cambridge 1975), Bull. Inst. Math. Appl. **12** (1976), 201–202.
18. M. L. Cartwright, *John Edensor Littlewood*, Bull. Inst. Math. Appl. **14** (1978), 87–90.
19. M. L. Cartwright, E. T. Copson, and J. Grieg, *Non-linear vibrations*, Advancement Sci. **6** No. 21 (1949), 64–75.
20. M. L. Cartwright and J. E. Littlewood, *On nonlinear differential equations of the second order: I. The equation* $\ddot{y} - k(1-y^2)\dot{y} + y = (b\mu k)\cos(\mu t + \alpha)$, *k large*, J. Lond. Math. Soc. **20** (1945), 180–189.
21. M. L. Cartwright and J. E. Littlewood, *On nonlinear differential equations of the second order: II. The equation* $\ddot{y}+kf(y)\dot{y}+g(y,k) = p(t) = p_1(t)+kp_2(t)$; $k > 0,\ f(y) \geq 1$, Annals of Math. (2) **48** (1947), 472–494.
Errata: Ann. of Math. (2) **49** (1948), 1010.
22. M. L. Cartwright and J. E. Littlewood, *Addendum to [18]*, Ann. of Math. (2) **50** (1949), 504–505.
23. M. L. Cartwright and J. E. Littlewood, *Some fixed point theorems*, with appendix by H. D. Ursell, Ann. of Math. (2) **54** (1951), 1–37.
24. M. L. Cartwright and J. E. Littlewood, *Some topological problems connected with forced oscillations*, Proc. International Congress of Mathematicians, Cambridge, Mass., 1950, Vol. 1, 429, A.M.S., Providence, R.I. 1952.
25. J. Gleick, *Chaos*, Viking, New York, 1987, pp. 44–53.
26. D. L. Herr, *Oscillations in certain nonlinear driven systems*, Proc. Inst. Radio Engrs **27** (1939), 396–402.

27. N. Levinson, *Transformation theory of nonlinear differential equations of the second order*, Annals of Math. **45** (1944), 723–737.
28. J. E. Littlewood, *On nonlinear differential equations of the second order: III. The equation $\ddot{y} - k(1-y^2)\dot{y} + y = (b\mu k)\cos(\mu t + \alpha)$ for large k and its generalizations*, Acta Math. **97** (1957), 267–308.
29. J. E. Littlewood, *On nonlinear differential equations of the second order: IV. The equation $\ddot{y} + kf(y)\dot{y} + g(y) = bkp(\phi)$, $\phi = t + \alpha$*, Acta Math. **98** (1957), 1–110.
30. J. E. Littlewood, *On van der Pol's equation with large k*, 161–165.
31. N. G. Lloyd, *On the non-autonomous van der Pol equation with large parameter*, Proc. Camb. Phil. Soc. **72** (1972), 213–227.
32. S. L. McMurran and J. J. Tattersall, *The mathematical collaboration of M. L. Cartwright and J. E. Littlewood*, Amer. Math. Monthly **103** (1996) No. 12, 833-845.
33. Radio Research Board, *A note on certain types of non-linear differential equations involved in radio engineering*, Math. Gazette **22** (1938), 217–218.
34. S. Smale, *Differential and Combinatorial Topology*, Princeton, 1965, 63–80.
35. B. van der Pol, *The nonlinear theory of electric oscillations*, Proc. Inst. Radio Engineers **22** (1934), 1051–1086.
36. B. van der Pol and J. van der Mark, *Frequency demultiplication*, Nature **120** (1927), 363-364.

UNIVERSITY OF REDLANDS, DEPARTMENT OF MATHEMATICS, REDLANDS, CA 92374
E-mail address: `mcmurran@uor.edu`

PROVIDENCE COLLEGE, PROVIDENCE, RI 02918
E-mail address: `tat@providence.edu`

Contemporary Mathematics
Volume **208**, 1997

One–sided Resonance for a Quasilinear Variational Problem

Adolfo J. Rumbos and Victor L. Shapiro

Abstract. We present some new existence results for a quasilinear elliptic problem with an unbounded driving force. The quasilinear elliptic operator is assumed to be variational, and is such that 0 acts like an isolated eigenvalue with a corresponding eigenfunction which does not change sign. The driving force is assumed to be in one–sided resonance around the eigenvalue 0, and a solvability condition of potential type is imposed. Variational methods are used to obtain existence.

1. Introduction

Let Ω be a bounded domain in $\mathbf{R}^N$, $N \geq 1$, with a continuous boundary, and satisfying the cone property, i.e, there exists a finite cone C such that each point x in Ω is a vertex of a finite cone C_x contained in Ω and congruent to C. Denote by D^α the differential operator

$$\frac{\partial^{|\alpha|}}{\partial x_1^{\alpha_1} \partial x_2^{\alpha_2} \cdots \partial x_N^{\alpha_N}}$$

where $\alpha = (\alpha_1, \alpha_2, \ldots, \alpha_N)$ is a multi–index consisting of nonnegative integers, and $|\alpha| = \sum_{j=1}^{N} \alpha_j$ denotes the order of D^α. In order to write nonlinear partial differential operators in a convenient form, we introduce as in [Br] the vector space $\mathbf{R}^{s_m}$ whose elements are of the form $\xi_m = \{\xi_\alpha \mid |\alpha| \leq m\}$. For each $u \in W^{m,p}(\Omega)$, define $\xi_m(u)(x)$ to be the vector in $\mathbf{R}^{s_m}$ given by $\{D^\alpha u(x) \mid |\alpha| \leq m\}$. (Observe that $D^{(0,0,\ldots,0)}u = u$.) In this paper we study the $2m$–th order quasilinear differential operator in generalized divergence form

$$Q(u) = \sum_{1 \leq |\alpha| \leq m} (-1)^{|\alpha|} D^\alpha A_\alpha(x, \xi_m'(u)), \tag{1}$$

1991 *Mathematics Subject Classification.* Primary: 35A15, 35J35; Secondary: 47H05.

Key words and phrases: Monotone operators, one–sided resonance, potential type, quasilinear elliptic, variational methods.

This paper is in final form and no version of it will be submitted for publication elsewhere.

where ξ'_m denotes the vector $\{\xi_\alpha \mid 1 \le |\alpha| \le m\}$ in $\mathbf{R}^{s_m - 1}$.

We will assume that Q has a variational structure in the sense that there exists a function $F: \Omega \times \mathbf{R}^{s_m-1} \to \mathbf{R}$ satisfying

(F–1) The map $x \to F(x, \xi'_m)$ is measurable for each $\xi'_m \in \mathbf{R}^{s_m-1}$, and the map $x \to F(x, \xi'_m)$ is continuously differentiable for a.e. $x \in \Omega$.

(F–2) There exist constants p and c_1, with $1 < p < \infty$ and $c_1 > 0$, and a nonnegative function $h \in L^1(\Omega)$ such that

$$|F(x, \xi')| \le h(x) + c_1 |\xi'|^p$$

for a.e. $x \in \Omega$ and all $\xi'_m \in \mathbf{R}^{s_m-1}$.

(F–3) $F(x, 0) = 0$ for a.e. $x \in \Omega$, and for each α, with $1 \le |\alpha| \le m$,

$$\frac{\partial F}{\partial \xi_\alpha}(x, \xi') = A_\alpha(x, \xi') \quad \text{for } (x, \xi') \in \Omega \times \mathbf{R}^{s_m-1}.$$

The functions $A_\alpha: \Omega \times \mathbf{R}^{s_m-1} \to \mathbf{R}$ defined in (F–3) above will be assumed to satisfy the following conditions

(A–1) There exists a constant c_2, with $c_2 > 0$, and a nonnegative function $\widetilde{h} \in L^{p'}(\Omega)$, where $p' = p/(p-1)$ and p is as in (F–2) above, such that

$$|A_\alpha(x, \xi'_m)| \le \widetilde{h}(x) + c_2 |\xi'_m|^{p-1} \qquad 1 \le |\alpha| \le m$$

for a. e. $x \in \Omega$ and for all $\xi'_m \in \mathbf{R}^{s_m-1}$.

(A–2) (*Ellipticity*) There exists a positive constant c_o such that

$$\sum_{1 \le |\alpha| \le m} A_\alpha(x, \xi'_m) \xi_\alpha \ge c_o \sum_{1 \le |\alpha| \le m} |\xi_\alpha|^p$$

for a. e. $x \in \Omega$ and for all $\xi'_m \in \mathbf{R}^{s_m-1}$, where p is as in (F–2).

(A–3) (*Monotonicity*) Let $\xi'_m = (\eta'_{m-1}, \zeta_m)$ be the division of ξ'_m into its mth order component and the corresponding $(m-1)$st order terms η'_{m-1}; i.e., $\eta'_{m-1} = \{\xi_\beta \mid 1 \le |\beta| \le m-1\} \in \mathbf{R}^{s_{m-1}-1}$, and $\zeta_m = \{\xi_\alpha \mid |\alpha| = m\}$. Put $A_\alpha(x, \xi'_m) = A_\alpha(x, \eta'_{m-1}, \zeta_m)$. Assume that for a.e. $x \in \Omega$ and each $\eta'_{m-1} \in \mathbf{R}^{s_{m-1}-1}$,

$$\sum_{|\alpha|=m} [A_\alpha(x, \eta'_{m-1}, \zeta_m) - A_\alpha(x, \eta'_{m-1}, \zeta^*_m)](\zeta_\alpha - \zeta^*_\alpha) > 0$$

for $\zeta_m \ne \zeta^*_m$ and a.e. $x \in \Omega$.

We define the following semilinear Dirichlet form

$$\mathcal{Q}(u, v) = \sum_{1 \le |\alpha| \le m} \int_\Omega A_\alpha(x, \xi'_m(u)) D^\alpha v \qquad \forall\, u, v \in W^{m,p}(\Omega). \tag{2}$$

In view of (A–1) we see that $\mathcal{Q}$ is well defined on $W^{m,p}(\Omega) \times W^{m,p}(\Omega)$.

Observe that by the definition of $\mathcal{Q}$ in (2) and (A–2) we get

$$\mathcal{Q}(u,u) \geq c_o \int_\Omega \sum_{1\leq|\alpha|\leq m} |D^\alpha u|^p \tag{3}$$

for all $u \in W^{m,p}(\Omega)$, so that $\liminf_{||u||_{L^p}\to\infty} \frac{\mathcal{Q}(u,u)}{||u||^p_{L^p}} \geq 0$. Define as in [S1, p. 1821]

$$\lambda_1 = \liminf_{||u||_{L^p}\to\infty} \frac{\mathcal{Q}(u,u)}{||u||^p_{L^p}}.$$

Since $\mathcal{Q}(u,u) = 0$ for u constant, we see that $\lambda_1 = 0$. On the other hand, for $v \in W^{m,p}(\Omega)$ such that $\int_\Omega v = 0$ we obtain from (3) and the Poincaré inequality (see [Ne, p. 32]) that

$$\mathcal{Q}(v,v) \geq \frac{c_o}{c} \int_\Omega |v|^p \tag{4}$$

where c is the best constant that appears in the Poincaré inequality:

$$\int_\Omega |v|^p \leq c \sum_{|\alpha|=1} \int_\Omega |D^\alpha v|^p \tag{5}$$

for all $v \in V = \{v \in W^{m,p}(\Omega) \mid \int_\Omega v = 0\}$; i.e., c is given by

$$\frac{1}{c} = \inf_{v\in V\setminus\{0\}} \frac{\sum_{|\alpha|=1}\int_\Omega |D^\alpha v|^p}{\int_\Omega |v|^p},$$

where V is the set $\{v \in W^{m,p}(\Omega) \mid \int_\Omega v = 0\}$. Define

$$\lambda_2 = \liminf_{||v||_{L^p}\to\infty} \frac{\mathcal{Q}(v,v)}{||v||^p_{L^p}} \qquad v \in V, \tag{6}$$

where $V = \{v \in W^{m,p}(\Omega) \mid \int_\Omega v = 0\}$. By (4) above we get that $\lambda_2 \geq c_o/c > 0$.

In this paper we deal with the solvability of the following problem

$$\begin{cases} Qu = -a(\cdot,u)|u|^{p-2}u^- + g(\cdot,u) + H & \text{a.e. in } \Omega \\ u \in W^{m,p}(\Omega) \end{cases} \tag{7}$$

where p is as in (F–2), $u^- = \max\{-u, 0\}$ is the negative part of u, $H \in W^{m,p}(\Omega)^*$, the dual of $W^{m,p}(\Omega)$, $g: \Omega \times \mathbf{R} \to \mathbf{R}$ is a function satisfying

(g–1) the Carathéodory conditions: the map $x \to g(x,s)$ is measurable for all $s \in \mathbf{R}$, and the map $s \to g(x,s)$ is continuous for a. e. $x \in \Omega$;

(g–2) g is bounded; that is, there exists constant M such that

$$|g(x,s)| \leq M \qquad \text{for all } s \in \mathbf{R} \text{ and a. e. } x \in \Omega;$$

and $a: \Omega \times \mathbf{R} \to \mathbf{R}$ is a function satisfying

(a–1) the Carathéodory conditions as in (g–1).

(a–2) $0 \leq a(x,s) \leq \gamma$ for a. e. $x \in \Omega$ and all $s \in \mathbf{R}$, where $0 < \gamma < \lambda_2$, with λ_2 being the constant given in (6).

(a–3) $\liminf_{s\to-\infty} a(x,s) = \gamma_1$ uniformly for a.e. $x \in \Omega$ and some $\gamma_1 > 0$. To be explicit, we mean that there exists $E \subset \Omega$ with $\text{meas}(E) = 0$ such that (i) $\liminf_{s\to-\infty} a(x,s) = \gamma_1$ for $x \in \Omega \backslash E$; and (ii) given $\varepsilon > 0$, there exists s_o such that for $s < s_o$, $a(x,s) > \gamma_1 - \varepsilon$ for $x \in \Omega\backslash E$.

By a solution to problem (7) we mean a weak or generalized solution; i.e., a function $u \in W^{m,p}(\Omega)$ satisfying

$$\mathcal{Q}(u,v) = -\int_\Omega a(x,u)|u|^{p-2}u^- v + \int_\Omega g(x,u)v + H(v), \tag{8}$$

for all $v \in W^{m,p}(\Omega)$.

We will prove the following result:

THEOREM 1. *Let $1 < p < \infty$ and let Ω be an open bounded connected set with continuous boundary and satisfying the cone property. Assume (F–1)–(F–3), and suppose that $\mathcal{Q}(u,v)$ is given by (2) where $A_\alpha(x,\xi_m')$ satisfies (A–1)–(A–3) for $1 \leq |\alpha| \leq m$. Let $g(x,s)$ satisfy (g–1)–(g–2), $a(x,s)$ satisfy (a–1)–(a–3), and $H \in W^{m,p}(\Omega)^*$, the dual of $W^{m,p}(\Omega)$. If*

$$\lim_{t\to\infty}\left\{\int_\Omega G(x,t) + H(t)\right\} = +\infty, \tag{9}$$

where $G(x,s) = \int_0^s g(x,t)\,dt$, then problem (7) has at least one solution.

In the special case in which g is of the form $g(x,s) = b(x)f(s)$ for $(x,s) \in \Omega\times\mathbf{R}$, where $b \in C(\Omega) \cap L^\infty(\Omega)$, $b(x) > 0$ for all $x \in \Omega$, $f: \mathbf{R} \to \mathbf{R}$ is continuous and satisfies $\lim_{s\to\infty} f(s) = f_+$ exists, with $f(s) < f_+$ for all $s \in \mathbf{R}$, one can show that if (7) has a solution then (9) holds. To see this, let $u \in W^{m,p}(\Omega)$ be a solution of (7) and take $v = 1$ in (8) to get

$$\int_\Omega a(x,u)|u|^{p-2}u^- = \int_\Omega b(x)f(u) + H(1).$$

So that by (a–2)

$$\int_\Omega b(x)f(u) + H(1) \geq 0.$$

Hence, using the assumptions $b > 0$ in Ω and $f(s) < f_+$ for all $s \in \mathbf{R}$, we get

$$f_+\int_\Omega b(x) + H(1) > 0$$

which can be shown to imply (9). Consequently, we obtain the following necessary and sufficient conditions for the solvability of problem (7):

THEOREM 2. *In addition to the hypotheses of Theorem 1, assume that* $g(x,s) = b(x)f(s)$ *for* $(x,s) \in \Omega \times \mathbf{R}$, *where* $b \in C(\Omega) \cap L^\infty(\Omega)$, $f\colon \mathbf{R} \to \mathbf{R}$ *is continuous, and* $b(x) > 0$ *for all* $x \in \Omega$. *If* $\lim_{s\to\infty} f(s) = f_+$ *exists, and*

$$f(s) < f_+ \qquad \text{for all } s \in \mathbf{R},$$

then condition (9) is necessary and sufficient for the solvability of problem (7).

REMARK 1. In the case that $p = 2$ in problem (7), let

$$g_1(x,s) = -a(x,s)s^- + g(x,s)$$

for a.e. $x \in \Omega$ and all $s \in \mathbf{R}$. Conditions (g–2) and (a–2) imply for g_1 that

$$0 \le \liminf_{|s|\to\infty} \frac{g_1(x,s)}{s} \le \limsup_{|s|\to\infty} \frac{g_1(x,s)}{s} \le \gamma < \lambda_2 - \lambda_1 \quad \text{for a.e. } x \in \Omega.$$

This is a one–sided resonance condition around the first eigenvalue in the case in which Q is a linear operator and $m = 1$. We are therefore well motivated to call problem (7) a one–sided resonance problem. We also note from (A–2) that if u is not constant, then

$$\liminf_{t\to\infty} \frac{Q(tu,tu)}{||tu||^p_{L^p}} > 0,$$

so that $\lambda_1 = 0$ acts as a simple eigenvalue.

REMARK 2. The solvability condition (9) in Theorem 1 was first introduced by Ahmad, Lazer, and Paul [ALP] to deal with resonance problems for bounded nonlinear perturbations of linear second order self–adjoint elliptic operators with Dirichlet boundary condition. A similar condition was used by Castro and Lazer [CL, Theorem 3, p. 148] to deal with nonlinear perturbations of the Neumann problem, in the case $p = 2$, in which the nonlinearity is assumed to be differentiable with derivative bounded from above. There is no result in the literature even for the case $p = 2$, other than the one presented here, that we know of, that treats this kind of condition with both an unbounded driving force and no boundedness assumption on the derivative of the driving force. The last example presented in Section 4 of this paper, which is covered by Theorem 1, has an unbounded driving force with a derivative, where it exists, with values going both to $\pm\infty$; i.e., the derivative is unbounded in both directions. In particular, for the case $p = 2$, this example is not covered by the Castro–Lazer result in [CL], but it is covered by our Theorem 1.

REMARK 3. There is a connection between problem (7) and the Neumann problem for Q. To see this connection we refer the reader to [S2, pp. 365–367].

REMARK 4. We shall present various examples of functions F which meet the conditions in (F–1)–(F–3), as well as examples of nonlinearities for which the solvability condition in (9) holds, in Section 4 of this paper.

2. Variational Setting and Preliminary Lemmas

Throughout this paper we will use the norm in $W^{m,p}(\Omega)$ given by

$$||u||^p_{m,p} = \sum_{|\alpha|\leq m} ||D^\alpha u||^p_{L^p}, \tag{10}$$

where $||\cdot||_{L^p}$ denotes the L^p norm. We will also be using the seminorm

$$|u|'_{m,p} = \left\{ \sum_{1\leq|\alpha|\leq m} ||D^\alpha u||^p_{L^p} \right\}^{1/p}. \tag{11}$$

Observe that in view of the Poincaré inequality [Ne, p. 32] the seminorm $|\cdot|'_{m,p}$ actually defines a norm in the space $V = \{v \in W^{m,p}(\Omega) \mid \int_\Omega v = 0\}$. Furthermore, in the space V, the norms defined by (10) and (11) are equivalent; in fact,

$$\frac{1}{1+c}||v||^p_{m,p} \leq |v|'^p_{m,p} \leq ||v||^p_{m,p} \qquad \forall v \in V, \tag{12}$$

where c is the constant that appears in the Poincaré inequality; see (5).

For $a(x,s)$ satisfying (a–1) and (a–2), set

$$A(x,s) = \begin{cases} \int_0^s a(x,t)|t|^{p-2}t dt & \text{for } s \leq 0 \\ 0 & \text{for } s > 0, \end{cases} \tag{13}$$

and for g satisfying (g–1) and (g–2), put

$$G(x,s) = \int_0^s g(x,t)dt \qquad \text{for all } s \in \mathbf{R}. \tag{14}$$

Let F be as given in (F–1)–(F–3). Define a functional $\mathcal{J}: W^{m,p}(\Omega) \to \mathbf{R}$ by

$$\mathcal{J}(u) = \int_\Omega F(x, \xi'_m(u)) - \int_\Omega A(x,u) - \int_\Omega G(x,u) - H(u) \tag{15}$$

for all $u \in W^{m,p}(\Omega)$. By virtue of (F–1)–(F–2), (g–1)–(g–2), and (a–1)–(a–2) we see that $\mathcal{J}$ is well defined and continuous. Observe also that if (F–2), (g–2) and (a–2) hold then $\mathcal{J}$ maps bounded sets in $W^{m,p}(\Omega)$ to bounded sets in $\mathbf{R}$. Moreover, using (F–3), (A–1), (g–2) and (a–2) we can show as in [Br, p. 35] that, in fact, $\mathcal{J} \in C^1(W^{m,p}(\Omega), \mathbf{R})$, and that its Fréchet derivative is given by

$$\mathcal{J}'(u)v = \mathcal{Q}(u,v) + \int_\Omega a(x,u)|u|^{p-2}u^- v - \int_\Omega g(x,u)v - H(v) \tag{16}$$

for all $u, v \in W^{m,p}(\Omega)$.

A critical point of $\mathcal{J}$ is a function $u \in W^{m,p}(\Omega)$ for which $\mathcal{J}'(u)v = 0$ for all $v \in W^{m,p}(\Omega)$. In view of (16) above, this corresponds precisely to the definition (8) of a solution for problem (7). Therefore, in order to prove Theorem 1 it suffices to show that $\mathcal{J}$ possesses at least one critical point. We will establish this fact by means of the Rabinowitz Saddle Point Theorem [Ra, Theorem 4.6, p. 24].

In order to put Theorem 1 in the setting of Theorem 4.6 in [Ra], we define $W = \text{span}\{1\}$. It is not hard to see that $W^{m,p}(\Omega) = W \oplus V$, where V is the set $\{v \in W^{m,p}(\Omega) \mid \int_\Omega v = 0\}$. By virtue of this decomposition, every element $u \in W^{m,p}(\Omega)$ can be uniquely written in the form

$$u = \widehat{u}(1) + u_2$$

where $\widehat{u}(1) \in \mathbf{R}$ and $u_2 \in V$.

LEMMA 1. *Assume (F–1)–(F–3), (A–1)–(A–2), (g–1)–(g–2), (a–1)–(a–3) and let $\mathcal{J}$ be given by (15). If condition (9) in Theorem 1 holds, then*

$$\mathcal{J}(w) \to -\infty \qquad \textit{as } ||w||_{m,p} \to \infty \textit{ in } W.$$

Proof: If $w \in W$ then $w = \widehat{w}(1)$ is constant, and therefore $\xi_m'(w) = 0 \in \mathbf{R}^{s_m - 1}$. Hence, since $F(x,0) = 0$ by (F–3),

$$\mathcal{J}(w) = -\int_\Omega A(x,w) - \int_\Omega G(x,w) - H(w). \tag{17}$$

Write $w = t$ for $t \in \mathbf{R}$, and consider first the case $t \leq 0$. Let $\gamma_2 > 0$ be such that $\gamma_2 < \gamma_1$, where γ_1 is as given in (a–3). By (a–3) and the definition of A in (13), there exist $R > 0$ and $C > 0$ such that

$$A(x,t) \geq \frac{\gamma_2}{p}|t|^p - C \tag{18}$$

for a.e. $x \in \Omega$ and all $t < -R$. So that, if $w = t < -R$, we obtain from (17), using (18) and (g–2), that

$$\mathcal{J}(t) \leq -\frac{\gamma_2}{p}|\Omega||t|^p + (M|\Omega| - H(1))t + C|\Omega|,$$

where $|\Omega|$ denotes the Lebesgue measure of Ω, which shows that

$$\mathcal{J}(t) \to -\infty \quad \text{as} \quad t \to -\infty \tag{19}$$

since $p > 1$.

On the other hand, if $w = t > 0$ then from (17) and the definition of $A(x,t)$ in (13) we obtain

$$\mathcal{J}(w) = -\left\{\int_\Omega G(x,t) + H(t)\right\},$$

which goes to $-\infty$ as $t \to \infty$ by condition (9). This combined with (19) proves the lemma. ■

LEMMA 2. *Assume (F–1)–(F–3), (A–1)–(A–2), (g–1)–(g–2), (a–1)–(a–2) and let $\mathcal{J}$ be given by (15). Then*

$$\mathcal{J}(v) \to \infty \qquad \textit{as } ||v||_{m.p} \to \infty \textit{ in } V,$$

where $V = \{v \in W^{m,p}(\Omega) \mid \int_\Omega v = 0\}$.

Proof: First observe that using (F–3) and Fubini's theorem, whose use in this case is justified by (A–1), we can write

$$\mathcal{J}(u) = \int_0^1 \mathcal{Q}(tu, u)dt - \int_\Omega A(x,u) - \int_\Omega G(x,u) - H(u) \tag{20}$$

for all $u \in W^{m,p}(\Omega)$, where $\mathcal{Q}$ is defined in (2).

By (a–2) and the definition of $A(x,s)$ in (13) we get that

$$|A(x,s)| \le \frac{\gamma}{p}|s|^p \quad \text{for a.e. } x \in \Omega \text{ and all } s \in \mathbf{R}. \tag{21}$$

Similarly, by (14) and (g–2) we get

$$|G(x,s)| \le M|s| \quad \text{for a.e. } x \in \Omega \text{ and all } s \in \mathbf{R}. \tag{22}$$

Using the estimates in (21) and (22) we then obtain from (20) that

$$\mathcal{J}(v) \ge \int_0^1 \mathcal{Q}(tv, v)dt - \frac{\gamma}{p}\int_\Omega |v|^p - M\int_\Omega |v| - C_3|v|'_{m,p}, \tag{23}$$

for $v \in V$, where C_3 is a constant that depends on the operator norm of H and the constant c in the Poincaré inequality (5).

Since $\gamma < \lambda_2$ by (a–2), we can find a $t_o \in (0,1)$ sufficiently small so that

$$\lambda_2 > \frac{\gamma}{1-t_o}. \tag{24}$$

Write the first term on the right hand side of (23) as

$$\int_0^1 \mathcal{Q}(tv,v)dt = \frac{\lambda_2 - \frac{\gamma}{1-t_o}}{\lambda_2}\int_0^1 \mathcal{Q}(tv,v)dt + \frac{\gamma}{(1-t_o)\lambda_2}\int_0^1 \mathcal{Q}(tv,v)dt. \tag{25}$$

Since $\mathcal{Q}(tv,v) = \frac{1}{t}\mathcal{Q}(tv,tv) \ge 0$ for all $t \in (0,1)$ by (A–2), we can estimate the second term on on the right hand side of (25) as follows

$$\frac{\gamma}{(1-t_o)\lambda_2}\int_0^1 \mathcal{Q}(tv,v)dt \ge \frac{\gamma}{(1-t_o)\lambda_2}\int_{t_o}^1 \mathcal{Q}(tv,v)dt. \tag{26}$$

By the definition of λ_2 in (6) there exists an $R > 0$ such that if $||v||_{L^p} \ge R/t_o$, then

$$\mathcal{Q}(tv,tv) \ge \frac{1-t_o}{1-t_o^p}\lambda_2 t^p ||v||^p_{L^p} \quad \text{for all } t \in [t_o, 1]. \tag{27}$$

On the other hand, if $||v||_{L^p} \le R/t_o$ then it follows from (A–2) that for $R_1 = R((1-t_o)\lambda_2)^{1/p}/t_o {c_o}^{1/p}(1-t_o^p)^{1/p}$ we have that if $|v|'_{m,p} \ge R_1$, then (27) holds. Therefore we see that (27) holds if $|v|'_{m,p} \ge R_1$.

Thus, it follows from (26) that

$$\frac{\gamma}{(1-t_o)\lambda_2}\int_0^1 \mathcal{Q}(tv,v)dt \geq \frac{\gamma}{p}||v||_{L^p}^p \tag{28}$$

for $v \in V$ such that $|v|'_{m,p} \geq R_1$.

We now use (25) and the estimate in (28) to obtain from (23) that

$$\mathcal{J}(v) \geq \frac{(1-t_o)\lambda_2 - \gamma}{(1-t_o)\lambda_2}\int_0^1 \mathcal{Q}(tu,u)dt - M\int_\Omega |v| - C_3|v|'_{m,p}, \tag{29}$$

for $v \in V$ such that $|v|'_{m,p} \geq R_1$.

Applying (A–2) to the first term on the right hand side of (29), and the Hölder and Poincaré inequalities to the second term, we get

$$\mathcal{J}(v) \geq \frac{(1-t_o)\lambda_2 - \gamma}{(1-t_o)\lambda_2 p}c_o|v|'^p_{m,p} - C_4|v|'_{m,p}, \tag{30}$$

for $v \in V$ such that $|v|'_{m,p} \geq R_1$, where C_4 is a constant that depends on C_3, M, the measure of Ω, p, and c.

Since $(1-t_o)\lambda_2 - \gamma > 0$ by (24), it follows from (30) that $\mathcal{J}(v) \to \infty$ as $|v|'_{m,p} \to \infty$. This fact in conjunction with (12) establishes the lemma. ∎

3. Proof of Theorem 1

In view of the fact $\mathcal{J}$ maps bounded sets in $W^{m,p}(\Omega)$ to bounded sets in $\mathbf{R}$, we conclude from Lemma 2 and (23) that $\mathcal{J}$ is bounded from below in V, and therefore $\inf_{v\in V}\mathcal{J}(v)$ is finite. By Lemma 1, there exists a $t_1 > 0$ large enough so that $\mathcal{J}(\pm t_1) < \inf_{v\in V}\mathcal{J}(v)$. Put $w_o = -t_1$ and $w_1 = t_1$, and let

$$\Gamma = \{h \in C([-t_1,t_1], W^{m,p}(\Omega)) \mid h(-t_1) = w_o \text{ and } h(t_1) = w_1\}.$$

Theorem 4.6 in [Ra, p. 24] implies that the value

$$\inf_{h\in\Gamma}\max_{t\in[-t_1,t_1]}\mathcal{J}(h(t))$$

will be a critical value of $\mathcal{J}$, and hence $\mathcal{J}$ will possess a critical point, provided that $\mathcal{J}$ satisfies the well known Palais–Smale condition:

(PS) *every sequence $\{u_n\} \subset W^{m,p}(\Omega)$ satisfying: (i) $\mathcal{J}(u_n)$ is bounded, and (ii) $\mathcal{J}'(u_n) \to 0$ in $W^{m,p}(\Omega)^*$ as $n \to \infty$, has a strongly convergent subsequence.*

Consequently, to complete the proof of Theorem 1 we need to establish that, under the hypotheses in Theorem 1, the functional $\mathcal{J}$ defined in (15) satisfies (PS) above. The verification of this fact is contained in two lemmas, the first of which is the following:

LEMMA 3. *Assume (F–1)–(F–3), (A–1)–(A–2), (g–1)–(g–2), (a–1)–(a–3) and let $\mathcal{J}$ be given by (15). Assume also that condition (9) in Theorem 1 holds. Suppose that $\{u_n\}$ is a sequence in $W^{m,p}(\Omega)$ satisfying: (i) $\mathcal{J}(u_n)$ is bounded, and (ii) $\mathcal{J}'(u_n) \to 0$ in $W^{m,p}(\Omega)^*$ as $n \to \infty$. Then $\{u_n\}$ is bounded in $W^{m,p}(\Omega)$.*

Proof: Assumptions (i) and (ii) in the hypotheses of the Lemma imply that there exists a constant K such that

$$|\mathcal{J}(u_n)| \le K \qquad \text{for all } n, \tag{31}$$

and that

$$|\mathcal{J}'(u_n)\varphi| \le \varepsilon_n ||\varphi||_{m,p} \qquad \text{for all } \varphi \in W^{m,p}(\Omega), \tag{32}$$

where $\varepsilon_n \to 0$ as $n \to \infty$.

Suppose to the contrary that $||u_n||_{m,p} \to \infty$ as $n \to \infty$. Put

$$v_n = \frac{u_n}{||u_n||_{m,p}} \qquad \text{for all } n.$$

Then $||v_n||_{m,p} = 1$ for all n, and we may therefore assume, passing to a subsequence if necessary, that

$$v_n \rightharpoonup v \quad \text{weakly in } W^{m,p}(\Omega), \tag{33}$$

$$v_n \to v \quad \text{strongly in } L^p(\Omega), \tag{34}$$

$$v_n(x) \to v(x) \quad \text{for a.e. } x \in \Omega. \tag{35}$$

Here we have used the facts that $W^{m,p}(\Omega)$ is a separable reflexive Banach space (see [Ad, p. 47]), and that the embedding $W^{m,p}(\Omega) \hookrightarrow L^p(\Omega)$ is compact (see for example [Ad, p. 144]).

Taking $\varphi = 1$ in (32), and using the fact that $\mathcal{Q}(u, \varphi) = 0$ for φ constant, we obtain from (16) that

$$\int_\Omega a(x, u_n)|u_n|^{p-2}u_n^- - \int_\Omega g(x, u_n) - H(1) \to 0 \quad \text{as } n \to \infty. \tag{36}$$

Dividing (36) by $||u_n||_{m,p}^{p-1}$ and letting $n \to \infty$ we get that

$$\lim_{n\to\infty} \int_\Omega a(x, u_n)|v_n|^{p-2}v_n^- = 0. \tag{37}$$

Here we have used (g–2) and the assumption $p > 1$.

Let $\Omega^- = \{x \in \Omega \mid v(x) < 0\}$. We claim that (37) implies that

$$\lim_{n\to\infty} \int_{\Omega^-} a(x, u_n)|v_n^-|^{p-1} = 0. \tag{38}$$

To see this observe that by (a–2) and Hölder's inequality

$$\left| \int_{\Omega\setminus\Omega^-} a(x, u_n)|v_n^-|^{p-1} \right| \le \gamma|\Omega|^{1/p} \left(\int_{\Omega\setminus\Omega^-} |v_n^-|^p \right)^{\frac{p-1}{p}}$$

so that, by (34),

$$\limsup_{n\to\infty} \left| \int_{\Omega\setminus\Omega^-} a(x, u_n)|v_n^-|^{p-1} \right| \le \gamma|\Omega|^{1/p} \left(\int_{\Omega\setminus\Omega^-} |v^-|^p \right)^{\frac{p-1}{p}} = 0,$$

from which (38) follows.

Next observe that by (a–3) and (35), for a.e. $x \in \Omega^-$,

$$\liminf_{n\to\infty} a(x, u_n(x))|v_n^-(x)|^{p-1} = \gamma_1 |v^-(x)|^{p-1},$$

since $u_n(x) = v_n(x)||u_n||_{m,p} \to -\infty$ for a.e. $x \in \Omega^-$. Thus, using Fatou's lemma, we obtain from (38) that

$$\gamma_1 \int_{\Omega^-} |v^-|^{p-1} \leq \liminf_{n\to\infty} \int_{\Omega^-} a(x, u_n)|v_n^-|^{p-1} = 0,$$

from which it follows that

$$\int_{\Omega^-} |v^-|^{p-1} = 0;$$

which implies that $\text{meas}(\Omega^-) = 0$. We therefore conclude that

$$v \geq 0 \quad \text{a.e. in } \Omega. \tag{39}$$

Write $v = \widehat{v}(1) + \widetilde{v}_2$ where $\widehat{v}(1) \in W$ and $\widetilde{v}_2 \in V$. We next show that $\widetilde{v}_2 = 0$. Taking $\varphi = v_n$ in (32), and letting $n \to \infty$, we get

$$\mathcal{Q}(u_n, v_n) + \int_\Omega a(x, u_n)|u_n|^{p-2}u_n^- v_n - \int_\Omega g(x, u_n)v_n - H(v_n) \to 0 \tag{40}$$

as $n \to \infty$, since $||v_n||_{m,p} = 1$ for all n. Dividing (40) by $||u_n||_{m,p}^{p-1}$, letting $n \to \infty$, and using (g–2) and the facts that $||v_n||_{m,p} = 1$ for all n and $p > 1$, we get that

$$\frac{\mathcal{Q}(u_n, v_n)}{||u_n||_{m,p}^{p-1}} + \int_\Omega a(x, u_n)|v_n^-|^p \to 0 \quad \text{as } n \to \infty. \tag{41}$$

The second term in (41) tends to 0 as $n \to \infty$ since both terms in (41) are nonnegative. We therefore conclude from (41) that

$$\lim_{n\to\infty} \frac{\mathcal{Q}(u_n, v_n)}{||u_n||_{m,p}^{p-1}} = 0. \tag{42}$$

On the other hand, by (A–2) and the semilinearity of $\mathcal{Q}$,

$$\mathcal{Q}(u_n, v_n) = \frac{1}{||u_n||_{m,p}} \mathcal{Q}(u_n, u_n) \geq \frac{c_o}{||u_n||_{m,p}} |u_n|'^{p}_{m,p}.$$

Thus

$$\frac{\mathcal{Q}(u_n, v_n)}{||u_n||_{m,p}^{p-1}} \geq c_o |v_n|'^{p}_{m,p},$$

from which we get, in conjunction with (42), that

$$\lim_{n\to\infty} |v_n|'_{m,p} = 0. \tag{43}$$

Write $v_n = \widehat{v}_n(1) + v_{n2}$, where $\widehat{v}_n(1) \in W$ and $v_{n2} \in V$ for all n, and observe that

$$|v_n|'_{m,p} = |v_{n2}|'_{m,p} \quad \text{for all } n.$$

Using this observation and (12), we conclude from (43) that

$$\lim_{n\to\infty} ||v_{n2}||_{m,p} = 0. \tag{44}$$

The fact that v_{n2} converges weakly to $\widetilde{v}_2$ by (33) and (34), and the lower semi–continuity of the norm $||\cdot||_{m,p}$, in conjunction with (44), imply

$$||\widetilde{v}_2||_{m,p} = 0,$$

and therefore that $\widetilde{v}_2 = 0$, which is what we wanted to show.

Consequently, $v = \widehat{v}(1) \in \mathbf{R}$ where $\widehat{v}(1) \geq 0$ by (39). In fact, we can show that $\widehat{v}(1) \neq 0$. To see this, use (34), (43), and the fact that $||v_n||_{m,p} = 1$ for all n, to get

$$\begin{aligned} 1 = ||v_n||^p_{m,p} &= \int_\Omega |v_n|^p + |v_n|'^p_{m,p} \\ &\to \int_\Omega |v|^p \quad \text{as } n \to \infty. \end{aligned}$$

So that $\widehat{v}(1) = 1/|\Omega|^{1/p}$, and therefore

$$\widehat{v}(1) > 0. \tag{45}$$

Write $u_n = \widehat{u}_n(1) + u_{n2} \in W \oplus V$ for all n. We claim that $\{u_{n2}\}$ is bounded in $W^{m,p}(\Omega)$. From (45) it follows that there exists some $n_1 \geq 1$ such that

$$\widehat{u}_n(1) > 0 \quad \text{for } n \geq n_1. \tag{46}$$

In the proof of the claim we will need the observation:

$$|u_n(x)|^{p-2}u_n^-(x)u_{n2}(x) \geq -|u_{n2}(x)|^p \quad \text{for a.e. } x \in \Omega \text{ and } n \geq n_1. \tag{47}$$

To establish (47) observe that the estimate holds trivially if either $u_{n2}(x) \geq 0$, or if $u_{n2}(x) < 0$ and $u_n(x) \geq 0$. It therefore suffices to consider the case

$$u_{n2}(x) < 0 \qquad \text{and} \qquad u_n(x) < 0.$$

In this case we have, for $n \geq n_1$,

$$\begin{aligned} 0 \leq u_n^-(x) &= -\widehat{u}_n(1) - u_{n2}(x) \\ &= -\widehat{u}_n(1) + |u_{n2}(x)| \leq |u_{n2}(x)|, \end{aligned}$$

since $\widehat{u}_n(1) \geq 0$ for $n \geq n_1$ by (46). Therefore

$$\begin{aligned} |u_n(x)|^{p-2}u_n^-(x)u_{n2}(x) &= -|u_n^-(x)|^{p-1}|u_{n2}(x)| \\ &\geq -|u_{n2}(x)|^{p-1}|u_{n2}(x)| = -|u_{n2}(x)|^p \end{aligned}$$

for $n \geq n_1$, which is the estimate in (47).

Next use (32) with $\varphi = u_{n2}$ and the fact that $\mathcal{Q}(u_n, u_{n2}) = \mathcal{Q}(u_{n2}, u_{n2})$ to get that there exists $n_2 \geq n_1$ such that

$$||u_{n2}||_{m,p} \geq \mathcal{Q}(u_{n2}, u_{n2}) + \int_\Omega a(x, u_n)|u_n|^{p-2}u_n^- u_{n2} - \int_\Omega g(x, u_n)u_{n2} - H(u_{n2})$$

for $n \geq n_2$. Consequently, by (47), (g–2), and Hölder's inequality,

$$||u_{n2}||_{m,p} \geq \mathcal{Q}(u_{n2}, u_{n2}) - \int_\Omega a(x, u_n)|u_{n2}|^p - C_5||u_{n2}||_{m,p}$$

for all $n \geq n_2$, where C_5 is a positive constant which depends on M, $|\Omega|$, c, p, and the operator norm of H. It then follows from (a–2) that

$$||u_{n2}||_{m,p} \geq \mathcal{Q}(u_{n2}, u_{n2}) - \gamma \int_\Omega |u_{n2}|^p - C_5||u_{n2}||_{m,p} \tag{48}$$

for all $n \geq n_2$.

Choose $\theta \in (0, 1)$ such that $\gamma < \theta\lambda_2$, and write the first term on the right hand side of (48) as

$$\mathcal{Q}(u_{n2}, u_{n2}) = \frac{\lambda_2 - \gamma/\theta}{\lambda_2}\mathcal{Q}(u_{n2}, u_{n2}) + \frac{\gamma}{\theta\lambda_2}\mathcal{Q}(u_{n2}, u_{n2}). \tag{49}$$

To show that $\{u_{n2}\}$ is bounded in $W^{m,p}(\Omega)$ assume by way of contradiction that

$$||u_{n2}||_{m,p} \to \infty \qquad \text{as } n \to \infty. \tag{50}$$

By virtue of the Poincaré inequality (5), there are only two possibilities to consider: either

$$||u_{n2}||_{L^p} \to \infty \qquad \text{as } n \to \infty, \tag{51}$$

for some subsequence which we again call $\{u_{n2}\}$, or $\{||u_{n2}||_{L^p}\}$ is bounded.

If (51) holds, then, by the definition of λ_2 in (6), there exists $n_3 \geq n_2$ such that

$$\mathcal{Q}(u_{n2}, u_{n2}) \geq \theta\lambda_2||u_{n2}||_{L^p}^p$$

for all $n \geq n_3$. Therefore, the second term on the right hand side of (49) can be estimated as

$$\frac{\gamma}{\theta\lambda_2}\mathcal{Q}(u_{n2}, u_{n2}) \geq \gamma||u_{n2}||_{L^p}^p$$

for all $n \geq n_3$, which in conjunction with (49) yields from (48) that

$$||u_{n2}||_{m,p} \geq \frac{\theta\lambda_2 - \gamma}{\theta\lambda_2}\mathcal{Q}(u_{n2}, u_{n2}) - C_5||u_{n2}||_{m,p} \tag{52}$$

for all $n \geq n_3$.

If, on the other hand, $\{||u_{n2}||_{L^p}\}$ is bounded, let $R > 0$ be such that $||u_{n2}||_{L^p} \leq R$ for all n. Choose, by virtue of (50), an $n_4 \geq n_3$ such that

$$|u_{n2}|'_{m,p} > R\left(\frac{\theta\lambda_2}{c_o}\right)^{1/p} \qquad \text{for all } \ n \geq n_4,$$

so that by (A–2) we obtain

$$\mathcal{Q}(u_{n2}, u_{n2}) \geq \theta\lambda_2 ||u_{n2}||^p_{L^p}$$

for all $n \geq n_4$. Hence, from (48) and (49) we conclude that (52) holds for all $n \geq n_4$.

Applying (A–2) in conjunction with (11) and (12) to the first term on the right hand side of (52), we obtain

$$||u_{n2}||_{m,p} \geq \frac{(\theta\lambda_2 - \gamma)c_o}{(1+c)\theta\lambda_2}||u_{n2}||^p_{m,p} - C_5||u_{n2}||_{m,p}$$

for all $n \geq n_4$, which contradicts (50) since $p > 1$ and $\gamma < \theta\lambda_2$. This contradiction shows that $\{u_{n2}\}$ is bounded in $W^{m,p}(\Omega)$.

Observe now that by (31)

$$\begin{aligned} K \geq \Bigg| \int_\Omega F(x, \xi'_m(u_n) - \int_\Omega A(x, u_n) - \int_\Omega [G(x, u_n) - G(x, \widehat{u}_n(1))] \\ - H(u_{n2}) - \int_\Omega G(x, \widehat{u}_n(1)) - H(\widehat{u}_n(1))\Bigg|. \end{aligned} \tag{53}$$

We claim that the fact that $\{u_{n2}\}$ is bounded in $W^{m,p}(\Omega)$ implies that the first three terms on the right hand side of (53) are bounded independently of n. To see this first observe that $\xi'_m(u_n) = \xi'_m(u_{n2})$ for all n, so that the growth condition for F in (F–2) and the fact that the $D^\alpha u_{n2}$ are bounded in L^p, for $1 \leq |\alpha| \leq m$, imply that

$$\left|\int_\Omega F(x, \xi'_m(u_n))\right| \leq K_1 \qquad \text{for all } n, \tag{54}$$

for some positive constant K_1. For the second term on the right of (53), observe that by (a–2) and the definition of $A(x,s)$ in (13)

$$|A(x, u_n)| \leq \frac{\gamma}{p}|u_n^-(x)|^p \qquad \text{for a.e. } x \in \Omega. \tag{55}$$

Now, since $\widehat{u}_n(1) > 0$ for $n \geq n_1$ by (46), we get that $u_n^- \leq u_{n2}^-$ for $n \geq n_1$. (To see this assertion note that $-u_n = -\widehat{u}_n(1) - u_{n2} \leq -u_{n2}$ for $n \geq n_1$, so that, if $-u_n > 0$, then $u_n^- = -u_n \leq -u_{n2} = u_{n2}^-$; on the other hand, if $-u_n \leq 0$ then $u_n^- = 0 \leq u_{n2}^-$.) We thus obtain from (55) that

$$\left|\int_\Omega A(x, u_n)\right| \leq \frac{\gamma}{p}\int_\Omega |u_{2n}^-|^p \leq \frac{\gamma}{p}\int_\Omega |u_{2n}|^p \text{ for } n \geq n_1,$$

which by the Poincaré inequality and the boundedness of $\{||u_{n2}||_{m,p}\}$ implies that

$$\left|\int_\Omega A(x,u_n)\right| \le K_2 \quad \text{for all } n \tag{56}$$

and some constant K_2. Next observe that by (g–2) and the Mean Value Theorem

$$\left|\int_\Omega [G(x,u_n) - G(x,\widehat{u}_n(1))]\right| \le M \int_\Omega |u_{n2}| \quad \text{for all } n,$$

so that by Hölder's inequality, the Poincaré inequality, and the boundedness of $\{||u_{n2}||_{m,p}\}$,

$$\left|\int_\Omega [G(x,u_n) - G(x,\widehat{u}_n(1))]\right| \le K_3 \quad \text{for all } n, \tag{57}$$

and some constant K_3.

Similarly, $\{H(u_{n2})\}$ is bounded since H is a bounded linear functional. Therefore, combining (54), (56) and (57), we get from (53) that

$$K \ge \left|\int_\Omega G(x,\widehat{u}_n(1)) + H(\widehat{u}_n(1))\right| - K_4 \quad \text{for all } n, \tag{58}$$

and some constant K_4. On the other hand, since

$$\widehat{u}_n(1) \to \infty \quad \text{as } n \to \infty,$$

by (34) and (45), we get from condition (9) that

$$\lim_{n\to\infty} \left\{\int_\Omega G(x,\widehat{u}_n(1)) + H(\widehat{u}_n(1))\right\} = \infty,$$

which contradicts (58). This contradiction establishes the lemma. ∎

The final lemma that we need in the verification of (PS) for $\mathcal{J}$ is the following:

LEMMA 4. *Assume (F–1)–(F–3), (A–1)–(A–3), (g–1)–(g–2), (a–1)–(a–2) and let $\mathcal{J}$ be given by (15). Suppose that $\{u_n\}$ is a sequence which is bounded in $W^{m,p}(\Omega)$, and such that $\mathcal{J}'(u_n) \to 0$ in $W^{m,p}(\Omega)^*$ as $n \to \infty$. Then $\{u_n\}$ has a convergent subsequence; that is, there exist $u \in W^{m,p}(\Omega)$ and a subsequence $\{u_{n_k}\}$ satisfying $\lim_{k\to\infty} ||u_{n_k} - u||_{m,p} = 0$.*

Proof: Assume that there exists a constant $K > 0$ such that

$$||u_n||_{m,p} \le K \qquad \text{for all } n = 1,2,3,\ldots, \tag{59}$$

and that

$$|\mathcal{J}'(u_n)\varphi| \le \varepsilon_n ||\varphi||_{m,p} \qquad \text{for all } \varphi \in W^{m,p}(\Omega), \tag{60}$$

where $\varepsilon_n \to 0$ as $n \to \infty$.

Since $\{u_n\}$ is bounded and $W^{m,p}(\Omega)$ is a separable reflexive Banach space [Ad, pg. 47], it follows from the compact embedding theorem for Sobolev spaces [Ad, pg. 144] that there exists a subsequence of $\{u_n\}$ (which for ease of notation we take to be the full sequence) and a function $u \in W^{m,p}(\Omega)$ satisfying

$$u_n \rightharpoonup u \quad \text{weakly in } W^{m,p}(\Omega) \text{ as } n \to \infty, \tag{61}$$

$$D^\beta u_n \to D^\beta u \quad \text{strongly in } L^p(\Omega) \text{ as } n \to \infty \text{ for } |\beta| \le m-1, \tag{62}$$

$$D^\beta u_n(x) \to D^\beta u(x) \quad \text{as } n \to \infty \text{ for a.e. } x \in \Omega \text{ and } |\beta| \le m-1. \tag{63}$$

Consequently, in order to prove the Lemma, it suffices to show that there exists a subsequence $\{u_{n_k}\}$ satisfying

$$D^\alpha u_{n_k} \to D^\alpha u \quad \text{strongly in } L^p(\Omega) \text{ as } k \to \infty \text{ for } |\alpha| = m. \tag{64}$$

We first observe that (59), (60), (61), and (62) imply that

$$\lim_{n\to\infty} \mathcal{Q}(u_n, u_n - u) = 0. \tag{65}$$

To see this, use (60) with $\varphi = u_n - u$, (16), and (59) to get that

$$\begin{aligned} \mathcal{Q}(u_n, u_n - u) &+ \int_\Omega a(x,u_n)|u_n|^{p-2}u_n^-(u_n - u) \\ &- \int_\Omega g(x,u_n)(u_n - u) - H(u_n - u) \to 0 \quad \text{as } n \to \infty. \end{aligned} \tag{66}$$

Next, use (a–2), Hölder's inequality, (59) and (62) on the second term of (66) to get that

$$\left| \int_\Omega a(x,u_n)|u_n|^{p-2}u_n^-(u_n - u) \right| \le \gamma K^{1-1/p} ||u_n - u||_{L^p} \to 0$$

as $n \to \infty$. So that

$$\int_\Omega a(x,u_n)|u_n|^{p-2}u_n^-(u_n - u) \to 0 \quad \text{as } n \to \infty. \tag{67}$$

Similarly, using (g–2), Hölder's inequality, and (62) on the third term in (66) yields

$$\left| \int_\Omega g(x,u_n)(u_n - u) \right| \le M|\Omega|^{1-1/p} ||u_n - u||_{L^p} \to 0$$

as $n \to \infty$; i.e.,

$$\int_\Omega g(x,u_n)(u_n - u) \to 0 \quad \text{as } n \to \infty. \tag{68}$$

By (61) we also get that

$$H(u_n - u) \to 0 \quad \text{as } n \to \infty. \tag{69}$$

Combining (67), (68), and (69) with (66) yields (65); i.e.,

$$\lim_{n\to\infty}\sum_{1\le|\alpha|\le m}\int_\Omega A_\alpha(x,\xi'_m(u_n))(D^\alpha u_n - D^\alpha u) = 0. \tag{70}$$

We can also show using (A–1), (61) and (62) that

$$\lim_{n\to\infty}\sum_{1\le|\alpha|\le m}\int_\Omega A_\alpha(x,\xi'_m(u))(D^\alpha u_n - D^\alpha u) = 0,$$

so that, in conjunction with (70),

$$\lim_{n\to\infty}\sum_{1\le|\alpha|\le m}\int_\Omega [A_\alpha(x,\xi'_m(u_n)) - A_\alpha(x,\xi'_m(u))](D^\alpha u_n - D^\alpha u) = 0. \tag{71}$$

Since (A–1) and (62) imply that

$$\lim_{n\to\infty}\sum_{1\le|\alpha|\le m-1}\int_\Omega [A_\alpha(x,\xi'_m(u_n)) - A_\alpha(x,\xi'_m(u))](D^\alpha u_n - D^\alpha u) = 0,$$

we get from (71) that

$$\lim_{n\to\infty}\sum_{|\alpha|=m}\int_\Omega [A_\alpha(x,\xi'_m(u_n)) - A_\alpha(x,\xi'_m(u))](D^\alpha u_n - D^\alpha u) = 0. \tag{72}$$

It is not hard to show, using (A–1), (59), (62), and Vitali's theorem [Ru, p. 143], that

$$\lim_{n\to\infty}\sum_{|\alpha|=m}\int_\Omega [A_\alpha(x,\eta'_{m-1}(u_n),\zeta_m(u)) - A_\alpha(x,\xi'_m(u))](D^\alpha u_n - D^\alpha u) = 0,$$

where $\xi'_m = (\eta'_{m-1},\zeta_m)$ is the decomposition of ξ'_m into its mth order component ζ_m and its corresponding lower order terms η'_{m-1} (see [S1, (3.21)]). It therefore follows from (72) that

$$\lim_{n\to\infty}\int_\Omega\sum_{|\alpha|=m}[A_\alpha(x,\xi'_m(u_n)) - A_\alpha(x,\eta'_{m-1}(u_n),\zeta_m(u))](D^\alpha u_n - D^\alpha u) = 0. \tag{73}$$

By (A–3) the integrand on the left hand side of (73) is nonnegative, so that for any measurable subset E of Ω, it is the case that

$$\lim_{n\to\infty}\sum_{|\alpha|=m}\int_E [A_\alpha(x,\xi'_m(u_n)) - A_\alpha(x,\eta'_{m-1}(u_n),\zeta_m(u))](D^\alpha u_n - D^\alpha u) = 0 \tag{74}$$

uniformly in E.

Let E be any Lebesgue measurable subset of Ω, and $\varepsilon > 0$ be given. From (A–2) and (74), there exists n_ε such that, if $n \geq n_\varepsilon$, then

$$
\begin{aligned}
\sum_{|\alpha|=m} \int_E |D^\alpha u_n|^p \leq \frac{1}{c_o} \int_E \sum_{|\alpha|=m} & A_\alpha(x, \xi'_m(u_n)) D^\alpha u \\
+ \frac{1}{c_o} \int_E \sum_{|\alpha|=m} & A_\alpha(x, \eta'_{m-1}(u_n), \zeta_m(u))(D^\alpha u_n - D^\alpha u) \\
+ \frac{1}{c_o} \int_E \sum_{1 \leq |\alpha| \leq m-1} & A_\alpha(x, \xi'_m(u_n)) D^\alpha u_n + \varepsilon
\end{aligned}
\tag{75}
$$

We now observe that the integrands on the right hand side of (75) are absolutely equi–integrable; that is, given $\varepsilon > 0$ there exists a $\delta > 0$ such that meas$(E) < \delta$ implies that the sum of the integrals is less than ε for each $n \geq n_\varepsilon$. We therefore get from (75) that

$$
\sum_{|\alpha|=m} \int_E |D^\alpha u_n|^p \leq 2\varepsilon \tag{76}
$$

for $n \geq n_\varepsilon$, provided that meas$(E) < \delta$. Now, for each n the integral in (76) is absolutely continuous in E, so that by choosing δ even smaller we can guarantee that the inequality in (76) holds for all n. Consequently, $\{|D^\alpha u_n|^p\}$ is absolutely equi–integrable for each $|\alpha| = m$.

In view of the fact that the sequences $\{|D^\alpha u_n|^p\}$ are absolutely equi–integrable for each $|\alpha| = m$, in order to prove (64), it suffices by Vitali's theorem [Ru, pg. 143] to show that there exists a subsequence $\{u_{n_k}\}$ satisfying

$$
\zeta_m(u_{n_k})(x) \to \zeta_m(u)(x) \quad \text{as } k \to \infty \text{ for a.e. } x \in \Omega. \tag{77}
$$

To prove (77) one proceeds as in [S1] by showing that there exists a subsequence $\{u_{n_k}\}_{k=1}^\infty$ satisfying

$$
\begin{aligned}
\lim_{k\to\infty} \sum_{|\alpha|=m} [A_\alpha(\cdot, \eta'_{m-1}(u_{n_k}), \zeta_m(u_{n_k})) - A_\alpha(\cdot, \eta'_{m-1}(u_{n_k}), \zeta_m(u))] \\
\times (D^\alpha u_{n_k} - D^\alpha u) = 0 \text{ a.e. in } \Omega
\end{aligned}
\tag{78}
$$

and $\{|\zeta_m(u_{n_k}(x))|\}$ is pointwise bounded for a.e. $x \in \Omega$. That is, there exists a function $\kappa \in L^\infty(\Omega)$ such that

$$
|\zeta_m(u_{n_k}(x))| \leq \kappa(x) \quad \text{for a.e. } x \in \Omega \text{ and } k = 1, 2, 3, \ldots \tag{79}
$$

The argument leading from (78) and (79) to (77) is by now standard in the theory of quasilinear boundary value problems. We therefore omit it, and refer the reader to [S1, pg 1829] or [Br, pg 30].

For the proofs of (78) and (79) we follow [S1, pp.1830–1836]. To establish (78) we observe from (A–3) that the integrand in (73) is nonnegative. Consequently, the integrand in (73) converges in the L^1–norm to zero. But then (78) follows from [Ru, Theorem 3.12, p. 80].

We next show that (79) holds.

Let Ω_1 be the set where, in view of (63),

$$\eta'_{m-1}(u_n(x)) \to \eta'_{m-1}(u(x)) \quad \text{for all } x \in \Omega_1,$$

where (78) holds, where

$$\xi'_m(u_{n_k}(x)),\ \xi'_m(u(x)),\ A_\alpha(x, \eta'_{m-1}(u_{n_k}(x)), \zeta_m(u(x))), \text{ and } \widetilde{h}(x) \text{ in (A–1)}$$

are finite valued for $1 \le |\alpha| \le m$ and $k = 1, 2, \ldots$, and also where (A–1) and (A–2) hold. Then Ω_1 has the same Lebesgue measure as Ω, and to establish (79) it is therefore sufficient to show that

$$\{|\zeta_m(u_{n_k}(x))|\}_{k=1}^{\infty} \quad \text{is pointwise bounded in } \Omega_1. \tag{80}$$

Suppose to the contrary that there exists a point $x_o \in \Omega_1$ and a subsequence $\left\{|\zeta_m(u_{n_{k_l}}(x))|\right\}_{l=1}^{\infty}$ for which

$$\lim_{l\to\infty} |\zeta_m(u_{n_{k_l}}(x_o))| = +\infty. \tag{81}$$

Let $\varepsilon > 0$ be such that $\varepsilon < 1$ and $p - \varepsilon > 1$. It follows from (A–2) that

$$c_o|\zeta_m(u_{n_{k_l}}(x_o))|^p \le \sum_{1\le|\alpha|\le m} A_\alpha(x_o, \xi'_m(u_{n_{k_l}}(x_o)))D^\alpha u_{n_{k_l}}(x_o). \tag{82}$$

Write (82) as

$$\begin{aligned} c_o|\zeta_m(u_{n_{k_l}}(x_o))|^p \le &\sum_{1\le|\beta|\le m-1} A_\beta(x_o, \xi'_m(u_{n_{k_l}}(x_o)))D^\beta u_{n_{k_l}}(x_o) \\ &+ \sum_{|\alpha|=m} A_\alpha(x_o, \xi'_m(u_{n_{k_l}}(x_o)))D^\alpha u_{n_{k_l}}(x_o), \end{aligned} \tag{83}$$

and observe that, by (A–1), (63), (81), and the fact that $\xi'_m(u_{n_{k_l}})$ is finite valued in Ω_1,

$$\lim_{l\to\infty} \frac{A_\beta(x_o, \xi'_m(u_{n_{k_l}}(x_o))D^\beta u_{n_{k_l}}(x_o)}{|\zeta_m(u_{n_{k_l}}(x_o))|^{p-\varepsilon}} = 0$$

for all β with $1 \le |\beta| \le m-1$. Therefore, dividing (83) by $|\zeta_m(u_{n_{k_l}}(x_o))|^{p-\varepsilon}$ and letting $l \to \infty$ we get

$$\begin{aligned} &c_o \limsup_{l\to\infty} |\zeta_m(u_{n_{k_l}}(x_o))|^\varepsilon \\ &\qquad \le \limsup_{l\to\infty} \frac{\sum_{|\alpha|=m} A_\alpha(x_o, \xi'_m(u_{n_{k_l}}(x_o)))D^\alpha u_{n_{k_l}}(x_o)}{|\zeta_m(u_{n_{k_l}}(x_o))|^{p-\varepsilon}} \end{aligned} \tag{84}$$

Next observe that, for fixed n,

$$\begin{aligned} A_\alpha(x, \xi'_m(u_n(x)))D^\alpha u_n = &\ A_\alpha(x, \xi'_m(u_n(x)))D^\alpha u \\ &+ A_\alpha(x, \eta'_{m-1}(u_n), \zeta(u))(D^\alpha u_n - D^\alpha u) \\ &+ [A_\alpha(x, \xi'_m(u_n)) - A_\alpha(x, \eta'_{m-1}(u_n), \zeta_m(u))] \\ &\qquad \times (D^\alpha u_n - D^\alpha u). \end{aligned} \tag{85}$$

Using (A–1), (63), (81) and the assumption $p-\varepsilon>1$ we obtain that

$$\lim_{l\to\infty}\frac{A_\alpha(x_o,\xi'_m(u_{n_{k_l}}(x_o)))D^\alpha u(x_o)}{|\zeta_m(u_{n_{k_l}}(x_o))|^{p-\varepsilon}}=0 \tag{86}$$

and

$$\lim_{l\to\infty}\frac{A_\alpha(x_o,\eta'_{m-1}(u_{n_{k_l}}(x_o)),\zeta_m(u(x_o)))(D^\alpha u_{n_{k_l}}(x_o)-D^\alpha u(x_o))}{|\zeta_m(u_{n_{k_l}}(x_o))|^{p-\varepsilon}}=0 \tag{87}$$

for all α with $|\alpha|=m$. Furthermore, from (78), (81), and the definition of Ω_1 we have that

$$\begin{aligned}\lim_{l\to\infty}\sum_{|\alpha|=m}&[A_\alpha(x_o,\xi'_m(u_{n_{k_l}}(x_o)))-A_\alpha(x_o,\eta'_{m-1}(u_{n_{k_l}}(x_o)),\zeta_m(u(x_o)))]\\ &\times(D^\alpha u_{n_{k_l}}(x_o)-D^\alpha u(x_o))\Big/|\zeta_m(u_{n_{k_l}}(x_o))|^{p-\varepsilon}=0.\end{aligned} \tag{88}$$

Consequently, using (85), (86), (87), and (88), we obtain from (84) that

$$c_o\lim_{l\to\infty}|\zeta_m(u_{n_{k_l}}(x_o))|^{\varepsilon}=0.$$

Since $c_o>0$, This gives a direct contradiction to (81). We therefore conclude that (80) must hold, and since $\Omega\backslash\Omega_1$ has Lebesgue measure zero, (79) must also hold. This concludes the proof of the lemma which gives the (PS) condition and establishes the proof of Theorem 1. ∎

4. Examples

We conclude this paper by presenting a few examples of quasilinear operators to which the results presented here apply. We will also consider examples of nonlinearities g for which the solvability condition in (9) holds. We begin with the case $m=1$ and the pth–Laplacian:

$$Q(u)=-\sum_{|\alpha|=1}D^\alpha(|\zeta_1(u)|^{p-2}D^\alpha u),$$

where $\zeta_1(u)$ is the gradient of u, and $p>1$. The potential $F\colon\mathbf{R}^N\to\mathbf{R}$ for this operator is given by

$$F(\zeta_1)=\frac{1}{p}\left(\sum_{|\alpha|=1}\zeta_\alpha^2\right)^{p/2}=\frac{1}{p}|\zeta_1|^p \tag{89}$$

for which (F–1) and (F–2) clearly hold. The function F in (89) above yields the functions A_α given by

$$A_\alpha(\zeta_1)=\begin{cases}|\zeta_1|^{p-2}\zeta_\alpha & \text{if } \zeta_1\neq 0\\ 0 & \text{if } \zeta_1=0\end{cases} \tag{90}$$

for all $|\alpha| = 1$, which clearly satisfy (A–1). The ellipticity condition (A–2) holds for the functions A_α defined in (90) above since

$$\sum_{|\alpha|=1} A_\alpha(\zeta_1)\zeta_\alpha = |\zeta_1|^p \geq \frac{1}{N} \sum_{|\alpha|=1} |\zeta_\alpha|^p.$$

The monotonicity condition (A–3) follows from the observation that the function F defined in (89) is strictly convex since $p > 1$, and consequently its gradient ∇F is strictly monotone; i.e.,

$$\sum_{|\alpha|=1} (|\zeta_1|^{p-2}\zeta_\alpha - |\zeta_1^*|^{p-2}\zeta_\alpha^*)(\zeta_\alpha - \zeta_\alpha^*) > 0$$

for $\zeta_1 \neq \zeta_1^*$.

Another example is provided by the function

$$F(\xi_m') = \sum_{1\leq|\alpha|\leq m} c_\alpha \frac{|\xi_\alpha|^p}{p}, \tag{91}$$

for some constants $c_\alpha > 0$ for $1 \leq |\alpha| \leq m$ and $p > 1$, which yields the functions A_α

$$A_\alpha(\xi_m) = \begin{cases} c_\alpha |\xi_\alpha|^{p-2}\xi_\alpha & \text{if } \xi_\alpha \neq 0 \\ 0 & \text{if } \xi_\alpha = 0 \end{cases} \tag{92}$$

for $1 \leq |\alpha| \leq m$.

It is not hard to see that the function F defined in (91) above satisfies (F–1) and (F–2), and that (A-1) and (A–2) hold for the A_α given in (92). To see that (A–3) also holds, observe that

$$\sum_{|\alpha|=1} c_\alpha(|\xi_\alpha|^{p-2}\xi_\alpha - |\xi_\alpha^*|^{p-2}\xi_\alpha^*)(\xi_\alpha - \xi_\alpha^*) > 0$$

for $\xi_m \neq \xi_m^*$, since $h(t) = |t|^{p-2}t$ is a monotonic function of t for $p > 1$; consequently, $(h(t) - h(s))(t - s) > 0$ if $t \neq s$.

For the bounded nonlinearity $g(s) = \arctan(s)$ one gets solvability of problem (7) for all those $H \in W^{m,p}(\Omega)^*$ satisfying

$$-H(1) < \frac{\pi}{2}\text{meas}(\Omega).$$

More generally, if $g(x, s) = b(x)f(s)$ where $b \in L^\infty(\Omega)$ satisfies $b(x) \geq b_o > 0$ for a.e. $x \in \Omega$, and $f: \mathbf{R} \to \mathbf{R}$ is a bounded continuous function, then problem (7) is solvable for $H \in W^{m,p}(\Omega)^*$ satisfying

$$-H(1) < f_+ \int_\Omega b(x), \tag{93}$$

where $f_+ = \liminf_{s\to+\infty} f(s)$. Condition (93) is the Landesman–Lazer condition [LL] similar to that used, for example, by Ahmad [Ah] to obtain existence in the one–sided resonance case for a linear Dirichlet problem. An existence result analogous to that obtained in [Ah] can also be obtained for the linear Neumann problem using the solvability condition in (93). The existence result given by Theorem 1 actually improves that obtained by means of the condition in (93). Hence we obtain a result which is new even for the linear problem with $p = 2$ and Neumann boundary conditions. To illustrate this consider the function $g(x,s) = b(x)f(s)$ where $b \in L^\infty(\Omega)$ satisfies $b \geq 0$ a.e., and f is a real valued continuous function defined as follows: $f(s)$ is bounded for $s \leq 0$, $f(s) = 0$ for $0 \leq s < 1$, and

$$f(s) = \begin{cases} 0 & \text{if } n-1 \leq s < n - 1/n; \\ -n2^n(s-n+1/n)/(2^n - n) & \text{if } n - 1/n \leq s < n - 1/2^n; \\ 2^n(s-n) & \text{if } n - 1/2^n \leq s < n; \end{cases}$$

for all integers n with $n \geq 2$. Then $\liminf_{s\to\infty} f(s) = -1$, so that, using condition (93) yields that problem (7) is solvable for $H \in W^{m,p}(\Omega)^*$ satisfying

$$-H(1) < -\int_\Omega b(x).$$

On the other hand, Theorem 1 implies that problem (7) is solvable for $H \in W^{m,p}(\Omega)^*$ satisfying

$$-H(1) < 0.$$

Which says that for this particular kind of nonlinearity, if $b(x) > 0$ for a.e. $x \in \Omega$, problem (7) is actually solvable for a larger class of distributions H.

5. References

[Ad] R. A. Adams, *Sobolev Spaces*, Academic Press, New York, 1975.

[Ah] S. Ahmad, *Nonselfadjoint Resonance Problems with Unbounded Perturbations*, Nonlinear Analysis, Theory, Methods, and Applications **10**, 2 (1986), 147–156

[ALP] S. Ahmad, A. C. Lazer, and J. L. Paul, *Elementary critical point theory and perturbations of elliptic boundary value problems at resonance*, Indiana Univ. Math. J., **25** (1976), 933–944.

[Br] F. E. Browder, *Existence theorems in partial differential equations*, Proc. Sympos. Pure Math. Vol. 16, Amer. Math. Soc., Providence, R.I., 1970, pp. 1–60.

[CL] A. Castro and A. C. Lazer, *Applications of a Max–Min principle*, Revista Colombiana de Matemáticas, **X** (1976), pp. 141–149.

[GT] D. Gilbarg and N. S. Trudinger, *Elliptic partial differential equations of second order*, 2nd ed., Springer–Verlag, Berlin, 1983.

[LL] E. M. Landesman and A. C. Lazer, *Nonlinear perturbations of a linear elliptic boundary value problem at resonance*, J. Math. Mech. **19** (1970), 609–623

[Ra] P. Rabinowitz, *Minimax Methods in Critical Point Theory with Applications to Differential Equations*, CBMS Regional Conf. Ser. in Math., No. 65, Amer. Math. Soc., Providence, R. I. 1986

[Ru] W. Rudin, *Real and Complex Analysis,* 2nd ed., McGraw–Hill, New York, 1974.

[S1] V. L. Shapiro, *Quasilinear ellipticity and the first eigenvalue* , Comm. in PDEs, **16** (1991), 1819–1855.

[S2] V. L. Shapiro, *Resonance and the second BVP* , *Trans. of the AMS*,**325** (1991), 363–387.

DEPERTMENT OF MATHEMATICS, POMONA COLLEGE, CLAREMONT, CALIFORNIA 91711
E–mail address: arumbos@pomona.edu

DEPERTMENT OF MATHEMATICS, UNIVERSITY OF CALIFORNIA AT RIVERSIDE, RIVERSIDE, CALIFORNIA 92521
E–mail address: shapiro@math.ucr.edu

Contemporary Mathematics
Volume **208**, 1997

Shock-Waves in General Relativity

by J. A. Smoller and J. B. Temple

ABSTRACT. In this paper we set out a general theory of shock-waves and we apply this to explicitly construct a class of spherically symmetric shock-wave solutions of the Einstein gravitational equations for a perfect fluid. Our construction entails matching a Friedmann-Robertson-Walker metric to an Oppenheimer-Tolman metric at a shock-wave interface across which the gravitational metric is Lipschitz continuous. In our dynamically matched solution, we imagine the F-R-W solution as an exploding inner core, and the boundary of this inner core is a shock surface that is driven by the expansion behind the shock out into the static O-S solution. Our solution solves the problem first posed by Oppenheimer and Snyder of extending their solution to the case of non-zero pressure. In the terminology of the classical theory of shock-waves, the interface of the O-S solution is a contact discontinuity, and this means that their solution is time-reversible. In contrast, our shock-wave solution, in which the pressure is non-zero, is an irreversible solution of the Einstein equations in which the irreversibility, the loss of information, and increase of entropy in the fluid puts time irreversibility into the dynamics of the gravitational field.

1. The Einstein gravitational field equations can be written in the form

$$G = \kappa T, \tag{1}$$

where G is the Einstein tensor, T is the stress-energy tensor (the source of the gravitational field), and $\kappa = 8\pi/c^2$, where c denotes the speed of light. In Einstein's theory, the gravitational field is identified with a Lorentzian metric g, a symmetric $(0,2)$ tensor which at any point can be diagonalized as $\mathrm{diag}(-1,1,1,1)$, and equation (1) describes, simultaneously, the evolution of g and sources in T.

In a given coordinate system, $x = (x^0, x^1, x^2, x^3)$, on 4-dimensional space-time, the components of G are given by $G_{ij} = R_{ij} - \frac{1}{2}Rg_{ij}$, where the Ricci tensor R_{ij} and the scalar curvature R are obtained from the Riemann curvature tensor $R^i_{jk\ell}$, computed from the metric g_{ij}. In this paper, we shall be concerned with a perfect fluid, whereby the stress energy tensor T takes the form

$$T_{ij} = (p + \rho c^2)u_i u_j + p g_{ij}, \tag{2}$$

1991 *Mathematics Subject Classification.* Primary: 83C05, 85A05; Secondary: 70F15, 76L05.

Key words and phrases. Black holes, Einstein equations, Oppenheimer-Snyder model, Riemann curvature tensor, Schwarzwild radius, shock-wave solutions.

This paper is in final form and no version of it will be submitted for publication elsewhere.

Research supported by the NSF, Contract No. DMS-G-95-01128
Research supported by the NSF, Contract No. DMS-95-00694

where ρ is the mass-energy density of the fluid (as measured in a frame moving with the fluid), p is the pressure, and (u^i) denotes the normalized unit 4-velocity vector of the fluid. When an equation of state $p = p(\rho)$ is specified, system (1) imposes 10 equations in 14 unknowns (10 metric components g_{ij}, the density ρ, and 3 velocity components). Four remaining equations can then be imposed in order to specify a coordinate system.

Note that (1) shows that the Einstein equations couple the gravitational field to the *undifferentiated* fluid variables. Now it is well known that, as a consequence of the fundamental geometric identities called the Bianchi identities, div $G \equiv 0$. (Here div denotes the covariant divergence computed with respect to the unique symmetric connection compatible with the metric g, [11].) The Einstein theory thus has the remarkable property that on solutions of (1), the equation

$$\operatorname{div} T = 0 \tag{3}$$

must hold, and (3) is equivalent to the conservation of energy and momentum. If g is the flat Minkowski metric of special relativity, "div" reduces to the ordinary divergence, and hence (2) and (3) reduce to the relativistic Euler equations, [8], which in turn reduce to the classical Euler equation for an inviscid fluid, in the limit of small velocities [11]. Thus in the setting of General Relativity, the Euler equations follow from (1) and (2), as a consequence of geometrical identities.

In the papers [8,9,10], we study shock-wave solutions of the Einstein equations (1), (2). This is done by solving a problem first posed by Oppenheimer and Snyder on black holes in 1939, [5], namely, how does one remove the assumption that the pressure be identically zero in the Oppenheimer-Snyder model? That is, the Oppenheimer-Snyder model, which is in virtually every textbook on the subject of general relativity, is an exact solution of the Einstein equations, that models the collapse of a star to a black hole. The simplifying assumptions that they make is that the pressure inside the star must be identically zero and the space outside the boundary of the star must be devoid of matter. In [8], we give a general technique for extending the Oppenheimer-Snyder model to the case of non-zero pressure. We then generalize these ideas and apply them to some interesting astrophysical problems, [9].

Our technique to obtain solutions of (1) consists of the matching of two well-known spherically symmetric solutions of (1) across a shock-wave interface. The inner metric is a standard model of an expanding universe, which we match to a well-known metric which models the interior of a star. What we obtain in [9] is a solution which models a shock-wave explosion into a singular, static isothermal sphere. This solution is a General Relativistic version of how stars are formed. In fact, this solution can model explosions on any scale: supernova explosions, or even the "big-bang." If we allow the sound speed $\sigma \to 0$, we then get models for spherical explosions in classical Newtonian gravity. In the paper [10], we demonstrate that our theory generates a large class of physically meaningful outgoing shock-waves that model blast waves in a general relativistic setting. We also obtain explicit formulas for the physical quantities that evolve according to the equations; these formulas are important for the numerical simulation of these solutions.

2. An important issue in general relativity is the problem of how time irreversibility should enter the theory. Indeed, all of the exact solutions of the Einstein equations (1) for classical sources assume the stress tensor for a perfect fluid, or for electro-magnetism, and it is well known that smooth solutions of the Einstein equations with these sources are time reversible. However, *it is our thesis here that time-irreversibility has been built into the theory of general relativity from the very start by way of the stress tensor for a perfect fluid, and that the irreversibility, loss of information and increase of entropy in the solutions of* (1) *is encoded in the structure of the shock-waves.*

The idea is that, in the case of a perfect fluid (2), the relativistic compressible Euler equations appear as a subsystem of the Einstein gravitational field equations (1) through the identity (3), and the relativistic Euler equations form a system of nonlinear hyperbolic conservation laws, the setting for the modern theory of shock-waves. Therefore, like the classical compressible Euler equations, solutions of the Einstein equations should be time reversible only up until the formation of shock-waves. For the classical Euler equations, the dissipation and increase of entropy at shock-waves can be interpreted physically as the non-negligible effects of viscosity that must appear because the inviscid Euler equations represent a *singular limit* of the Navier-Stokes equations as the viscosity is taken to zero: however small the viscosity parameters, the dissipative effects of viscosity are not negligible in the surface layer of a shock, and this dissipation is fully accounted for in the shock-waves that appear in the zero viscosity limit. One of the complications in the theory of relativity is that there is no simple model like the classical Navier-Stokes equations that plays a similar role, because the introduction of second order parabolic terms to model dissipation introduces an infinite speed of propagation similar to that observed in the linear heat equation. However, it is important to emphasize that the irreversibility, loss of information and increase of entropy at shock-waves in a generic system of conservation laws div $T = 0$ is explicitly determined at the level of the zero viscosity limit, and this applies even when the precise form of the viscosity is not known. The conclusion one must draw from the theory of shock-waves is that shock-waves will always form in solutions of a system of conservation laws except in very special cases when the fluid is everywhere expansive [1,3,6]. We believe that this result carries over to the Einstein equations for a perfect fluid as well. Indeed, in locally Lorentzian coordinates at a point, (coordinates in which $g_{ij} = \eta_{ij} = \text{diag}(-1, 1, 1, 1)$, and $g_{ij,k} = 0$ for all $i, j, k = 0, \ldots, 3$), the covariant divergence equals the classical divergence; and thus, to within higher order errors, the perfect fluid in general relativity satisfies the relativistic Euler equations in special coordinates at each point. Thus if shock-waves never form in T, then in essence, the gravitational field has the effect of removing all of the compressive action of the Euler equations, and thereby has the effect of eliminating dissipation in fluids! This we find physically implausible, and thus we maintain that shock-waves are as fundamental to solutions (1), (2) as they are to the classical Euler equations. This suggests that when viscosity and heat conduction are neglected, (as they are in the case of a perfect fluid), it is the shock-waves in the sources T that fundamentally introduce irreversibility into the dynamics of the gravitational field. This point of view seems to have been essentially overlooked [2,4]. In [8], we organize the general theory of matching two metric solutions of (1) across an interface from the point of view of the theory of shock-waves, and we apply this theory to explicitly construct spherically symmetric shock-waves solutions of the Einstein equations. This construction solves the problem first posed by Oppenheimer and Snyder in 1939 of extending their model to the case of non-zero pressure.

If the fluid variables shock, the entries in the stress tensor T become discontinuous, and thus by the Einstein equation, the metric loses continuity in the second derivative. In Section 3 of [8], we present a self-contained treatment of the general theory of shock-waves for the Einstein equations by considering the general problem of allowing the metric to be only Lipschitz continuous across a hypersurface Σ in space-time. This treatment reorganizes and extends previous results of Israel. (See also [4] and references therein where this topic is referred to by the heading *Junction Conditions*). The result is the following theorem:

THEOREM 1. *Let Σ denote a smooth, 3-dimensional shock surface in space-time with space-like normal vector* $\mathbf{n}$. *Assume that the components g_{ij} of the gravitational metric g are smooth on either side of Σ, (continuous up to the boundary on either side separately), and Lipschitz continuous across Σ in some fixed coordinate system. Then the following statements are equivalent:*

(i) *$[K]=0$ at each point of Σ.*
(ii) *The curvature tensors $R^i_{jk\ell}$ and G_{ij}, viewed as second order operators on the metric components g_{ij}, produce no delta function sources on Σ.*
(iii) *For each point $P\in\Sigma$ there exists a $C^{1,1}$ coordinate transformation defined in a neighborhood of P, such that, in the new coordinates, (which can be taken to be the Gaussian normal coordinates for the surface), the metric components are $C^{1,1}$ functions of there coordinates.*
(iv) *For each $P\in\Sigma$, there exists a coordinate frame that is locally Lorentzian at P, and can be reached within the class of $C^{1,1}$ coordinate transformations.*

Moreover, if any one of these equivalencies hold, then the Rankine-Hugoniot jump conditions

$$[G]^{\sigma}_{i}n_{\sigma}=0,$$

(which express the weak form of conservation of energy and momentum across Σ when $G=\kappa T$), hold at each point on Σ.

Here $[K]$ denotes the jump in the second fundamental form (extrinsic curvature) K across Σ, (this being determined by the metric separately on each side of Σ because g_{ij} is only Lipschitz continuous across Σ), and by $C^{1,1}$ we mean that the first derivatives are Lipschitz continuous. Theorem 1 should be credited mostly to Israel, who obtained results (i)–(iii) in Gaussian normal coordinates. Our contribution was to identify the covariance class of $C^{1,1}$ transformations, and to thereby obtain precise coordinate independent statements for (ii) and (iii), as well as the equivalence with (iv). As a consequence of this, we obtain the result that the Ricci scalar curvature R *never* has delta function sources at a Lipschitz continuous matching of the metrics, as well as the following theorem that partially validates the statement that shock-wave singularities in the source free Einstein equations $R_{ij}=0$ or $G_{ij}=0$ can only appear as coordinate anomalies, and can be transformed away by coordinate transformation:

THEOREM 2. *If a smooth shock surface Σ forms in weak solutions of $R_{\alpha\beta}=0$ or $G_{\alpha\beta}=0$ posed in some given coordinate system y, such that the y-components*

$g_{\alpha\beta}$ of the metric tensor g are Lipschitz continuous across Σ, and C^k functions of y on either side of Σ (continuous up to the boundary on either side separately), then there exists a regular $C^{1,1}$ coordinate transformation taking $y \to x$, such that, the components g_{ij} of g in the x-coordinates are actually C^k functions of x in a neighborhood of each point on the surface Σ.

Note that when the metric g is only Lipschitz continuous across a surface Σ, the Einstein tensor G, which contains second derivatives of g, can in general contain delta function sources, and then Σ should be interpreted as a surface layer and not as a shock-wave. In the presence of spherical symmetry, however, the weak form of conservation of energy and momentum, (4), implies the absence of surface layers, so long as the areas of the spheres of symmetry match up on Σ. This is implied in the following theorem, proved in [8].

THEOREM 3. *Let g and $\bar{g}$ be 2 spherically symmetric metrics:*

$$g\colon\ ds^2 = -a(t,r)\,dt^2 + b(t,r)\,dr^2 + c(t,r)d\Omega^2,$$

$$\bar{g}\colon\ d\bar{s}^2 = -\bar{a}(\bar{t},\bar{r})\,d\bar{t}^2 + \bar{b}(\bar{t},\bar{r})\,d\bar{r}^2 + \bar{c}(\bar{t},\bar{r})\,d\Omega^2,$$

where $d\Omega^2 = d\theta^2 + \sin^2\theta\, d\theta$. Assume that there exists a smooth map: $\psi(t,r) \to (\bar{t},\bar{r})$ defined in a neighborhood of a shock interface Σ, such that the metrics agree on Σ. Assume too that $c(t,r) = \bar{c}(\psi(t,r))$, the normal $\bar{n}$ to Σ is non-null, and that the derivative of c in the direction $\bar{n}$ is non-zero. Then the following are equivalent:

(a) *$g \cup \bar{g}$ is $C^{1,1}$ in some neighborhood of Σ.*
(b) $[G^i_j]n_i = 0,\ j = 1,2.$
(c) $[G^{ij}]n_i n_j = 0.$
(d) $[K] = 0.$

In the above theorem and in what follows, $[f]$ denotes the jump in the quantity f across the shock interface and $C^{1,1}$ denotes the class of functions with Lipschitz continuous first derivatives. Note that the condition $c(t,r) = \bar{c}(\psi(t,r))$ implies that the areas of the spheres of symmetry in the barred and unbarred coordinates agree on Σ. The condition $\bar{n}(c) \neq 0$ implies that the areas of the spheres of symmetry change monotonically in the normal direction if $c(t,r) = r^2$, e.g., and if this condition failed, it would imply that the shock speed would exceed the speed of light.

We conclude that in the presence of spherical symmetry one condition, (c), is equivalent to the jump conditions (4); i.e., to the fact that the weak form of conservation of energy and momentum holds. This is an important fact in our extension of the (OS) model to non-zero pressure.

3. We now discuss how we explicitly construct spherically symmetric shock-wave solutions of the Einstein equations for a perfect fluid. To do this, we shall match different metrics, Lipschitz-continuously across a surface of discontinuity for the fluid variables (a shock-wave). In what follows, we shall always assume that the fluid is comoving relative to the metric; thus for diagonal metrics, $u^i = 0$, $i = 1,2,3$, and $u^0 = (-g_{00})^{-1/2}$.

In our shock-wave model, the *inner* metric is the Friedmann-Robertson-Walker (FRW) metric

$$\text{(FRW)} \qquad ds^2 = -dt^2 + R(t)^2 \left[\frac{dr^2}{1-kr^2} + r^2\, d\Omega^2 \right],$$

where k is a constant, $k = 0, 1$, or -1, and

$$d\Omega^2 = d\theta^2 + \sin^2\theta\, d\phi^2.$$

This metric is spherically symmetric and describes a homogeneous, isotropic universe (no preferred point and no preferred direction), [11]. The space expands or contracts in time according to the function $R(t)$, the cosmological scale factor. The (FRW) metric is used to model an expanding universe, and there is, generally speaking (e.g. if $R(t) > 0$ for some t), a singularity ($R(t) = 0$) in "backwards" time. The function $R(t)$ determines the "red-shift" factor for distant astronomical objects. If we are given an equation of state of the form $p = p(\rho)$, $R(t)$ is determined by the Einstein equations (1), (2), under the assumption that $\rho = \rho(t)$.

The *outer* metric in our model is the Oppenheimer-Tolman (OT) metric:

$$\text{(OT)} \qquad d\bar{s}^2 = -B(\bar{r})\, d\bar{t}^2 + \left(1 - \frac{2\mathcal{G}M(\bar{r})}{\bar{r}}\right)^{-1} d\bar{r}^2 + \bar{r}^2\, d\Omega^2.$$

(We use barred coordinates here to distinguish from the unbarred coordinates in the FRW metric.) If we are given an equation of state of the form $\bar{p} = \bar{p}(\bar{\rho})$, then using Einstein's equations, we get the following differential equations for the unknown quantities $\bar{\rho}$, B, and M (see [11]):

$$\text{(4)} \qquad \frac{B'}{B} = \frac{-2\bar{p}'}{\bar{p}+\bar{\rho}}, \qquad M'(\bar{r}) = 4\pi\bar{r}^2\bar{\rho}(\bar{r}),$$

and (the Oppenheimer-Volkoff equation),

$$\text{(5)} \qquad -\bar{r}^2\bar{p}' = \mathcal{G}M(\bar{r})\bar{\rho}(\bar{r}) \left(1 + \frac{\bar{p}}{\bar{\rho}}\right) \left(1 + \frac{4\pi\bar{r}^3\bar{p}}{M}\right) \left(1 - \frac{2\mathcal{G}M(\bar{r})}{\bar{r}}\right)^{-1}.$$

Here $M(\bar{r})$ denotes the "total mass" inside a ball of radius $\bar{r}$. The (OT) metric is a time independent spherically symmetric solution of the Einstein equations (1), (2); it is used to model the interior of a star, [11].

We now discuss the Oppenheimer-Snyder (OS) model, [5]. In the (OS) model, the *outer* metric is the empty space Schwarzschild metric ([11]):

$$d\bar{s}^2 = -\left(1 - \frac{2\mathcal{G}M}{\bar{r}}\right) d\bar{t}^2 + \left(1 - \frac{2\mathcal{G}M}{\bar{r}}\right)^{-1} d\bar{r}^2 + \bar{r}^2\, d\Omega^2,$$

where M is a positive constant. This metric describes the gravitational field outside of a ball in $\mathbb{R}^3$, assuming that there is no matter in the exterior of the ball; i.e., $T_{ij} \equiv 0$. The OS result is the following: Take $p \equiv 0$ in the FRW metric. Then there exists a coordinate transformation $(t, r) \to (\bar{t}, \bar{r})$, where $Rr = \bar{r}$, such that under this identification of coordinates the (FRW) and Schwarzschild metrics agree and are Lipschitz continuous across a 3-dimensional interface, the surface of a star. In (r, t) coordinates, the surface of the star is $r = a$ (fixed in the FRW metric). Then given $\rho(0)$ and M, a is determined by $M = \frac{4\pi}{3}\rho(0)$ and $k = \frac{2M\mathcal{G}}{a^3}$. The functions $\rho(t)$ and $R(t)$ are determined by the Einstein equations. On the other hand, in $\bar{r}$-$\bar{t}$ coordinates the surface collapses to a black hole: $\bar{r}(\bar{t}) \to \bar{r} = 2\mathcal{G}M$, and $\rho(\bar{t}) \to \infty$ as $\bar{t} \to \infty$. This (OS) result was the first example of gravitational collapse.

The total mass inside the interface is fixed since mass does not cross the interface. The (OS) solution is thus a contact discontinuity, [10] (as opposed to a true fluid-dynamical shock-wave). The function $R(t)$ vanishes at some finite T; thus, (cf. [11]) "a fluid sphere of initial density $\rho(0)$ and zero pressure will collapse from rest to a state of infinite density in the finite time T."

In the shock-wave model in [8,9,10], we also take the inner metric to be the (FRW) metric but with a non-zero pressure function $p = p(\rho)$, $\rho = \rho(t)$, and we take the outer metric to be the (OT) metric, with non-zero pressure function $\bar{p} = \bar{p}(\bar{\rho})$, $\bar{\rho} = \bar{\rho}(\bar{r})$. We show that we can define a coordinate mapping $(t, r) \to (\bar{t}, \bar{r})$, where $Rr = \bar{r}$, such that under this identification of coordinates, the (FRW) and (OT) metric match Lipschitz continuously along a 3-dimensional interface: the shock surface. That is, the following theorem holds [8]:

THEOREM 4. *Under the above procedure, there exists a shock-wave solution for arbitrary equation of state $p = p(\rho)$ in the* (FRW) *space, and arbitrary equation of state $\bar{p} = \bar{p}(\bar{\rho})$ in the* (OT) *space.*

It is instructive to outline the matching procedure. Thus, we first take $\bar{r} = Rr$, so that the spheres of symmetry have the same area. In order to match the $d\bar{r}^2$ term in the (OT), namely $\left(1 - \frac{2\mathcal{G}M(\bar{r})}{\bar{r}}\right)$, to the corresponding term in (the coordinate transformation of) the (FRW) metric, we find the explicit relation

$$M(\bar{r}) = \frac{4\pi}{3}\rho(t)\bar{r}^3; \tag{6}$$

this defines the shock surface. In order to match B, the $d\bar{t}^2$ term in (OT), to the corresponding term in the (FRW) metric we find that $\bar{t} = \bar{t}(t, r)$ solves a linear partial differential equation with initial conditions on the shock surface (6) chosen to match B on (6). The result is the following theorem, [8].

THEOREM 5. *The shock surface* (6) *is non-characteristic for this partial differential equation away from the Schwarzschild radius.*

Thus the transformation $\bar{t} = \bar{t}(t, r)$ exists, but is not explicit. So the coordinate transformation is defined; we shall not need any explicit information about $\bar{t}$.

Now there arises the following problem: If the metrics match Lipschitz-continuously across a shock surface Σ, does the weak form of the conservation of energy and momentum hold across the shock surface; i.e., does the following (jump condition) hold:

$$\left[T^{ij}\right] n_i = 0 \tag{7}$$

where $\bar{n} = \left(n^i\right)$ is the normal vector to Σ, cf. Figure 1.
Now in view of (1), (7) holds if and only if

$$\left[G^{ij}\right] n_i = 0 \tag{8}$$

That is, assume div $G = 0$ on each side of Σ, and g is Lipschitz continuous across Σ. Does this imply that (8) holds? The answer is no, in general. To understand this, we consider the extrinsic curvature (second fundamental form) K

$$K\colon T_p\Sigma \to T_p\Sigma$$
$$\bar{x} \to \nabla_{\bar{x}}\bar{n},$$

where $T_p\Sigma$ is the tangent space to Σ at p. Then Israel, [2], has proved the following theorem.

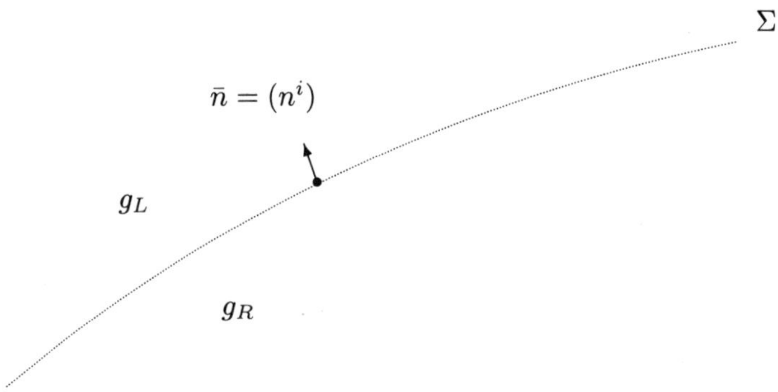

FIGURE 1

THEOREM 6. *If the extrinsic curvature K is continuous across Σ, then* (8) *holds.*

This theorem is generalized in Theorem 3, above, where we see that in the presence of spherical symmetry the jump conditions (7) reduce to the single condition

$$[T^{ij}]\, n_i n_j = 0. \tag{9}$$

In fact, in the presence of spherical symmetry, (9) holds if and only if the extrinsic curvature is continuous across Σ.

4. In this section, we shall apply the general results discussed in §3, to obtain explicit shock-wave solutions of the Einstein equations, which model shock-waves exploding into the general relativistic version of a static, singular, isothermal sphere. We shall also describe how this solution can be used in stellar dynamics to model the birth of a star.

Before proceeding with the construction of our exact shock-wave solution, it is instructive to see what (9) implies in the Oppenheimer-Snyder model discussed above. Thus, in this case, we have $\bar{p} \equiv 0 \equiv \bar{\rho}$, so (9) gives

$$p\rho\big[(1-\theta)p+\rho\big] = 0. \tag{10}$$

Thus if $\rho > 0$, then $p \equiv 0$, because $\theta \leq 1$; see [10]. Thus, in the (OS) model, there is no solution for non-zero pressure!

Now the idea in [8,9,10] is to fix the (OT) outer solution: $\bar{\rho}$, $\bar{p}$, and $M(\bar{r})$, where $\bar{p} = \bar{p}(\bar{\rho})$. Then we seek equations for the unknown quantities $R(t)$ and $\rho(t)$ in the inner (FRW) solution. These equations are (cf. [11]):

$$\dot{R}^2 = \frac{8\pi\mathcal{G}}{3}\rho R^2 - k, \tag{11}$$

$$p = -\frac{d}{dt}\left(\rho R^3\right)/3R^2\dot{R}. \tag{12}$$

Note that if we are given an equation of state $p = p(\rho)$, then the system (11), (12) becomes a system of two equations in the two unknowns R and ρ, which has a unique solution, given any initial condition. In this case, generally speaking, we cannot be sure that the conservation equation (9) will hold. We thus take a different approach, allow p to be another variable, and we take the three (autonomous) equations (10), (11), and (12) to determine the three unknowns, R, ρ, and p.

The procedure outlined here leaves open a few important questions; for example

1. Is $p \geq 0$?
2. Do $p(t)$ and $\rho(t)$ give a reasonable equation of state $p = p(\rho)$; for example, is the sound speed less than 1, the (normalized) speed of light?
3. Are p and ρ larger behind the shock-wave?
4. Do the Lax characteristic conditions hold (cf. [3,6])?

In general, the answer to these questions depends on the initial conditions, as well as on the given outer (OT) equation of state $\bar{p} = \bar{p}(\bar{\rho})$. However, when $\bar{p} = \bar{\sigma}\bar{\rho}$, and $k = 0$, we obtain an *explicit* solution, by assuming that the inner (FRW) equation of state is of the form $p = \sigma\rho$; (cf [11]).

We now construct an explicit (OS) type shock-wave with $p \equiv 0$, which satisfied the conservation constraint (10). To do this, choose $k = 0$ in the (FRW) metric, and choose the equation of state in the outer (OT) metric to be of the form

$$\bar{p} = \bar{\sigma}\bar{\rho}, \quad 0 < \bar{\sigma} < 1.$$

We have the following theorem, [9].

THEOREM 7. *Under these assumptions, the* (FRW) *equation of state* $p = \sigma\rho$, *where* σ *is a constant is determined by the algebraic equation*

$$\sigma = \frac{\bar{\sigma}(\bar{\sigma}+7)}{3(1-\bar{\sigma})}.$$

The solution is a shock-wave which satisfies the conservation constraint (9) , *and is given explicitly by*

$$M(\bar{r}) = 4\pi\gamma\bar{r}, \quad B(\bar{r}) = \bar{r}^{4\bar{\sigma}/(1+\bar{\sigma})}, \quad \rho(\bar{r}) = \frac{\gamma}{\bar{r}^2},$$

where

$$\gamma = \frac{1}{4\pi\mathcal{G}}\left(\frac{\bar{\sigma}}{1+6\bar{\sigma}+\bar{\sigma}^2}\right),$$

and

$$R(t) = R_0\left(\frac{\rho(t)}{\rho_0}\right)^{-1/3(1+\sigma)}, \quad \rho(t) = \frac{3\gamma}{\left[\sqrt{18\pi\mathcal{G}\gamma}\,(1+\sigma)(t-t_0)+\bar{r}_0\right]^2}.$$

The shock position is given by

$$\bar{r}(t) = \sqrt{18\pi\mathcal{G}\gamma}\,(1+\sigma)(t-t_0)+\bar{r}_0,$$

or

$$r(t) = \bar{r}_0 R_0^{-1}\left(\bar{r}(t)/\bar{r}_0\right)^{(1+3\sigma)/(3+3\sigma)}.$$

The above solution is an outgoing shock-wave and satisfied $\rho = 3\bar{\rho}$ on the shock surface. Note too that there is a singularity in backward time

$$t_* = t_0 - \frac{\bar{r}_0}{\sqrt{18\pi\mathcal{G}\gamma}\,(1+\sigma)},$$

and as $t \to t_*$, $\bar{r} \to 0$; ρ, $\bar{\rho}$, p and $\bar{p}$ all tend to infinity; and R and r tends to zero. If we take this as a cosmological model, then $t = t_*$ represents the initial big-bang singularity in which a shock-wave emerges from $\bar{r} = 0$. The solution describes an expanding shock-wave, (a blast wave), as t increases; cf. Figure 2.

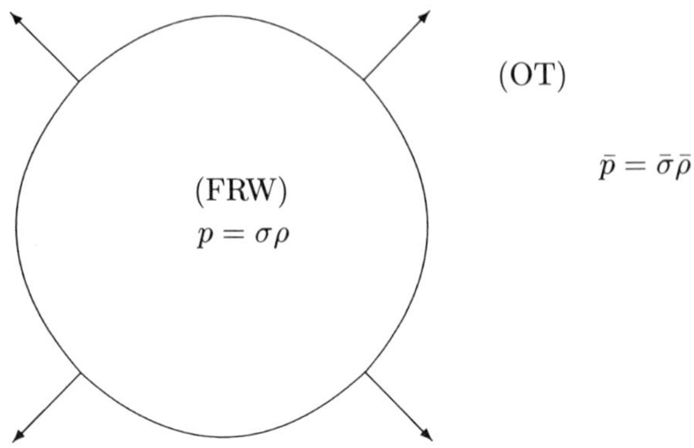

FIGURE 2

Note that the outer (OT) solution, with $p = \bar{\sigma}\bar{\rho}$, implies that $\bar{\rho}(\bar{r}) = \gamma/\bar{r}^2$, so $\bar{\rho} \to \infty$ as $\bar{r} \to 0$; this is referred to as a static, singular isothermal sphere. (For our shock-wave solution, the density is finite after some initial time $t > 0$, since $\bar{r} = 0$ is not in the domain of the solution after this initial time.)

The outer static solution is called a *static isothermal sphere* because the metric entries are time independent, and the constant sound speed models a gas at constant temperature. It is *singular* because it has an inverse-square density profile, and thus the density and pressure tend to ∞ at the center of the sphere. The Newtonian version of a static singular isothermal sphere is well known and is important in theories of how stars form from gaseous clouds [4]. The idea in the Newtonian case goes as follows: a star begins as a diffuse cloud of gas, which slowly contracts under its own gravitational force by radiating energy out through the gas cloud as gravitational potential energy is converted into kinetic energy. This contraction continues until the gas cloud reaches the point where the mean free path for transmission of light is small enough that light is scattered, instead of transmitted, through the cloud. The scattering of light within the gas cloud has the effect of equalizing the temperature within the cloud. At this point the gas begins to drift toward the most compact configuration of the density that balances the pressure when the equation of state is isothermal; namely, it drifts toward the configuration of a static, singular, isothermal sphere. Since this solution of the Newtonian case is also an inverse square in density and pressure, the density tends to infinity at the center of the sphere, and this ignites thermonuclear reactions. The result is a shock-wave explosion emanating from the center of the sphere, and this signifies the birth of a star. The explicit solution, which we present here, is an exact, *general relativistic version* of such a shock-wave explosion.

Finally, we have the following theorem; [9].

THEOREM 8. *There exist numbers $\sigma_1 < \sigma_2$, $0 < \sigma_1 = .458 < .744 = \sigma_2 < 1$ such that*

(i) *The Lax characteristic condition holds iff $0 < \sigma < \sigma_1$, and*
(ii) *The shock speed is less than the speed of light iff $0 < \sigma < \sigma_2$.*

We conclude that if $\sigma_1 < \sigma < \sigma_2$, a new type of shock-wave appears, in which the shock speed exceeds the characteristic (sound) speeds on each side of the shock. Note that $\sqrt{\sigma_1} \approx .677$, and $\sqrt{\sigma_2} = .863$. Thus, a fluid with sound speed near $\sqrt{\sigma_2} = .863$ can drive shock waves having speeds arbitrarily close to the speed of light.

5. Our dynamical shock-wave solution of the Einstein equation opens up intriguing possibilities in cosmology, as well as in the classical theory of shock-waves. For cosmology, our solution opens up the possibility, (with an explicit model), that there is a shock-wave at the edge of the universe. Since shock-waves are time irreversible solutions of the equations, because of the increase of entropy (in a generalized sense, see [3,6]), we infer from the mathematical theory of shock-waves that many solutions must decay time asymptotically to the same shock-wave. Thus, in contrast to the pure (FRW) solution, or to the (OS) solution, in our model we should not expect a unique time reversal of the solution all the way back to the initial big-bang singularity.

In the recent paper [10], we extend our results and show that our theory generates a large class of physically meaningful multi-dimensional shock-waves that model blast waves in a General Relativistic setting. In addition, we obtain formulas for the physical quantities that evolve according to the equations. These formulas are important for the numerical simulation of these solutions.

We believe that the new techniques which we have introduced have opened up a great many exciting possibilities for obtaining *multi-dimensional* shock-wave solutions for the relativistic Euler equations, as well as for the classical Euler equations by taking the limit $c \to \infty$. The implementation of this program is just beginning, because we have only recently obtained *explicit, closed-form* solutions of the equations. In addition, the rather "clean" form of the differential equations, which we have just recently derived, will enable us to easily implement them on a computer, and we can track the global evolution of the solutions, as t increases. This clearly opens up the possibility of finding new shock-wave phenomena. For example, we will be able to study such questions as: do singularities form in the equations as $t \to \infty$? (If so, are they related to "gravitational collapse"?) Do the fluid variables (pressure, density, sound speeds) remain well-behaved?

Finally, because no derivatives fall on the fluid variables in Einstein's equations (1), it suggests that the Einstein equations represent a sort of "hyperbolic regularization" of the compressible Euler equations, in a manner somewhat analogous to the role played by the Hamilton-Jacobi equations for scalar conservation laws. In [8], we showed that the *smoothness* of the metric translates over to *conservation* of the sources, and we continue to investigate the possibility that there could actually be numerical advantages to computing with Einstein-like regularizations of the Euler equations — analogous to the viscous regularizations that have played a fundamental role to date.

References

1. Glimm, J., *Solutions in the large for nonlinear hyperbolic systems of equations*, Comm. Pure Appl. Math, **18**, 697–715, (1965).
2. Israel, W., *Singular hypersurfaces and thin shells in general relativity*, It Nuovo Cimento, **54B**, 1–14, (1966).
3. Lax, P., *Hyperbolic systems of conservation laws, II*, Comm. Pure Appl. Math., **10**, 537–566, (1957).
4. Misner, C., Thorne, K., and Wheeler, J., *Gravitation*, Freeman (1973).
5. Oppenheimer, J. and Snyder, H., *On continued gravitational contraction*, Phys. Rev., **56**, 455–459, (1939).
6. Smoller, J., *Shock-Waves and Reaction-Diffusion Equations*, Springer-Verlag, (1983).
7. Smoller, J. and Temple, B., *Global solutions of the relativistic Euler equations*, Comm. Math. Phys., **156**, 67–99, (1993).
8. Smoller, J. and Temple, B., *Shock-wave solutions of the Einstein equations*, Arch. Rat. Mech. Anal., **128**, 249–297, (1994).
9. Smoller, J. and Temple, B., *Astrophysical shock-wave solutions of the Einstein equations*, Phys. Rev. D, 2733–2743, (1995).
10. Smoller, J. and Temple, B., *General relativistic shock-waves that extend the Oppenheimer-Snyder model*, Arch. Rat. Mech. Anal., (to appear).
11. Weinberg, S., *Gravitation and Cosmology*, Wiley, (1972).

University of Michigan (J.A.S.)
Ann Arbor, MI (48109)

University of California, Davis (J.B.T.)
Davis, CA (95616)

Contemporary Mathematics
Volume **208**, 1997

DYADIC HARMONIC ANALYSIS

WILLIAM R. WADE

ABSTRACT. We discuss progress made on the problems of uniqueness, convergence, and summability of Walsh series. We pay special attention to the role Professor V.L. Shapiro and his students played in this development. Our bibliography contains 191 entries.

1. INTRODUCTION

1.1. The Walsh system. Let $\mathbf{N} := \{0, 1, \dots\}$ represent the set of natural numbers. Define a function $r(x)$ on the unit interval $\mathbf{I} := [0, 1)$ by

$$r(x) := \begin{cases} 1 & x \in [0, 1/2) \\ -1 & x \in [1/2, 1) \end{cases}$$

and extend it to all of $[0, \infty)$ by periodicity of period 1. The *Rademacher system* is the collection of functions defined by $r_k(x) := r(2^k x)$ for $k \in \mathbf{N}$. This system is *orthonormal* in $L^2(\mathbf{I})$, i.e.,

$$\int_0^1 r_k(x) r_j(x)\, dx = \begin{cases} 1 & \text{if } k = j \\ 0 & \text{if } k \neq j, \end{cases}$$

but not complete.

The *Walsh system* $\{w_n\}_{n=0}^\infty$, the completion of the Rademacher system, is defined as follows. For each $n \in \mathbf{N}$ let $n = \sum_{k=0}^\infty n_k 2^k$ represent the dyadic expansion of n, i.e., each $n_k = 0$ or 1. Then the nth *Walsh function* w_n is defined by

$$(1) \qquad w_n(x) := \prod_{k=0}^\infty r_k^{n_k}(x), \qquad x \in \mathbf{I}.$$

1991 *Mathematics Subject Classification.* Primary 42C10; Seconday 43A75.

Key words and phrases. Walsh series, Dyadic Group, Sets of uniqueness, convergence of Walsh-Fourier series, Sets of divergence, Quasimeasures, Sidon inequalities, $(C, 1)$ summability of Walsh–Fourier series.

This article, an expanded version of a talk given at the Southern California Seminar on Analysis and Differential Equations, November 1995, is dedicated to Professor V.L. Shapiro on the occasion of his retirement from teaching at the University of California, Riverside.

This paper is in final form and will be submitted no where else.

Thus $w_0 \equiv 1$,

$$w_1(x) = \begin{cases} 1 & x \in [0, 1/2) \\ -1 & x \in [1/2, 1) \end{cases}$$

$$w_2(x) = \begin{cases} 1 & x \in [0, 1/4) \\ -1 & x \in [1/4, 1/2) \\ 1 & x \in [1/2, 3/4) \\ -1 & x \in [3/4, 1) \end{cases}$$

and

$$w_3(x) = \begin{cases} 1 & x \in [0, 1/4) \\ -1 & x \in [1/4, 3/4) \\ 1 & x \in [3/4, 1). \end{cases}$$

Among orthonormal systems, the Walsh system satisfies some unusual properties. For each $n > 0$, it is evident that the Walsh function w_n is piecewise constant on $\mathbf{I}$, has range $\{1, -1\}$, and if $m \in \mathbf{N}$ satisfies $2^m \le n < 2^{m+1}$, then w_n changes signs once on every *dyadic interval* of the form $I(k, m) := [k2^{-m}, (k+1)2^{-m})$, for $k = 0, 1, \dots, 2^m - 1$. Since the product of any two Walsh functions is a Walsh function, it follows that the Walsh system is *orthonormal* on $L^2(\mathbf{I})$, i.e.,

$$\int_0^1 w_n(x) w_m(x)\, dx = \begin{cases} 1 & \text{if } n = m \\ 0 & \text{if } n \ne m. \end{cases}$$

It is also easy to see by induction that the *Walsh Dirichlet kernels*

$$D_k(x, t) := \sum_{j=0}^{k-1} w_j(x) w_j(t)$$

have nonnegative 2^nth partial sums; in fact if $I_n(x) := I(k, n)$, where k is determined by the property $x \in I(k, n)$, then

$$D_{2^n}(x, t) = \begin{cases} 2^n & t \in I_n(x) \\ 0 & t \notin I_n(x) \end{cases} \tag{2}$$

for all $n \in \mathbf{N}$.

These properties actually characterize the Walsh system in the following sense. Waterman ([166] and [167]) proved that if r_k is ANY sequence of real functions on $\mathbf{I}$ dominated almost everywhere by 1, and if the product system w_n defined by (1) is an orthogonal system with essential range $\{1, -1\}$ whose L^2 norms are constant for sufficiently large indices, then $\{w_n\}$ is a measure preserving transformation of the Walsh system. Levizov [68] proved that any orthonormal system whose functions f_n have exactly n sign changes on $\mathbf{I}$, have range $\{1, -1\}$, and satisfy $f_n(0) = 1$ is (a reordering of) the Walsh system. And Price [103] proved that the Walsh system is the only orthonormal system whose functions alternate signs on finer and finer partitions of $\mathbf{I}$ with nonnegative Dirichlet kernels of order 2^n.

Walsh series can be used to approximate any function $f \in L^1(\mathbf{I})$. To see why this is true, suppose $S = \sum_{k=0}^{\infty} a_k w_k$ is a Walsh series which converges to f uniformly, or in some L^p norm, $p \ge 1$. Then

$$\int_0^1 f(x) w_j(x)\, dx = \int_0^1 \left(\sum_{k=0}^{\infty} a_k w_k(x) \right) w_j(x)\, dx = \sum_{k=0}^{\infty} a_k \int_0^1 w_k(x) w_j(x)\, dx = a_j$$

by orthogonality. Thus to find a Walsh series which converges to some function f uniformly (or in some L^p norm), we must use the *Walsh–Fourier series* of f, i.e., the Walsh series of the form

$$(Sf)(x) := \sum_{k=0}^{\infty} \widehat{f}(k) w_k(x), \quad \text{where} \quad \widehat{f}(k) := \int_0^1 f(x) w_k(x)\, dx. \tag{3}$$

(The numbers $\widehat{f}(k)$ are called *Walsh–Fourier coefficients* of f.)

Does this series actually converge to f? Yes, provided we use the 2^nth partial sums. Indeed, by (3) and (2),

$$\begin{aligned}
(S_{2^n} f)(x) &:= \sum_{k=0}^{2^n-1} \widehat{f}(k) w_k(x) \\
&= \sum_{k=0}^{2^n-1} \left(\int_0^1 f(t) w_k(t)\, dt \right) w_k(x) \\
&= \int_0^1 f(t) D_{2^n}(x,t)\, dt \\
&= 2^n \int_{I_n(x)} f(t)\, dt.
\end{aligned}$$

Since the Lebesgue measure of $I_n(x)$ is 2^{-n}, and $I_n(x)$ shrinks to x as $n \to \infty$, it follows from well known results about differentiating an indefinite integral that $S_{2^n} f \to f$ uniformly when f is continuous and periodic on $\mathbf{I}$, and $S_{2^n} f \to f$ almost everywhere when $f \in L^1(\mathbf{I})$. (Notice, then, that the Walsh system is complete. For if $\widehat{f}(k) = 0$ for all $k \in \mathbf{N}$, then $f = \lim_{n\to\infty} S_{2^n} f = 0$ almost everywhere on $\mathbf{I}$.)

These simple calculations introduce two of the main questions which interest people who study Walsh series:

1) [UNIQUENESS] Given a Walsh series S, what conditions guarantee that S is Sf, the Walsh–Fourier of some $f \in L^1(\mathbf{I})$?
2) [CONVERGENCE] Given $f \in L^1(\mathbf{I})$, what conditions guarantee that Sf, the Walsh–Fourier series of f, converges to f?

Of course, answers to these questions depend on what one means by "converges" (e.g., pointwise, uniform, norm, almost everywhere) and which method of convergence one uses (e.g., the full sequence of partial sums, some subsequence of partial sums, or some summability method). As we saw above, answers to these questions are quite easy to obtain if uniform convergence and dyadic partial sums are used. We shall survey other answers to these questions below. We shall pay special attention to the role Professor Shapiro and his students have played in this development.

Before we proceed farther we must apologize to the reader who works in classical harmonic analysis on the circle group, and to most of the pioneers who have gone before us. Although many of the results mentioned below have older trigonometric analogues, we have not included much information of this type. We have done this not to slight the older area, but to keep this article down to a manageable size. Further information of this type can be found in the almost encyclopedic monograph by Schipp, Wade, Simon, and Pál [122]. This work also includes a broader sample of the work done in dyadic harmonic analysis than is given here together with many general references on the subject. For a less comprehensive, more classically oriented reference, see Golubov, Efimov, and Skvortsov [43].

1.2. Early history. Although Euler, D'Alembert, and Daniel Bernoulli stumbled onto concrete examples as early as 1748, the study of orthonormal systems began in 1811 with Fourier's treatise on the theory of heat which introduced trigonometric Fourier series to the mathematical community. Guided by a connection between trigonometric series and the distribution of heat, Fourier conjectured that "any" function could be approximated by its Fourier series. Research on this conjecture, one of the central themes of analysis during the nineteenth century, lead to many discoveries and clarifications including the function concept, the distinction between absolute, uniform, and pointwise convergence, set theory and transfinite numbers, the Riemann integral, and the need for something like the Lebesgue integral. At the same time, the connection between the trigonometric system and partial differential equations lead Sturm and Liouville to introduce a class of orthonormal systems which also contributed to the understanding of these concepts.

For a time it was thought that any continuous function could be uniformly approximated by such series. When that position (even in the trigonometric case) was shown to be untenable by Du Bois Reymond in 1873, the search for orthonormal series with good approximation properties was broadened to include non Sturm–Liouville systems. Haar [47], writing a dissertation under Hilbert's direction which was published in 1909, found such a system.

Since the Haar system is unbounded and the theory of bounded orthonormal systems is much nicer than that of unbounded types, it was natural to seek a bounded system which also had good convergence properties. Walsh [164], writing a dissertation under G.D. Birkhoff at Harvard University which was published in 1923, used linear combinations of Haar functions to introduce a bounded orthonormal system which is a rearrangement of the Walsh system defined above. As we have already seen, this system has good convergence properties if attention is restricted to the 2^nth partial sums.

The next milestone occurred in 1933 when Paley [102] noticed that the Walsh system could be defined by using products of Rademacher functions (as we did above). He showed that the Walsh system is the completion of the Rademacher system, and proved that the Walsh–Fourier series of a function $f \in L^p(\mathbf{I})$ converges to f in L^p norm, when $1 < p < \infty$. Because the Rademacher system plays a central role in the study of independent, identically distributed random variables, this paper forever linked Walsh theory with probability theory.

The final step toward a general theory took place in 1949 when Fine [27], a student of Zygmund at the University of Pennsylvania, showed that the Walsh system could be identified with the characters of a compact abelian group, i.e., the study of Walsh series is a special case of the general theory of harmonic analysis on groups. This is where dyadic harmonic analysis begins.

2. Why study Walsh series?

2.1. Analogy with the trigonometric system. The Walsh system, the best piecewise constant orthonormal approximation to the trigonometric system, is analogous to the trigonometric system only simpler. (Let $t = 2\pi x$ and compare the graphs of $w_0(x)$ with $\cos(0 \cdot t)$, of $w_1(x)$ with $\sin(1 \cdot t)$, of $w_2(x)$ with $\sin(2 \cdot t)$, and of $w_3(x)$ with $\cos(1 \cdot t)$.) Consequently, many results for Walsh and trigonometric series are the same. This suggests at least three reasons for studying Walsh series.

1) Trigonometric results can be discovered by considering the Walsh case first. This has been happening with greater frequency as the study of Walsh series has grown more mature. Recent examples include: Garnett and Jones [34] (Hardy

spaces), Kochneff, Sagher,and Zhou [62] (lacunary series), Schipp [121] (Sidon inequalities), Weisz [169] (strong summability), and Aubertin and Fournier [5] (integrability conditions).

2) Some aspects of Fourier analysis might be better taught by considering the Walsh case first. This would convey the overall aims of the discipline in the simplest possible setting. An excellent example of this is the Carleson–Hunt technique for proving almost everywhere convergence of Fourier series. The basic idea, using local Fourier coefficients to systematically discard a number of exceptional sets where the maximal function is not of restricted weak type (p, p), is the same for both proofs. However, in the Walsh case one need not deal with Hilbert transforms and instead of seven exceptional sets, in the Walsh case there are only three.

3) Since the Walsh system is the simplest bounded orthonormal system, it is an excellent place to look for counterexamples. Enflo [25], for example, used the Walsh system to solve the basis problem by constructing a separable Banach space which had no basis. (Schipp [119] showed that a class of such Banach spaces can be defined using VMO, functions of vanishing mean oscillation based on Walsh series.)

2.2. Connections with probability theory. Another reason to study the Walsh system is its connection with probability theory. Jerison and Rabson [55] were the first to notice that the sequence of 2^nth partial sums of any Walsh series forms a martingale (with respect to the σ algebras generated by dyadic intervals of length 2^{-n}). Thus not only is the Walsh system the completion of the Rademacher system, it also serves as a prototype for a large class of martingales. This connection has benefited both sides. The fundamental theorem of martingales was first proved for the Walsh system (Paley [102]), and the Walsh system has been used to characterize Banach spaces which possess Burkholder's UMD property (Wenzel [172]). Conversely, martingale theory has provided Walsh series with a dyadic analogue of the classical Hardy spaces (see Garsia [35]). (The classical Hardy spaces are of no use for Walsh series because there is no connection between boundary values of analytic functions and Walsh series.)

We shall denote the dyadic Hardy spaces by $H^p(\mathbf{I})$, $0 < p < \infty$. They are defined to be the space of functions (or, in the case $0 < p < 1$, distributions) f whose *dyadic maximal function* $f^* := \sup_{n \in \mathbf{N}} |S_{2^n} f|$ belong to $L^p(\mathbf{I})$. Although the spaces $L^p(\mathbf{I})$ and $H^p(\mathbf{I})$ are equivalent for $1 < p < \infty$, $H^p(\mathbf{I})$ is a proper subset of $L^p(\mathbf{I})$ for each $0 < p \leq 1$. In particular, many results which fail to hold for $L^1(\mathbf{I})$ come alive again when the smaller space $H := H^1(\mathbf{I})$ is used. In a similar way, the dual of H, BMO, serves as a replacement space for $L^\infty(\mathbf{I})$ in many situations.

2.3. Applications. Applications provide an even more compelling reason to study Walsh series. Because the Walsh system is similar to the trigonometric system, many of the standard applications of the classical Fourier transform also work for the Walsh transform. These include filtering, image enhancement, data compression, and approximation. Moreover, since the Walsh system is an orthogonal system, it can be used in transmission of data. (Harmuth, [48] and [49], has written two books on this subject.) But since the Walsh system is so simple (no complex arithmetic, the values ± 1 can easily be simulated using open/closed switches, hence Walsh transforms can easily be built as hardware), these applications can be carried out more quickly (and with more primitive computers) in the Walsh case than in the trigonometric case. For this reason, the Walsh system is preferred when weight and space are at a premium, e.g., the Mariner spacecraft used the Walsh transform instead of the Fourier transform.

There are, however, a whole class of applications which have no connection with the trigonometric system but rely instead on peculiarities of the Walsh system itself. Here are six examples. Whelchel and Guinn [173] introduced a method for collecting data across several channels at once which simulates taking the Walsh transform of the data. (For certain situations, e.g., spectroscopy and radio astronomy, this reduces background noise in weak signals.) Lanning and Johnson [67] have used the Walsh transform to provide a fast, simple, accurate means of dividing well log data into groups corresponding to different rock layers. (The fact that the Walsh functions have jump discontinuities played a crucial role here.) Optican and Richmond [100] have shown how two dimensional Walsh functions can be used to decipher the neural code which carries visual information from eyes to brain in primates. Bethke [10] has used Walsh functions to analyze genetic algorithms, iterative procedures for optimizing nondifferentiable functions when traditional methods either fail to converge, or converge badly at suboptimal points. Bohn [15] has examined the problem of error propagation resulting from truncation of infinite dimensional operational matrices used to form algebraic combinations of Walsh Fourier coefficients which lead to solutions of differential equations in certain types of dynamical systems. (He actually derived recursive expressions for evaluating the residual terms associated with such a truncation.) And Daly [24] has used Walsh functions to find a stopping rule that gives the optimal histogram bin width to approximate a probability density function. (This procedure, which minimizes the L^2 error between the histogram and the probability density function, requires no prior knowledge of the density function and seems to compare favorably with two older methods by Scott, and by Freedman and Diaconis.)

Applications of a more speculative nature include statistical measurements by Maqusi [75] and a substitution for string theory by Vladimirov, Volovich, and Zelonov [154].

2.4. Connections with abstract harmonic analysis. A final reason to study Walsh series is their connection with harmonic analysis on groups and fields. Fine [27] and Vilenkin [153] independently discovered that the Walsh system is the set of characters of the *dyadic group* $\mathbf{G}$, i.e., the set of sequences $(x_0, x_1, \dots)$ where each $x_j = 0$ or 1 and addition is performed coordinatewise, modulo 2, with no carrying. This group is a prototype for the entire class of zero dimensional compact abelian metrizable groups (see Vilenkin [153]). Hence Walsh functions are a key ingredient in the study of harmonic analysis on nonclassical compact abelian groups.

The Walsh functions occupy an even more central position in the theory of harmonic analysis on locally compact fields. To describe the situation, recall that given any prime p, there are two fields which can be formed from the doubly infinite sequences of the form

$$\{(\dots, x_{-1}, x_0, x_1, \dots) : 0 \le x_j < p \text{ such that } x_j = 0 \text{ for large negative } j\} :$$

the p–adic field (whose elements are of characteristic p) and the p–series field (whose elements are of characteristic zero). It turns out that a nondiscrete locally compact topological field with reasonable topology can be one of only three types: a classical field (e.g., the real line), a p-adic field, or a p-series field for some prime p (see Taibleson [142]).

What is the connection with Walsh series? The simplest p-series field, the 2-*series field* $\mathbf{F}$, is a prototype for all p-series fields, and the dyadic group $\mathbf{G}$ is the ring of integers of $\mathbf{F}$. (Notice that $\mathbf{G}$ can be considered a subset of $\mathbf{F}$ by taking each element $(x_0, x_1, \dots) \in \mathbf{G}$ to $(\dots, 0, 0, x_0, x_1, \dots) \in \mathbf{F}$.)

These discoveries have reinforced the analogy between the Walsh system and the trigonometric system, since **G** plays the role of the circle group, and **F** plays the role of the real line. They have also affected the range and scope of techniques available to dyadic harmonic analysis in a profound way.

To see how this works, notice that *Fine's map*,

$$\phi : (\dots, x_0, x_1, \dots) \longmapsto \sum_{k=\infty}^{\infty} x_k 2^{k-1},$$

takes **F** onto $[0, \infty)$ and **G** onto $[0, 1]$. This allows the topological and algebraic structures of **F** to be transferred to **I** and $[0, \infty)$. For example, define the *dyadic sum* of two numbers $x, y \in \mathbf{I}$ by

$$x \dot{+} y := \phi(\phi^{-1}(x) + \phi^{-1}(y)),$$

and define a function f to be *W–continuous* on **I** if $f \circ \phi$ is continuous on **G**. (Every continuous function is W–continuous, but not conversely.)

What advantage is there for introducing these nonclassical concepts on **I**? Since the characters of **G** are given by $\psi_n \equiv w_n \circ \phi$, these concepts allow greater freedom when adapting proofs from the trigonometric case to the Walsh case. Here are two examples: 1) Many arguments for classical Fourier series use trigonometric identities (equivalently, the exponential properties of the complex characters e^{ix}) to simplify expressions. Although these properties are not valid for the Walsh functions under the usual addition on **I**, they are valid if we use dyadic addition: for each $k \in \mathbf{N}$ $w_k(x \dot{+} y) = w_k(x)w_k(y)$ almost everywhere on **I**. (This holds because the ψ_n's are characters of **G** and ϕ takes addition on the group to dyadic addition on **I**.) 2) Many trigonometric proofs, appealing to powerful tools of functional analysis of the Banach space of continuous functions, use the fact that the trigonometric system is continuous and periodic (equivalently, the complex characters e^{ix} are continuous on the circle group). Although these techniques are invalid for the Walsh system (because for each $k > 0$, w_k is not continuous on **I** under the usual topology), they are valid if we use W–continuity: each w_k is W–continuous on **I**. Why not work directly on **G** and avoid these ad hoc concepts on **I**? Because **I** has a natural linear ordering which is useful in many ways, e.g., for defining indefinite integrals. Thus much work is done on **I** where the best of both worlds is available.

3. Uniqueness

3.1. Sets of Uniqueness. Let

$$S(x) := \sum_{k=0}^{\infty} a_k w_k(x)$$

represent an arbitrary Walsh series. It is easy to see that if $S_n \to f$ uniformly as $n \to \infty$, then f is W–continuous and $S = Sf$. How far can the hypothesis about uniform convergence be weakened?

The first results in this direction concentrated on the case when $f \equiv 0$. Walsh [164] proved that if E_0 is a finite set and $S_n \to 0$ uniformly on compact subsets of $\mathbf{I} \setminus E_0$, then $a_k = 0$ for all $k \in \mathbf{N}$. Next, Vilenkin [153] obtained the same conclusion under the hypothesis that $S_n \to 0$ everywhere on **I**.

At this point it was natural to ask the following question. Can the hypothesis about pointwise convergence in Vilenkin's result be relaxed on certain sets? If so, how big can these sets be?

To discuss these questions, we introduce a concept borrowed from trigonometric series. A set $E \subseteq \mathbf{I}$ is called a *set of uniqueness* if every Walsh series $S = \sum_{k=0}^{\infty} a_k w_k$ which satisfies $S_n(x) \to 0$ for all $x \in \mathbf{I} \setminus E$ also satisfies $a_k = 0$ for all $k \in \mathbf{N}$. If a set is not a set of uniqueness, it is called a *set of multiplicity*. In this language, Vilenkin's result cited above says that the empty set is a set of uniqueness. Are there any others?

Fine [27] and Shneider [126] showed that any countable set is a set of uniqueness. On the other hand, Shneider showed that every set of positive Lebesgue measure is a set of multiplicity, i.e., if $|E_0| > 0$, then there is a nonzero Walsh series S which converges to zero everywhere off E_0. Thus sets of multiplicity can be thick in a measure theoretic sense. They can also be quite thick in a topological sense: Coury [21] showed that sets of multiplicity can be dense in $\mathbf{I}$.

How thick can a set of uniqueness be? From the naive point of view taught in a first course in measure theory, one might guess that sets of uniqueness coincide with sets of Lebesgue measure zero. Shneider [126] showed that the picture is very much more complicated. On the one hand, he proved that the Cantor "middle halves" set is a set of uniqueness. His argument, similar to Rajchman's proof of the trigonometric analogue, uses formal products of Walsh polynomials and Walsh series to obtain sufficient conditions for sets of uniqueness for Walsh series. These conditions can be summarized succinctly using the language of functional analysis: a set E is a set of uniqueness if there is a sequence of functions whose Walsh–Fourier series are uniformly convergent, which vanish on E and converge to 1 in the weak* topology generated by the space of pseudofunctions. Lippman and Wade [70] have shown that these sets of uniqueness (now called *elementary sets of uniqueness*) can be used as building blocks for uniqueness in the following sense: every closed set of uniqueness is a countable union of elementary sets of uniqueness. (This problem was posed by Professor Shapiro to Gary Lippman while he was a graduate student at the University of California, Riverside. Lippman is Shapiro's 10th Ph.D. student, and Wade is his 7th).

On the other hand, Shneider [126] proved that not all sets of measure zero are sets of uniqueness. Indeed, using infinite products of Rademacher polynomials, Shneider constructed a nonzero Walsh series which converges almost everywhere to zero. This situation is not confined to Lebesgue measure. Skvortsov [135] has shown that given any positive, increasing h on $[0, \infty)$ with $h(0+) = 0$, there exist sets of multiplicity which have h measure zero. Thus among the uncountable sets of measure zero there are both sets of uniqueness and sets of multiplicity.

Putting conditions on the partial sums, or on the coefficients, does not seem to help things much (unless, of course, those conditions are strong enough to insure norm convergence). Schipp [113] showed that there exist nonzero Walsh series which converge almost everywhere to zero whose partial sums are actually nonnegative. Skvortsov [136] showed that there exist nonzero Walsh series which converge almost everywhere to zero such that the ℓ^2 norm of its coefficients diverge as slowly as one wishes. (Navasardyan [93] has extended this result to double Walsh series.) Gevorkyan [41] obtained a similar result for Walsh series whose coefficients decay as slowly as one wishes, provided the ℓ^2 norm is still infinite. (For information about uniqueness of Walsh series whose coefficients belong to ℓ^p for $1 < p \leq 2$, see Gevorkyan [39].)

Can one combine sets of uniqueness to get other sets of uniqueness? Not always. It is easy to see that the union of two sets of uniqueness may not be a set of uniqueness (see Zygmund [191], for example). Shneider [126], however, proved that a finite union of closed sets of uniqueness is a set of uniqueness. Wade [156] extended this to countable unions of closed sets of uniqueness, and Kholshchevnikova [59] showed that this result still holds if the sets of uniqueness are only relatively closed (with respect to the union). In particular, the union of two disjoint $\mathcal{G}_\delta$ sets of uniqueness is a set of uniqueness.

We have seen that sets of uniqueness must be thin in a measure theoretic sense. They must also be thin in a topological sense. Kholshchevnikova [60], solving a long standing open problem, showed that if E is a set of uniqueness which is the union of a $\mathcal{G}_\delta$ set and a set of the first category, then E is of the first category. In particular, any Borel set of uniqueness which is both analytic and coanalytic is of the first category.

If the coefficients decay rapidly enough, however, then the sets of uniqueness can get quite thick. Shapiro [124] showed that given any sequence of numbers $\epsilon_k \downarrow 0$ as $k \to \infty$, there is a set $E \subset \mathbf{I}$ whose Lebesgue measure is as close to 1 as one wishes such that if $S = \sum_{k=0}^{\infty} a_k w_k$ converges to zero off E and $|a_k| \leq \epsilon_k$ for all $k \in \mathbf{N}$, then S is the zero series. Gevorkyan [38] proved that such a set of uniqueness can have full measure. Gevorkyan [40] also has obtained further results of this type which compare one growth condition with another.

An even stronger result is true if enough of the coefficients are zero. Stechkin and Ul'yanov [138] proved that if the Lebesgue measure of a set $E \subset \mathbf{I}$ is $< 1/2$, then E is a set of uniqueness for all Rademacher series with respect to any regular method of summability. A weaker multidimensional version of this result has been obtained by Mushegyan [92]. He proved that if M_i is regular in $\mathbf{R}^d$ (i.e., $M_i \subseteq \mathbf{R}^d$ and given any integer lattice point $\mathbf{n} \in \mathbf{N}^d$ there is a subsequence $\{i_k\}$ such that $\mathbf{n} \in M_{i_k}$ for all k) and if P_i is a sequence of Rademacher polynomials of the form

$$P_i(\mathbf{x}) = \sum_{\mathbf{n} \in M_i} a_{\mathbf{n}}^{(i)} r_{\mathbf{n}}(\mathbf{x})$$

which satisfy $P_i(\mathbf{x}) \to 0$ for all points $\mathbf{x}$ in some subset of the d–dimensional cube $\mathbf{I}^d$ of measure $> 1 - 2^{-d}$, then $a_{\mathbf{n}}^{(i)} \to 0$ as $i \to \infty$ for each $\mathbf{n} \in \mathbf{N}^d$. Tetunashvili [151] has also investigated uniqueness of multiple Rademacher series.

A Rademacher series is just a Walsh series with gaps. What happens to the result of Stechkin and Ul'yanov when the gaps become smaller? Roider [107] has shown that when the gaps occur at integers with large *Vielfalt* (namely, if there is an N such that $\widehat{f}(k) = 0$ whenever $k = 2^{n_1} + \cdots + 2^{n_m}$ for some $m > N$), then every $f \in L^1(\mathbf{I})$ which assumes only integer values is a Walsh polynomial.

Despite a fairly long history, concrete examples of sets of uniqueness for Walsh series with unrestricted coefficients are scarce, and there are still many open questions. Wade [159], using Shneider's elementary sets of uniqueness, showed that $H(n)$ sets (a particular class of nowhere dense sets of Lebesgue measure zero) are sets of uniqueness. (Although this produced some new examples of sets of uniqueness, we still do not know whether the Cantor "middle thirds" set is a set of uniqueness for Walsh series.) Yoneda [178], building on this foundation, showed that any closed subgroup of the dyadic group of Haar measure zero is a set of uniqueness. Yoneda [183] has also shown that the class of $H(n)$ sets is closed under translations, dilations, and unions.

Although it is not known if Yoneda's result about subgroups holds when "Borel" replaces "closed," he did use it to characterize a particular subclass of sets of uniqueness. To formulate his results, recall that a subset E of $\mathbf{G}$ is a *Dirichlet set* if

$$\liminf_{n\to\infty} \sup_{x\in E} |1 - \psi_n(x)| = 0.$$

Yoneda [178] proved that given $E \subset \mathbf{G}$, E is a Dirichlet set if and only if it is a subset of some closed subgroup of Haar measure zero. In particular, every Dirichlet set is a set of uniqueness for Walsh series. Yoneda [186] improved this observation by obtaining an analogue of Salem's result: every Dirichlet set E is actually a *strong U-set*, i.e., every Walsh series $\sum_{k=0}^{\infty} a_k\psi_k$ which converges to zero off E also satisfies

$$\sup_k |a_k| = \liminf_{k\to\infty} |a_k|.$$

Yoneda [182] has also identified the following structural condition sufficient for Dirichlet sets (analogous to a theorem of Kahane for trigonometric series): if $\mathcal{N}_n(E)$ represents the number of dyadic intervals of rank n which intersect E, and $E \subset \mathbf{G}$ satisfies

$$\liminf_{n\to\infty} \frac{\mathcal{N}_n(E)}{n} = 0,$$

then E is a Dirichlet set. On the other hand, there exist Dirichlet sets which satisfy $\mathcal{N}_n(E)/n \to \infty$ as $n \to \infty$. In particular, there exist Dirichlet sets of positive Hausdorff dimension.

Yoneda [186] used Dirichlet sets to generate sets of uniqueness for classes of Walsh series described by how rapidly their coefficients decay. He proved that given a nonnegative function ψ on $[0,\infty)$ which is continuous at 0, every Dirichlet set E is a set of uniqueness for the class of Walsh series which satisfies

$$\sum_{k=2^n}^{2^{n+1}-1} \psi(|a_k|) = O(1), \qquad \text{as} \quad n \to \infty.$$

This includes the classes $\sum_{k=2^n}^{2^{n+1}-1} |a_k|^2 = O(1)$ and $\sum_{k=2^n}^{2^{n+1}-1} a_k\psi_k(x) = O(1)$ uniformly in x, as $n \to \infty$. This last observation is best possible in the following sense. Given any sequence $\epsilon_n \downarrow 0$ which satisfies $1/\epsilon_n = o(n)$ as $n \to \infty$, there is a Dirichlet set which is a set of multiplicity for the class of Walsh series which satisfies

$$\frac{1}{2^n} \sum_{k=0}^{2^n-1} |a_k|^2 = o(\epsilon_n) \qquad \text{as} \quad n \to \infty.$$

Sets of uniqueness for multiple Walsh series are also thin. Lukomskiĭ [73] has identified a condition on the cross sections of a set $E \subset \mathbf{I}^d$, $d \geq 2$, sufficient for E to be a set of uniqueness for d-dimensional Walsh series. He used it to build sets of uniqueness in higher dimensions from lower ones: a set E is a set of uniqueness for d-dimensional Walsh series, $d \geq 1$, if and only if $E \times \mathbf{I}$ is a set of uniqueness for $(d+1)$-dimensional Walsh series. It follows that every locally convex surface $E \subset \mathbf{I}^d$, $d \geq 2$, is a set of uniqueness for d-dimensional Walsh series.

Finally, in the single variable case there is an intriguing connection between sets of uniqueness and number theory. In classical number theory, the set T of algebraic integers $\theta > 1$ whose conjugates lie in the unit disk $\{z \in \mathbf{C} : |z| \leq 1\}$ are divided into two classes: Pisot numbers, those whose conjugates lie in the open disk $\{z \in \mathbf{C} : |z| < 1\}$, and Salem numbers, the rest of the numbers in the set T. Salem and Zygmund showed that for trigonometric series, a Cantor set

$$C(\theta) := \left\{ \sum_{n=1}^{\infty} \epsilon_n \theta^{-n} : \epsilon_n = 0, \text{ or } 1 \right\}$$

is a set of uniqueness if and only if θ is a Pisot number. It is not known whether this result holds for the Walsh case. In fact, it is not yet known whether the Cantor middle thirds set is a set of uniqueness for Walsh series.

If we look at these results on the dyadic group, however, there are some positive results. The field structure of the dyadic field $\mathbf{F}$ can be described by identifying its elements as coefficients of a field of Laurent series in a variable $1/t$, where addition and multiplication are done formally and the coefficients are viewed as elements of the binary field $GF(2) := \{0, 1\}$. Since $\mathbf{F}$ is a formal Laurent series field, the set T of Pisot numbers and Salem numbers in $\mathbf{F}$ is already well known (see Bateman and Duquette [7].) Indeed, fix $c > 1$ and define the absolute value of an $x \in \mathbf{F}$ by

$$|x| = c^j,$$

where j is the smallest index for which the coefficient x_j of x is nonzero. A basis of $\mathbf{F}$ is given by $\{e_j\}$, where the coefficients of $e_j \in \mathbf{F}$ are described by $e_{jk} = 0$ for $j \neq k$ and $e_{jj} = 1$. Let $R := B[t]$ represent the ring of polynomials over the binary field $B := GF(2)$ and let $R' := B(t)$ represent the ring of rational functions formed by quotients of elements in R. Then T is the collection of elements $\theta \in \mathbf{F}$ with $|\theta| > 1$, such that $\sum \theta_j t^{-j}$ is algebraic and integral (over the ring R), and all of its remaining conjugates (in R') have absolute value ≤ 1.

Since $\mathbf{G} \subset \mathbf{F}$ (in fact, $\mathbf{G}$ is the ring of integers in $\mathbf{F}$), we can also define $\theta^{-n} \in \mathbf{F}$ for $\theta \in \mathbf{G}$ and $n \in \mathbf{N}$. Thus we can write down a dyadic group analogue of the Cantor sets defined above:

$$F(\theta) := \left\{ \sum_{n=1}^{\infty} \epsilon_n \theta^{-n} : \epsilon_n = 0 \text{ or } 1 \right\}$$

where $\theta \in \mathbf{F}$ and $|\theta| > c^2$. Unfortunately, these sets and uniqueness have no connection with number theory. Each of them is a subgroup of $\mathbf{G}$ of Haar measure zero, hence (see Yoneda [178]), a set of uniqueness.

How can we recapture the connection between sets of uniqueness and number theory on the dyadic group? Aubertin [3] showed that one answer to this question is to modify the sets $F(\theta)$. For each $\theta \in \mathbf{F}$, $|\theta| > c^2$, he defined a class of Cantor like sets by

$$E(\theta) := \left\{ \sum_{n=2}^{\infty} \epsilon_n \theta^{-n} : \epsilon_n \in I_\theta \right\},$$

where $I_\theta := \{e_0, e_1, \ldots, e_{2L(\theta)-1}\}$ and $L : \mathbf{F} \to \mathbf{Z}$ is defined by $L(x) = j$ if and only if $|x| = c^j$. (The sum starts with $n = 2$ so that the sets $E(\theta)$ are subsets of the dyadic group $\mathbf{G}$.) Aubertin proved that each set $E(\theta)$ is a perfect set of Haar measure zero and that $E(\theta)$ is a set of uniqueness if and only if $\theta \in T$.

On the surface, this result is disappointing because there is no distinction between Salem and Pisot numbers which appears in the trigonometric case. However, it reveals an even sharper departure from the classical case in at least two ways. First, since T is a countable set, we see that in a set theoretical sense, most of the sets $E(\theta)$ are not sets of uniqueness. Secondly, Grandet-Hugot [45] proved that the Pisot numbers in T do not form a closed subset but in fact are dense in $\mathbf{F}$. Thus unlike the trigonometric case, a set $E(\theta)$ will not remain a set of multiplicity under small perturbations of the parameter θ.

3.2. Dyadic sets of uniqueness. For Walsh series, it is much more natural to consider dyadic partial sums than full partial sums. Thus a set E is called a *dyadic set of uniqueness* (for some collection of Walsh series $\mathcal{A}$) if the only Walsh series S (in $\mathcal{A}$) which satisfies

$$\lim_{n\to\infty} S_{2^n}(x) = 0, \qquad x \notin E, \tag{4}$$

is the zero series. Since every convergent Walsh series has convergent dyadic partial sums, there are fewer dyadic sets of uniqueness than sets of uniqueness. Thus the problem of identifying dyadic sets of uniqueness is more manageable.

The idea of looking at dyadic sets of uniqueness originated with Crittenden and Shapiro [23], who proved that a set $E \subset \mathbf{G}$ is a dyadic set of uniqueness for the collection of Walsh series which satisfies the *C–S condition*, i.e.,

$$\lim_{n\to\infty} 2^{-n} S_{2^n}(x) = 0 \qquad \text{for all } x \in \mathbf{G}, \tag{5}$$

if and only if E contains no perfect subsets. Thus a Borel set E is a dyadic set of uniqueness for the collection of Walsh series satisfying the C–S condition if and only if E is countable. Their proof relied on applying a "dyadic" differentiation result to the first formal integral of the Walsh series S. This connection between uniqueness and dyadic differentiation, hinted at in Fine [27] and developed further by Skvortsov [132], was brought to maturity by Yoneda (see Wade and Yoneda [163]). Indeed, it turns out that (unlike the trigonometric case) every Walsh series is a Walsh-Fourier Stieltjes series in the following sense: given ANY Walsh series $S = \sum_{k=0}^{\infty} a_k w_k$ there is a *quasimeasure* (i.e., a finitely additive set function μ defined on the algebra generated by the dyadic intervals) such that

$$\mu(I_n(x)) = 2^{-n} S_{2^n}(x), \qquad n \in \mathbf{N},\ x \in \mathbf{I},$$

and

$$a_k = \widehat{\mu}(k) := \int_0^1 w_k \, d\mu \qquad k \in \mathbf{N}.$$

When viewed from this vantage point, the Crittenden–Shapiro result looks very natural: the C–S condition (5) becomes $\mu(I_n(x)) \to 0$ as $n \to \infty$, a continuity condition, and condition (4) becomes

$$\lim_{n\to\infty} \frac{\mu(I_n(x))}{|I_n(x)|} = 0,$$

a differentiability condition. It follows that the hard part of the Crittenden–Shapiro theorem is equivalent to the following analogue of a classical theorem of Dini: if a quasimeasure μ is everywhere continuous and its derivative is zero except possibly at countably many points, then μ is identically zero.

It should be pointed out that the Crittenden-Shapiro theorem is false unless some growth condition is imposed. Indeed, consider the Walsh series $D := \sum_{k=0}^{\infty} w_k$. By (2), $D_{2^n}(x) \to 0$ for all $x \neq 0$, but D is evidently not the zero series. Notice that D fails to converge at only one point: $x = 0$. What confuses things here is that D also fails to satisfy the C–S condition at only one point too. It was thought for many years that the C–S condition could not be relaxed at one point and that (4) could not be relaxed to include any uncountable set E, but Wade and Yoneda [163] proved that if S is a Walsh series such that some dyadic subsequence $S_{2^{n_k}}$ converges almost everywhere to zero, as $k \to \infty$, and if

$$\limsup_{k\to\infty} |S_{2^{n_k}}(x)| < \infty$$

for every $x \in \mathbf{G}$, then S is the zero series. Thus it is finiteness, not convergence, which plays a crucial role in uniqueness of dyadic partial sums. (See also Wade [162].)

The simple but elegant result of Crittenden–Shapiro must have been surprising when it first appeared. It opened up a line of investigation which remains active today. Here are a few more results along this line.

Professor Shapiro, noticing an analogy between dyadic partial sums of Walsh series and Cesàro means of trigonometric series, conjectured that a result of Broman concerning sets of uniqueness for Cesàro means of trigonometric series also held for dyadic sets of uniqueness. Wade [157] verified this conjecture for nonnegative Walsh series, showing that for the collection of Walsh series which satisfy $S_{2^{n_k}} \geq 0$ for some subsequence of integers $\{n_k\}$ and

$$\sum_{k=0}^{\infty} \sum_{j=2^k}^{2^{k+1}-1} 2^{k(\alpha-1)} a_j^2 < \infty, \tag{6}$$

a closed set $E \subset \mathbf{G}$ is a dyadic set of uniqueness if and only if E is of α-capacity zero. Yoneda [179] solved the original conjecture, showing that the nonnegativity hypothesis was superfluous. In fact, under a growth condition even more general than (6), he obtained a structural characterization of dyadic sets of uniqueness, describing them by how frequently they intersect successive layers of dyadic intervals. These results have been generalized further in Yoneda [188].

Yoneda (see [180], [181], [184], [185], and [189]) has also studied dyadic sets of uniqueness and multiplicity under a variety of other growth conditions, some weaker, and some stronger that the C–S condition. In each case, he obtains structural information or estimates the size of these sets. For example, in [180] he investigated uniqueness of the collection of Walsh series whose coefficients satisfy the stronger condition

$$\lim_{n\to\infty} \frac{1}{n} \sum_{k=0}^{n-1} |a_k|^2 = 0.$$

He proved that any perfect set of measure zero is a set of multiplicity for this collection, but there exist perfect sets of uniqueness for Walsh series satisfying a slightly stronger growth condition.

3.3. Uniqueness for pointwise convergent Walsh series. What happens to these results when the limit is nonzero? Fine [27], after proving that every countable set is a set of uniqueness for Walsh series, conjectured that if the dyadic partial sums of a Walsh series S converge pointwise, except possibly at countably many points, to an integrable function f, then $S = Sf$. This conjecture was solved some fifteen years later, independently, by Arutunyan and Talalyan [2] and by Crittenden and Shapiro [23]. Their methods and the final form of their results were quite different.

Arutunyan and Talalyan's proof, by contradiction, concentrated first on the Haar case where localization was more natural. Their results were valid for any Walsh series S whose coefficients $a_k \to 0$ as $k \to \infty$ provided some dyadic subsequence $S_{2^{n_k}}$ converged to f, as $k \to \infty$, for all but countably many $x \in \mathbf{I}$.

Crittenden and Shapiro's proof relied on the formal integration of Walsh series and a delicate Baire category argument. Their results used a weaker growth condition, the C–S condition (see (5) above), and replaced convergence by the finiteness conditions

$$f(x) \leq \liminf_{n\to\infty} S_{2^n}(x) \quad \text{and} \quad \limsup_{n\to\infty} |S_{2^n}(x)| < \infty \tag{7}$$

for all but countably many $x \in \mathbf{G}$.

This result was the first of its kind. Lindahl [69] simplified the categorical part of the argument by focussing on the dyadic rationals. Conditions (7) have been modified or weakened by Skvortsov [133], and Wade [158] and [162]. The growth condition on the right side of (7) also played a prominent role in the uniqueness for almost everywhere convergence (see Wade and Yoneda [163]).

Grubb [46] has streamlined these arguments even further. His new proof, which avoids set theoretical and topological subtleties altogether, has three main steps: 1) if the left side of (7) holds then one can assume that the quasimeasure μ associated with S is nonnegative; 2) any quasimeasure which is nonnegative must be a finite Borel measure, and if S satisfies the C–S condition, then the singular part of μ must be zero; 3) if μ is a finite Borel measure with no singular part, then S must be a Walsh–Fourier series. This approach is so simple it generalizes all the way to certain compact 0-dimensional metric spaces (not even a group structure is needed in this setting).

3.4. Integrability conditions. Some uniqueness results seek conditions on the coefficients of a Walsh series S which guarantee that S is a Walsh–Fourier series of some $f \in L^1(\mathbf{I})$ AND converges to f in some sense. The first theorem of this kind was obtained by Yano [174]. To state his results, we introduce the following notation. The *first difference* of a sequence $\{a_k\}$ is the sequence $\Delta a_k := a_k - a_{k+1}$ and the *second difference* of $\{a_k\}$ is the sequence $\Delta^2 a_k := \Delta(\Delta(a_k))$. Yano proved that if the coefficients of a Walsh series S converge to zero and are quasi–convex, i.e., if $a_k \to 0$ and

$$\sum_{k=0}^{\infty} k|\Delta^2 a_k| < \infty,$$

then S converges everywhere except possibly at $x = 0$ to an $f \in L^1(\mathbf{I})$, and S is the Walsh–Fourier series of f. (This shows that the coefficients of a Walsh–Fourier series can decay as slowly as one wishes.) Fomin [30] (and, independently, Siddiqi [127]) proved that under these conditions, S converges in L^1 norm if and only if $a_n \log n \to 0$ as $n \to \infty$.

Since a convex sequence which converges must be monotone, it is natural to ask whether these results extend to Walsh series whose coefficients $a_k \downarrow 0$. Rubinshtein [108] showed that although the limit of such a series exists on $\mathbf{I}$ and is always (improperly) Riemann integrable, it may not be Lebesgue integrable. These limits do not always behave like the trigonometric case. For example, Coury [20] proved the limit of such a series can take on negative values.

Siddiqi [127] was the first to notice that an old multiplier result for trigonometric series is also valid for Walsh series: if $a_k \to 0$ and

$$\sum_{k=1}^{\infty} |\Delta a_k| \log k < \infty, \tag{8}$$

then $S = \sum_{k=0}^{\infty} a_k w_k$ is the Walsh–Fourier series of some $f \in L^1(\mathbf{I})$ and converges in L^1 norm. Applying Abel's transform to this result, it follows that if $a_k \downarrow 0$ and

$$\sum_{k=1}^{\infty} \frac{a_k}{k} < \infty, \tag{9}$$

then S converges to some integrable f in L^1 norm. (For the Vilenkin version of these results, see Bloom and Fournier [12]. For a two–dimensional version of these results, see Mòricz, Schipp, and Wade [89].)

Baĭarstanova [8] refined condition (9) by showing that if $b_k \downarrow 0$ as $k \to \infty$ then $\sum_{k=0}^{\infty} a_k w_k$ is a Walsh–Fourier series for all sequences $\{a_k\}$ which satisfy

$$a_k \downarrow 0, \quad a_k \leq b_k, \qquad k = 1, 2, \dots,$$

if and only if

$$\sum_{k=1}^{\infty} \frac{b_k}{k} < \infty.$$

Her proof relied on deriving necessary and sufficient conditions for a Walsh series whose coefficients satisfy a stair-like condition to be a Walsh–Fourier series. Namely, she proved that if $\{n_k\}$ satisfies

$$n_k = 2^{2^k} + 2^{2^k - 2} + 2^{2^k - 4} + \cdots + 1, \qquad k = 0, 1, \dots$$

or

$$n_k = 2^k + 2^{k-2} + 2^{k-4} + \cdots + 2^{\alpha_k}, \qquad \alpha_k = 0 \quad \text{or} \quad 1, \qquad k = 0, 1, \dots,$$

and if $a_k \to 0$ is a sequence of real numbers which are locally constant in the sense that

$$a_m = a_{n_k}, \qquad n_{k-1} < m \leq n_k, \; k = 1, 2, \dots, \tag{10}$$

then $\sum_{k=0}^{\infty} a_k w_k$ is the Walsh–Fourier series of some integrable function if and only if

$$\sum_{p=1}^{\infty} \left(\sum_{k=p}^{\infty} |\Delta a_{n_k}|^2 \right)^{1/2} \log \frac{n_{p+1}}{n_p} < \infty.$$

The quasi–convex result of Yano has been generalized in several other directions. A sequence $\{a_k\}$ is said to belong to the Fomin class $\mathcal{F}_p$, for some $1 \leq p < \infty$, if

$$\sum_{k=0}^{\infty} 2^{k(1-1/p)} \left(\sum_{j=2^k}^{2^{k+1}-1} |\Delta a_j|^p \right)^{1/p} \infty. \tag{11}$$

The class $\mathcal{F}_\infty$ is the set of sequences which satisfy

$$\sum_{k=0}^{\infty} 2^k \sup_{2^{k-1}\le j<2^k} |\Delta a_j| < \infty$$

and the class $\mathcal{F}$ is the set of sequences in $\mathcal{F}_p$ for some $p > 1$ which satisfy $a_k \to 0$ as $k \to \infty$. Notice that by Hölder's inequality, if a sequence satisfies condition $\mathcal{F}_p$ for some $p > 1$ then it satisfies condition $\mathcal{F}_q$ for all $q < p$. In particular, $\mathcal{F}_\infty$ is the strongest of these conditions and $\mathcal{F}_1$ is the weakest. Notice also that a sequence satisfies $\mathcal{F}_1$ if and only if it is of bounded variation. In particular, any Walsh series whose coefficients belong to one of these spaces must converge everywhere on **I** except possibly at $x = 0$.

This limit is not necessarily integrable when the coefficients belong to the largest of these spaces $\mathcal{F}_1$ (see Rubinshtein [108] or Baĭarstanova [8]). However, Bloom and Fournier [12] proved that this limit is always integrable when the coefficients belong to $\mathcal{F}$. (Since every convex sequence which converges to zero satisfies condition $\mathcal{F}_\infty$, this result contains Yano's result.) They also proved that if f belongs to the generalized Lipschitz space $\Lambda(1/p, p, 1)$ for some $1 \le p < \infty$, then the following conditions are equivalent:

$$\int_{\mathbf{G}} \frac{|f(x)|}{|x|}\, dx < \infty, \tag{12}$$

$$\sum_{N=1}^{\infty} \frac{|S_N f(0)|}{N} < \infty,$$

and

$$\sum_{k=0}^{\infty} |S_{2^k} f(0)| < \infty.$$

Using this characterization, they derived the following integrability condition: If $a_k \to 0$ as $k \to \infty$ and the formal product of an adapted function with the Walsh series $S := \sum_{k=0}^{\infty} a_k w_k$ is the Walsh–Fourier series of some $f \in \Lambda(1/p, p, 1)$, $1 \le p < \infty$, then S is a Walsh–Fourier series if and only if (12) holds.

Mòricz has examined L^1 convergence when it is already given that the series is a Walsh–Fourier series. He showed that if $f \in L^1(\mathbf{I})$ and if

$$\lim_{\lambda\to 1+} \limsup_{n\to\infty} \sum_{k=n}^{[\lambda n]} k^{p-1} |\Delta \widehat{f}(k)|^p \tag{13}$$

is zero [79] or finite [80] for some $p > 1$, then $S_n f$ converges to f in $L^1(\mathbf{I})$ norm if and only if

$$\lim_{n\to\infty} a_{n+1} \|D_n\|_1 = 0.$$

Thus $S_n f \to f$ in $L^1(\mathbf{I})$ norm if $\widehat{f}(k) \log k \to 0$ as $k \to \infty$ and either

$$\lim_{n\to\infty} \frac{1}{n} \sum_{k=1}^{n} k^p |\Delta \widehat{f}(k)|^p < \infty$$

or

$$\sum_{k=0}^{\infty} k^{p-1}|\Delta \widehat{f}(k)|^p < \infty$$

for some $p > 1$. (For a multiple Walsh series version of this result, see Mòricz [86].)

A Sidon inequality is an upper estimate of the L^1 norm of $\sum_{k=0}^{n} c_k D_k$ in terms of its coefficients c_k. The first of these was discovered by Sidon in 1939, who showed that for the trigonometric case,

$$\frac{1}{n+1}\left\|\sum_{k=0}^{n} c_k D_k\right\|_1 \leq \max_{0\leq k\leq n} |c_k|.$$

Sidon inequalities have been used by Mòricz and Siddiqi [91] to study Nördlund means of Walsh–Fourier series.

Telyakovskiĭ [150] discovered that Sidon inequalities can be used to obtain integrability conditions because by Abel's transform

$$\sum_{k=0}^{n} a_k w_k = \sum_{k=0}^{n-1} \Delta a_k D_k + a_n D_n.$$

Hence L^1 estimates of partial sums of Walsh series follow from Sidon inequalities using the coefficients $c_k \equiv \Delta a_k$. This discovery has generated many integrability conditions for Walsh series. (For a complete, scholarly survey of these results for both the trigonometric and Vilenkin cases, see Fridli [32].)

The first results along these lines for Walsh series were obtained by Mòricz and Schipp [87], who established the Sidon inequality

$$\frac{1}{n+1}\left\|\sum_{k=0}^{n} c_k D_k\right\|_1 \leq C_p \left(\frac{1}{n+1}\sum_{k=0}^{n} |c_k|^p\right)^{1/p}$$

for $1 < p < \infty$, where C_p is an absolute constant which does not depend on the coefficients c_k. They used this to obtain independently Bloom and Fournier's integrability condition, namely, if $a_k \to 0$ and (11) holds for some $p > 1$, then S is the Walsh–Fourier series of some $f \in L^1(\mathbf{I})$. Mòricz and Schipp [88] also investigated the double Walsh series version of this result on the *unit square* $\mathbf{I}^2$. They proved that if a_{jk} are convex and converge to zero, then the *rectangular partial sums*

$$S_{mn} := \sum_{j=0}^{m}\sum_{k=0}^{n} a_{jk} w_j w_k$$

converge to an $f \in L^1(\mathbf{I}^2)$ as $\min\{m, n\} \to \infty$, and they converge to f in L^1 norm if and only if $a_{jk} \log(j+2)\log(k+2) \to 0$ as $\min\{j, k\} \to \infty$.

Schipp [121] improved condition (11) by replacing the ℓ^p norm in the Sidon inequality above by the Hardy space sequence norm. Specifically, he proved that there is an absolute constant $C > 0$ such that given any coefficients c_k and any $n \in \mathbf{N}$,

$$\frac{1}{2^n}\left\|\sum_{k=0}^{2^n-1} c_k D_k\right\|_1 \leq C \left\|\{c_k\}_{k=0}^{2^n-1}\right\|_h, \tag{14}$$

where h is the Hardy norm for sequence spaces. Although this generalizes the "Fomin" integrability classes of Bloom and Fournier (see (11) above), it is not easy to apply because, unlike the ℓ^p norm, the Hardy norm of a sequence depends on the coefficients c_k in a complicated way.

The latest result along these lines is due to Fridli [31]. He has obtained the best possible Sidon inequality for a rearrangement invariant space X, which turned out to be an Orlicz space. It is noteworthy that his results can be expressed directly in terms of the coefficients and also contain the classical multiplier result that a Walsh series whose coefficients converge to zero and satisfy (8) above is a Walsh–Fourier series and converges in L^1 norm. This cannot be said of any of the earlier convergence classes identified by Sidon inequalities. Using an Orlicz norm in place of the Hardy norm (the dyadic Hardy norm is NOT rearrangement invariant), Fridli identified a condition on the coefficients, weaker than both Schipp's condition (14) and Mòricz's condition (13), sufficient to conclude that S is a Walsh–Fourier series. His estimates include a shifted version of the Sidon inequalities, again directly in terms of the coefficients c_k, which is sharper than anything before it. Among several consequences of his work is the following uniqueness result. If $a_k \to 0$ satisfies

$$\sum_{k=0}^{\infty} \sum_{j=2^k+1}^{2^{k+1}} |\Delta a_j| \left(1 + \log^+ \frac{\Delta a_j}{2^{-k} \sum_{\ell=2^k+1}^{2^{k+1}} |\Delta a_\ell|}\right) < \infty, \tag{15}$$

then S is the Walsh–Fourier series of some $f \in H$ if and only if (9) holds.

There is another generalization of Yano's classical result which contains the multiplier result cited above. Using entirely different techniques, Aubertin and Fournier [4] proved that if $a_k \downarrow 0$ and

$$\sum_{k=0}^{\infty} \left(\sum_{j=1}^{\infty} \left(\sum_{\ell=j2^k}^{(j+1)2^k-1} |\Delta a_\ell| \right)^2 \right)^{1/2} < \infty, \tag{16}$$

then S is a Walsh–Fourier series, and S converges in L^1 if and only if $a_k \log k \to 0$ as $k \to \infty$. Since (15) and (16) are incomparable (see Fridli [31]), perhaps there is one more result out there which unifies the integrability conditions of both Fridli and Aubertin–Fournier.

3.5. Uniqueness via Cesàro means. Morgenthaler [77] discovered that Cesàro means can be used to characterize Walsh–Fourier series. He proved that a Walsh series $S = \sum_{k=0}^{\infty} a_k w_k$ is the Walsh–Fourier series of some $f \in L^p(\mathbf{I})$, for $1 < p < \infty$, if and only if the *Cesàro means*

$$\sigma_n := \sum_{k=0}^{n-1} \left(1 - \frac{k}{n}\right) a_k w_k$$

of S are uniformly bounded in L^p norm. This result has been extended to (C, α) means on Besov spaces by Tateoka and Wade [149]. For the case $p = 1$, the result still holds, but Walsh–Fourier–Stieltjes series must be used in place of Walsh–Fourier series (see Fine [29]).

In the early sixties, Professor Shapiro asked his third Ph.D. student, Richard Crittenden, whether a Walsh analogue of a theorem of M. Riesz held, namely, whether uniqueness held for Walsh series which were pointwise Cesàro summable to an integrable function f. This problem turned out to be very difficult to solve. Although Crittenden [22] obtained some partial results in this direction, it was more than a decade before the original question of Shapiro was answered in the affirmative.

The first step toward a resolution of this problem was taken by Wade [155], who proved uniqueness for Cesàro summable Haar series. Due to the wavelet–like behavior of Haar functions which makes localization arguments much simpler, the techniques of Arutunyan and Talalyan [2] could be modified to apply to the Cesàro case. It was shown that under a suitable growth condition, if some subsequence σ_{n_j} of Cesàro means of a Haar series S converges almost everywhere to an integrable function f, and if

$$\limsup_{j\to\infty} |\sigma_{n_j}(x)| < \infty$$

for all but countable many $x \in \mathbf{I}$, then S is the Haar–Fourier series of f.

The next important step was taken by Shaginyan [123], who showed that for Walsh series there is a close relationship between convergence and summability. He proved that given ANY Walsh series S with Cesàro means σ_n, if

$$\limsup_{n\to\infty} |\sigma_{2^n}(x)| < \infty \tag{17}$$

for all x in some measurable set E, then S_{2^n} converges almost everywhere on E and its limit f satisfies

$$\liminf_{n\to\infty} \sigma_{2^n}(x) \leq f(x) \leq \limsup_{n\to\infty} \sigma_{2^n}(x)$$

for almost every $x \in E$. In particular, for dyadic partial sums, convergence and summability of Walsh series are equivalent up to sets of measure zero.

The final step was taken by Skvortsov [134] who used Shaginyan's result to prove that if S is a Walsh series which satisfies the C–S condition everywhere and (17) off a countable set, and if $\sigma_{2^n} \to f$ almost everywhere, where f is Perron integrable, then S is the Perron–Walsh–Fourier series of f. It follows that if S is Cesàro summable off a countable set to some finite–valued, Lebesgue integrable function f and the coefficients of S satisfy $a_k \to 0$ as $k \to \infty$, then S is the Walsh–Fourier series of f.

4. Convergence

4.1. Absolute convergence. The space $\mathcal{A}$ of absolutely convergent Walsh–Fourier series shares several properties with its trigonometric analogue. Namely, $\mathcal{A}$ is a Banach algebra under pointwise addition and multiplication. The Walsh–Fourier coefficient map is an isometric isomorphism from $\mathcal{A}$ onto the convolution Banach algebra ℓ^1. And, Agaev [1] proved that if $f \in \mathcal{A}$ is bounded away from zero, then $1/f$ also belongs to $\mathcal{A}$.

Most conditions which control absolute convergence of Fourier series are related to the *moduli of continuity*

$$\omega^{(p)}(f;\delta) := \sup_{y\in\mathbf{G},|y|<\delta} \|f(\cdot+y) - f(\cdot)\|_p, \qquad 1 \leq p \leq \infty.$$

The first result along these lines for Walsh series was obtained by Fine [27], who proved that if $f \in \operatorname{Lip} \alpha$ for some $\alpha > 1/2$, i.e., if f is continuous on $\mathbf{G}$ and

$\omega(f;\delta) := \omega^{(\infty)}(f;\delta) = O(\delta^\alpha)$ as $\delta \to 0$, then Sf converges to f absolutely on $\mathbf{I}$. Yoneda [177] localized this result for certain summability methods. Neither of these results holds for $\alpha \le 1/2$.

Fine's result has been generalized in several directions. Yoneda [177] showed that if

$$\sum_{k=1}^{\infty} \frac{\omega^{(2)}(f;1/k)}{\sqrt{k}} < \infty,$$

then Sf converges absolutely. It follows that if f is continuous and of bounded variation, then Sf converges absolutely when

$$\sum_{k=1}^{\infty} \frac{\sqrt{\omega(f;1/k)}}{k} < \infty. \tag{18}$$

In connection with this last condition, it is interesting to note that Bochkarev [14] proved that if ω is ANY modulus of continuity which satisfies (18), then for ANY bounded orthonormal system in $L^2(\mathbf{I})$ there exists an absolutely continuous function f whose classical modulus of continuity satisfies $\omega(f;\delta) = O(\omega(\delta))$, as $\delta \to 0$, but whose Fourier series does not converge absolutely.

Onneweer introduced the following measurement of the oscillation of a function:

$$\mathcal{F}_p(f) := \sum_{k=0}^{\infty} \left(\sum_{j=0}^{2^k} \sup_{x,y \in z_{j,k}+\mathbf{G}_k} |f(x)-f(y)|^p \right)^{1/p},$$

where for each k, $\{z_{j,k} + \mathbf{G}_k : j = 0,\ldots,2^k\}$ represents the collection of distinct cosets of the subgroup $\mathbf{G}_k := \{x \in \mathbf{G} : x = (0,0,\ldots,0,x_k,x_{k+1},\ldots)\}$ in $\mathbf{G}$. Onneweer [95] showed that if $\mathcal{F}_p(f)$ converges for some $f \in L^p(\mathbf{I})$, where $1 \le p \le 2$, then Sf converges absolutely. It follows that if $f \in \mathrm{Lip}\ \alpha$ for some $\alpha > 0$, and if f is of p *-generalized bounded fluctuation* (i.e., if the terms of the series $\mathcal{F}_p(f)$ are uniformly bounded), then Sf converges absolutely.

A function f is said to belong to the generalized Lipschitz spaces $\mathrm{Lip}\ (\alpha, L^p)$ for some $1 \le p < \infty$ if $\omega^{(p)}(f;\delta) = O(\delta^\alpha)$ as $\delta \to 0$. Onneweer [96] has shown that if $f \in \mathrm{Lip}\ (\alpha, L^p) * \mathrm{Lip}\ (\alpha, L^q)$ for some $1 \le p \le 2$, $q > 1$, $0 < \alpha, \beta \le 1$, and $(\alpha+\beta)p > 1$, then Sf converges absolutely. This is also true (see Ladhawala [65] and Butzer and Wagner [17]) when $f \in \mathrm{Lip}\ (\alpha, L^p)$ for some $\alpha, p > 1$. (Yes, unlike the classical case, dyadic Lipschitz spaces are nontrivial even when $\alpha > 1$. This is due to the "totally disconnected" nature of $\mathbf{G}$.)

Absolute convergence can be characterized by the *Stechkin rate of best approximation* in $L^2(\mathbf{I})$:

$$\mathcal{E}_n(f) := \inf\{\|f-P\|_2 : P \text{ is a Walsh polynomial with at most } n \text{ terms}\}.$$

Indeed, an $f \in L^2(\mathbf{I})$ has an absolutely convergent Walsh–Fourier series if and only if

$$\sum_{k=1}^{\infty} \frac{\mathcal{E}_k(f)}{\sqrt{k}} < \infty$$

(see Schipp, Wade and Simon [122]). Tateoka [148] has shown that absolute convergence still holds if this condition is weakened to "f belongs to the Besov space $\mathcal{B}^{1/p}_{p,1}$ for some $0 < p \le 2$."

4.2. Uniform convergence. Although (as we saw in 1.1) the 2^nth partial sums of the Walsh–Fourier series of a continuous function converge uniformly on $\mathbf{I}$, this may not be the case for the full sequence of partial sums. Indeed, Walsh [164] proved that given any $a \in \mathbf{I}$, there is a continuous function f such that $(Sf)(a)$ diverges. Therefore, one can ask: what more is needed (besides continuity of f) to conclude that Sf converges uniformly? This question, which has been considered by many authors, can be answered in a variety of ways.

Some of the answers focus on properties of the coefficients. By modifying the trigonometric proofs, one can show that if f is continuous and either $\sum_{k=1}^{n} k|\widehat{f}(k)| = o(n)$ or $\widehat{f}(k) \geq 0$, then Sf converges uniformly on $\mathbf{I}$. If one allows the signs of the coefficients to change, even weaker conditions apply. For example, Watari [165] proved that if

$$\sum_{k=0}^{\infty} a_k^2 (\log k)^{1+\epsilon} < \infty$$

for some $\epsilon > 0$, then $\sum_{k=0}^{\infty} \pm a_k w_k$ converges uniformly for almost all choices of signs.

Other answers focus on how much f wiggles. Suppose f is continuous on $\mathbf{I}$ and periodic. Walsh [164] proved that if f is also of bounded variation, then Sf converges uniformly on $\mathbf{I}$. (The same argument extends to W–continuous functions of bounded variation.) Onneweer [97] showed that this conclusion still holds if bounded variation is replaced by

$$\lim_{n\to\infty} \sum_{k=1}^{2^n-1} \frac{1}{k} \left| f\left(x \dotplus \frac{k}{2^n}\right) - f\left(x \dotplus \frac{k+1/2}{2^n}\right) \right| = 0. \tag{19}$$

Mòricz [82] has extended this result to multiple Walsh series. He found sufficient conditions for L^p convergence of the rectangular partial sums $S_{mn}f$ when $f \in L^p(\mathbf{I}^2)$ for some $1 \leq p \leq \infty$, as well as uniform convergence of $S_{mn}f$ when f is W–continuous and of bounded ϕ–variation on $\mathbf{I}^2$.

The results of Onneweer have been generalized further. Onneweer and Waterman [98] proved that if f is W–continuous and of 1–generalized bounded fluctuation (see 4.1 above for a definition), then Sf converges uniformly. Onneweer and Waterman [99] showed that this result still holds if generalized bounded fluctuation is replaced by harmonic bounded fluctuation. Mòricz [85] has extended these results to multiple Walsh–Fourier series.

Skvortsov [137] proved that if f is uniformly continuous on the interval $\mathbf{I}$, if ω_n is a positive sequence which satisfies $n\omega_n \to \infty$ as $n \to \infty$ such that for every x and n satisfying $0 \leq x < x + 2^{-n} < 1$ it is also the case that

$$f(x + 2^{-n}) - f(x) \geq -K\omega_n$$

for some positive constant K, then (19) holds. Hence it follows from Onneweer [97] that the Walsh-Fourier series of such an f converges uniformly on $\mathbf{I}$. This is a Walsh analogue of the Nevai test.

Fine [27] showed that Sf converges uniformly on $\mathbf{I}$ when f is continuous and $\omega(f;\delta) = o(\log^{-1}(1/\delta))$, as $\delta \to 0$. In particular, Sf converges uniformly when $f \in \operatorname{Lip} \alpha$ for some $\alpha > 0$. Schipp [111] has shown that Fine's result is best possible in the following sense. There are continuous functions which satisfy $\omega(f;\delta) = O(\log^{-1}(1/\delta))$, as $\delta \to 0$, such that Sf diverges at at least one point. Schipp also proved that there exist continuous functions whose Walsh–Fourier series converge everywhere but not uniformly.

Fine's result is true for multiple Walsh series. After estimating the d–dimensional rectangular partial sums

$$\|S_{\mathbf{n}}f - f\|_{\boldsymbol{X}}$$

for the Banach spaces $\boldsymbol{X} = \mathcal{C}(\mathbf{I}^d)$ and $\boldsymbol{X} = L^1(\mathbf{I}^d)$, Getsadze [37] proved that if $f \in \boldsymbol{X}$ and

$$\omega^{\boldsymbol{X}}(f, \delta_{i_1}, \dots, \delta_{i_k}) = o\left(\prod_{j=1}^{k} \log 1/\delta_{i_j}\right)^{-1},$$

as $\delta_{i_j} \to 0$, for all subsets $\{i_1, \dots i_k\}$ of $\{1, 2, \dots, d\}$, then $S_{\mathbf{n}}f$ converges to f in $\boldsymbol{X}$ as $\|\mathbf{n}\| \to \infty$. As a matter of fact, he showed this result holds for any orthonormal system which satisfies the Olevskiĭ condition.

4.3. Pointwise convergence. Localization holds for Walsh series, i.e., convergence of a Walsh–Fourier series Sf at a specific point a is governed by the behavior of f in a neighborhood of a. Thus many conditions for pointwise convergence of single variable Walsh–Fourier series are local in nature. Walsh [164] proved that if f is of bounded variation on $[0, 1]$ and continuous at a point a, then $(Sf)(a)$ converges to $f(a)$. Fine [27] proved that if an integrable function f satisfies

$$\int_{a-\delta}^{a+\delta} \left| \frac{f(t) - b}{t \dotplus a} \right| dt < \infty$$

for some constant b and some $\delta > 0$, then $(Sf)(a)$ converges (to b). And, Yano [176] proved that if $\widehat{f}(k) = O(k^{-\delta})$, as $k \to \infty$, for some $\delta > 0$, and if either $|f(a \dotplus h) - f(a)| = o(\log^{-1}(1/h))$, as $h \to 0$, or

$$\frac{1}{I_n(a)} \int_{I_n(a)} |f(a \dotplus x) - f(a)|\, dx = o\left(\frac{1}{n}\right),$$

as $n \to \infty$, then $(Sf)(a)$ converges to $f(a)$. These tests are known by their trigonometric analogues: Jordan, Dini, and Hardy–Littlewood.

Mòricz [84] has modified the Jordan test to handle functions of bounded fluctuation which are W–continuous. He proved that if f is of bounded fluctuation (defined in section 4.1) and $f(x \dotplus t) \to f(x)$ as $t \to 0+$, then $(S_n f)(x) \to f(x)$ as $n \to \infty$. Moreover, this convergence is uniform when $f(x \dotplus t) \to f(x)$ uniformly in x as $t \to 0+$.

The bounded variation condition of Walsh has been generalized by Yoneda and Tevzadze. Yoneda [187] proved that if f is of bounded p–variation for some $1 \le p < \infty$, i.e., if

$$\sup_{0=x_0<x_1<\cdots<x_n=1^-} \left\{\sum_{k=1}^{n} |f(x_k) - f(x_{k-1})|^p\right\}^{1/p} < \infty,$$

then $(Sf)(x)$ converges to $f(x)$ when f is continuous at x and diverges when f is not continuous at x. Tevzadze [152] has announced that if $[a, b] \subset [0, 1]$ and the modulus of variation $v(f; n)$ of f satisfies

$$\sum_{n=1}^{\infty} \frac{v(f; n)}{n^2} < \infty,$$

then $(Sf)(x)$ converges on (a,b) at each point x of continuity of f, converges to the right–hand limit $f(x+0)$ at each dyadic rational $x \in (a,b)$, and converges uniformly on compact subsets of (a,b) when f is continuous on $[a,b]$. This contains Yoneda's result since the hypotheses are met when f is of bounded p–variation for some $1 \le p < \infty$.

What happens near a point of discontinuity? Something similar to Gibbs phenomenon for the trigonometric system. To state these results, let f be a function of bounded variation and set

$$(Af)(x_0) := \frac{f(x_0+0)+f(x_0-0)}{2}.$$

For the trigonometric system, recall that

$$\begin{aligned} G(x_0) &:= \limsup_{\substack{x\to x_0 \\ n\to\infty}} \frac{1}{|(Af)(x_0)-f(x_0-0)|} \mid (S_n)f(x)-(Af)(x_0) \mid \\ &= \frac{2}{\pi}\int_0^\pi \frac{\sin t}{t}\,dt = 1.179\ldots \end{aligned}$$

Balashov and Skvortsov [6] showed that up to sets of measure zero, a similar situation prevails for the Walsh case, but with a worse Gibbs constant. Namely, they proved that $G(x_0) = 3/2$ for almost every $x_0 \in \mathbf{I}$. What happens at other x_0's? They proved that

$$\frac{4}{3} \le G(x_0) \le \frac{3}{2} \tag{20}$$

for any dyadic irrational x_0, with $4/3$ best possible as a lower bound (for example, there is an f such that $G(1/3) = 4/3$). They also proved that, unlike the trigonometric case, Gibbs phenomenon holds for Cesàro means as well. The corresponding constants in (20) are 1 and $1 + (12+2\sqrt{34})^{-1} \approx 1.05608$, respectively, with the latter achieved for almost every x_0. Thus, as one might expect, Gibbs phenomenon is less pronounced for Cesàro means than for partial sums.

Analogy with the trigonometric case completely falls apart if attention is restricted to functions with jump discontinuities at dyadic rationals. Indeed, Qian [101] has shown that neither the partial sums $S_n f$ nor the (C,α) means of Sf (for any $\alpha > 0$) exhibit Gibbs phenomenon at dyadic rational points of discontinuity of the first kind when f is of bounded variation.

Not much is known about pointwise convergence of multiple Walsh–Fourier, and with good reason. Although localization still holds for double Walsh–Fourier series in $L^p(\mathbf{I}^2)$, $p > 1$, it breaks down in a significant way for functions in 3 or more variables. Indeed, Bloshanskaya and Bloshankiĭ [13] proved that if $f \in L^p(\mathbf{I}^2)$, $p > 1$, vanishes on some open set V, then $S_{mn} f \to 0$ almost everywhere on V. If $d \ge 3$, however, then given any set $E \subset \mathbf{I}^d$ of positive Lebesgue measure which is not dense in $\mathbf{I}^d$, there is a continuous function f such that

$$\limsup_{\|\mathbf{n}\|\to\infty} |S_{\mathbf{n}} f| = \infty$$

on some subset E_0 of E of positive measure. Tateoka [145] has shown that localization can be recaptured using Abel means.

4.4. Almost everywhere convergence. Integrability by itself is not enough to conclude that a Walsh–Fourier series converges almost everywhere. Stein [139] proved that there exist integrable functions whose Walsh–Fourier series diverge everywhere on $\mathbf{I}$. Cobzas [19] went even further by showing that the set of functions whose Walsh–Fourier series diverge unboundedly on a superdense subset of $\mathbf{I}$ is itself superdense in $L^1(\mathbf{I})$. Thus if one wants almost everywhere convergence, one must insist on conditions stronger than integrability.

Such conditions divide themselves nicely into two classes. The first class focuses on the variability of the function. Yano [174] proved that if

$$\frac{1}{h}\int_0^h |f(x+t)-f(x)|\,dt = O\left(\log^{-1}(1/h)\right),$$

as $h \to 0$, or if

$$\int_0^1\int_0^1 \frac{|f(x+t)-f(x-t)|^p}{t}\,dt\,dx < \infty \tag{21}$$

holds for some $1 < p < \infty$, then Sf converges to f almost everywhere on $\mathbf{I}$. Schipp [114] showed that the first result still holds if $+$ is replaced by $\dot{+}$. Schipp [117] obtained the same conclusion for functions f which satisfy any one of the following three conditions:

$$\sum_{k=0}^{\infty}\int_0^1\int_0^1 |f(x\dot{+}t)-f(x)|D_{2^k}(t)\,dx\,dt < \infty,$$

$$\sum_{k=1}^{\infty}\frac{1}{k}\omega^{(1)}(f;1/k) < \infty,$$

or

$$\omega^{(1)}(f;\delta) = O(\log^{-1-\epsilon}(1/\delta)), \quad \text{as } \delta \to 0.$$

Schipp [114] also proved that this last condition is best possible in the following sense: there is an integrable f satisfying $\omega^{(1)}(f;\delta) = O(\log\log^{-1}(1/\delta))$, as $\delta \to 0$, whose Walsh–Fourier series diverges almost everywhere.

The second class of conditions sufficient to conclude that a Walsh–Fourier series is almost everywhere convergent involves integrability conditions. This line of investigation began when Billard [11] adapted Carleson's trigonometric argument to the Walsh setting, proving that if $f \in L^2(\mathbf{I})$, then Sf converges to f almost everywhere. (Gosselin [44] has a simpler proof of this result based on C. Fefferman's graph theory technique.) This result was improved by Tateoka [144] and again by Sjölin [131], who proved that the conclusion still holds if $f \in L\log^+ L\log^+\log^+ L(\mathbf{I})$. (A more natural proof of these results based on martingale trees can be found in Schipp, Wade, and Simon [122].) In particular, Sf converges to f almost everywhere when $f \in L^p(\mathbf{I})$ or $f \in L(\log^+ L)^p(\mathbf{I})$ for some $p > 1$.

On the other hand, Moon [76] proved that for each $q < 1$, there is an $f \in L(\log^+\log^+ L)^q$ whose Walsh–Fourier series diverges everywhere, and Ladhawala and Pankratz [66] proved that there is an f in the dyadic Hardy space H whose Walsh–Fourier series diverges almost everywhere. Thus the question of almost everywhere convergence is still open for the spaces $L(\log^+ L)^p$ for $0 < p \le 1$.

There is another class of functions which contains $L^p(\mathbf{I})$, $p > 1$, whose Walsh–Fourier series converge almost everywhere. Using the trigonometric techniques of

Taibleson and Weiss [143], Chao [18] introduced the block spaces $\mathcal{B}_p$ and proved that if $f \in \mathcal{B}_p$ for some $p > 1$, then Sf converges to f almost everywhere. Since each block space $\mathcal{B}_p$, $p > 1$, is a complete quasi–normed linear space which lies between $L^1(\mathbf{I})$ and $L^p(\mathbf{I})$, $p > 1$, this result includes one of the corollaries of Sjölin's work cited above. It does not include the results about the logarithm spaces since no block space is comparable to any other Orlicz space except L^1. The block spaces do, however, include all functions of finite entropy and all functions which satisfy the Dini condition (21).

The situation gets better if some partial sums are skipped. For example, in section 1.1 we proved that if $f \in L^1(\mathbf{I})$ then $S_{2^n}f$ converges almost everywhere as $n \to \infty$. Ladhawala and Pankratz [66] showed that for functions in the dyadic Hardy space, this can be improved. Indeed, if n_k is *lacunary* (i.e., $n_{k+1}/n_k \geq q > 1$ for $k \in \mathbf{N}$) and $f \in H$, then $S_{n_k}f \to f$ almost everywhere on $\mathbf{I}$. Konyagin [63] has shown that if the n_k's increase rapidly enough, this also works for integrable functions. He proved that if $f \in L^1(\mathbf{I})$ and $\{n_k\}$ is an increasing sequence of integers such that n_{k+1} can be divided by $2^{[\log n_k]+1}$, then $S_{n_k}f \to f$ almost everywhere on $\mathbf{I}$. Both proofs boil down to showing that the maximal functions

$$f^\star := \sup_{k \in \mathbf{N}} |S_{n_k} f|$$

are of weak type $(1,1)$.

Although there are some positive results for multiple Walsh–Fourier series, the picture is still pretty muddy. Mòricz [83] proved that if $f \in L^2(\mathbf{I}^2)$, then the square partial sums $S_{nn}f$ converge to f almost everywhere on $\mathbf{I}^2$. This proof can readily be adapted to handle the d–dimensional case. Mòricz [81] has also shown that if $\widehat{f}(j,k)$ is of bounded variation and converges to zero, then $\|S_{mn}f - f\|_r \to 0$, as $\min\{m,n\} \to \infty$ for $0 < r < 1$. (For the case $r = 1$, see Mòricz, Schipp, and Wade [89].)

As in the one–dimensional case, the situation gets better for dyadic partial sums. Sanadze and Kheladze [109] proved that if $f \in L(\log^+ L)^{2d-1}(\mathbf{I}^d)$ then the rectangular partial sums of Sf converge almost everywhere on $\mathbf{I}^d$ provided all but one of the indices are lacunary. In particular, the dyadic rectangular partial sums of a multiple Walsh–Fourier series of any $f \in L^p(\mathbf{I}^d)$, $p > 1$, converge almost everywhere on $\mathbf{I}^d$.

These results are not valid for $p = 1$. In fact, unlike the trigonometric case, not even the condition $\sup L_{n_k} < \infty$ is sufficient for convergence of the square partial sums. Lukomskiĭ [74] proved there exist such indices and an $f \in L^1(\mathbf{I}^2)$ such that $S_{n_k,n_k}f$ diverges almost everywhere as $k \to \infty$.

Mòricz [83] has shown that if f is W–continuous on $\mathbf{I}^2$ and the two–dimensional analogue of (13) is finite, then $S_{mn} \to f$ in L^r, $0 < r < 1$, and if in addition, $f \in L\log^+ L(\mathbf{I}^2)$, then $S_{mn} \to f$ almost everywhere on the unit square. On the other hand, Getsadze [36] proved that there is a continuous function f such that the rectangular partial sums of the double Walsh-Fourier series of f diverges almost everywhere on $\mathbf{I}^2$.

Harris [52] proved that the situation is just as bad for other methods of convergence. He showed that no matter what piecewise smooth non-rectangular region is summed over (in particular, triangular or spherical partial sums), there exist functions $f \in L^p(\mathbf{G}^2)$ for $1 \leq p < 2$ whose double Walsh-Fourier series diverge almost everywhere and in L^p norm. The key step was to reduce the problem of convergence/divergence to a multiplier problem which generalized a result of Wo-Sang Young [190].

Harris's result is not rearrangement invariant. Kemkhadze [56] proved that given any $f \in L^p(\mathbf{I}^d)$, $p > 1$, the multiple Walsh–Fourier series Sf can be arranged so that its spherical partial sums converge to f almost everywhere on $\mathbf{I}^d$.

4.5. Sets of divergence. A subset E of $\mathbf{I}$ or the dyadic group $\mathbf{G}$ is called a *set of divergence* for a class of functions $\mathcal{L}$ if there exists a function $f \in \mathcal{L}$ whose Walsh–Fourier series diverges at every point in E.

The study of sets of divergence for Walsh series began when Professor Shapiro asked his fourth Ph.D. student, Larry Harper, to look for a dyadic analogue of a theorem of Salem and Zygmund which characterizes sets of divergence for a particular class of trigonometric series in terms of capacity. Harper [50] developed a theory of capacity on the group $\mathbf{G}$, identified the correct growth condition, and proved that given $0 \le \alpha < 1$, a closed set $E \subset \mathbf{G}$ is a set of divergence for the class of functions whose Walsh–Fourier coefficients satisfy

$$\sum_{k=0}^{\infty} \sum_{j=2^k}^{2^{k+1}-1} 2^{k(1-\alpha)} |\widehat{f}(j)|^2 < \infty \tag{22}$$

if and only if E is of α–capacity zero. (Notice by Parseval's identity that each function which satisfies (22) belongs to $L^2(\mathbf{I})$.) The necessity part of Harper's result was generalized from L^2 to L^p, $1 \le p \le 2$, by Kobayashi [61]. He proved that if $f \in L^p(\mathbf{I})$, for some $1 \le p < \infty$, and if

$$\sum_{k=0}^{\infty} \sum_{j=2^k}^{2^{k+1}-1} 2^{k(\alpha-1)/p} \widehat{f}(j) w_j(x)$$

diverges on a closed set $E \subset \mathbf{G}$, then E is of α–capacity zero (when $1 \le p \le 2$) and of $(\alpha + \epsilon)$–capacity zero (when $p > 2$).

According to Sjölin's result (cited in section 4.4 above), sets of divergence for $L^p(\mathbf{G})$ for $p > 1$ must be of Haar measure zero. What about the converse of this statement? Kheladze [57], and Harris and Wade [53], showed that every set of Haar measure zero is a set of divergence for the class $L^p(\mathbf{I})$, when $1 < p < \infty$. The case $p = \infty$, which turned out to be considerably more difficult, was obtained only recently by Bugadze [16]. (A multidimensional version of Kheladze's theorem has been obtained by Sanadze and Kheladze [110].)

If analogy holds with the trigonometric system, sets of divergence for continuous functions should also be characterized to be of Haar measure zero. Thus far, this conjecture has remained unresolved. Schipp [111] showed that every singleton is a set of divergence for $\mathcal{C}(\mathbf{G})$. (Since Harris and Wade [53] proved that the countable union of sets of divergence is a set of divergence, it follows that any countable set is a set of divergence for $\mathcal{C}(\mathbf{G})$.) Schipp [112] was the first to construct an uncountable set of divergence for $\mathcal{C}(\mathbf{G})$. This feat was repeated by Simon [128] and Harris and Wade [53]. Onneweer [97] identified a nice sufficient condition, proving that if E is of logarithmic capacity zero, then E is a set of (bounded) divergence for $\mathcal{C}(\mathbf{G})$.

The original problem of characterizing sets of divergence for $\mathcal{C}(\mathbf{G})$ remains open. The best result thus far is due to Harris [51], a student of Wade, who proved that any compact subset of $\mathbf{G}$ which is of Haar measure zero is a set of divergence for $\mathcal{C}(\mathbf{G})$. Since the countable union of sets of divergence is a set of divergence, it follows that any $\mathcal{F}_\sigma$ set of Haar measure zero is a set of divergence for $\mathcal{C}(\mathbf{G})$.

The set theoretic structure of sets of divergence has been investigated by Lukashenko [71], who proved that every $\mathcal{G}_\delta$ set is a set of divergence for $L^1(\mathbf{G})$, but there are $\mathcal{G}_\delta$ sets (and $\mathcal{F}_\sigma$ sets) which are not sets of divergence for $\mathcal{C}(\mathbf{G})$. Lukashenko also proved that there is a function $f \in \mathcal{C}(\mathbf{G})$ such that

$$E := \{x \in \mathbf{G} : S_n f(x) \text{ diverges}\}$$

is a $\mathcal{G}_{\delta\sigma}$ set but is neither a $\mathcal{G}_\delta$ set nor an $\mathcal{F}_\sigma$ set. Furthermore, E cannot be written in the form

$$E = (E_1 \setminus E_2) \bigcup E_3$$

for E_1 a $\mathcal{G}_\delta$ set and E_2, E_3 at most countable, or for E_1 an $\mathcal{F}_\sigma$ set and E_2, E_3 of the first category on $\mathbf{G}$.

Kheladze [58] has studied sets of divergence for multiple Vilenkin series, in particular, multiple Walsh–Fourier series. He has given an outline of a proof to the following result. If $\{B_k\}_{k\in\mathbf{N}}$ is a sequence of bounded subsets of $\mathbf{N}^d$ which satisfy

$$B_k \subset B_j \qquad k < j, \text{ and } \bigcup_{k\in\mathbf{N}} B_k = \mathbf{N}^d,$$

then given any subset E of $\mathbf{I}^d$ of d-dimensional Lebesgue measure zero, there is a function $f \in L^p(\mathbf{I}^d)$, $1 \le p < \infty$, such that

$$\limsup_{k\to\infty} \Big| \sum_{\mathbf{n}\in B_k} \widehat{f}(\mathbf{n})\psi_{\mathbf{n}}(x) \Big| = \infty$$

for every $x \in E$. Thus every set of measure zero is a set of divergence for rectangular, spherical, and triangular multiple Walsh-Fourier series.

5. Summability

5.1. Definition. Although there are many methods of summability by which one can make sense out of the " limit" of a divergent series, we will concentrate on Cesàro summability, i.e., on convergence of the Cesàro means

$$\sigma_n f := \frac{S_1 f + S_2 f + \cdots + S_n f}{n} \equiv \sum_{k=0}^{n-1} \left(1 - \frac{k}{n}\right) \widehat{f}(k) w_k.$$

Since it is easy to check that

$$(\sigma_n f)(x) = \int_0^1 f(x \dot{+} y) K_n(y)\, dy,$$

where the *Fejér kernels* K_n are defined by

$$K_n(x) := \sum_{k=0}^{n-1} \left(1 - \frac{k}{n}\right) w_k(x),$$

results about summability frequently depend on information about Fejér kernels.

It is well known (and easy to see) that if L is an extended real number and $(S_n f)(x) \to L$, then $(\sigma_n f)(x) \to L$ as $n \to \infty$, but not conversely. Thus results which fail to hold for $S_n f$ might be true for $\sigma_n f$. This point of view has a long and fruitful history for the trigonometric system, and many of these results go over to the dyadic case. There is an extra twist, however, for the Walsh system. The Fejér kernels are not nonnegative, but only quasipositive (see Yano [175]). This makes the theory of summability for Walsh series somewhat trickier than it is for trigonometric series. Moreover, because the Walsh system takes on only the values ± 1, there is a limitation on how fast the Cesàro means of a nonconstant function can converge. For example, if $\|\sigma_{2^n} f - f\|_X = o(2^{-n})$, as $n \to \infty$, for f in some homogeneous Banach space X, then f is constant (see Schipp, Wade, and Simon [122], p. 191).

5.2. Summability in norm. Fine [27] proved that if f is W–continuous, then $\sigma_n f \to f$ uniformly on $\mathbf{I}$. Yano [174] localized this result by proving that if $f(x) \to L$ as $x \to a$, then $\sigma_n(a) \to L$ as $n \to \infty$. These results hold if continuity is taken on the group also.

Morgenthaler [77] proved that if $f \in L^1(\mathbf{I})$, then $\sigma_n f \to f$ in L^1 norm. This result was extended to the case L^p, $1 \leq p < \infty$ by Kobayashi [61].

Esfahanizadeh and Siddiqi [26] proved that if f is bounded then the de la Vallée Poussin means $\tau_n f \equiv \sigma_{2n} f - \sigma_n f$ of Walsh-Fourier series satisfy

$$|\tau_n f(x) - f(x)| \leq 7E_n(f, L^\infty),$$

where E_n is the best rate of approximation by Walsh polynomials of order n to f in the L^∞ norm. It follows that if $f \in \text{Lip}\ \alpha$, then $\tau_n f(x) - f(x) = O(n^{-\alpha})$ as $n \to \infty$.

5.3. Almost everywhere summability. Fine [28] proved that if $f \in L^1(\mathbf{I})$, then $\sigma_n f \to f$ almost everywhere on $\mathbf{I}$. In 1967, Professor Shapiro asked whether this result could be generalized to higher dimensions along the lines of the Marcienkiewicz–Zygmund result which states that rectangular sums of double trigonometric Fourier series $\sigma_{m,n} f$ of integrable functions converge almost everywhere on the unit square provided the indices m, n converge to ∞ in some positive cone, i.e., do not get too close to either of the positive axes. At that time it was determined that Fine's proof could not be altered to fit the multidimensional setting.

A major breakthrough occurred when Schipp [118] obtained a new proof of the one–dimensional result. His proof, based on an estimate of the Fejér kernels which involved a linear combination of translates of Dirichlet kernels of order 2^j, showed that the maximal operator

$$\sigma^* f := \sup_{n>0} |\sigma_n f|$$

is of weak type $(1, 1)$. The crucial step used the Calderon–Zygmund decomposition to break f into a good piece g and bad pieces b_k supported on dyadic intervals I_k. Since I_k and the "support" J_j of some translate of D_{2^j} are both dyadic intervals, I_k and J_j are either disjoint or one is a subset of the other. In the disjoint case or in the case when I_k is a subset of J_j, the integral of $b_k D_{2^j}$ is zero because D_{2^j} is constant on J_j (see (2) above) and each b_k is of mean zero. Thus, many of the bad functions can be completely ignored when estimating σ^*.

Adapting this argument to higher dimensions, Wade [161] obtained estimates on the growth of arbitrary rectangular partial sums of double Walsh–Fourier series and introduced Shapiro's question to the open literature. Mòricz, Schipp, and Wade [90] answered the question for dyadic Cesàro means of double Walsh–Fourier series: if $f \in L^1(\mathbf{I}^2)$, and $\alpha \geq 0$, then $\sigma_{2^n, 2^m} f \to f$ almost everywhere on $\mathbf{I}^2$ as $\min\{n, m\} \to \infty$ provided $|n - m| \leq \alpha$. Recently, Weisz [171] showed that the restriction to dyadic means was unnecessary. If $f \in L^1(\mathbf{I}^2)$, and $\alpha > 0$, then $\sigma_{n,m} f \to f$ almost everywhere on $\mathbf{I}^2$ as $\min\{n, m\} \to \infty$ and $\alpha^{-1} \leq m/n \leq \alpha$.

Mòricz, Schipp, and Wade [90] also examined Cesàro summability of multiple Walsh–Fourier series without the cone condition. They introduced the hybrid Hardy space $H^{\#}(\mathbf{I}^2)$ (the set of f which are dyadic Hardy in one variable and integrable in the other) and used it to prove that if $f \in L \log^+ L(\mathbf{I}^2)$, then $\sigma_{n,m} f$ converges to f almost everywhere on $\mathbf{I}^2$ as $\min\{n, m\} \to \infty$.

Weisz [170] has improved this result further by replacing $L \log^+ L(\mathbf{I}^2)$ by $H^{\#}(\mathbf{I}^2)$. He introduced a new atomic decomposition of the dyadic Hardy spaces on $\mathbf{I}^2$ (in terms of the quadratic variation of a function) and used it to estimate the maximal Cesàro means σ^* on a large class of Lorenz spaces. Consequences of these estimates include:

1) σ^* is bounded from H^p to L^p for $4/5 < p \leq \infty$ and of weak type $(H^{\#}, L^1)$.
2) The maximal operator associated with the dyadic Cesàro means is bounded from $H^p(\mathbf{I}^2)$ to $L^p(\mathbf{I}^2)$ for $2/3 < p \leq \infty$.

Consequence 1), which generalizes a one variable result of Fujii when $p = q = 1$, shows that if $f \in H^{\#}(\mathbf{I}^2)$, then $\sigma_{n,m} f$ converges to f almost everywhere on $\mathbf{I}^2$ as $\min\{n, m\} \to \infty$.

5.4. Strong summability. Since $\sigma_n f \to f$ almost everywhere on $\mathbf{I}$ for $f \in L^1(\mathbf{I})$, it is clear that

$$\lim_{n\to\infty} \frac{1}{n} \sum_{k=0}^{n-1} (S_k f)(x) - f(x) = 0$$

for almost every $x \in \mathbf{I}$. It is natural to ask whether the stronger condition

$$\lim_{n\to\infty} \frac{1}{n} \sum_{k=0}^{n-1} |(S_k f)(x) - f(x)| = 0 \tag{23}$$

holds for almost every $x \in \mathbf{I}$. (This question was first addressed in the trigonometric case by Hardy and Littlewood.)

Sunouchi ([140] and [141]) was the first to examine this question for Walsh series. He proved that if $f \in L^p(\mathbf{I})$ for some $1 < p \leq 2$, then

$$\lim_{n\to\infty} \frac{1}{n} \sum_{k=0}^{n-1} |(S_k f)(x) - f(x)|^p = 0 \tag{24}$$

almost everywhere on $\mathbf{I}$. His proof was based on the following observation. Since $\sigma_n f \to f$ almost everywhere for all $f \in L^1(\mathbf{I})$, one can replace f by $\sigma_n f$ in (24). Hence it suffices to prove that

$$U_r f := \left(\sum_{k=1}^{\infty} \frac{|S_k f - \sigma_k f|^r}{k} \right)^{1/r}$$

belongs to $L^p(\mathbf{I})$ when f does. Sunouchi studied both this operator and the closely related

$$Tf := \left(\sum_{k=0}^{\infty} |S_{2^k} f - \sigma_{2^k} f|^2 \right)^{1/2}.$$

He proved that there are constants A_p and B_p, depending only on $p > 1$, such that

$$A_p\|f\|_p \le \|\Lambda f\|_p \le B_p\|f\|_p$$

for both $\Lambda \equiv U_r$, $r \ge 2$, and $\Lambda \equiv T$. In particular, the operators U_r, $r \ge 2$, and T are bounded endomorphisms on $L^p(\mathbf{I})$ when $1 < p < \infty$.

What happens to these results when $p \le 1$? Wade [160] showed that for the case $0 < p < 1$, the p norm of $U_2 f$ can be estimated by $\|Tf\|_1$, the p norm of Tf can be estimated by $\|\mathcal{Q}f\|_1$ (where $\mathcal{Q}f$ is the martingale square function), and the p norm of $\mathcal{Q}f$ can be estimated by $\|U_2 f\|_1$. A much more comprehensive result was obtained by Simon [129], who proved that the operator T is bounded from the dyadic Hardy space H to $L^1(\mathbf{I})$, but U_r is not for any $r \ge 1$.

These inequalities have been used by Tateoka to study the generalized Littlewood-Paley function

$$(\mathbf{g}\,f)(x) := \left(\sum_{n=1}^{\infty} n\epsilon_n|\sigma_{n+1}f(x) - \sigma_n f(x)|^2\right)^{1/2},$$

where ϵ_k is a sequence of positive numbers which satisfy

$$cn \le \sum_{k=1}^{n} \epsilon_k \le Cn \qquad n = 1, 2, \ldots$$

for positive constants c and C. Igari [54] had proved that the map $f \to \mathbf{g}\,f$ is not of weak type (1,1). Tateoka [146] obtained a number of inequalities which can be considered in some sense replacements for the weak type (1,1) inequalities. Specifically, he shows that there exists absolute constants C and C_p such that

$$|\{\mathbf{g}f > y\}| \le C\frac{\|f\|_H}{y} \qquad y > 0,$$

$$\|\mathbf{g}f\|_p \le C_p\|f\|_H \qquad 0 < p < 1,$$

$$\|f\|_{H^p} \le C_p\|\mathbf{g}f\|_1 \qquad 0 < p < 1,$$

and

$$\|\mathbf{g}f\|_1 \le C\|\mathcal{Q}f\|_{L\log^+ L} + C$$

where $\mathcal{Q}f$ is the martingale square function.

Rodin [104] has investigated what happens to Sunouchi's theorem about strong summability when $p > 2$. He obtained BMO estimates analogous to (23), and estimates in any Orlicz space whose norm is generated by a function of exponential type.

The multidimensional version of Sunouchi's theorem has been investigated by Sharashenidze [125]. He announced that if $f \in L(\log^+ L)^{d-1}(\mathbf{I}^d)$, then (23) holds almost everywhere. However, if $d \ge 2$, then there exist $f \in L(\log^+\log^+ L)^{d-1}(\mathbf{I}^d)$ for which (23) fails to hold on some set of positive measure.

Sunouchi's theorem was improved by Schipp [115], who showed that (24) holds almost everywhere for any $p > 0$ when $f \in L^1(\mathbf{I})$. Rodin [105] has an extremely nice and general result along these lines. If $f \in L^1(\mathbf{I})$ and $A > 0$, then

$$\lim_{n\to\infty} \frac{1}{n}\sum_{k=0}^{n} e^{A|(S_k f)(x) - f(x)|} = 1$$

for almost every $x \in \mathbf{I}$.

Rodin [106] has a double Walsh series analogue of Schipp's strong summability result. It turns out that in the two variable case, in order for (24) to hold for all $0 < p < \infty$ one also needs to assume that $f \in L \log^+ L\,(\mathbf{I}^2)$.

Since for an arbitrary $f \in L^1(\mathbf{I})$ the partial sums $S_n f$ do not tend to f, the only way that (24) holds is that the number of indices n where $S_n f - f$ is large must be relatively small incomparison to n. Since (24) gets stronger as p gets bigger, it is natural to wonder whether these results hold if p is replaced by a sequence p_n which tends to infinity as $n \to \infty$. Schipp [116] showed that this is not the case. Given any sequence $p_n \uparrow \infty$, there is an $f \in L^1(\mathbf{I})$ such that

$$\limsup_{n\to\infty} \frac{1}{n} \sum_{k=0}^{n-1} |(S_k f)(x) - f(x)|^{p_n} = \infty$$

almost everywhere.

Fridli and Schipp [33] have discovered an intriguing connection between Sidon inequalities and strong summability which holds in any homogeneous Banach space. From this and Fridli's rearrangement invariant Sidon inequality, they proved that if ϕ is monotone increasing on $(0, \infty)$ and $\phi(0+) = 0$, then

$$\lim_{n\to\infty} \frac{1}{n} \sum_{k=0}^{n-1} \phi\left(|(S_k f)(x) - f(x)|\right) = 0$$

uniformly for any $f \in \mathcal{C}(\mathbf{I})$ if and only if ϕ of exponential type, namely, there exists an $A > 0$ such that $\phi(t) \le e^{At}$ for $t \ge 0$.

The L^1 norm analogue of strong summability has also been investigated. Simon [130] proved that

$$\lim_{n\to\infty} \frac{1}{\log n} \sum_{k=1}^{n} \frac{\|S_k f - f\|_1}{k} = 0$$

for all $f \in H$. This result has been carried over to multiple Walsh–Fourier series by Weisz [168], who used the martingale Hardy space generated by the diagonal maximal function. It is interesting to note that his investigations lead to new results about trigonometric series as well.

What happens to (23) if we pass to some subsequence of S_k's? Using an estimate for the Lebesgue constants associated with any triangular method of summability, Glukhov [42] proved that if $\{n_k\}$ is a convex sequence of natural numbers which satisfies

$$\log n_k = O(\sqrt{k}) \qquad k \to \infty,$$

then

$$\lim_{m\to\infty} \frac{1}{m} \sum_{k=0}^{m-1} |f(x) - S_{n_k} f(x)| = 0 \tag{25}$$

for all f continuous on $\mathbf{I}$. For the trigonometric case when $\{n_k\}$ is convex, the condition $\log n_k = O(\sqrt{k})$ is both necessary and sufficient for (25). This is not true for Walsh systems, since

$$\lim_{m\to\infty} \frac{1}{m} \sum_{k=0}^{m-1} |f(x) - S_{2^k} f(x)| = 0$$

for all f continuous on $\mathbf{I}$.

Glukhov also proved that, as in the trigonometric case, the Lebesgue constant estimate can be used to show that (C,α) sums all have the same growth in L^p norm. Indeed, given $\alpha > 0$, there exist absolute constants C_1 and C_2 such that

$$C_1\|f-\sigma_n f\|_p \le \|f-\sigma_n^{(\alpha)} f\|_p \le C_2\|f-\sigma_n f\|_p$$

for all $f \in L^p(\mathbf{I})$, $1 \le p \le \infty$.

Kowalski [64] has also investigated these questions. He showed that if $\{n_k\}$ is a convex increasing sequence of integers which satisfies $\log n_k = O(\sqrt{k})$, as $k \to \infty$, then there is an absolute constant C which depends only on the sequence $\{n_k\}$ such that

$$\left(\frac{1}{m}\sum_{k=0}^{m-1} |(S_{n_k} f)(0)|^2\right)^{1/2} \le C\|f\|_\infty$$

for all $f \in L^\infty(\mathbf{G})$. It follows that

$$\frac{1}{m}\sum_{k=0}^{m-1} |(S_{n_k} f)(0)| \le C\|f\|_\infty$$

for all $f \in L^\infty(\mathbf{G})$, and

$$\lim_{m\to\infty} \frac{1}{m}\sum_{k=0}^{m-1} |f - S_{n_k} f|^2 = 0,$$

uniformly on $\mathbf{G}$ for all f continuous on $\mathbf{G}$.

Baĭarstanova [9] has studied the Cesàro means of subsequences $S_{n_k} f$ of Walsh–Fourier series without the convexity condition. She proved that if $n_k \ge k$ and $\log n_k = O(\sqrt{k})$, as $k \to \infty$, then Cesàro means of $\{S_{n_k} f\}$ converge to f uniformly on $\mathbf{I}$ for any $f \in \mathcal{C}(\mathbf{I})$, i.e., (25) holds if the absolute values are erased.

References

1. G. Agaev, *A Wiener type theorem for series of Walsh functions*, Dokl. Akad. Nauk SSSR **142** (1962), 751–753.
2. F.G. Arutunyan and A.A. Talalyan, *On uniqueness of Haar and Walsh series*, Izv. Akad. Nauk SSSR **28** (1964), 1391–1408.
3. B. Aubertin, *Algebraic elements and sets of uniqueness in the group of integers of a p-series field*, Canad. Math. Bull. **29** (1986), 177–184.
4. B. Aubertin and J. Fournier, *An integrability theorem for Walsh series*, Studia Math. **107** (1993), 33–59.
5. B. Aubertin and J. Fournier, *Integrability theorems for trigonometric series*, Boll. Un. Mat. Ital. B (8) **7** (1994), 775–789.
6. L.A. Balashov and V.A. Skvortsov, *Gibbs constants for partial sums of a Walsh-Fourier series and their* $(C,1)$*-means*, Trud. Mat. Inst. Akad. Nauk SSSR **164** (1983), 37–48.
7. P.T. Bateman and A.L. Duquette, *The analogue of the Pisot–Vijayaraghavan numbers in fields of formal power series*, Illinois J. Math. **6** (1966), 594–606.
8. A.S. Baĭarstanova, *On a class of series with monotone coefficients*, Izv. Vyssh. Ucheb. Zaved. Mat., No. 12 (1983), 3–7.
9. A.S. Baĭarstanova, $(C,1)$ *summability of partial sums of Walsh–Fourier series*, Functional Analysis, Differential Equations and their Applications, Kazakh. Gos. Univ., Alma–Ata, 1987, pp. 7–11.
10. A.D. Bethke, *Genetic algorithms as function optimizers*, Dissertation, University of Michigan (1980).

11. P. Billard, *Sur la convergence presque partout des séries de Fourier–Walsh des fonctions de l'espace* $L^2(0,1)$, Studia Math. **28** (1967), 363–388.
12. W.R. Bloom and J.J.F. Fournier, *Smoothness conditions and integrability theorems on bounded Vilenkin groups*, J. Austral. Math. Soc. **45** (1988), 46–61.
13. S.K. Bloshanskaya and I.L. Bloshankiĭ , *Generalized and weak generalized localization for multiple Fourier–Walsh series of functions in* L_p, $p \geq 1$, Russian Acad. Sci. Dokl. Math. **48** (1994), 359–364.
14. S.V. Bochkarev, *A method of averaging in the theory of orthogonal series and some problems in the theory of bases*, Trudy Mat. Inst. Steklov, No. 146 (1978), 1–92.
15. E.V. Bohn, *Recursive evaluation of Walsh coefficients for multiple integrals of Walsh series*, Automatica-J. IFAC **20** (1984), 243–246.
16. B.M. Bugadze, *On divergence of Walsh–Fourier series of bounded functions on sets of measure zero*, Mat. Sbornik **185** (1994), 119–127.
17. P.L. Butzer and H.J. Wagner, *Walsh series and the concept of a derivative*, Applic. Anal. **3** (1973), 29–46.
18. J.-A. Chao, *Convergence of generalized Walsh–Fourier series*, Tamkang J. Math. **17** (1986), 123–128.
19. S. Cobzas, *Double condensation of singularities for Walsh–Fourier series*, Rev. D'Anal. Num. et de Th.L'Approx. **21** (1992), 119–129.
20. J.E. Coury, *Walsh series with coefficients tending monotonically to zero*, Pac. J. Math. **54** (1974), 1–16.
21. J.E. Coury, *A class of Walsh M–sets of measure zero (corrigendum)*, J. Math. Analysis **52** (1975), 669–671.
22. R.B. Crittenden, *A theorem on the uniqueness of the* $(C,1)$ *summability of Walsh series*, Dissertation, University of Oregon (1964).
23. R.B. Crittenden and V.L. Shapiro, *Sets of uniqueness on the group* 2^ω, Annals of Math. **81** (1965), 550–564.
24. J.E. Daly, *The construction of optimal histograms*, Comm. Stat. Theory Meth. **17** (1988), 2921–2931.
25. P. Enflo, *A counterexample to the approximation property in Banach spaces*, Acta Math. **139** (1973), 308–317.
26. J. Esfahanizadeh and A.H. Siddiqi, *On the approximation of functions by the de la Vallée-Poussin means of their Walsh-Fourier series*, Aligarh Bull. Math. **8** (1978), 59–64.
27. N.J. Fine, *On the Walsh functions*, Trans. Amer. Math. Soc. **65** (1949), 372–414.
28. N.J. Fine, *Cesàro summability of Walsh–Fourier series*, Proc. Nat. Acad. Sci. USA **41** (1955), 588–591.
29. N.J. Fine, *Fourier–Stieltjes series of Walsh functions*, Proc. Amer. Math. Soc. **86** (1957), 246–255.
30. G.A. Fomin, *Certain properties of orthogonal series*, Int. Conf. Appl. Funct. Anal. in Theory of Approx. Kalinin (1970), 165–169.
31. S. Fridli, *An inverse Sidon type inequality for the Walsh system*, J. Math. Anal. Appl. **193** (1995), 715–736.
32. S. Fridli, *Integrability and* L^1*–convergence of trigonometric and Walsh series*, Annales Univ. Sci. Budapest Eötvös, Sect. Math. (to appear).
33. S. Fridli and F. Schipp, *Strong summability and Sidon type inequalities*, Acta Sci. Math. (Szeged) **60** (1995), 277–289.
34. J.B. Garnett and P.W. Jones, *BMO from dyadic BMO*, P acific J. Math. **99** (1982), 351–371.
35. A.M. Garsia, *Martingale Inequalities*, W.A. Benjamin, Inc., Reading Massachusetts, 1973.
36. R.D. Getsadze, *A continuous function with multiple Fourier series in the Walsh–Paley system that diverges almost everywhere*, Math. USSR Sbornik **56** (1987), 262–278.
37. R.D. Getsadze, *On the convergence and divergence of multiple Fourier series with respect to the Walsh–Paley system in spaces* C *and* L, Analysis Math. **13** (1987), 29–39.
38. G.G. Gevorkyan, *Uniqueness sets for Haar and Walsh systems*, Akad. Nauk Armyan. SSR Dokl. **73** (1981), 91–96.
39. G.G. Gevorkyan, *On* U_p^**-sets for the Haar, Walsh, and trigonometric systems*, Izv. Akad. Nauk Armyan. SSR **20** (1985), 325–348.
40. G.G. Gevorkyan, *On classification of* $U(S)$ *sets of the Walsh system*, Mathematics, No. 5, Erevan. Gos. Univ., Erevan, 1986, pp. 42–49.

41. G.G. Gevorkyan, *On coefficients of null-series and on sets of uniqueness of trigonometric and Walsh systems*, Analysis Math. **14** (1988), 219–251.
42. V.A. Glukhov, *On summability of Walsh-Fourier series*, Ukrain. Mat. Ž. **38** (1986), 303–309.
43. B. Golubov, A. Efimov, and V. Skvortsov, *Walsh Series and Transforms*, Kluwer Academic Publ., Dordrecht, 1991.
44. J. Gosselin, *On convergence of Walsh–Fourier series for* $L^2[0,1]$, Studies in Analysis, Advances in Math., Suppl. Studies, 4, Academic Press, 1979, pp. 223–232.
44. M. Grandet–Hugot, *Éléments algébraiques remarquables dans un corps de séries formelles*, Acta Arith. **14** (1967/68), 177–184.
46. D.J. Grubb, *Sets of uniqueness in compact, 0–dimensional metric groups*, Trans. Amer. Math. Soc. **301** (1987), 239–249.
47. A. Haar, *Zur Theorie der orthogonal en Funktionen systeme*, Mat. Ann. **69** (1910), 331–371.
48. H.F. Harmuth, *Transmission of Information by Orthogonal Functions*, Springer–Verlag, Berlin, 1972.
49. H.F. Harmuth, *Sequency Theory*, Academic Press, New York, 1977.
50. L. Harper, *Capacities of sets and harmonic analysis on the group* 2^ω, Trans. Amer. Math. Soc. **126** (1967), 303–315.
51. D.C. Harris, *Compact sets of divergence for continuous functions on a Vilenkin group*, Proc. Amer. Math. Soc. **98** (1986), 436–440.
52. D.C. Harris, *Almost everywhere divergence of multiple Walsh-Fourier series*, Proc. Amer. Math. Soc. **101** (1987), 637–643.
53. D.C. Harris and W.R. Wade, *Sets of divergence on the group* 2^ω, Trans. Amer. Math. Soc. **240** (1978), 385–392.
54. S. Igari, *An extension of the interpolation theorem of Marcinkiewicz*, Tôhoku Math. J. **15** (196), 343–358.
55. M. Jerison and G. Rabson, *Convergence theorem sobtained from induced homomorphisms of a group algebra*, Ann. of Math. **63** (1956), 176–190.
56. G.G. Kemkhadze, *A remark on the convergence of spherical partial sums of Walsh–Fourier series of functions of class* L^p, $p > 1$, Dokl. Akad. Nauk SSSR **222** (1975), 277–280.
57. Sh.V. Kheladze, *Divergence everywhere of Fourier series in bounded Vilenkin systems*, Sakharth. SSR Mecn. Akad. Math. Inst. Shrom **58** (1978), 224–242.
58. Sh.V. Kheladze, *On divergence of multiple Fourier series on a given set of measure zero*, Trud. Tbilis. Mat. Inst. **82** (1986), 142–146.
59. N.N. Kholshchevnikova, *A generalization of a theorem of Bary for the Walsh system*, Mat. Sbornik **183** (1992), 3–12.
60. N.N. Kholshchevnikova, *On the category of U–sets for series in the Walsh system*, Mat. Zametki **53** (1993), 129–151.
61. M. Kobayashi, *Capacities of sets and harmonic analysis on the group* 2^ω, Tôhoku Math. J. (2) **21** (1969), 419–433.
62. E. Kochneff, Y. Sagher, and K. Zhou, *BMO estimates for lacunary series*, Ark. Mat. **28** (1990), 301–310.
63. S.V. Konyagin, *On a subsequence of Walsh–Fourier sums*, Mat. Zametki **54** (1993), 69–75.
64. T. Kowalski, *On some properties of partial sums of Walsh-Paley-Fourier series*, Analysis Math. **14** (1988), 49–63.
65. N.R. Ladhawala, *Absolute summability of Walsh–Fourier series*, Pacif. J. Math. **65** (1976), 103–108.
66. N.R. Ladhawala and D.C. Pankratz, *Almost everywhere convergence of Walsh–Fourier series of* H^1 *functions*, Studia Math. **59** (1976), 85–92.
67. E.N. Lanning and D.M. Johnson, *Automated identification of rock boundaries: An application of the Walsh transform to geophysical well–log analysis*, Geophysics **48** (1983), 197–205.
68. S.V. Levisov, *Some properties of the Walsh system*, Mat. Zametki **27** (1980), 715–720.
69. R. Lindahl, *A differentiation theorem for functions defined on the dyadic rationals*, Proc. Amer. Math. Soc. **30** (1971), 349–352.
70. G.E. Lippman and W.R. Wade, *Pseudofunctions and uniqueness on the group of integers of a p–series field*, Acta Math. Acad. Sci. Hungar. **35** (1980), 1–12.
71. S. Yu. Lukashenko, *The structure of divergence sets for Walsh-Fourier series*, Analysis Math. **10** (1984), 23–41.

72. S.F. Lukomskiĭ, *On some classes of sets of uniqueness for multiple Walsh series*, Mat. Sbornik **180** (1989), 937–945.
73. S.F. Lukomskiĭ, *On a U-set for multiple Walsh series*, Analysis Math. **18** (1992), 127–138.
74. S.F. Lukomskiĭ, *On divergence almost everywhere of Fourier–Walsh quadratic partial sums of integrable functions*, Mat. Zametki **56** (1994), 57–62.
75. M. Maqusi, *Applied Walsh Analysis*, Heyden, Ltd., London, 1981.
76. K.H. Moon, *An everywhere divergent Fourier–Walsh series of the class* $L(\log^+ \log^+ L)^{1-\epsilon}$, Proc. Amer. Math. Soc. **50** (1975), 309–314.
77. G.W. Morgenthaler, *Walsh–Fourier series*, Trans. Amer. Math. Soc. **84** (1957), 472–507.
78. F. Mòricz, *On the convergence of double orthogonal series*, Analysis Math. **2** (1976), 287–304.
79. F. Mòricz, *On* L^1*-convergence of Walsh–Fourier series I*, Rend. Circ. Mat. Palermo **38** (1989), 411–418.
80. F. Mòricz, *On* L^1*-convergence of Walsh–Fourier series II*, Acta Math. Hung. **58** (1991), 203–210.
81. F. Mòricz, *Double Walsh–Fourier series with coefficients of bounded variation*, Zeit. für Anal. Anwend. **10** (1991), 3–10.
82. F. Mòricz, *On the uniform convergence and* L^1*-convergence of double Walsh–Fourier series*, Studia Math. **102** (1992), 225–237.
83. F. Mòricz, *Pointwise convergence of double Walsh–Fourier series*, Analysis **12** (1992), 121–137.
84. F. Mòricz, *A quantitative Dirichlet–Jordan test for Walsh–Fourier series*, Proc. Amer. Math. Soc. **118** (1993), 143–149.
85. F. Mòricz, *Rate of convergence for double Walsh–Fourier series of functions of bounded fluctuation*, Arch. Math. (Basel) **60** (1993), 256–265.
86. F. Mòricz, L^1*-convergence of double cosine- and Walsh–Fourier series*, J. Anal. Math. **62** (1994), 115–130.
87. F. Mòricz and F. Schipp, *On the integrability and* L^1 *convergence of Walsh series with coefficients of bounded variation*, J. Math. Anal. Appl. **146** (1990), 99–109.
88. F. Mòricz and F. Schipp, *On the integrability and* L^1 *convergence of double Walsh series*, Acta Math. Hung. **57** (1991), 371–380.
89. F. Mòricz, F. Schipp, and W.R. Wade, *On the integrability of double Walsh series with special coefficients*, Mich. Math. J. **37** (1990), 191–201.
90. F. Mòricz, F. Schipp, and W.R. Wade, *Cesàro summability of double Walsh–Fourier series*, Trans. Amer. Math. Soc. **329** (1992), 131–140.
91. F. Mòricz and A.H. Siddiqi, *Approximationby Nördlund means of Walsh–Fourier series*, J. Approx. Theory **70** (1992), 375–389.
92. G.M. Mushegyan, *Uniqueness of multiple series in the Rademacher system*, Akad. Nauk Armyan. SSR Dokl. **80** (1985), 152–156.
93. K.A. Navasardyan, *On null-series by double Walsh system*, Izv. Nat. Akad. Nauk Arm. Mat. **29** (1994), 50–68.
94. C.W. Onneweer, *On uniform convergence of Walsh–Fourier series*, Pacif. J. Math. **34** (1970), 117–122.
95. C.W. Onneweer, *Absolute convergence of Fourier series*, Duke Math. J. **39** (1972), 599–609.
96. C.W. Onneweer, *Absolute convergence of Fourier series on groups, II.*, Duke Math. J. **41** (1974), 679–688.
97. C.W. Onneweer, *Sets of divergence in the group of integers of a p–adic or p–series field*, Annales Univ. Sci. Budapest Eötvös, Sect. Math. **21** (1978), 13–20.
98. C.W. Onneweer and D. Waterman, *Uniform convergence of Fourier series on groups, I.*, Mich. Math. J. **18** (1971), 265–273.
99. C.W. Onneweer and D. Waterman, *Fourier series of functions of harmonic bounded fluctuation on groups*, J.d'Anal. Math. **27** (1974), 79–93.
100. L.M. Optican and B.J. Richmond, *Temporal encoding of two dimensional patterns by single units in primate inferior temporal cortex, III*, J. Neurophysiology **57** (1987), 162–178.
101. Qian Fang Xu, *On a property of the sums of Walsh–Fourier series*, Acta Math. Sinica **29** (1986), 293–298.
102. R.E.A.C. Paley, *A remarkable system of orthogonal functions*, Proc. Lond. Math. Soc. **34** (1932), 241–279.

103. J.J. Price, *Orthonormal sets with non–negative Dirichlet kernels*, Trans. Amer. Math. Soc. **95** (1960), 356–362.
104. V.A. Rodin, *The space BMO and strong means of Walsh–Fourier series*, Mat. Sbornik **182** (1991), 1463–1478.
105. V.A. Rodin, *Totik's conjecture for the Chrestenson and Walsh systems*, Analysis Math. **18** (1992), 295–305.
106. V.A. Rodin, *Strong means and oscillation of multiple Walsh–Fourier series*, Mat. Zametki **56** (1994), 102–117.
107. B. Roider, *Über Stufenfunktionen Endlicher Vielfalt*, Doctoral Dissertation, University of Innsbruck, Austria, 1969.
108. A.I. Rubinshtein, *The A–integral and series with respect to the Walsh system*, Uspekhi Mat. Nauk **18** (1963), 191–197.
109. D.K. Sanadze and Sh. V. Kheladze, *The convergence of multiple Walsh–Fourier series*, Soobshch. Akad.Nauk Gruz. SSR **80** (1975), 285–288.
110. D.K. Sanadze and Sh. V. Kheladze, *Convergence and divergence of multiple Walsh–Fourier series*, Sakharth. SSR Mecn. Akad. Mat. Inst. Shrom. **55** (1977), 93–106.
111. F. Schipp, *Bermerkung zur Divergenz der Walsh–Fourierreihen*, Annales Univ. Sci. Budapest Eötvös, Sect. Math. **11** (1968), 53–58.
112. F. Schipp, *On the strong approximation of Walsh–Fourier series*, MTA III. Oszt. Közl. (in Hungarian) **19** (1969), 101–111.
113. F. Schipp, *Über Walsh–Fourierreihen mit nichtnegativen Partialsummen*, Annales Univ. Sci. Budapest Eötvös, Sec. Math. **12** (1969), 43–48.
114. F. Schipp, *Über die Divergenz der Walsh–Fourierreihen*, Annales Univ. Sci. Budapest Eötvös, Sec. Math. **12** (1969), 49–62.
115. F. Schipp, *Über die starke Summation von Walsh–Fourierreihen*, Acta Math. Sci. (Szeged) **30** (1969), 77–87.
116. F. Schipp, *Bermerkung zur starken Summation der Walsh–Fourierreihe*, Acta Math. Acad. Sci. Hungar. **20** (1969), 263–274.
117. F. Schipp, *Über die Konvergenz von Reihen nach Produkt systemmen*, Acta Math. Sci. (Szeged) **35** (1973), 13–16.
118. F. Schipp, *Über Geweissen Maximal operatoren*, Annales Univ. Sci. Budapest Eötvös, Sec. Math. **18** (1976), 189–195.
119. F. Schipp, *VMO spaces not having a Schauder basis*, Analysis Math. **9** (1983), 313–322.
120. F. Schipp, *On boundedly divergent Walsh–Fourier series*, Acta Math. Hung. **56** (1990), 361–367.
121. F. Schipp, *Approximation Theory*, (G.A. Anastassiou, Ed.), Lecture Notes in Pure and Applied Mathematics, Vol. 138, Dekker, New York, 1991, pp. 421–426.
122. F. Schipp, W.R. Wade, and P. Simon (with assistance by J. Pàl), *Walsh Series: An Introduction to Dyadic Harmonic Analysis*, Adam Hilger, Ltd., Bristol and New York, 1990.
123. L.A. Shaginyan, *On indefinite limits and sets of limit functions of Walsh series*, Mat. Sbornik **95** (1974), 263–271.
124. V.L. Shapiro, *$U(\epsilon)$–sets for Walsh series*, Proc. Amer. Math. Soc. **16** (1965), 867–870.
125. T.V. Sharashenidze, *Multiple Walsh–Fourier series*, Soobshch. Akad. Nauk Gruz. SSR **81** (1976), 565–567.
126. A.A. Shneider, *Uniqueness of expansions with respect to the Walsh system of functions*, Mat. Sbornik **24** (1949), 279–300.
127. A.H. Siddiqi, *Walsh–Fourier series with monotone coefficients*, Aligarh Bull. Math. **1** (1971), 25–29.
128. P. Simon, *Verallgemeinerten Walsh–Fourier Reihen*, Annales Univ. Sci. Budapest Eötvös, Sect. Math. **16** (1973), 103–113.
129. P. Simon, *(L^1, H)–type estimations for some operators with respect to the Walsh–Paley system*, Acta Math. Hung. **46** (1985), 307–310.
130. P. Simon, *Strong convergence of certain means with respect to the Walsh–Fourier series*, Acta Math. Hung. **49** (1987), 425–431.
131. P. Sjölin, *An inequality of Paley and convergence a.e. of Walsh–Fourier series*, Arkiv för Mat. **8** (1969), 551–570.
132. V.A. Skvortsov, *Differentiation with respect to nets and the Haar system*, Mat. Zametki **4** (1968), 33–40.

133. V.A. Skvortsov, *Some generalizations of uniqueness theorems for Walsh series*, Mat. Zametki **13** (1973), 367–372.
134. V.A. Skvortsov, *Uniqueness theorems for Walsh series that are summable by the $(C, 1)$ method*, Vestnik Mosk. Univ. Ser. I Mat. Mekh. **31** (1976), 73–80.
135. V.A. Skvortsov, *The h-measure of M-sets for the Walsh systems*, Mat. Zametki **21** (1977), 335–340.
136. V.A. Skvortsov, *The rate of convergence to zero of coefficients of zero-series in the Haar and Walsh systems*, Izv. Akad. Nauk SSSR, Ser. Mat. **41** (1977), 703–716.
137. V.A. Skvortsov, *An analogue of the Nevai test for Walsh–Fourier series*, Vestnik Mosk. Univ. Ser. Mat. Mekh., No. 4 (1987), 61–63.
138. S.B. Stechkin and P.L. Ul'yanov, *On sets of uniqueness*, Izv. Akad. Nauk SSSR, Ser. Mat. **26** (1962), 211–222.
139. E.M. Stein, *On limits of sequences of operators*, Annals Math. **74** (1961), 140–170.
140. G. Sunouchi, *On the Walsh-Kaczmarz series*, Proc. Amer. Math. Soc. **2** (1951), 5–11.
141. G. Sunouchi, *Strong summability of Walsh–Fourier series*, Tôhoku Math. J. **16** (1964), 228–237.
142. M.H. Taibleson, *Fourier analysis on local fields*, Princeton Univ. Press, Princeton, N.J., 1975.
143. M.H. Taibleson and G. Weiss, *Certain function spaces connected with almost everywhere convergence of Fourier series*, Proc. Conf. A. Zygmund's 80th Birthday (Chicago,1980), Wadsworth Publ. Co., 1982.
144. J. Tateoka, *On almost everywhere convergence of Walsh–Fourier series*, Proc. Japan Acad. **44** (1968), 647–650.
145. J. Tateoka, *On the localization property of the double Walsh–Fourier series*, Studia Sci. Math. Hung. **11** (1976), 407–411.
146. J. Tateoka, *On some inequalities for Walsh–Fourier series*, Acta Sci. Math. **52** (1988), 387–392.
147. J. Tateoka, *The modulus of continuity and the best approximation over the dyadic group*, Acta Math. Hung. **59** (1992), 115–120.
148. J. Tateoka, *On the characterization of Hardy–Besov spaces on the dyadic group and its applications*, Studia Math. **110** (1994), 127–148.
149. J. Tateoka and W.R. Wade, *On strong approximation and summability by Cesàro means on Besov spaces over the 2-series field*, Acta Sci. Math. Sci. (Szeged) **60** (1995), 685–703.
150. S.A. Telyakovskiĭ , *On a sufficient condition of Sidon for the integrability of trigonometric series*, Mat. Zametki **14** (1973), 317–328.
151. Sh.T. Tetunashvili, *On the uniqueness of multiple Rademacher series*, Soobshch. Akad. Nauk Gruzin. SSR **123** (1986), 33–35.
152. V.I. Tevzadze, *On the convergence of Walsh-Fourier series at a point*, Soobshch. Akad. Nauk Gruz. SSR **125** (1987), 249–251.
153. N. Ya. Vilenkin, *A class of complete orthonormal series*, Izv. Akad. Nauk SSSR Ser. Mat. **11** (1947), 363–400.
154. V.S. Vladimirov, I.V. Volovich, and E.I. Zelonov, *p-Adic Analysis and Mathematical Physics*, Series in Soviet and East European Mathematics, Vol. 10, World Scientific Publ., Singapore, New Jersey, London, and Hong Kong, 1994.
155. W.R. Wade, *Uniqueness theory for Cesàro summable Haar series*, Duke Math. J. **38** (1971), 211–227.
156. W.R. Wade, *Summing closed U-sets for Walsh series*, Proc. Amer. Math. Soc. **29** (1971), 123–125.
157. W.R. Wade, *Sets of uniqueness and α-capacity on the group 2^ω*, Trans. Amer. Math. Soc. **208** (1975), 309–315.
158. W.R. Wade, *Growth conditions and uniqueness for Walsh series*, Mich. Math. J. **24** (1977), 153–155.
159. W.R. Wade, *Sets of uniqueness for the group of integers of a p-series field*, Canad. J. Math. **31** (1979), 858–866.
160. W.R. Wade, *L^r inequalities for Walsh series, $0 < r < 1$*, Acta Sci. Math. (Szeged) **46** (1983), 233–241.
161. W.R. Wade, *A growth estimate for Cesàro partial sums of multiple Walsh–Fourier series*, Colloq. Math. Soc. Janos Bolyai **49** (1985), 975–991.
162. W.R. Wade, *A unified approach to uniqueness of Walsh series and Haar series*, Proc. Amer. Math. Soc. **99** (1987), 61–65.

163. W.R. Wade and K. Yoneda, *Uniqueness and quasi–measures on the group of integers of a p–series field*, Proc. Amer. Math. Soc. **84** (1982), 202–206.
164. J.L. Walsh, *A closed set of normal orthogonal functions*, Amer. J. Math. **45** (1923), 5–24.
165. C. Watari, *Multipliers for Walsh–Fourier series*, Tôhoku Math. J. **16** (1964), 239–251.
166. D. Waterman, *W–systems are the Walsh functions*, Bull. Amer. Math. Soc. **75** (1969), 139–142.
167. D. Waterman, *On systems of functions resembling the Walsh system*, Mich. Math. J. **29** (1982), 83–87.
168. F. Weisz, *Strong summability of two–dimensional Walsh–Fourier series*, Acta Sci. Math.(Szeged) **60** (1995), 779–803.
169. F. Weisz, *Strong convergence theorems for two–parameter Walsh–Fourier and trigonometric–Fourierseries*, Studia Math. **117** (1996), 173–194.
170. F. Weisz, *Cesàro summability of two–parameter Walsh–Fourier series*, J. Approx. Theory (to appear).
171. F. Weisz, *Cesàro summability of two–dimensional Walsh–Fourier series*, Trans. Amer. Math. Soc. (to appear).
172. J. Wenzel, *Mean convergence of vector–valued Walsh series*, Math. Nachr. **162** (1993), 117–124.
173. J.E. Whelchel and D.F. Guinn, *Fast Fourier–Hadamard transform and its use in signal representation and classification*, EASON (1968), 561–573.
174. Sh. Yano, *On Walsh series*, Tôhoku Math.J. **3** (1951), 223–242.
175. Sh. Yano, *On approximation by Walsh functions*, Proc. Amer. Math. Soc. **2** (1952), 962–967.
176. Sh. Yano, *A convergence test for Walsh-Fourier series*, Tôhoku Math. J. **6** (1954), 226–230.
177. K. Yoneda, *On the absolute convergence of Walsh–Fourier series*, Math. Japonica **18** (1973), 71–78.
178. K. Yoneda, *Perfect sets of uniqueness on the group* 2^ω, Canad. J. Math. **34** (1982), 759–764.
179. K. Yoneda, *Sets of uniqueness for a certain class* $\mathcal{M}_\epsilon$ *on the dyadic group*, Proc. Amer. Math. Soc. **89** (1983), 279–284.
180. K. Yoneda, *Sets of multiplicity on the dyadic group*, Acta Math. Hung. **41** (1983), 195–200.
181. K. Yoneda, *On generalized uniqueness theorems for Walsh series*, Acta Math. Hung. **43** (1984), 209–217.
182. K. Yoneda, *A sufficient condition for a set to be a Dirichlet set on the dyadic group*, Math. Japonica **29** (1984), 45–50.
183. K. Yoneda, *Some properties of* $H^{(s)}$*–sets defined on the dyadic group*, Math. Japonica **29** (1984), 29–33.
184. K. Yoneda, *Uniqueness theorems for a certain small class of Walsh series*, Math. Japonica **30** (1985), 321–326.
185. K. Yoneda, *Summing generalized closed U–sets for Walsh series*, Proc. Amer. Math. Soc. **94** (1985), 110–114.
186. K. Yoneda, *Dirichlet sets and some uniqueness theorems for Walsh series*, Tôhoku Math. J. **38** (1986), 1–14.
187. K. Yoneda, *A note on convergence of Walsh–Fourier series of a function on a class* V_α, Acta Math. Hung. **49** (1987), 77–81.
188. K. Yoneda, *A generalized capacity and a uniqueness theorem on the dyadic group*, Proc. Amer. Math. Soc. **102** (1988), 52–56.
189. K. Yoneda, *Uniqueness theorems for Walsh series under a strong condition*, Acta Math. Hung. **61** (1993), 1–15.
190. Wo–Sang Young, *A note on Walsh Fourier series*, Proc. Amer. Math. Soc. **59** (1976), 305–310.
191. A. Zygmund, *Trigonometric Series*, Cambridge Univ. Press, New York, 1959.

Mathematics Department, University of Tennessee, Knoxville, TN 37996-1300

Selected Titles in This Series

(*Continued from the front of this publication*)

(See the AMS catalog for earlier titles)